PERSPECTIVES IN

BIOANALYSIS

VOLUME 2

Cover image: Ian Tattersall, American Museum of Natural History

PERSPECTIVES IN BIOANALYSIS

NEW HIGH THROUGHPUT TECHNOLOGIES FOR DNA SEQUENCING AND GENOMICS

EDITOR

KEITH R. MITCHELSON

CAPITALBIO CORPORATION:
NATIONAL ENGINEERING RESEARCH CENTRE
FOR BEIJING BIOCHIP TECHNOLOGY
18 LIFE SCIENCE PARKWAY
CHANGPING DISTRICT
BEIJING 102206
CHINA

AND

MEDICAL SYSTEMS BIOLOGY RESEARCH CENTER
TSINGHUA UNIVERSITY SCHOOL OF MEDICINE
BEIJING 100084
CHINA

VOLUME 2

ELSEVIER

AMSTERDAM – BOSTON – HEIDELBERG – LONDON – NEW YORK – OXFORD
PARIS – SAN DIEGO – SAN FRANCISCO – SINGAPORE – SYDNEY – TOKYO

Elsevier
Radarweg 29, PO Box 211, 1000 AE Amsterdam, The Netherlands
Linacre House, Jordan Hill, Oxford, OX2 8DP, UK

First edition 2007
Reprinted 2008

British Library Cataloguing in Publication Data
A catalogue record for this book is available from the British Library

Library of Congress Cataloging-in-Publication Data
A catalog record for this book is available from the Library of Congress

ISBN: 978-0-444-52223-8
ISSN: 1871-0069

Printed and bound in *China*

08 09 10 10 9 8 7 6 5 4 3 2

Contents

Contributors

Numbers in parentheses indicate the pages where the authors' contributions can be found.

P. Adams (303), Department of Mathematics, University of Queensland, St. Lucia, Queensland 4072, Australia

J. J. Austin (357), Department of Environmental Biology, University of Adelaide, North Terrace, Adelaide 5005, South Australia, Australia

Dirk van den Boom (97), Sequenom Corporation, 3595 John Hopkins Court, San Diego, CA 92121, USA

I. Braslavsky (209), Department of Physics and Astronomy, Clippinger 251B, Ohio University, Athens, OH 45701, USA

D. E. Bryant (303), Department of Mathematics, University of Queensland, St. Lucia, Queensland 4072, Australia

J. C. Carter (303), Leukaemia Foundation Queensland Laboratories, Queensland Institute of Medical Research, Herston, Queensland 4006, Australia

J. Cheng (45), Capitalbio Corporation: National Engineering Research Center for Beijing Biochip Technology, Beijing, 18 Life Science Parkway, Changping District, Beijing 102206, China; and the Medical Systems Biology Research Center, Tsinghua University School of Medicine, Beijing 100084, China

D. A. E. Cochran (303), Agen Biomedical Limited, Durbell Street, Acacia Ridge, Queensland 4110, Australia

J. R. Edwards (187), Columbia Genome Center, Columbia University College of Physicians and Surgeons, Russ Berrie Medical Science Pavilion, 1150 St. Nicholas Avenue, New York, NY 10032, USA

M. Ehrich (97), Sequenom Corporation, 3595 John Hopkins Court, San Diego, CA 92121, USA

C. W. Fuller (119), GE Healthcare, 800 Centennial Avenue, Piscataway, NJ 08855, USA

P. S. Gooding (357), Agricultural Division – AGRF, Plant Genomics Centre, University of Adelaide, Hartley Grove, Waite Campus PMB1, Glen Osmond, SA 5064, Australia

D. B. Hawkes (3,303), AGRF, Institute of Molecular Bioscience, University of Queensland, St. Lucia, Queensland 4072, Australia

B. Hebert (209), Department of Physics, McGill University, Rutherford Physics Building 228, 3600 University Street, Montreal, Quebec H3A 2T8, Canada

J. Herschleb (265), Laboratory for Molecular and Computational Genomics, UW Biotechnology Center, Laboratory of Genetics and Department of Chemistry, University of Wisconsin, 425 Henry Mall, Madison 53706, USA

F. Hillenkamp (97), Institute for Medical Physics and Biophysics, University of Münster, Robert-Koch-Str. 31, Münster D-48149, Germany
T. P. Jarvie (153), 454 Life Sciences Corporation, 20 Commercial Street, Branford, CT 06405, USA
J. Ju (187), Columbia Genome Center, Columbia University College of Physicians and Surgeons, Russ Berrie Medical Science Pavilion, 1150 St. Nicholas Avenue, New York, NY 10032, USA
J. M. Keith (303), Institute of Molecular Bioscience and Department of Mathematics, University of Queensland, St. Lucia, Queensland 4072, Australia
D. H. Kim (187), Columbia Genome Center, Columbia University College of Physicians and Surgeons, Russ Berrie Medical Science Pavilion, 1150 St. Nicholas Avenue, New York, NY 10032, USA
J. R. Knight (153), 454 Life Sciences Corporation, 20 Commercial Street, Branford, CT 06405, USA
G. A. Kowalchuk (357), Netherlands Institute of Ecology, PO Box 40, 6666 ZG Heteren, The Netherlands
S. Kumar (119), GE Healthcare, 800 Centennial Avenue, Piscataway, NJ 08855, USA. Present Address: 21 Muirhead Court, Belle Mead, NJ 08502, USA
J. W. Lee (245), Oak Ridge National Laboratory, Oak Ridge, TN 37831-6194, USA
M. Margulies (153), 454 Life Sciences Corporation, 20 Commercial Street, Branford, CT 06405, USA
A. McGrath (327), Australian Genome Research Facility, University of Queensland, St. Lucia, Queensland 4072, Australia
A. Meller (245), The Department of Physics and Biomedical Engineering, Boston University, 44 Cummington Street, Boston, MA 02215, USA
A. E. Men (3), AGRF, Institute of Molecular Bioscience, University of Queensland, St. Lucia, Queensland 4072, Australia
K. R. Mitchelson (3,303), Capitalbio Corporation: National Engineering Research Centre for Beijing Biochip Technology, 18 Life Science Parkway, Changping District, Beijing 102206, China; and Medical Systems Biology Research Center, Tsinghua University School of Medicine, Beijing 100084, China
D. C. Schwartz (265), Laboratory for Molecular and Computational Genomics, UW Biotechnology Center, Laboratory of Genetics and Department of Chemistry, University of Wisconsin, 425 Henry Mall, Madison 53706, WI, USA
J. F. Simons (153), 454 Life Sciences Corporation, 20 Commercial Street, Branford, CT 06405, USA
J. R. Stephen (357), Agricultural Division – AGRF, Plant Genomics Centre, University of Adelaide, Hartley Grove, Waite Campus PMB1, Glen Osmond, SA 5064, Australia
R. Turakulov (3), AGRF, Walter and Eliza Hall Institute for Medical Research, 1G Royal Parade, Parkville, Victoria 3050, Australia

Q. Xiong (45), Capitalbio Corporation: National Engineering Research Center for Beijing Biochip Technology, 18 Life Science Parkway, Beijing 102206, China
S. Zhou (265), Laboratory for Molecular and Computational Genomics, UW Biotechnology Center, Laboratory of Genetics and Department of Chemistry, University of Wisconsin, 425 Henry Mall, Madison 53706, WI, USA

Preface

Since the independent invention of DNA sequencing by Sanger and by Gilbert 30 years ago, it has grown from a small-scale technique capable of reading several kilobase-pair of sequence per day into today's multibillion dollar 'industry', with large Sequencing Centers for large-scale delineation of entire genomes, and supporting DNA sequencing activity at some level at virtually all Universities and larger hospitals throughout the world. We are now in a "post-genomic era" with possibly more than 150 billion base-pair of sequence information held at international Bioinformatics Centers, yet DNA sequencing continues as a major diagnostic and research activity in many areas of life science and medicine. This growth has spurred the development of new sequencing technologies that do not involve either electrophoresis or Sanger sequencing chemistries. Sequencing by synthesis (SBS) involves multiple parallel micro-sequencing addition events occurring on a surface, where data from each round is detected by imaging. The recent plan to sequence a complete Neandertal genome (*Homo sapiens neanderthalensis*) by the 454 Life Sciences and the Max Planck Institute using SBS underlines the relevance of these new sequencing technologies to post-genomic science. This will be the second entire human genome project, and it is expected to radically advance knowledge of the human genomic biology and provide profound new insights into genetic diseases in man.

This volume brings together some of the new developments in DNA sequencing technology. Reviews of complementary developments in Sanger and SBS sequencing chemistries (Kumar and Fuller), capillary electrophoresis and micro-device integration (Xiong and Cheng), MS sequencing (Ehrich *et al.*) and applications (Mitchelson *et al.*) set the framework. Reviews of new developments in SBS technology (Margulies *et al.*), the chemistry of nucleotide-dye SBS sequencing (Edwards *et al.*), and steps toward realizing the sequencing of single DNA molecules by cyclic synthesis (Hebert and Braslavsky), nano-pore sequencing (Lee and Meller) and optical mapping of arrayed DNA (Schwartz *et al.*) indicate the latest advances. Finally, bioinformatics tools for genome assembly (McGrath), sequencing ancient and environmental DNA samples (Kowalchuk *et al.*) and support for SBS sequence assembly (Keith *et al.*) discuss many issues relevant to new applications using SBS sequencing. The authors hope this volume will provide stimulus to both students and researchers interested in this vital field of chemistry and technological innovation.

July 2006 Keith Mitchelson

Enabling Technologies

Chapter 1

Overview: Developments in DNA Sequencing

Keith R. Mitchelson,[1,2] David B. Hawkes,[3] Rustam Turakulov[4] and Artem E. Men[3]

[1]*Capitalbio Corporation: National Engineering Research Centre for Beijing Biochip Technology, 18 Life Science Parkway, Changping District, Beijing 102206, China*
[2]*Medical Systems Biology Research Center, Tsinghua University School of Medicine, Beijing 100084, China*
[3]*AGRF, Institute of Molecular Bioscience, University of Queensland, St. Lucia, Queensland 4072, Australia*
[4]*AGRF, Walter and Eliza Hall Institute for Medical Research, 1G Royal Parade, Parkville, Victoria 3050, Australia*

Contents

PERSPECTIVES IN BIOANALYSIS, VOLUME 2
ISSN 1871-0069 DOI: 10.1016/S1871-0069(06)02001-5

1. INTRODUCTION

The origins of genome era arguably commenced with the Noble prize winning work of Frederick Sanger who developed methods for utilizing *in vitro* DNA synthesis in the presence of dideoxyribonucleotides to generate a partial ladder of DNA fragments that differ by single nucleotide base steps, allowing determination of the sequence of the DNA polymer (Sanger *et al.*, 1977a, 1977b, 1982), as well as his methods of cloning in single-stranded bacteriophage to aid rapid DNA sequencing (Sanger *et al.*, 1980). Sanger sequencing has undergone a remarkable evolution since its radioactive slab-gel beginnings more than 30 years ago to a current practice of electrophoretic sizing on parallel arrays of microcapillaries and chip-based ultra-microcapillaries. Throughout the entire era of primary genome discovery its pre-eminence is because of its robustness with a wide range of different DNA sequence motifs and its unique ability to generate both long reads and highly accurate base calls.

To many scientists the report of initial draft of the sequences of the human genome in 2001 by the International Human Genome Sequencing Consortium (IHGSC, 2001) and the subsequent report of the entire euchromatin genome in 2004 (IHGSC, 2004) ended the decades long "genomic era". This era commenced in earnest in the early 1990s with initial reports from collaborative projects to sequence the *E. coli* genome (Yura *et al.*, 1992) and the *C. elegans* genome (Sulston *et al.*, 1992). Planning for the massive undertaking to sequence the human genome (Collins and Galas, 1993) also commenced. These large sequencing projects (the Human Genome Project was the largest biological program ever undertaken) commenced using Sanger sequencing, and employed electrophoretic size separations of sequentially terminated primer extensions from fragmented genomic templates. This era also saw a large increase in the efficiency of sequencing as new technologies for sequence reading – capillary array sequencers, liquid handling robotics and improved fluorescent dye chemistries and signal detectors were developed to provide for the worldwide demand for better and more cost-effective DNA sequencing.

We have now ostensibly entered the "post-genomic era" with the accumulation of significantly more than 150 billion base-pair of sequence information (142 Gb at June 2006), which provides an unprecedented opportunity to investigate deeply into the information revealed by the known genomes, and yet DNA sequencing will continue as a major and growing research and diagnostic activity, for biomedicine, to understand the basis of disease, to gain understanding of the depth of information encoded in the primary genomes, for forensic, biosafety and species identification purposes, and for comparative, archeological, taxonomic and evolutionary studies, to name but a few of the myriad applications.

The *de novo* sequencing of the genomes of many economically and medically important species is currently being undertaken (Benson *et al.*, 2005; GeneBank, 2006), as well as an accelerated re-sequencing of known mammalian genomes to discover the genetic variation lying behind phenotypic diversity and disease susceptibility (The International HapMap Consortium, 2003, 2005; The ENCODE Project Consortium, 2004). Each of these sequencing activities undoubtedly will continue apace as more and more variant genomes are examined to determine the genetic component(s) to a myriad of common diseases and conditions in man and in model animals.

Sequencing technology based on Sanger sequencing – capillary electrophoresis (CE) is still the gold standard and many large and small laboratories have access to this technology. However, for the planned large-scale screening and re-sequencing projects the current technology is either too expensive, or its capacity is too small. However, recently Blazej *et al.* (2006) have suggested that the continued use of Sanger sequencing may still be viable and that it is essential that the cost efficiency and scalability limits of this technique are taken to its ultimate limit (Figure 1). Blazej and colleagues propose that a fully integrated microfluidic genome sequencing system should achieve this aim and also lead to significant infrastructure and labor savings as well as template and reagent requirements being reduced an additional 100-fold from current array CE sequencer levels.

To address the need for a major improvement in sequencing throughput, both companies and governments are pushing to develop novel technologies that will bring the cost of sequencing to between $100,000 (http://grants.nih.gov/grants/guide/rfa-files/RFA-HG-05-003.html) and $1,000 (http://grants.nih.gov/grants/guide/rfa-files/RFA-HG-05-004.html) per whole human genome, compared to today's $3 million price. In the main, these new technologies move away from Sanger sequencing to modular unitary base addition onto multiple templates arrayed onto solid surfaces, using microfluidics and new approaches to sequence addition and ultra-sensitive signal reading technologies to capture the signals from the densely arrayed reactions (10^4–10^6 features per cm^2), and thus bring about enormous increases in productivity and volume of quality data generation.

Several excellent recent reviews (Marziali and Akeson, 2001; Shendure *et al.*, 2004; Metzker, 2005; Mitchelson, 2005; Church *et al.*, 2006) provide comprehensive information about the advantages and problems faced in development of these new sequencing technologies, and discuss polymerase colony (polony) sequencing, nanopore sequencing and sequencing by hybridization (SBH), CE and microchip sequencing and sequencing of repeated elements, and also provide very detailed information on the different DNA termination strategies and on chemistries for signal detection. This introductory chapter will draw attention to recent advances in these new sequence technologies, and also indicates some of the surprising and exciting applications in which these technologies provide real advantage.

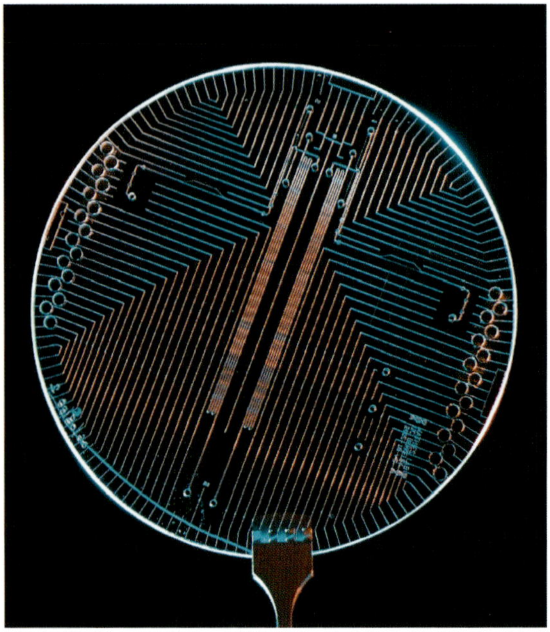

Fig. 1. A microfabricated nanoliter-scale "Bioprocessor Sequencing Factory". The microfabricated device integrates all three Sanger sequencing steps, thermal cycling, sample purification and CE. Importantly, a combination of glass and polydimethylsiloxane (PDMS) wafers was used to construct different functional elements including 250-nl reactors, affinity-capture purification chambers, high-performance CE channels and pneumatic valves and pumps onto a single microfabricated device. The "lab-on-a-chip-level" of integration requires only 1 fmol of DNA template for complete Sanger sequencing of up to 550 continuous bases with 99% accuracy. The performance of this miniaturized DNA sequencer provides a new benchmark for reducing the cost and determining the efficiency limits of Sanger sequencing of read lengths required for *de novo* sequencing of human and other complex genomes. Reprinted from Blazej *et al.* (2006). Copyright (2006), reprinted with permission from National Academy of Sciences (USA).

1.1. Biotechnological implications of ultra-high-throughput sequencing capability

Several different technological approaches are being developed to increase sequence reading throughput, while simultaneously reducing the cost of sequencing by several orders of magnitude. These developing (and prospective) technologies are being undertaken by both companies and academic researchers (see Table 1), and include new single-molecule sequencing (SMS) technologies and instruments employing the "clonal single-molecule array" developed by Solexa Inc., and "single-molecule sequencing by synthesis" developed by Helicos BioSciences corporation (see Table 1 for web site links), improvement of "sequencing by synthesis (SBS)" using pyrosequencing by the 454 Life Science Corporation (Margulies *et al.*, 2005a, 2007), and both sequencing-by-ligation

Table 1. Web-sites providing information on different aspects of new DNA sequencing technology and sequencing output

Technologies	Web-sites
Array capillary electrophoresis instrumentation and cyclic dye terminator chemistry	http://home.appliedbiosystems.com/ http://amershambiosciences.com http://www.gehealthcare.com/auen/index.html http://www.cchem.berkeley.edu/ramgrp/alpha/
MALDI-TOF mass spectrographic sequencing. SNuPE and small oligomer fragmentation sequencing	http://www.sequenom.com/ http://www.methexis-genomics.com/ http://www.nuvelo.com/
"Sequencing by synthesis" and sequential polymerization enzymology	http://www.pyrosequencing.com/ http://www.biotage.com/ http://www.agencourt.com/
Massively parallel "sequencing by synthesis". Array chemistry and advanced signal detection technologies. Advanced base-calling software	http://www.454.com/ http://www.solexa.com/ http://www.helicos.com/ http://www.visgen.com/
Nanofluidic barrier technology. Nanopore technologies. Sequencing of single DNA molecules	http://www.nanofluidics.com/ http://www.ionian-technologies.com/ http://www.usgenomics.com/ http://www.mcb.harvard.edu/branton/index.htm http://www.ornl.gov/
Alternative DNA sequencing tools. DNA-barrier breaking enzymes, sequencing enhancers, enzyme re-engineering	http://www.molecularstaging.com/ http://www.fidelitysystems.com/ http://www.stratagene.com/ http://www2.mrc-lmb.cam.ac.uk/groups/ph1/pub.html http://www.maxygen.com/ http://www.scripps.edu/news/sk/sk2005/sk05barbas.html http://www.venterinstitute.org/ http://www.deltadot.com/
DNA and oligonucleotide arraying technologies. Optical masking, nanobead arrays	http://www.nimblegen.com/ http://www.affymetrix.com/index.affx http://www.illumina.com/
Laboratory-on-a-chip technologies, nanoengineering, nanofluidic and nanoanalytical technologies	http://www.capitalbio.com http://www.microchipbiotech.com/ http://www.nanofluidics.com/ http://www.biotrove.com/ http://www.lab-on-a-chip.com/home/index.asp http://thebigone.stanford.edu/

Table 1 (*continued*)

Technologies	Web-sites
Fundamental advances in dye chemistries, CE equipment and micro/nanofabricated device engineering	http://www.cchem.berkeley.edu/~cab/ http://depts.washington.edu/chemfac/dovichi/ http://www.chem.harvard.edu/faculty/whitesides.html
Microarrays for specific "sequencing by hybridization" of known SNPs and tiling of known genome regions. Non-coding RNA gene detection, promoter detection.	http://www.capitalbio.com http://www.affymetrix.com http://www.nimblegen.com/
Physical and electrical analysis of DNA sequence	http://www.integratednano.com/ http://www.handylab.com/ http://www.ornl.gov/
Genomic analysis technologies	http://www.keygene.com/ http://www.broad.mit.edu/ http://www.opgen.com/
Genomic sequence data bases Advanced access tools for genomic analysis and sequence analysis	URLs http://genamics.com/ http://www.geospiza.com/ http://www.technelysium.com.au/chromas.html http://www.nucleics.com/ http://www.eecs.berkeley.edu/~gene//
MicroRNAs analysis and microRNA genomics tools	http://www.mekentosj.com/ http://www.rosettagenomics.com/ http://www.wi.mit.edu/
GenBank	http://www.ncbi.nlm.nih.gov/Genbank/index.html
European Bioinformatics Institute	http://www.ebi.ac.uk/
DOE Office of Science	http://www.doegenomes.org/
International HapMap Project	http://www.hapmap.org/
Human Genome Project	http://www.sanger.ac.uk/HGP/

(Shendure *et al.*, 2005) and fluorescent *in situ* sequencing (FISSEQ) (Mitra *et al.*, 2003), while advances have also been made to greatly reduce the costs of microchip sequencing, which although utilizing Sanger chemistries provide substantially longer read lengths than the above solid-phase sequencers, and have sequence turnaround in minutes rather than hours while using minute volumes of reagents.

2. ADVANCED SEQUENCING TECHNOLOGIES

Worldwide, approximately US$3 billion was spent in 2005 on all aspects of genomic sequencing – the analyzer equipment and software and on sequencing reagents and enzymes. The vast majority of this sequence output was determined using CE technology with Sanger sequencing, which commensurately has been developing progressively over the past 20 years. Following the commercial release of the "Sequencer GS20" array pyrosequencer by the 454 Life Science Corporation (Margulies *et al.*, 2005a, 2007), this industrial space will be increasingly occupied by sequencing technologies that do not use electrophoresis, although for many conventional, sequencing projects, Sanger sequencing employing fragment separation will remain the preferred technology, if its advantage of long sequence reads will be used in combination with the alternate ultra-high-throughput sequencing technologies that generate short-read lengths (Desany *et al.*, 2005).

2.1. Capillary electrophoresis and Sanger sequencing

CE offers high resolution and high throughput, automatic operation and data acquisition, with on-line detection of dyes bound to DNA extension products – for reviews see Kan *et al.* (2004) and Mitchelson (2003). Operational advances such as graduated electric fields and automated thermal ramping programs as the run progresses can result in higher base resolution and longer sequence reads. Advanced base-calling algorithms and DNA marker additives that utilize known fragment sizing landmarks can also help to improve fragment base calling, increasing call accuracy and read lengths by 20–30% (http://www.nucleics.com). Yet, despite the high efficiency of CE sequencers, the delineation of the human genome and its implication for genome-wide analysis for personalized medicine is driving the development of devices and chemistries capable of massively increased sequence throughput compared to the throughput capable on conventional CE sequencers. Miniaturization of CE onto chip-based devices provides some increased efficiencies – a significant improvement in the speed of long sequence reads and improved automation of handling and analysis. However, the new array based sequencing devices also promise a quantum increase in efficiency, which microchip sequencers will find hard to match. Each of these new devices provides an extremely high throughput, high-quality data and low-process costs, yet currently their reads are between 30 and 100 bp average length.

2.2. High-throughput capillary-array sequencing

A CE instrument comprises two electrolyte chambers linked by a thin silica capillary 50–100 μm in diameter, or as a fine microchannel on a silicon chip. The thin capillary rapidly dissipates heat generated by the large electric fields, stabilizing band resolution. An in-line detector positioned close to the capillary outlet acquires data from the size-fractionated molecules. Typically, dyes attached to the DNA fragments are detected using laser-induced fluorescence (LIF). During the recent accumulation phase of genome research, electrophoresis-based capillary-array sequence analysis developed rapidly becoming the paradigm for DNA sequencing and providing simultaneous multi-parallel analyses. Capillary-array electrophoresis (CAE) is a multiplied version of conventional CE, with up to 100 parallel capillaries, or channels on miniature CE chips arranged radially (Paegel *et al.*, 2002a, 2003), such that each one can simultaneously analyze individual samples. Sequencers such as the Megabace 4500 system (GE Healthcare) with 384 capillaries process sequence data from long 800 bp reads, equivalent to some 2.8 Mb per day. Novel new block copolymer separation media for CE and chip electrophoresis devices represent a new paradigm for longer and faster sequence analysis (Doherty *et al.*, 2004). Sparsely cross-linked "nanogels" composed of sub-colloidal polymer structures of covalently linked, linear polyacrylamide chains act as novel replaceable DNA sequencing matrixes, particularly for microchip electrophoresis (Barron, 2006). The physical network stability provided by the internally cross-linked structure of the nanogels results in substantially longer average read lengths compared to conventional LPA matrix. More conventional sequencing developments involving Sanger sequencing include improvement in chip CE equipment and design (Kan *et al.*, 2004; Xiong and Cheng, 2007), while Jovanovich and colleagues (http://microchipbiotech.com/) and Blazej *et al.* (2006) intend to push Sanger-based sequencing toward its performance limit in a fully automated, bench-top system. The heart of these systems will be a microchip-based device that can label and process DNA fragments from individual microbeads or solution in low-volume reactions, followed by ultra-fast separation and analysis on microfabricated CE channels.

2.3. Signal detection dyes and detectors

Dye-tagged nucleotides with higher sensitivity and better spectral discrimination than earlier fluorophore dyes are being developed, in concert with new enzyme systems to efficiently incorporate them into DNA (Metzker, 2005; Kumar and Fuller, 2007). Highly sensitive photometric devices such as CCD cameras can detect extremely low numbers of signal molecules, despite constraints on the ease of detection of single molecules due to photobleaching of dyes (Eggeling *et al.*, 2006). Micromachined sheath-flow cuvettes that precisely control both capillary alignment and matrix flow, and confocal LIF systems with one lens, for both excitation and detection optics, that scan continuously across a bundle of capillaries held in register can provide both longer reads and more accurate base-calling. Alternative detection methods such as time-resolved fluorescence decay, electrochemical detection, chemi-luminescence and near-infrared (NIR) detectors can also be incorporated into CE devices, yet none have emerged yet as competitors to fluorescent dyes except for IR dye sequencing systems from LiCor (www.licor.com). Recent development of efficient incorporation systems for these alternative dyes will make their use more widespread. Dye-tagged nucleotides with longer linker lengths and charge matching can also improve the incorporation of these bulky molecules into nascent DNA (Kumar and Fuller, 2007). Finn *et al.* (2003) created a series of charge-modified, dye-labeled 2′,3′-dideoxynucleoside-5′-triphosphate terminators, which possess a net positive charge and migrate in the opposite direction to dye-labeled Sanger fragments during electrophoresis. The charge-modified nucleotides are efficiently incorporated by a number of DNA polymerases. Post-sequencing purification is not required to remove unreacted nucleotides prior to electrophoresis, and DNA sequencing reaction mixtures can be loaded directly onto a separating medium. New nucleotides dye labeled at the terminal phosphate (Kumar *et al.*, 2005; Edwards *et al.*, 2007) will also play a role in the new single nucleotide-addition sequencing (SBE) systems. Chain extension halts after each addition step allowing excitation and fluorescent signal detection, and further extension is prevented until the dye is removed, regenerating a terminal phosphate.

2.4. Microchip electrophoresis

We also draw the reader's attention to reviews by Bruin (2000) and Xiong and Cheng (2007) and to research by Krishnan *et al.* (2001) of recent advances in the miniaturization and automation of nano and microliter volume reaction processing and their roles in the improvement of DNA amplification processes and for alternative approaches to sequencing. The microchip CE analysis format has significant advantages over conventional CE. It is up to 10 times faster and uses sub-microliter volumes of analysis reagents. Pal and colleagues (2005) elegantly demonstrated the extremely rapid DNA analysis using an integrated microchip system incorporating both DNA amplification and CE separation of products and identified sequence-specific hemagglutinin A subtype for the A/LA/1/87 strain of influenza virus. This system integrated fluidic and thermal components such as heaters, temperature sensors and addressable valves to control two nanoliter reactors in series and is suitable for a variety of genetic analyses. Significant advances in the amplification and detection of single DNA template molecules on integrated devices are providing unprecedented levels of sensitivity.

New constriction-microchannel designs from Paegel *et al.* (2002a, 2002b, 2003) improve fragment resolution and increase the scope for longer path lengths, permitting single base resolution over longer fragments (see also, Figure 2h). The miniaturization of chip CE systems with nanochannels $<1\,\mu m$ allows analysis to be undertaken on limiting numbers of molecules held pico- and nanomolar concentrations, with amplification and detection of signals from single template molecules (Xiong and Cheng, 2007). Microfabricated multi-reflection absorbance cells for microchip-based CE are being built with 5- to 10-fold enhanced sensitivity over single-pass devices. These schemes are being built into devices with hundreds of capillaries to achieve high-speed and extremely high-throughput and detection sensitivity.

2.5. Capillary electrophoretic sequencing on microcapillary chips

Doherty *et al.* (2004) demonstrated the improvement of microchip-based DNA sequencing read-lengths and base-call accuracy with nanogel matrixes in a high-throughput microfabricated DNA sequencing device consisting of 96 separation channels densely fabricated on a 6-in. glass wafer. Aborn *et al.* (2005) also described the development of a 768-lane microfabricated system in large-format ($25\,cm \times 50\,cm$) 384-lane arrays for high-throughput *de novo* genomic DNA

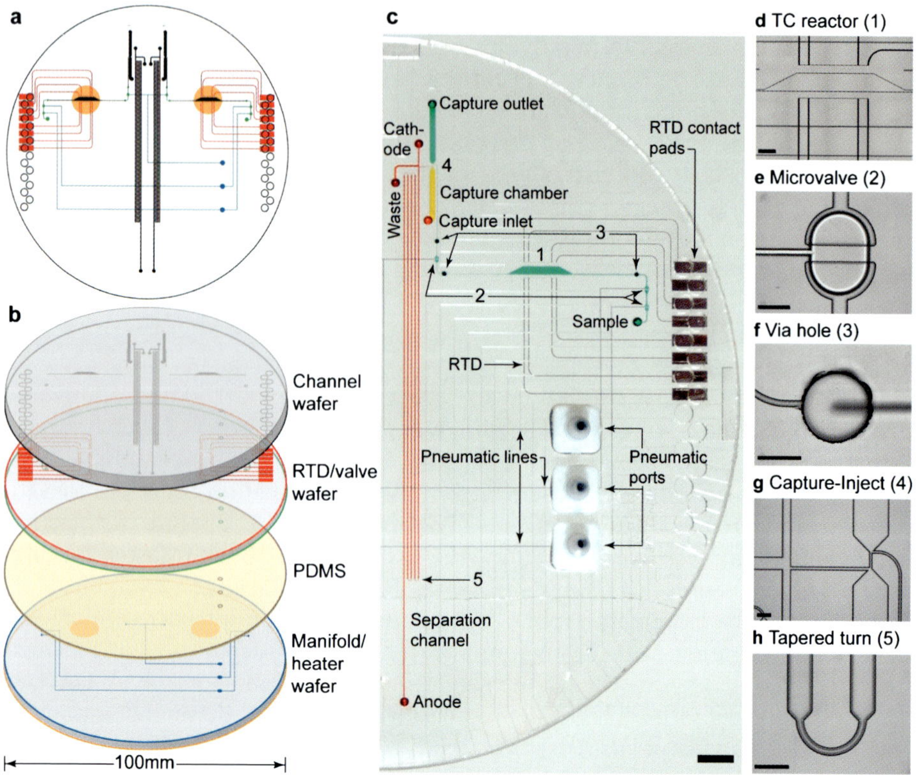

Fig. 2. An integrated nanoliter-scale nucleic acid bioprocessor for Sanger DNA sequencing developed by the Mathies group. (a) Top view of the assembled bioprocessor containing two sets of thermal cycling reactors, purification/concentration chambers, CE channels (black), RTDs (red), microvalves/pumps (green), pneumatic manifold channels (blue) and surface heaters (orange). (b) Expanded view, showing microdevice layers. Rim colors indicate the surface on which the respective features are fabricated. The top two glass wafers are thermally bonded and then assembled with a featureless PDMS membrane and manifold wafer. (c) A photograph of the microdevice, showing one of the two complete nucleic acid processing systems. Colors indicate the location of sequencing reagent (green), capture gel (yellow), separation gel (red) and pneumatic channels (blue) (scale bar, 5 mm). Notations d–h correspond to the following component microphotographs: (d) a 250-nl thermal cycling reactor with RTDs (scale bar, 1 mm); (e) a 5-nl displacement volume microvalve; (f) a 500-μm diameter via hole; (g) capture chamber and cross injector; (h) a 65-μm wide tapered turn (scale bars, 300 μm). All features are etched to a depth of 30 μm. Reprinted from Blazej *et al.* (2006). Copyright (2006), reprinted with permission from National Academy of Sciences (USA).

sequencing. The two 384-lane plates are alternatively cycled between electrophoresis and regeneration and achieve a total of $>172{,}000$ bases, at 99% accuracy (quality score Phred 20) for each run of a 384-lane plate. This corresponds to a throughput >4 Mb of raw Phred 20 sequence per day. This microcapillary format allows operation at "$1/32\times$ "Sanger chemistry, and tests suggest that sample can be further reduced to $1/256\times$ Sanger chemistry in the microdevice. Yet, this microcapillary device still used conventional microliter-scale processing that generates a 1000-fold more product than is needed for analysis (Shi *et al.*, 1999). Together, these advances directly address the cost model requirements for the next step beyond capillary array instruments, while retaining the long read lengths of Sanger chemistries. Other nanoreactors, with serial electrodes that provide for high "sweeping field" separation using low voltage supplies suited to hand held devices, can achieve PCR-amplification in 15 min and CE analysis in 2 min (Krishnan *et al.*, 2001; Pal *et al.*, 2005). The nanoreactors can be interfaced to either microelectrophoresis chips or capillary gel tubes via micromachined capillary connectors or zero-dead volume unions, and signals are detected using NIR fluorescence detector. Moreover, new low-voltage closed-loop CE devices also offer the promise of hand-held or readily transportable analysis instruments. Microdevices that require no operator intervention and that integrate sample purification, sample amplification, amplicon product purification and DNA sequencing by CE have been developed (Paegel *et al.*, 2002a). Unincorporated dye terminators are electrically separated from sequencing products under high voltage into a waste channel, prior to diverting the sample into a separation capillary for size resolution and sequence analysis.

Further advances with these types of Sanger sequencing devices may permit complete CE analysis in an easily transportable format (a DARPA request), a capacity that the large new SBS array devices with ultra-sensitive optical systems currently do not allow. In this regard, Blazej *et al.* (2006) recently reported on the construction and application of an efficient, nanoliter-scale microfabricated bioprocessor (Figure 2) integrating all three Sanger sequencing steps of thermal cycling, sample purification and CE. The design had a number of novel features that aided miniaturization into an integrated device – the use of electrophoretic and pneumatic forces for sample movement improved sample transfer through holes into channels, and the transitioning from a monolithic substrate to a hybrid glass – PDMS assembly was also critical to the function of the device (Figure 2a–c). Multi-layer construction also enabled a much greater design complexity and permitted the exchange of materials across fluidic and pneumatic lines (Skelley *et al.*, 2005), and was necessary for parallel processing. The wafer-scale device was constructed to form a single microfabricated instrument with 250-nl reactor chambers, affinity-capture purification chambers, high-performance CE channels, and pneumatic valves and pumps. This device involved "lab-on-a-chip-level integration" (Krishnan *et al.*, 2001; Hansen and Quake, 2003) of each of these functional components and was shown to be capable of undertaking complete Sanger sequencing from only 1 fmol of DNA template. Their associated development of optimized capture and resolution gels also aided the single base separation. The volume of affinity-capture gel (Paegel *et al.*, 2002b) used to pre-concentrate the sequencing products was scaled down to 250 nl to eliminate excess sample. The resolution of the commonly used dye-terminator samples was improved by extension of the separation channel from 16 to 30 cm, which increased the resolving power of the system, producing error rates of ~1 in a million between 100 and 300 bases read. The device was capable of single base length fractionating and thus continuous reading of up to 556 bases with 99% accuracy. The lengths of these reads are thus still superior to the best of SBS devices and are realistically the lengths required for *de novo* sequencing of mammalian and other complex genomes.

Blazej *et al.* (2006) note additional potential improvements to their system may be possible, and that appropriately tuned separation gels (Doherty *et al.*, 2004) could result in uniformly resolving peaks over greater amplicon lengths and extend the currently >99% accuracy range. Improvement of injection techniques combined with increased scanner sensitivity could extrapolate the ultimate minimal template quantity to a conservative ~100 amol, within a fabricated 25-nl reactor (requiring further reduction in volume 10-fold), and would represent a 400-fold reduction from the current Sanger sequencing reagent consumption and require 800-fold less DNA template (Karlinsey *et al.*, 2005). The practical limits to any reduction in reaction volume will be determined by the sensitivity of the detection system. The ability to achieve single-molecule detection (SMD) coupled with the miniaturization technologies described above will be required to achieve the optimal requirements for the analysis and manipulation of samples on a single molecule scale. Dittrich and Manz (2005) present the unique benefits of single fluorescent molecule detection in microfluidic channels, which may be central to the reduction of Sanger sequencing chemistries. The integrated device of Blazej *et al.* (2006) provided a new performance benchmark for the evaluation of the feasibility of miniaturizing high-throughput Sanger sequencing and for determination of the costs of this established technology against newly emerging low-consumable cost solid-phase-array sequencing systems (Margulies *et al.*, 2007).

2.6. Sequencing by mass spectrometry

Mass spectrometry is used to determine the sequence of a polynucleotide by analysis of the atomic masses of a series of polynucleotide sub-fragments derived by the partial and uniform fragmentation, or by the uniform extension of the polynucleotide (see this volume, Ehrich *et al.*, 2007). The polynucleotide sub-fragments are released from a solid phase and analyzed by matrix-assisted laser desorption/ionisation time-of-flight mass spectrometry (MALDI-TOF MS). The techniques include partial enzymatic cleavage, chemical cleavage, base specific cleavage of PCR products and primer extension Sanger sequencing (single nucleotide primer extension, SNuPE) methodologies. Nucleotide analogs are used widely during DNA sequence analysis by mass spectrometry – to modify polynucleotide electro-chemistry and stabilize the N-glycosidic linkages, to block base loss and subsequent random backbone cleavage or to introduce site-specific backbone weakness for controlled fragmentation at particular types of base. Wolfe *et al.* (2002, 2003) have employed analogs such as 7-deaza analogs, 5′-amino-5′-deoxy- and 5′-amino-2′,5′-dideoxy-analogs, and stable mass isotope tags to substitute for particular nucleotides. These nucleotide analogs introduce differential mass properties or differential stability into the DNA sub-fragments which improve the mass separations. Similarly, acids and bases can cleave dideoxy and amine-modified analog backbones generating sequencing ladders, which may be analyzed by mass spectrometry.

3. SOLID-PHASE ARRAY SEQUENCING DEVICES

The demand for ultra-high-throughput re-sequencing for personal medicine is driving the development of non-electrophoretic platforms, capable of sequencing multiple single (amplified) molecules held at defined locations in parallel. These integrated platforms employ technologies ranging from microfluidic arrays and solid-phase chemistries, SBS chemistries, ultra-sensitive optics and CCD data capture (see Table 1). Newly commercialized nanosequencing technologies already offer markedly increased capacity in sequence output, as required for rapid genome-sequencing projects and for large genome re-sequencing programs.

3.1. Ultra-sensitive detectors and sequencers

Hebert and Braslavsky (2007) provide a detailed description of equipment being developed to achieve longer reads from very large numbers of single DNA molecules by improving the sensitivity of SMD of fluorescence resonance energy transfer on a total internal reflection microscope. Simultaneous with these developments, versatile and improved nucleotide analogs with better incorporation kinetics and better fluorescent signal output are being developed (Kumar and Fuller, 2007). Different positions of dye labeling have also been explored. For example, a number of terminal phosphate-labeled nucleotides with three or more phosphates and with varied length linkers attached between the terminal phosphate and the dye have been synthesized (Kumar *et al.*, 2005; Kumar and Fuller, 2007; Edwards *et al.*, 2007). These nucleotides have utility as substrates for DNA sequencing where the base addition is unitary, until the reporter is eliminated from the terminal position and a terminal phosphate is regenerated. Williams and colleagues at LiCOR (http://www.licor.com/) are developing charge-switch technology to detect the release of reaction products when nucleotides are incorporated into single DNA strands, while Metzker's (2005) group are also working on developing novel fluorescent, photolabile nucleotide terminators for SBS. Edwards *et al.* (2007) also describe the use of reversibly terminating dye-tagged nucleotides for single nucleotide extension SBS, as well as making improvements to DNA polymerases that will support their accurate incorporation into DNA. One focus of their research is novel chemistry that allows a fluorescent molecule attached to a nucleotide to be detected and then removed with a flash of light after its addition to a growing DNA molecule, as well as integrated sequencing systems by combining terminating nucleotides and array platforms.

3.2. Sequencing by synthesis

"Sequencing by synthesis" is a method common to primer extension methods such as SNuPE and pyrosequencing, in which a unitary base addition chemistry that allows single nucleotide additions to growing chains to be monitored on each oligonucleotide feature, simultaneously with the addition of one of the four differentially labeled terminating nucleotides. Church and colleagues

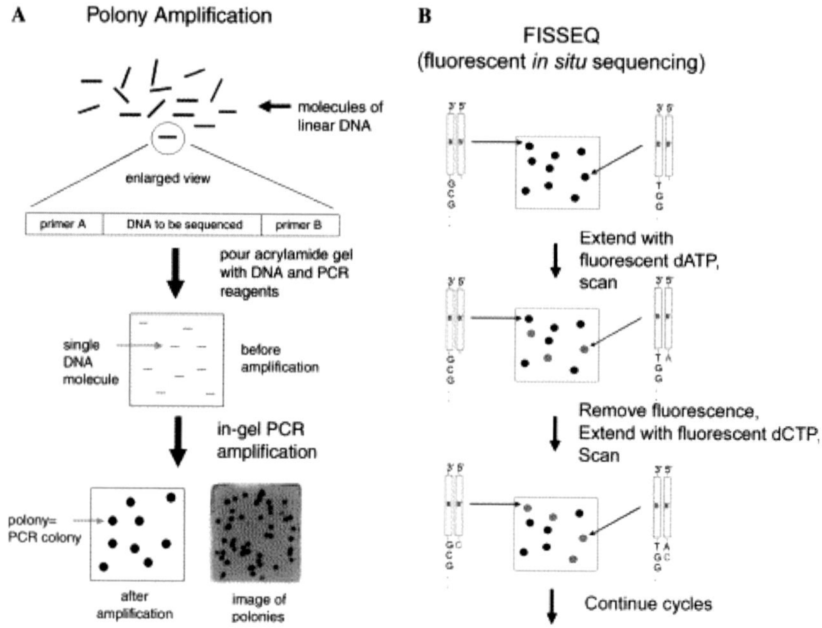

Fig. 3. (A) Polony amplification. A library of linear DNA molecules with universal priming sites is PCR amplified within a polyacrylamide gel. Each single template molecule gives rise to a polymerase colony or "polony". (B) Fluorescent *in situ* sequencing. Polonies are denatured, and a sequencing primer is annealed. Polonies are sequenced by serial additions of a single fluorescent nucleotide. Reprinted from Mitra *et al.* (2003). Copyright (2003), reprinted with permission from Elsevier.

(Mitra *et al.*, 2003; Shendure *et al.*, 2005; Zhang *et al.*, 2006a) initiated the integration of solid-state DNA sequencing using polymerase colonies ("polonies") and cycles of fluorescent dNTP incorporation with high-signal sensitivity that allow multiple polonies to be sequenced in parallel. Large-scale arrays of discrete "polonies" can be extended cyclically, bringing cost-effective genome-scale single-array sequencing (Figure 3). "FISSEQ" involves the addition of the one nucleotide, the extended fragments are then all detected simultaneously using CCD optics, the terminating moiety and fluoro tag are then removed chemically from each attached nucleotide readiness for the addition of the next nucleotide in the following cycle. The series of base additions is only interrupted for signal scanning (and for data-acquisition) and for chemical treatment of the slides to remove dye-signals prior to the next extension step. An alternative "sequencing by ligation" (Shendure *et al.*, 2005) technology has been rapidly developed and an advanced version is currently undergoing commercial development by the Applied Biosystems subsidiary, Agencourt Biosciences/Agencourt Personal Genomics (http://www.agencourt.com/) as the "Supported Oligo Ligation Detection" (SOLiD) process for massively parallel sequencing by stepwise ligation.

The technology company the 454 Life Sciences Corporation (Margulies *et al.*, 2007) have developed a solid-phase parallel microarray system of microfluidic

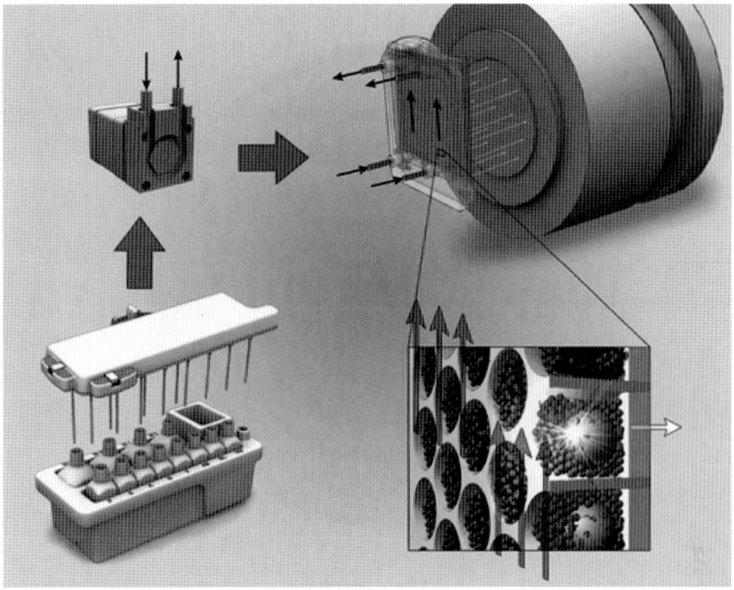

Fig. 4. The GS20 Sequencer. The bead-based pyrosequencing process involves the flow of sequencing reagents containing buffers and nucleotides in a fixed sequential order across the PicoTiterPlate™ device during a sequencing run. Each of the hundreds of thousands of beads (each with millions of clonal copies of a DNA fragment) is located in fixed micro well and all are sequenced in parallel. The addition of one (or more) nucleotide(s) results in a reaction that generates a light signal, which is then recorded by the CCD camera. The strength of the light signal is proportional to the number of nucleotides incorporated. For example, a short homopolymer region incorporates a number of nucleotides in proportion to its length, in a single nucleotide flow. The speed of sequencing and process chemistry is simplified by use of native dNTPs. Copyright (2005), reprinted with permission from the 454 Life Science Corporation.

wells in which 3×10^4 features per cm^2 (captured wells) are used to capture single-stranded fragments of sheared genomic DNA attached to microbeads, which are then PCR-amplified *in situ* by emulsion PCR, resulting in some 10^7 copies of a unique fragment attached at each bead (Figure 4). The DNA sequence is read by cycles of incorporation of single native deoxynucleotide triphosphates (dA, dG, dC or dT) at each feature and detected by pyrosequencing (Hyman, 1988; Nyren *et al.*, 1993; Ronaghi, 2001). Gharizadeh *et al.* (2004) noted the interference from primer–dimers and loop structures that give rise to false sequence signals during pyrosequencing could be improved by employing Sequenase polymerase, and homopolymeric regions could be read through for more than five T bases. Interestingly, Eriksson *et al.* (2004) reported that the new analog, 7-deaza-2′-deoxyadenosine-5′-triphosphate (c7dATP) has low substrate specificity for luciferase, while the inhibition of apyrase was reduced significantly, and read lengths up to 100 bases were obtained by pyrosequencing for several templates from fungi, bacteria and viruses. In order to prevent the incorporation of erroneous bases, limiting amounts of each dNTP are added at each addition cycle. This measure however can contribute to dephasing the synchronicity of nucleotide addition across all copies of a template attached to a bead. Homopolymer tracts are sinks for incorporation of the low amount of dNTP, and failure to complete extension across all copies of a homopolymer may result in mixed or dephased signals in later addition rounds and cause bead dropout.

Despite these technicalities, the outcomes from the "Genome Sequencer GS20 (Mb)" pyrosequencing platform have developed rapidly. In August 2003, the 454 Life Sciences Corporation announced the sequence the 35 kb genome sequence of adenovirus by pyrosequencing and by 2004 could achieve the sequence of entire bacterial genomes of several Mb length (Margulies *et al.*, 2005a; Andries *et al.*, 2005) with some 20 Mb of total sequence generated on a single PicoTitre plate within 3.5 h. By the end of 2005, the sequence of simple eukaryote genomes such as yeast (12 Mb) and other small eukaryote microorganism genomes could be undertaken using one Sequencer GS20 machine within one week. Poinar *et al.* (2006) also recently reported using the platform to sequence some 14 Mb of ancient DNA from preserved mammoth tissues indicating the rapid development of applications for ultra-high-throughput short-read sequence (see Section 5). Recently, the 454 Life Sciences Corporation introduced an updated version of the GS20 v1.02, with improved single-read accuracy, new gasket formats, software algorithms with additional applications and an LIMS interface. They also announced development of a new version of their sequencer, the GS100, for analysis of larger genomes than the current GS20, with an expected release in 2007. The broad swathe of applications to which SBS technology can be applied has been growing rapidly as researchers explore the benefits of a low-cost platform that provides extremely deep-sequence coverage of small libraries. It appears very suited to identifying genetic variation in mixed samples due to its high depth of coverage (Thomas *et al.*, 2006). This coverage depth to some extent overcomes some of the limitations of the shorter reads, particularly with nucleic acids that are short, or low in repetition, such as RNA species, fragmented DNAs and the simpler genomes of microorganisms. Goldberg *et al.* (2006) also evaluated the integration of GS20 data with conventional Sanger whole shotgun sequencing data for genomic assembly, concluding there was improved cost-effectiveness using a hybrid sequencing approach by combining standard capillary sequencer Sanger WGS data and GS20 data to generate higher-quality lower-cost assemblies of microbial genomes.

Ju and colleagues (Meng *et al*, 2006; Edwards *et al*, 2007) have developed an approach to DNA SBS using reversible fluorescent nucleotide terminators to address the limitations of current DNA sequencing techniques. The photocleavable fluorescent nucleotide analog, 3′-*O*-allyl-dGTP-PC-Bodipy-FL-510 has been developed as a reversible terminator for SBS. The nucleotide is incorporated by DNA polymerase efficiently into an extending DNA strand, where it terminates the polymerase reaction. Following the unitary addition, the fluorophore is photocleaved quantitatively by irradiation at 355 nm and the allyl group is rapidly and efficiently removed by using a Pd-catalyzed reaction under DNA-compatible conditions to regenerate a free 3′-OH group on the ribose, which reinitiates the polymerase addition reaction. Successive cycles of such addition cleavage-reactivation steps could be used successfully to sequence a homopolymeric region of a DNA template (Meng *et al.*, 2006). This reversible-termination SBS technology promises to be a viable approach for high-throughput DNA sequencing.

Recently, Aksyonov *et al.* (2006) reported a new DNA SBS method in which the sequences of DNA templates were obtained by determining the number of nucleotides extended within the primers at each array spot in sequential DNA polymerase-catalyzed nucleotide incorporation reactions, using single fluorescein-labeled dNTP species. The fluorescein label can be destroyed following the readout of each addition step by a photo-stimulated reaction. Self-quenching was avoided by diluting the labeled dNTP with unlabeled reagent.

3.3. Single DNA molecule sequencing

The need for analysis of single DNA molecules has stimulated the development of technologies with a lead-time of 3–5 years, which have unique single-molecule sensitivity. Several solid-state methods for single DNA molecule sequencing have been reported recently, again with promise of highly parallel, genome-scale efficiencies. Several corporations (see Table 1) are developing state-of-the-art array instruments for sequencing of individual molecules of DNA or cDNA (RNA). The technologies developed by Solexa currently involve amplified single template molecules and use several innovations. The first involves a zero-mode waveguide (Levene *et al.*, 2003), which confines optical excitation and detection to the few zeptoliters of fluid surrounding the polymerase at the interface between the attached DNA molecules and the surface of the chip (Figure 5); the second innovation involves the development of cluster DNA amplification whereby DNA molecules were modified by two adapters attached at either ends, and then replicated *in situ* via a bridging process between surface attached complementary adapters; and the third, the use of extremely sensitive CCD low-intensity imagers (Jansen *et al.*, 1989), which can capture low-intensity signals from single molecule sequence extension events. These detectors can capture images with densities of 10^8 pixels per cm^2. Single DNA molecule imaging can potentially achieve simultaneous analysis of up to 100,000 distinct target molecules every second (Bennett *et al.*, 2005).

Helicos Corp is developing a procedure called "true single molecule sequencing by synthesis" (tSMS). The procedure involves working directly on fragments of genomic DNA, eliminating DNA amplification (Braslavsky *et al.*, 2003). The use of single DNA molecules means that they can be packed closely on the solid surface, with an entire human genome arrayed across a single glass substrate chip. The tSMS technology relies on cyclic SBS, using some 1.2 billion strands of DNA attached to a quartz slide, by directly interrogating each of the single molecules after each nucleotide addition step. The tSMS process does not employ template molecule amplification. Despite the need for highly sensitive detectors and greater statistical coverage to confirm sequence, this lack of amplification provides a number of benefits, including no PCR bias thus it has potentially fewer errors, and no dephasing issues as individual molecules are

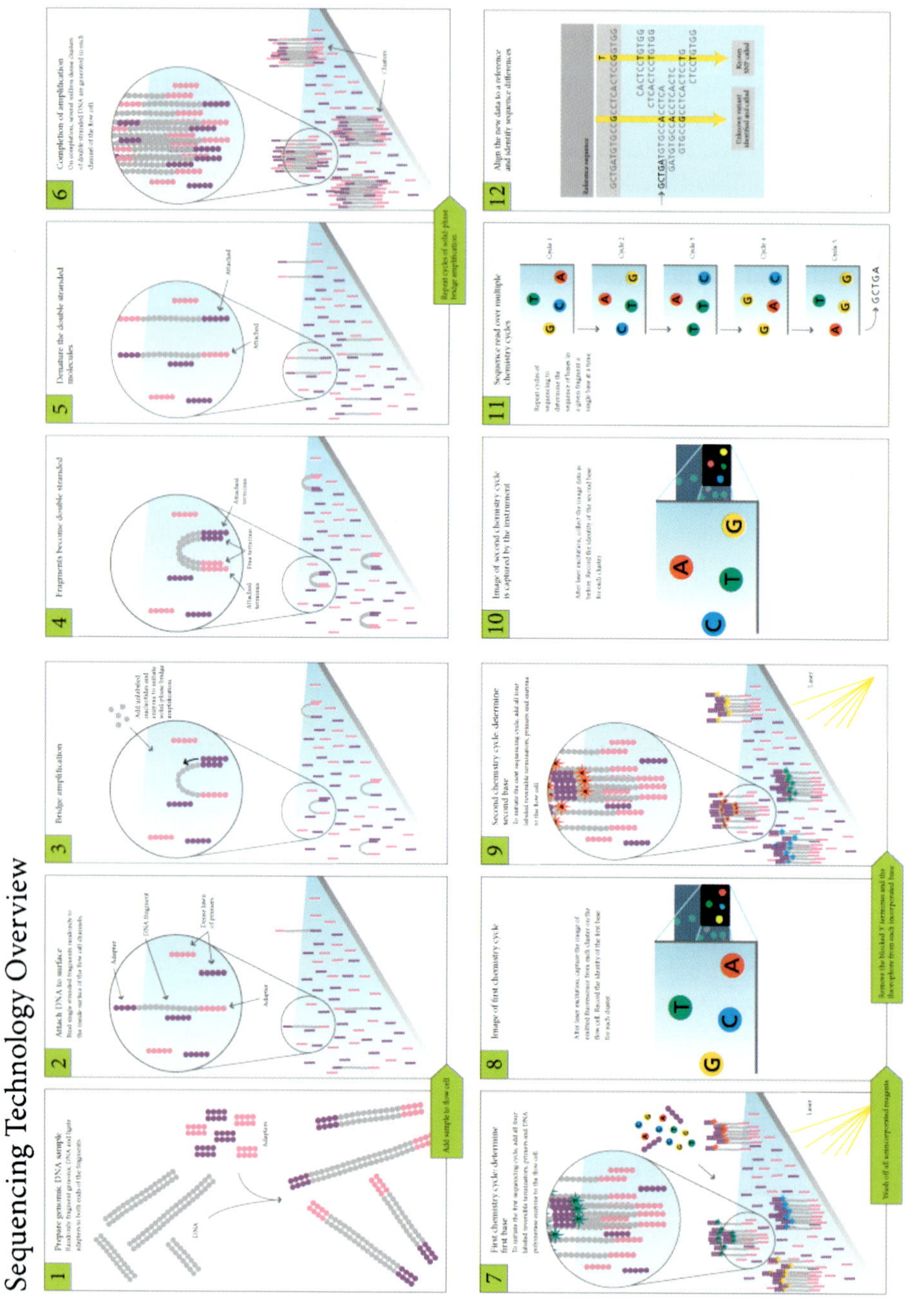
Sequencing Technology Overview
1 Prepare genomic DNA sample
2 Attach DNA to surface
3 Bridge amplification
4 Fragments become double stranded
5 Denature the double stranded molecules
6 Completion of amplification
7 First chemistry cycle: determine first base
8 Image of first chemistry cycle
9 Second chemistry cycle: determine second base
10 Image of second chemistry cycle is captured by the instrument
11 Sequence read over multiple chemistry cycles
12 Align the new data to a reference and identify sequence differences

either read in one round, or are read in a subsequent round of the cyclic process. The high template-packing density on the slide surface, with up to 10^8 molecules per cm^2 projected (Kartalov and Quake, 2004) will provide the largest amount of sequence information per data image. This high-density and high-throughput sequencing will allow the detection of rare genomic mutations and polymorphisms, as well as rare transcripts if cDNAs are arrayed. The method is expected to have running costs around 1000 times less than Sanger sequencing. Recently, Helicos announced the successful sequencing of the 6.4 kb long M13 phage genome including short homopolymeric sequences. These analyzers are expected to develop sequencing rates of 10^9 or more base reads per day, the equivalent of a billion-lane sequencer that reads the sequence of each molecule at the speed of the addition reaction. Although currently the efficiency and uniformity of extension is poor, it is expected that if each molecule could be extended by an average 50 nucleotides, it will allow parallel discovery and detection of genetic variation on 10^8 molecules that can be aligned to known reference sequence (such as the human genome). Currently, 'SBS' methods that generate short reads between 25 and 100 bp may permit *de novo* sequencing of entire genomes of low repetition. Methods to eliminate repetitive DNA from genomes have also emerged as complementary technologies: techniques such as methyl DNA depletion (Emberton *et al.*, 2005) and high-Cot fractionation (Braun *et al.*, 1978; Peterson *et al.*, 2002) can effectively enrich for the unique genome fraction of organisms ranging from mammals to those with highly repetitive genomes such as crops and other plant species. These fractionation techniques require highly fragmented DNA, which conveniently is a size range compatible with the short-read array sequencing platforms.

Genome assembly is a key outcome associated intimately with the manner of genomic sequencing, which is reviewed by McGrath (2007) in this volume. The vast majority of DNA sequencing is still performed using Sanger methods, while array pyrosequencing is still a relatively new technology. This gives rise to questions of whether sequencing data from both technologies be combined and assembled together? The 454 Life Sciences Corporation suggests that the same assembly tool allow flowgrams and chromatograms to be assembled together

---

Fig. 5. Solexa's genetic analysis technology is based on massively parallel short-read sequencing, using its Clonal Molecular Array technology (Steps 1–6) and novel reversible terminator-based sequencing chemistry (Steps 7–11). The approach relies on attachment of randomly fragmented genomic DNA to a planar optically transparent surface (Steps 1–2) and solid-phase amplification (Steps 3–6) to create an ultra-high-density sequencing flow cell with >10 million clusters per cm^2, each containing $\sim$1000 copies of template. These templates are sequenced using a very robust four-color DNA SBS technology that employs reversible terminators with removable fluorescence (Steps 7–11). This approach ensures high accuracy and avoidance of artefacts with homopolymeric repeats. High sensitivity fluorescence detection is achieved using laser excitation and total internal reflection optics (Step 10). Short sequence reads are aligned against a reference genome and genetic differences called using a specially developed data pipeline (Step 12). Alternative sample preparation methods allow the same system to be used for a range of other genetic analysis applications, including gene expression. Copyright (2006), reprinted with permission from the Solexa Corporation.

(Desany *et al.*, 2005), and that this combined technology approach can improve assemblies that suffer from cloning biases such as unclonable regions. Further support for incorporating flowgram data into existing tools is underway, for example, with the release of the Staden package 1.6.0. Recent announcements from the 454 Life Sciences Corporation indicate that they intend to develop a software with features to address many of these issues. These include new algorithms that improve sequence assembly and contig building, and algorithms that will also facilitate sequencing of large number of short DNA fragments, like serial analysis of gene expression (SAGE) tags, cap analysis of gene expression (CAGE) tags and microRNA analysis.

3.4. Hybridization re-sequencing

The unparalleled success in the application of competitive hybridization to discriminate gene expression at tens of thousands of genes simultaneously on microarrayed assemblies of (addressed) complementary oligonucleotides, displayed on slide surfaces (Schena *et al.*, 1995; Bertone *et al.*, 2005), is presently being coupled with rapid improvements in reducing feature size, in higher printing density and in detector resolution and sensitivity (www.capitalbio.com) which have lead to a renewed interest in the use of massively parallel hybridization analysis to genotype or "re-sequence" and "tile" selected unique (non-repetitive) regions of the human genome for purposes as diverse as comparative hybridization (Wilson *et al.*, 2006), identification of conserved non-coding RNA genes (Washietl *et al.*, 2005) and the distribution of methylated DNA in different cells (Schumacher *et al.*, 2006). By virtue of the impracticality of representing all possible variants of a large genome region for detection and also by its inability to screen long repeat motifs and homopolymer regions, this technology is *in sensu stricto* limited to detection of unique copy regions that contain SNP variants. Nonetheless, its speed of data generation coupled with the widespread availability of high-resolution arrayers and analyzers (www.capitalbio.com; www.affymetrix.com) will engender defined SBH sequencing applications, such as for large-scale single chromosome re-sequencing (Garnis *et al.*, 2005), for re-sequencing of large sets of known developmental genes or disease-linked genes (Bertone *et al.*, 2004; Selzer *et al.*, 2005), as well as for more limited SNP typing 're-sequencing' applications (Hardenbol *et al.*, 2005). Another advantage of hybridization sequencing by microarray is the ability to repeat the analysis, and indeed to target genomic regions of interest and variation with confirmatory analysis. These features, along with the robustness of the platform and ease of sample preparation will continue to attract users of the technology.

4. FUTURE TECHNOLOGIES

Until recently, there were few serious contenders to challenge the supremacy of Sanger sequencing with respect to data quality, read-length and flexibility of application. However recent broadening of the applications of microarrays and the emergence of several ultra-throughput SBS platforms to market reality has changed the spectrum of future technologies to emerging technologies. Single-molecule technologies that promise large gains in sequencing performance remain in accelerating development.

4.1. Nanopore membranes

Recently, Nakane *et al.* (2004) described the detection of the ion current signal during single DNA molecule capture at an α-Hemolysin nanopore in a lipid bilayer membrane (Figure 6). The successful capture of an analyte molecule in the pore results in a reduction of the potential to +10 mV for a short period. The exit of the probe from the pore is then prevented by the bound analyte, and impedance remains at blocked channel values. When the potential is reversed to −60 mV this provides the force to withdraw the probe from the pore and after time the probe disassociates from the analyte and an open channel (reverse) current is restored. More speculative SMS technologies involving nanopore membranes (Deamer and Akeson, 2000; Rhee and Burns, 2006) are also under development (Lee and Meller, 2007 this volume). While most development of nanopore techniques involves DNA sequence transducers to detect an electrical or ionic signal from individual DNA molecules (Nakane *et al.*, 2004), Lee intends to examine stretched DNA molecules made to pass between ultra-sharp electrodes spaced 2–5 nm apart to distinguish between the four types of nucleotides based on differences in a physical phenomenon called electron tunneling (Lee and Meller, 2007).

In contrast, the Meller group intends to use nanopores to simultaneously detect electrical and fluorescent signals. The method employs ultra-fast optical reading of Design DNA polymers (Meller *et al.*, 2005; Lee and Meller, 2007). Here, each nucleotide in the original DNA is substituted with a group (composed of a unique binary sequence of 3–16 nucleotides). Fluorescently tagged oligonucleotides that are complementary to this sequence are hybridized to the converted DNA. A nanopore is then used to unwrap the hybrid, and different fluorescent signals are detected briefly (1 ms) as they emerge from process, and achieve read rates of ~1 Mb/s (Lee and Meller, 2007). Other evidence collected by Mathé *et al.* (2005) suggests that the translocation process through a narrow pore involves base tilting and stretching of ssDNA molecules inside the confining pore and is strand-orientation dependent. In other studies, Viasnoff *et al.* (2006) found that rapid cooling (>100 °C/ms) locks DNA molecules into a unique alternative conformation that is retained over weeks at room temperature, and that this state can be probed using fluorescent energy transfer. These observations suggest that non-equilibrium DNA switch states may be more amenable to nanotechnology applications (such as for SMS). Fologea *et al.* (2005) observed that voltage biased solid-state nanopores can detect and characterize individual single-stranded DNA molecules of fixed length by operating a nanopore detector at pH values greater than approximately 11.6. They found that at this pH a large component of the DNA molecules are unfolded and can access the pore in this state.

Others intend to amplify the nucleotide signals by producing conical gold-coated nanopores in a synthetic membrane to control DNA transport and then

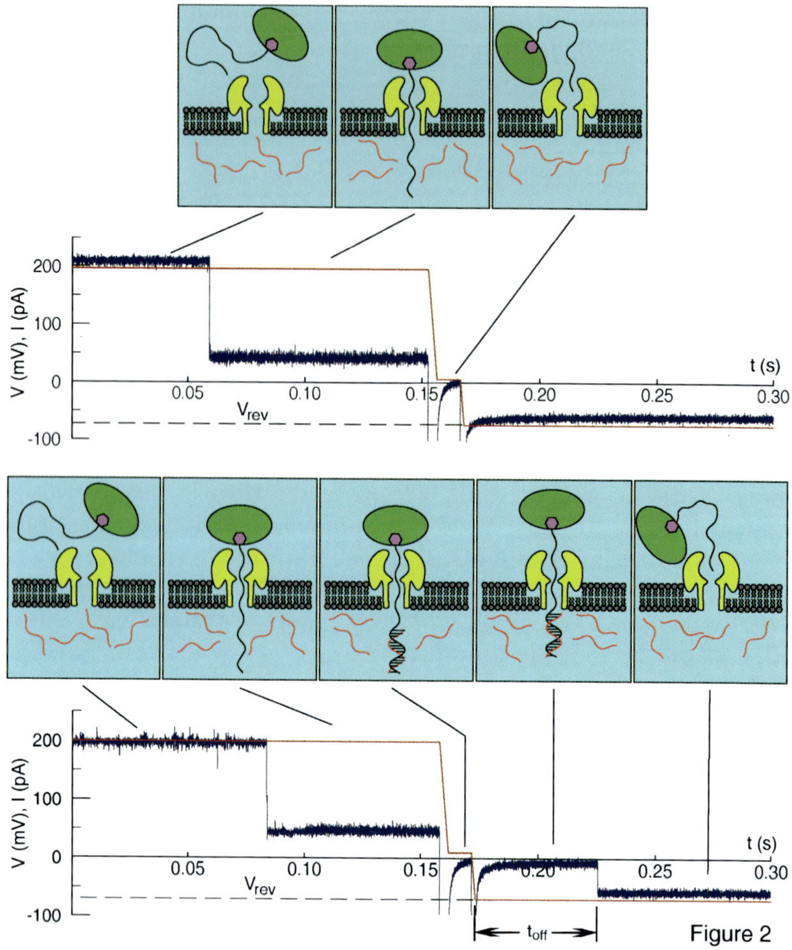

Fig. 6. Cartoon representing ion current signal detection during single DNA molecule capture at an α-Hemolysin nanopore in a lipid bilayer membrane. *Upper*: experimental data from the unsuccessful capture of an analyte molecule. *Lower*: experimental data from the successful capture of an analyte molecule. After the probe is captured in the pore, the potential is reduced to +10 mV for a short period. The exit of the probe from the pore is then prevented by the bound analyte, and impedance remains at blocked channel values. The potential is then reversed to −60 mV, this provides the force to withdraw the probe from the pore, and after time t_{off} the probe disassociates from the analyte and an open channel (reverse) current is restored. Analysis of many different dissociation event lifetimes (t_{off}) at different reverse potentials (V_{rev}) can identify the different analyte characteristics. Reprinted from Nakane *et al.* (2004). Copyright (2004), reprinted with permission from The BioPhysical Society and Highwire Press.

detect different signals from the four types of chemically modified nucleotides introduced into the DNA. Karhanek *et al.* (2005) reported an alternative to nanopore membranes where single DNA molecules labeled with nanoparticles can be detected as they translocate through a finely pulled nanopipette tip by their blockades of ionic current. SMS promises to radically improve DNA sequencing as it is potentially 10,000 times faster than present production systems that rely on single lanes. It can potentially start directly with genomic DNA, reducing the need for sample preparation. SMS read lengths are also potentially significantly longer than those obtained from gel electrophoresis systems. Longer read lengths will simplify sequence reconstruction and reduce the total number of runs required to get full coverage of the genome, although at its present state of development only short reads are possible (Lee and Meller, 2007). Uniquely, because fragments from single alleles of chromosomes will be used, SMS sequencing potentially can directly detect haplotypes over several linked polymorphisms.

4.2. Direct electrical detection of DNA synthesis

Pourmand *et al.* (2006) reported the development of electrical biosensors for direct electrical detection of enzymatically catalyzed DNA synthesis by induced surface-charge perturbation. The incorporation of a complementary deoxynucleotide (dNTP) into a self-primed single-stranded DNA attached to the surface of a gold electrode evokes an electrode surface-charge perturbation, which can be detected as a transient current by a voltage-clamp amplifier. It is proposed that the electrode detects proton removal from the 3′-hydroxyl group of the DNA molecule during phosphodiester bond formation and this phenomenon can be potentially exploited at polarizable interfaces for evoking surface-charge perturbations specific for the addition of individual nucleotide species. The use of electrical biosensors are of interest for genotyping and re-sequencing, because of their potential high speed and for the elimination of the need for DNA labeling and optical detection, as well as their potential for miniaturization and automation.

5. APPLIED SHORT-READ GENOMIC SEQUENCING

This section hopes to capture some of the new high-throughput sequencing applications, as researchers explore questions that could not be asked economically by Sanger sequencing. Particularly, the momentous announcement that the second human genome to be sequenced will be that of the extinct Neanderthals using the pyrosequencing technology of the 454 Life Sciences Corporation.

5.1. Genotyping by re-sequencing

The new class of surface-array sequencing instruments provides greatly increased and parallel sequencing capacity, sufficient for "whole genome analysis". For the biomedical profiling of individuals, extremely dense polymorphism maps need to be created, involving the directed re-sequencing of targeted unique sequence regions. Although DNA hybridization microarrays and genome-tiling arrays (see Table 1) produced by Affymetrix, CapitalBio, ParAllele BioScience and other companies offer one form of re-sequencing – pre-determined genomic re-sequencing by hybridization (re-SBH) that is limited to the analysis of known (and common) polymorphisms that are represented on the arrays. Detection of novel or rare variants is possible if the alternative alleles are also present in the arrayed set of hybridization probes.

5.1.1. Polony genotyping

Re-sequencing that also profiles different patients for responses to drugs and genetic predisposition to disease, and for the monitoring of newly developed mutations in cells (http://cancergenome.nih.gov) across the entire genome for the molecular identification of nascent cancers is also an important new application for high-throughput sequencing technologies. Zhang and colleagues (2006b) reported a method for *in vitro* multi-locus long-range haplotyping of human chromosome molecules, based on DNA polymerase colony (polony) technology. They immobilized thousands of intact chromosome molecules within a polyacrylamide gel on a microscope slide and could perform multiple amplifications from single molecules and then sequence defined polymorphic loci on located chromosomes. Long-range haplotypes spanning distances over hundreds of megabase of different human chromosomes could be examined. This type of detailed information will catapult human medicine into a new era, where accurate monitoring is achievable within several days, rather than months.

5.1.2. Pyrosequencer genotyping

Recently, Keygene (www.keygene.com) reported development of an application on the pyrosequencer Sequencer GS20 platform to enable large-scale SNP discovery and analysis in higher eukaryotic organisms. The technology, which they call complexity reduction of polymorphic sequences, or "CRoPS", helps to identify large-scale polymorphism in organisms with low levels of germplasm polymorphism or with highly repeated genomes that are difficult to analyze. Use is made in the CRoPS process of AFLP technology (Vos *et al.*, 1995) to prepare tagged complexity-reduced libraries of two or more genetically diverse samples, which are then sequenced at 5- to 10-fold redundancy with the 454 Life Sciences Corporation's Sequencer GS20. Typically this yields a total of over 20 Mb of sequence, each read of about 100 bases. The resulting sequences are then clustered and bioinformatics tools are used to inspect the sequence contigs for two or more classes of differences in common elements between the two varieties. In effect, it identifies additional sequence polymorphism at the level of the gene in a significant portion of the unique sequence regions of the genome. In another application of short-read array sequencing, repeated sequences from conifer genomes are being segregated on the basis of their level of repetition in the genome (Cot analysis), and different repetition groups are sequenced to very-high-coverage levels, allowing the evolution of these repeated sequence elements to be investigated in depth.

5.1.3. Polymorphism ratio sequencing

Blazej *et al.* (2003) also recently reported the development of polymorphism ratio sequencing (PRS) that combines new DNA labeling and sample pooling schemes with high-throughput DNA sequencing to create a sensitive new assay for SNP discovery, for rapid genotyping, and for accurate determination of allele frequency with a multiplexed set of samples. The PRS method uses dideoxy-terminator extension ladders generated from a sample and reference template labeled with different fluorescent dyes on two bases A + C or G + T. These labeled ladders are co-injected into a CE separation capillary for comparison of their relative signal intensities. Sample and reference produce coincident peaks where there is identity of sequence, and non-coincident peaks when there is a polymorphism. Insertions, substitutions and deletions can be readily detected. Because only two bases are labeled, the peaks are more well spaced, and polymorphism is easily seen, even with poor sequence regions. Titration of multiplexed DNA samples was used to determine the limit of minor allele frequency detection, which could be seen at 5%. Thus, PRS is a sensitive and robust polymorphism detection method for analysis of multiplexed samples and is compatible with any four-color fluorescence DNA sequencer.

5.1.4. BEAMing

Dressman *et al.* (2003) described a new high-throughput method for the detection of uncommon variations in individual genes or transcripts. Each DNA molecule in a collection of many similar molecules is attached onto the surface of a single magnetic particle, this single sequence is then amplified by emulsion PCR and by rolling-circle amplification (Li *et al.*, 2006) to create a bead with up to 30,000 copies of the DNA, identical to the original bound molecule. The population of beads thus represents a one-to-one population of the starting DNA molecules, but each is highly amplified on its unique bead location. Variation within the original population of DNA molecules can then be simply assessed by counting fluorescently labeled particles via flow-cytometry. The targets were labeled by hybridization of a target sequence detector, then they were identified by a specific fluorescently labeled antibody. They called this approach "BEAMing" (beads, emulsion, amplification and magnetics). Millions of individual DNA molecules can be rapidly assessed, and moreover, specific variants can be isolated by flow-sorting for further analysis. BEAMing can be used for the identification and quantification of rare mutations as well as to study variations in particular gene sequences or transcripts in specific populations or tissues. Although this is a complementary technology, it uses several methods in common with array SBS, and then selects rare molecules for further analysis. A combination of the two technologies is readily imagined, where selected rare-DNA beads are immobilized in a PicoTitre plate and sequenced.

5.2. PaleoGenomics

Several types of sequence analysis, which have been difficult to undertake conventionally, have recently generated rapidly growing interest. The isolation and sequence determination of DNA from mummified organisms and other forms of ancient preserved tissues have been pioneered by Pääbo and colleagues

(Pääbo *et al.*, 2004; Noonan *et al.*, 2005; Krause *et al.*, 2006). Several particular problems beset such sequencing – one is the degraded state of the ancient DNA and second, the likelihood of contamination with DNA from extant organisms from the environment containing the sample tissue. A description of issues surrounding such studies with preserved DNA is described by Pääbo *et al.* (2004), and further detailed in this volume for both preserved and environmental samples (Kowalchuk *et al.*, 2007). Recently, Poinar *et al.* (2006) reported the sequencing of random genomic elements from mammoth tissues preserved for ~20,000 years in the Siberian permafrost in a "metagenomics approach". These were determined in high-throughput manner using the short-read array GS20 (the 454 Life Sciences Corporation) system despite some >50% contamination of the mammoth DNA with fragments from the genomes of decay organisms. The process of unique single-strand capture onto nanobeads and emulsion PCR amplification of that single strand used by the 454 Life Sciences Corporation, allowed the mammoth (and contaminant) sequences to be readily determined. These different sequences could then be easily distinguished from one another by reference to detailed databases containing modern elephant reference sequence and those of reference microorganisms.

Although Cooper and Poinar (2000), and in this volume, Kowalchuk *et al.* (2007) indicated the measures necessary for usual work with ancient DNA samples with highly degraded DNA and for validation of findings, this metagenomic approach – using emulsion PCR and pyrosequencing for the determination of all the genomic sequence found in ancient tissues – marks a technical breakthrough for paleogenomics. The short DNA fragment size found in such preserved tissue samples is also often an ideal size for the average 100 bp read length of pyrosequencing. Other approaches have also been successful, but these used pre-amplification of mitochondrial genomic elements prior to sequencing (Noonan *et al.*, 2005). This pre-amplification approach however is usually particularly difficult with the short nuclear genomic fragments typically found in such samples. Similarly, the paucity of ancient DNA fragments recovered from preserved tissues stands against use of complementary techniques such as hybridization enrichment onto capture arrays to remove contaminating environmental DNAs. Interestingly, the wealth of information generated on both the target species and on its ancient contaminants by this metagenomic sequencing approach is likely to define a new direction in paleogenomic sequencing.

5.3. Neanderthal genomics

In July 2006, scientists at the Max Planck Institute for Evolutionary Anthropology and the 454 Life Sciences Corporation announced (http://www.eva.mpg.de/) the intention to sequence the complete genome of the closest (extinct) relative of humans – the Neanderthal. Comparative sequencing has already shown that approximately 99% of the *Homo sapiens* genome is identical to the chimpanzee genome, our closest living relative. It is estimated from pilot sequencing of about 1 Mb of Neanderthal genome (from a 45,000-year-old Croatian fossil) that it shares about 96% of the 1% difference (with chimpanzee) with *Homo sapiens*, making its genome significantly more similar to that of human than chimpanzee. The Neanderthal shares the remaining 4% of the difference with the chimpanzee, allowing a further comparison between Neanderthals and higher primates. Project director Professor Svante Pääbo suggested that "the analysis of the estimated 4% of genome variation that Neanderthal shares with the chimpanzee will help us to understand the evolution of characteristics specific to the *Homo sapiens* and perhaps even aspects of cognitive function."

This genome project will be completed in about 2 years, yet compared to the first human genome sequence, which accumulated with increasing speed as sequencing technology developed to provide it, will still represent a major force for further technical advance (Margulies *et al.*, 2007). The pyrosequencing platform represented by the GS20 with about 25 Mb sequence per run is being expanded to a 100 Mb capacity and paired-end sequencing will be implemented. This project also represents a major reconstruction of our approaches to genomics and builds on the significant body of knowledge of genomes acquired through Sanger sequencing platforms. Traditionally, great care was needed to ensure purity of genomic materials and absence of all contaminants except the target. With ancient DNA, additional care is needed to avoid modern contamination (see Kowalchuk *et al.*, 2007), yet the natural contamination of the Neanderthal genomic material with decay organisms and environmental organisms as well as the chemical modification and fragmentation of DNA requires combined metagenomic and paleogenomic approaches. Further, the limited amount of sample means that maximum information must be extracted from all material, and measures must be implemented to improve data recovery that will require development of new bioinformatics tools for extraction of validated data and for statistical interpretation of sequences from the multiple genomes. This need has already been illustrated in the pioneering pyrosequencing of preserved mammoth genomic sequences (Poinar *et al.*, 2006) that identified a significant number of co-contaminant species and in itself represents a technical bioinformatic *tour de force*.

5.4. MetaGenomics

The new ultra-high-capacity sequencing systems will also usher in an era of renewed microbiological analysis for both biosafety applications, industrial process and quality control uses, and will foster the emergent study of whole ecosystems at the level of their microbial biota – "metagenomics" (Venter *et al.*, 2004). In the former of these applications, the definition of industrial species will involve determining whole genome sequence in addition to physiological classifications, and will lead to new insights into metabolic capabilities. The demand for ongoing monitoring of the genetic identity of important industrial strains for optimal production will also use whole genome sequencing at regular intervals. Andries *et al.* (2005) illustrated another important application for genomic sequencing in the development of new drugs. Mutants resistant to the new drug were sequenced to identify newly acquired polymorphisms in unknown genes. Such analysis could provide early leads for second tier drugs that are targeted at different enzymes or functions.

The capability to sequence mixed biota to sufficient depth will allow entirely new species of microorganisms to be identified in the environment in the absence of classical culture of the organism. Complementary use of isothermal *in vitro* methods such as strand displacement amplification (SDA) and rolling-circle amplification for amplification of a specific DNA sequence or even whole genomes to concatamer lengths (Hutchison *et al.*, 2005) will be used to capture the genomic content. Environments will be defined in terms of the enzyme and gene systems resident within them, rather than by purely classical terms of identifiable culturable organisms defined by a limited set of physiological parameters (Remington *et al.*, 2005). Even with the more simple genomes of bacteria, the question of how many genes define the genome of a bacterial species is not fully known. In some species, new genes continue to be discovered even after sequencing of perhaps hundreds of genomes per species (Medini *et al.*, 2005). A bacterial species might be ultimately defined by its "pan-genome", which is composed of a "core genome" containing genes present in all strains, and an additional "dispensable genome" containing genes present in two or more strains, as well as genes unique to single strains. The resulting pan-genome of a bacterial species might be several orders of magnitude larger than the single genome of any individual strain.

Gladitz *et al.* (2005) also used a pan-genomic and metagenomic approach in studies on the similarity of codon usage when comparing novel gene sequences found in clinical isolates of *H. influenzae.* The genes were compared against reference sets of different prokaryotic, eukaryotic and viral genomes. About 65% of the novel sequences identified in the *H. influenzae* isolates have codon use similar to other *Haemophilus* sp., indicating a pool of variants of common genes. However the remaining 35% of the novel sequences are more similar to other reference genomes, suggesting these sequences entered the *H. influenzae* gene pool more recently. Bacteria such as *H. influenza* grow with other bacteria in biofilms, with the possibility that the higher number of bacterial transformations seen in biofilms is a factor permitting a process of horizontal transfer genes between species. Definition of these variant genes in the different genomes of microorganisms and bacteria will continue to be a formidable task that will require comprehensive and rapid new sequencing technologies to cater for the expected high volume of new sequencing.

5.5. SAM sequencing of repetitive DNAs

Sequencing Aided by Mutation (SAM) (Keith *et al.*, 2004a, 2004b) is a radically different method that overcomes sequencing difficulties caused by problematic and repetitive motifs – regions where local sequence characteristics hinder existing sequencing technologies and subsequent sequence assembly. Problematic motifs and regions are mutated in a random manner, sufficiently to introduce diversity into repeated motifs. The random locations of the mutations cause the copies of the region to have different sequences. The DNA sequences are determined from a low number of the altered copies, which are then analyzed using Bayesian methods to reconstruct the original wild-type sequence, with efficiencies and accuracy similar to conventional sequencing (see this volume,

Keith *et al.*, 2007). The method may have advantages for the assembly of short sequence reads during *de novo* genome sequencing, including very short reads <50 bp in length (Cochran *et al.*, 2006). SBS methods such as pyrosequencing are also particularly prone to error at homopolymer regions. This is because the quantification of the strength of dNTP addition signals does not permit homopolymers longer than 6–7 bp to be distinguished accurately. Another source of error is because limiting amounts of individual natural dNTPs are used in each extension cycle to minimize mis-incorporation effects that occur at higher concentrations. Thus, one major drawback with pyrosequencing is the incomplete extension through long homopolymer repeats, leading to loss of register on the many copies of the template, and causing read dropout on individual beads. The introduction of random mutations during the SAM process can reduce the effective homopolymer lengths into a series of shorter homopolymers, which are then more tractable to pyrosequencing (Keith *et al.*, 2007). This measure could potentially result in both longer and more accurate sequencing of homopolymer regions, as well as reduced bead dropout.

5.6. Transcriptome and expressed RNA sequence analysis

Massively parallel signature sequencing involves using microbeads with 32-mer tags comprising strings of eight 4-mer "words" that hybridize amplified cDNA library targets (Brenner *et al.*, 2000a, 2000b; Reinartz *et al.*, 2002). The beads each hybridize about 100,000 molecules of a unique cDNA, and the beads are then sorted on an FACS analyzer and distributed in a micromachined flow-cell into a planar array. The sequence of each immobilized cDNA bead is then "read" successively by the restriction enzyme *BbvI* after linking of encoded "adapters" to the end of each exposed specific cDNA overhang and by a second set of fluorescently labeled "decoder" oligonucleotides. The system has large capacity for sequence analysis, with some 1 million cDNA-coated beads being simultaneously analyzed in a single machine experiment within several days, with over about 20 bases read per bead. This technology provides a "signature" for each cDNA rather than full-length sequence analysis. Massively parallel signature sequencing (MPSS) has no requirement that genes be identified and characterised prior to conducting an experiment as it is based on sequencing of a signature region, and it is sufficiently sensitive to routinely detect a few molecules of mRNA per cell. The system has particular utility for the quantitative analysis of expression data of mRNA and other transcribed RNA sequences accumulated from representative, but uncharacterized cDNA libraries (Nakano *et al.*, 2006). It can measure the gene expression level under defined conditions and tissues and provide information about potentially novel transcripts such as antisense transcripts, regulatory intergenic transcripts and alternative splice (isoform) transcripts. In addition, a modified version of MPSS could be used to perform deep profiling of small RNAs (http://mpss.udel.edu). MPSS can provide many advantages in generating the type of complete transcriptional data sets needed to facilitate hypothesis-driven experiments.

This utility has been employed by groups defining data sets of representative differences between cell lines and disease states, for in-depth digital analysis of the accumulated data. For example, Jongeneel *et al.* (2003) described the characteristic transcriptomes of two cultured cell lines, HB4a (normal breast epithelium) and HCT-116 (colon adenocarcinoma), using MPSS. Their comprehensive snapshot showed that the number of genes expressed at one copy per cell or more in either of the lines was between 10,000 and 15,000 and the majority of the transcripts could be mapped to known genes and their polyadenylation variants. Of the known genes that could be identified from their signature sequences, about 8500 were expressed by both cell lines, whereas 6000 showed cellular specificity. More recently Liu *et al.* (2006) compared seven different human embryonal cell lines (stem cells and carcinomas) as part of a database establishment study. This detailed level of cellular expression identity can have an enormous predictive role in systems analysis of cellular functions (Janes *et al*, 2005; Brandman *et al*, 2005; Bornholdt, 2005).

Importantly, Oudes *et al.* (2005) undertook comparison of two high-throughput methods frequently used to profile transcriptomes – DNA hybridization analysis by Affymetrix GeneChip Array and MPSS. Each method has certain strengths and weaknesses and this study undertook a direct comparison between both the analysis systems. The transcriptome of two lineage-related prostate cancer cell lines, LNCaP and C4-2, was analyzed to identify genes associated with prostate cancer progression. All together, the LNCaP and C4-2 cell lines were found to express 10,308 genes. Figure 7 illustrates the data acquisition by the two methods. Both technologies detected genes that the other

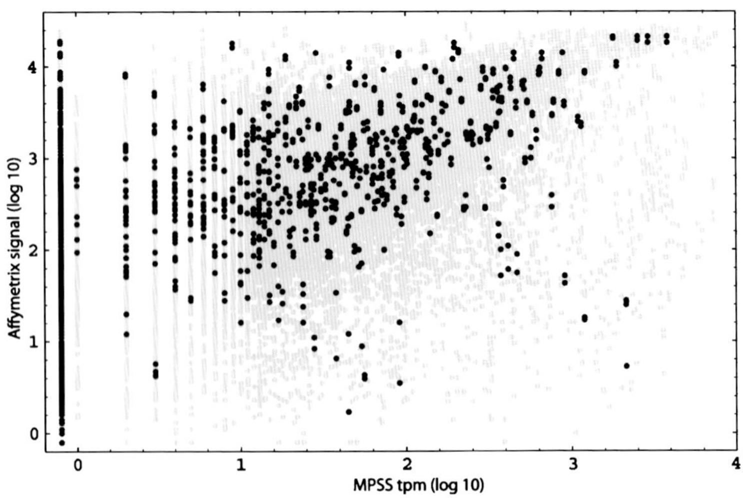

Fig. 7. Comparison of Affymetrix and MPSS gene expression signals for the analysis of two prostate cancer cell lines, LNCaP and C4-2. Unique gene expression signals were found with the two analysis systems. In cell line LNCaP, 3180 genes were only detected by Affymetrix array and 1169 genes were only detected by MPSS. Similarly, in cell line C4-2, 4121 genes were only detected by Affymetrix and 1014 genes were only detected by MPSS. Reprinted from Oudes *et al.* (2005). Open access article no. doi: 10.1186/1471-2407-5-86. The electronic version can be found at: http://www.biomedcentral.com/1471-2407/5/86

did not. These data provide evidence that transcriptome profiling using a single methodology does not fully assess the full spectrum of expression of all genes in a cell line and a combination of transcription profiling technologies such as DNA array and MPSS provides a more robust means to assess the spectrum of gene expression.

Although genome sequencing provides primary sequence of an organism, various search tools are used to identify putative genes and their regulatory elements – start and stop codons of genes, splice junctions, poly adenylation sites, etc. Yet this annotation is often incomplete as many genes do not conform to normal or general rules of identity, and many non-coding and small RNA genes have characteristics that are not yet identified by search algorithms. Strategies such as MPSS sequencing of the transcriptome can provide primary evidence for the presence of a gene, or for another type of transcribed locus of the genome. Shah *et al.* (2006) have used MPSS transcriptome sequencing to identify and annotate the genome from 73% of computationally predicted genes in the *Theileria parva* schizont life-cycle stage. They identified some 83 putative genes, >100 codons overlooked by annotation software and 139 potentially incorrect gene models (either with truncated ORFs or with overlooked exons) by interfacing signature locations with stop codon maps. Their study illustrates an important role for mass random re-sequencing using tools such as MPSS or other short-read platforms for improving the annotation of small, gene-rich microbial or eukaryotic genomes where primary genome sequence data are not as well studied as is mammalian genomes.

Zhu *et al.* (2003) described the use of "digital polony exon profiling", a technology for studying complex alternative pre-messenger RNA splicing based on single-molecule analysis. It allows monitoring of the combinatorial diversity of exon inclusion in individual transcripts. A polony mini-sequencing strategy was then used for resolution of single nucleotides in a sequential manner such that individual splicing variants could be quantified. Digital polony exon profiling could be used to investigate the roles of alternately spliced messenger RNAs in different tissue states. Analysis of the mechanism of alternative pre-messenger RNA splicing is important for understanding the generation of protein diversity.

Increasingly ultra-high-throughput short-read sequencing technologies are being used to sequence nucleic acids that are not amenable to Sanger sequencing (Meyers *et al.*, 2006). The determination of the sequence of small RNAs and microRNAs by the MPSS (Mineno *et al.*, 2006; Nakano *et al.*, 2006) and recently by the GS20 pyrosequencer platforms (Henderson *et al.*, 2006; Girard *et al.*, 2006) is an application that these two technologies are well suited for, as the "sequence dead space" inherent in electrophoretic separation of Sanger sequencing fragments is absent from the output of both platforms, and reads are obtained from very close to the 5′-terminus of the target. The increasing importance of analysis of small nucleic acid regions and motifs, either expressed or genomic sequences for the investigation of the genetic basis of diseases (Plasterk, 2006; Mattick and Makunin, 2006), suggests increased use of these and other new non-electrophoretic sequencing technologies in the near future.

5.7. MPSS and genome analysis

Yet another powerful use to which MPSS has been applied, is for the mass re-sequencing of a genome-wide library of DNase I hypersensitive (HS) sites (Crawford *et al.*, 2006). Mapping of DNase I HS sites in nuclear chromatin is a powerful method of identifying many different types of regulatory elements active in that tissue, and can be used to understand how genes are regulated in tissues at various stages of development and in disease states. Crawford and colleagues sequenced close to 230,000 short tags and found 14,190 clusters of sequences that occur in close proximity to each other in the genome. About 80% of these DNase HS sites map uniquely within one or more annotated genomic regions that contain highly conserved regulatory sequences, including CpG island regions 2 kb upstream of genes. About 10% of the DNase HS sites were cell-type specific, suggesting regulatory elements that specify cell type or development can be identified. This application of MPSS might be used more generally to study chromatin structures involved in dictating cell function and fate.

5.8. Optical mapping

Optical mapping (Ashton *et al.*, 1999; Greulich, 2005; Reslewic *et al.*, 2005) is an ingenious complementary technology that aids the mapping and reconstruction of genomes and employs an advanced nanotechnological development of the concepts of ordered restriction mapping introduced by Smith and Birnstiel (1976) and later by Schwartz and colleagues in the 1990s (Reed *et al.*, 1998). Here, large genomic DNA molecules are prepared for analysis using methods similar to that used for pulsed field gel electrophoresis (PFGE). The long DNA molecules are then extended linearly over the derivatized glass surface of a microfluidic optical chip device (Ashton *et al.*, 1999; also see this volume Zhou *et al.*, 2007). As the molecules flow through the optical chip, individual DNA molecules elongate and attach to the surface. The immobilized DNA molecules are digested with sequence-specific restriction endonucleases, typically at 6-bp recognition sequences. The DNA molecules are under slight tension, and the ends retract upon cleavage leaving gaps which can be visualized, and the fragments sized precisely using vibration-free precision optical measurement equipment. The information can be used to assemble overlapping sets of single molecules, identified by an ordered assemblage of closely similar DNA restriction fragments, including identifying errors in standard sequence assembly (Reslewic *et al.*, 2005; Zhou *et al.*, 2007). During the sequence finishing process optical maps can confirm the sequence assembly by imaging fragments in a range between 5 and 45 kb, underscoring the unique "molecular cytogenetics" niche in resolution occupied by the optical mapping system. Software searches for similar fragment patterns in different single molecules and assembles them into a consensus map representing the whole genome. The consensus maps show $> 50\times$ coverage of each genomic marker, and this high level of redundancy results in highly accurate, ordered maps. Thus optical mapping technology enables whole genome analysis that is complementary to existing SNP microarray-genotyping systems and advanced genomic re-sequencing technologies. Where current technologies are best at detecting small changes in DNA sequence, optical mapping detects genome rearrangements, including insertions/deletions ("indels"), translocations, inversions, repeats and gene amplification events (Giacalone *et al.*, 2000). The array of aligned molecules can also be interrogated for the presence of specific gene markers using genomic hybridization to surface arrayed molecules.

6. SUMMARY

In 2002, the entire human genome was randomly sequenced within 6 month by Celera Genomics Corp with sequencing performed at an average rate of $\sim 9 \times 10^9$ bases/year with a factory of some 200 or more 96-capillary array electrophoresis sequencers. Recent improvements in the design of radial CAE microplate capillary channels and new matrix polymers can extend electrophoretic resolution and throughput, each by an order of magnitude or more above the sequencers used by Celera. Each 384-channel microchip sequencer can read ~4 Mb per 24 hr. If an equivalent number of microchip sequencers were used today, their capacity would translate into 1.4×10^9 bases/year $\times 200 = 2.88 \times 10^{11}$ bases/year, or about 10 human genomes sequenced to 10-fold coverage. Yet, this capacity is insufficient for planned levels of whole genome re-sequencing and *de novo* sequencing. Alternatively, new surface-array-based sequencing approaches that employ unitary base addition chemistries promise to make large-scale whole genome analysis a routine activity for smaller laboratories. Short-sequencing reactions over 30–120 nucleotides using new polymerase enzymes and sequencing chemistries such as pyrosequencing, cluster amplification and SBS on 'Clonal Single-Molecule Array', 'tSMS' and 'FISSEQ' methods, combine the enormous parallelism of microarrays with unitary base addition and with ultra-high-resolution optical systems for signal capture. Physical methods for large fragment sizing and sequencing using nanopore technologies and short fragments by fragmentation sequencing using mass spectrometry are also under rapid development of throughput, accuracy and sensitivity.

The recent completion of a high-quality sequence of the human genome poses the challenge to understand the functional elements that it encodes, beyond the protein sequences that can be identified computationally. Comparative genomic analysis offers a powerful approach for finding such elements by identifying sequences that have been highly conserved during evolution. Margulies *et al.* (2005b) propose an initial strategy for detecting regions, which were highly conserved during evolution by generating low-redundancy sequence (~2-fold redundancy) from a collection of 16 eutherian mammals, above the 7 mammals for which genome sequence data are already available in mid-2006. Margulies *et al.* (2006) also show that multi-sequence alignment methods are much better at aligning (and identifying) the available orthologous sequence from phylogenetically diverse vertebrates and contain significant amounts of both exonic and highly conserved non-exonic sequences that are the goal of such comparative sequencing programs. Although the data employed in these reports were from published sequence emanating from genomic Sanger sequencing programs, the potential to utilize sequence derived from SBS analysis for comparative alignment should be explored.

The recent announcement of the planned sequencing of a complete Neanderthal genome (*Homo sapiens neanderthalensis*) by the 454 Life Sciences Corporations and the Max Planck Institute using SBS, underlines the importance of new sequencing technologies to post-genomic science. This project will be a major milestone for human genomic science. It represents the second human species to be sequenced, this time using extremely limited amounts of unique genetic material, where only an ultra-efficient technology could be conceived as making it possible. The scale of this project also represents a major stimulus for the development of the 454 Life Sciences Corporation's pyrosequencing technology and other short-read technologies. Attention to the development of specialist software and complementary technologies that utilize the data to the fullest will be other areas of rapid development. This will further stimulate interest from conventional life sciences in these platforms, as the ability to utilize the technologies and their data output is broadened.

In that regard, another application of the pyrosequencing is for areas requiring the ability to sequence limited amounts of tumor material and for providing sufficient depth of coverage so as to identify rare genetic mutations occurring in a small proportion of the sample. By comparison the sensitivity of conventional Sanger CE sequencing of tumor biopsies is limited by the stromal contamination and by genetic heterogeneity within the cancer, and insufficient depth of coverage can be achieved economically to identify all mutations. Thomas *et al.* (2006) showed that GS20-based pyrosequencing can detect rare cancer-associated sequence variations by independent and parallel sampling of multiple representatives of a given DNA fragment, facilitating the accurate molecular diagnosis of cancer specimens within heterogeneous cellular genomes. New single molecule sequence approaches continue to be explored. Greenleaf and Block (2006) described a novel approach in which the sequence of DNA could be determined in principle by a motion-based method involving an ultra-stable optical trapping device capable of recording pauses in the motion of RNA polymerase when a limiting nucleotide limits the rate of transcription. Patterns of pauses could be aligned and a DNA sequence determined.

See also: DNA sequencing, capillary array electrophoresis, silicon-chip microanalytical device, pyrosequencing, sequencing by synthesis (SBS), single molecule sequencing, sequencing by hybridization, tiling microarrays, sequencing by mass spectrometry, molecular pore sequencing, sequencing of ancient DNA, transcriptome sequencing.

REFERENCES

Aborn, J. H., El-Difrawy, S. A., Novotny, M., Grismondi, E. A., Lam, R., Matsudaira, P., McKenna, B. K., O'Neil, T., Streechon, P. and Erlich, D. J. (2005). A 768-lane microfabricated system for high-throughput DNA sequencing. Lab. Chip 5, 669–674.

Aksyonov, S. A., Bitter, M., Bloom, L. B., Reha-Krantz, L. J., Gould, I. R., Hayes, M. A., Kiernan, U. A., Niederkofler, E. E., Pizziconi, V., Rivera, R. S., Williams, D. J. and Williams, P. (2006). Multiplexed DNA sequencing-by-synthesis. Anal. Chem 348, 127–138.

Andries, K., Verhasselt, P., Guillemont, J., Gohlmann, H. W., Neefs, J. M., Winkler, H., Van Gestel, J., Timmerman, P., Zhu, M., Lee, E., Williams, P., de Chaffoy, D., Huitric, E., Hoffner, S., Cambau, E., Truffot-Pernot, C., Lounis, N. and Jarlier, V. (2005). A diarylquinoline drug active on the ATP synthase of *Mycobacterium tuberculosis*. Science 307, 223–237.

Ashton, C., Mishra, B. and Schwartz, D. C. (1999). Optical mapping and its potential for large-scale sequencing projects. Trends. Biotechnol. 17, 297–302.

Barron, A. E. (2006). DNA sequencing and genotyping. Electrophoresis 27, 3687–3688.

Bennett, S. T., Barnes, C., Cox, A., Davies, L. and Brown, C. (2005). Toward the $1000 human genome. Pharmacogenomics 6, 373–382.

Benson, D. A., Karsch-Mizrachi, I., Lipman, D. J., Ostell, J. and Wheeler, D. L. (2005). GenBank. Nucleic Acids Res. 33(Database issue), D34–D38.

Bertone, P., Gerstein, M. and Snyder, M. (2005). Applications of DNA tiling arrays to experimental genome annotation and regulatory pathway discovery. Chromosome Res. 13, 259–274.

Bertone, P., Stolc, V., Royce, T. E., Rozowsky, J. S., Urban, A. E., Zhu, X., Rinn, J. L., Tongprasit, W., Samanta, M., Weissman, S., Gerstein, M. and Snyder, M. (2004). Global identification of human transcribed sequences with genome tiling arrays. Science 306, 2242–2246.

Blazej, R. G., Kumaresan, P. and Mathies, R. A. (2006). Microfabricated bioprocessor for integrated nanoliter-scale Sanger DNA sequencing. Proc. Natl. Acad. Sci. USA 103, 7240–7245.

Blazej, R. G., Paegel, B. M. and Mathies, R. A. (2003). Polymorphism ratio sequencing: a new approach for single nucleotide polymorphism discovery and genotyping. Genome Res. 13, 287–293.

Bornholdt, S. (2005). Systems biology. Less is more in modeling large genetic networks. Science 310, 449–451.

Brandman, O., Ferrell, J. E., Jr., Li, R. and Meyer, T. (2005). Interlinked fast and slow positive feedback loops drive reliable cell decisions. Science 310, 496–498.

Braslavsky, I., Hebert, B., Kartalov, E. and Quake, S. R. (2003). Sequence information can be obtained from single DNA molecules. Proc. Natl. Acad. Sci. USA 100, 3960–3964.

Braun, B. A., Schanke, K. E. and Graham, D. E. (1978). Isolation of discrete repetitive sequence classes from Xenopus DNA by high temperature reassociation. Nucleic Acids Res. 5, 4283–4304.

Brenner, S., Johnston, M., Bridgham, J., Golda, G., Lloyd, D. H., Johnson, D., Luo, S., McCurdy, S., Foy, M., Ewan, M., Roth, R., George, D., Eletr, S., Albrecht, G., Vermaas, E., Williams, S. R., Moon, K., Burcham, T., Pallas, M., DuBridge, R. B., Kirchner, J., Fearon, K., Mao, J. and Corcoran, K. (2000a). Gene expression analysis by massively parallel signature sequencing (MPSS) on microbead arrays. Nat. Biotechnol. 18, 630–634.

Brenner, S., Williams, S. R., Vermaas, E. H., Storck, T., Moon, K., McCollum, C., Mao, J. I., Luo, S., Kirchner, J. J., Eletr, S., DuBridge, R. B., Burcham, T. and Albrecht, G. (2000b). *In vitro* cloning of complex mixtures of DNA on microbeads: physical separation of differentially expressed cDNAs. Proc. Natl. Acad. Sci. USA 97, 1665–1670.

Bruin, G. J. (2000). Recent developments in electrokinetically driven analysis on microfabricated devices. Electrophoresis 21, 3931–3951.

Church, G., Shendure, J. and Porreca, G. (2006). Sequencing thoroughbreds. Nat. Biotechnol. 24, 139.

Cochran, D. A. E., Lala, G. H., Keith, J. M., Adams, P., Bryant, D. and Mitchelson, K. R. (2006). Sequencing by aligning mutated DNA fragments. In: W. L. Xing and J. Cheng (Eds), The Frontiers of Biochip Technologies (pp. 231–245). Springer Publishers, New York.

Collins, F. and Galas, D. (1993). A new five-year plan for the U.S. Human Genome Project. Science 262, 43–46.

Cooper, A. and Poinar, H. N. (2000). Ancient DNA: do it right or not at all. Science 289, 1139.
Crawford, G. E., Holt, I. E., Whittle, J., Webb, B. D., Tai, D., Davis, S., Margulies, E. H., Chen, Y., Bernat, J. A., Ginsberg, D., Zhou, D., Luo, S., Vasicek, T. J., Daly, M. J., Wolfsberg, T. G. and Collins, F. S. (2006). Genome-wide mapping of DNase hypersensitive sites using massively parallel signature sequencing (MPSS). Genome Res. 16, 123–131.
Deamer, D. W. and Akeson, M. (2000). Nanopores and nucleic acids: prospects for ultrarapid sequencing. Trends Biotechnol. 18, 147–151.
Desany, B., Alvarez, P., Rounsley, S., Brockman, W., Altman, B., Benitez, M., Clark, C., Knight, J., Perbost, C., Vijayadamodar, G., Butler, J., Kleber, M., Zody, M. C., Nusbaum, C., Irzyk, G., Du, L. and Jaffe, D. B. (2005). Assembly of *Neurospora crassa* genome using 454 sequencing technology and new assembler. In: Genomes, Medicine and the Environment Conference, Hilton Head, SC, October 17–19 (http://www.454.com/downloads/news-events/publications/GMEC_454.pdf).
Dittrich, P. S. and Manz, A. (2005). Single-molecule fluorescence detection in microfluidic channels – the Holy Grail in muTAS? Anal. Bioanal. Chem. 382, 1771–1782.
Doherty, E. A. S., Kan, C. W., Paegel, B. M., Yeung, S. H. I., Cao, S. T., Mathies, R. A. and Barron, A. E. (2004). Sparsely cross-linked "nanogel" matrixes as fluid, mechanically stabilized polymer networks for high-throughput microchannel DNA sequencing. Anal. Chem. 76, 5249–5256.
Dressman, D., Hai, Y., Traverso, G., Kinzler, K. W. and Vogelstein, B. (2003). Transforming single DNA molecules into fluorescent magnetic particles for detection and enumeration of genetic variations. Proc. Natl. Acad. Sci. USA 100, 8817–8822.
Edwards, J. R., Kim, D. H. and Ju, J.-Y. (2007). An integrated system for DNA sequencing by synthesis. In: K. R. Mitchelson (Ed), New High Throughput Technologies for DNA Sequencing and Genomics (pp. 187–205). Elsevier, Amsterdam.
Eggeling, C., Widengren, J., Brand, L., Schaffer, J., Felekyan, S. and Seidel, C. A. (2006). Analysis of photobleaching in single-molecule multicolor excitation and Förster resonance energy transfer measurements. J. Phys. Chem. A Mol. Spectrosc. Kinet. Environ. Gen. Theory 110, 2979–2995.
Ehrich, M., Hillenkamp, F. and van den Boom, D. (2007). Comparative sequence analysis by MALDI-TOF mass spectrometry – utilizing the known to discover the new. In: K. R. Mitchelson (Ed), New High Throughput Technologies for DNA Sequencing and Genomics (pp. 97–117). Elsevier, Amsterdam.
Emberton, J., Ma, J., Yuan, Y., San Miguel, P. and Bennetzen, J. L. (2005). Gene enrichment in maize with hypomethylated partial restriction (HMPR) libraries. Genome Res. 15, 1441–1446.
The ENCODE Project Consortium (2004). The ENCODE (ENCyclopedia Of DNA Elements) Project. Science 306, 636–640.
Eriksson, J., Gharizadeh, B., Nourizad, N. and Nyren, P. (2004). 7-Deaza-2′-deoxyadenosine-5′-triphosphate as an alternative nucleotide for the pyrosequencing technology. Nucleosides Nucleotides Nucleic Acid 23(10), 1583–1594.
Finn, P. J., Bull, M. G., Xiao, H., Phillips, P. D., Nelson, J. R., Grossmann, G., Nampalli, S., McArdle, B. F., Mamone, J. A., Flick, P. K., Fuller, C. W. and Kumar, S. (2003). Efficient incorporation of positively charged 2′, 3′-dideoxynucleoside-5′-triphosphates by DNA polymerases and their application in 'direct-load' DNA sequencing. Nucleic Acids Res. 31, 4769–4778.
Fologea, D., Gershow, M., Ledden, B., McNabb, D. S., Golovchenko, J. A. and Li, J. (2005). Detecting single stranded DNA with a solid state nanopore. Nano Lett. 5, 1905–1909.
Garnis, C., Davies, J. J., Buys, T. P., Tsao, M. S., MacAulay, C., Lam, S. and Lam, W. L. (2005). Chromosome 5p aberrations are early events in lung cancer: implication of glial cell line-derived neurotrophic factor in disease progression. Oncogene 24, 4806–4812.
GeneBank. (2006). (http://www.ncbi.nlm.nih.gov/Genbank/index.html).

Gharizadeh, B., Eriksson, J., Nourizad, N., Nordstrom, T. and Nyren, P. (2004). Improvements in pyrosequencing technology by employing Sequenase polymerase. Anal. Biochem. 330(2), 272–280.
Giacalone, J., Delkobette, S., Gibaja, V., Ni, L., Skiadas, Y., Qi, R., Edington, J., Lai, Z., Gebauer, D., Zhao, H., Anantharaman, T., Mishra, B., Brown, L. G., Saxena, R., Page, D. C. and Schwartz, D. C. (2000). Optical mapping of BAC clones from the human Y chromosome DAZ locus. Genome Res. 10, 1421–1429.
Girard, A., Sachidanandam, R., Hannon, G. J. and Carmell, M. A. (2006). A germline-specific class of small RNAs binds mammalian Piwi proteins. Nature 442, 199–202.
Gladitz, J., Shen, K., Antalis, P., Hu, F. Z., Post, J. C. and Ehrlich, G. D. (2005). Codon usage comparison of novel genes in clinical isolates of *Haemophilus influenzae*. Nucleic Acids Res. 33, 3644–3658.
Goldberg, S. M. D., Johnson, J., Busam, D., Feldblyum, T., Ferriera, S., Friedman, R., Halpern, A., Khouri, H., Kravitz, S. A., Lauro, F. M., Li, K., Rogers, Y.-H., Strausberg, R., Sutton, G., Tallon, L., Thomas, T., Venter, E., Frazier, M. and Venter, J. C. (2006). A Sanger/pyrosequencing hybrid approach for the generation of high-quality draft assemblies of marine microbial genomes. Proc. Natl. Acad. Sci. USA 10.1073/pnas.0604351103.
Greenleaf, W. J. and Block, S. M. (2006). Single-molecule, motion-based DNA sequencing using RNA polymerase. Science 313, 801.
Greulich, K. O. (2005). Single-molecule studies on DNA and RNA. Chem. Phys. Chem. 6, 2458–2471.
Hansen, C. and Quake, S. R. (2003). Microfluidics in structural biology: smaller, faster and better. Curr. Opin. Struct. Biol. 13, 538–544.
Hardenbol, P., Yu, F., Belmont, J., Mackenzie, J., Bruckner, C., Brundage, T., Boudreau, A., Chow, S., Eberle, J., Erbilgin, A., Falkowski, M., Fitzgerald, R., Ghose, S., Iartchouk, O., Jain, M., Karlin-Neumann, G., Lu, X., Miao, X., Moore, B., Moorhead, M., Namsaraev, E., Pasternak, S., Prakash, E., Tran, K., Wang, Z., Jones, H. B., Davis, R. W., Willis, T. D. and Gibbs, R. A. (2005). Highly multiplexed molecular inversion probe genotyping: over 10,000 targeted SNPs genotyped in a single tube assay. Genome Res. 15, 269–275.
Hebert, B. and Braslavsky, I. (2007). Single-molecule fluorescence microscopy and its applications to single-molecule sequencing by cyclic synthesis. In: K. R. Mitchelson (Ed), New High Throughput Technologies for DNA Sequencing and Genomics (pp. 207–244). Elsevier, Amsterdam.
Henderson, I. R., Zhang, X., Lu, C., Johnson, L., Meyers, B. C., Green, P. J. and Jacobsen, S. E. (2006). Dissecting *Arabidopsis thaliana* DICER function in small RNA processing, gene silencing and DNA methylation patterning. Nature Genet. 38(6), 721–725.
Hutchison, C. A., III., Smith, H. O., Pfannkoch, C. and Venter, J. C. (2005). Cell-free cloning using {phi}29 DNA polymerase. Proc. Natl. Acad. Sci. USA 102(48), 17332–17336.
Hyman, E. D. (1988). A new method of sequencing DNA. Anal. Biochem. 174, 423–436.
The International HapMap Consortium (2003). The International HapMap Project. Nature 426, 789–796.
The International HapMap Consortium (2005). A haplotype map of the human genome. Nature 437, 1299–1320.
International Human Genome Sequencing Consortium (IHGSC) (2001). Initial sequencing and analysis of the human genome. Nature 409, 860–921.
International Human Genome Sequencing Consortium (IHGSC) (2004). Finishing the euchromatic sequence of the human genome. Nature 431, 931–945.
Janes, K. A., Albeck, J. G., Guadet, S., Sorger, P. K., Lauffenburger, D. A. and Yaffe, M. B. (2005). A systems model of signaling identifies a molecular basis set for cytokine-induced apoptosis. Science 310, 1646–1653.
Jansen, E. H., Buskens, C. A. and van den Berg, R. H. (1989). A sensitive CCD image system for detection of chemiluminescent reactions. J. Biolumin. Chemilumin. 3, 53–57.
Jongeneel, C. V., Iseli, C., Stevenson, B. J., Riggins, G. J., Lal, A., Mackay, A., Harris, R. A., O'Hare, M. J., Neville, A. M., Simpson, A. J. and Strausberg, R. L. (2003). Comprehensive sampling of gene expression in human cell lines with massively parallel signature sequencing. Proc. Natl. Acad. Sci. USA 100, 4702–4705.

Kan, C. W., Fredlake, C. P., Doherty, E. A. and Barron, A. E. (2004). DNA sequencing and genotyping in miniaturized electrophoresis systems. Electrophoresis 25, 3564–3588.

Karhanek, M., Kemp, J. T., Pourmand, N., Davis, R. W. and Webb, C. D. (2005). Single DNA molecule detection using nanopipettes and nanoparticles. Nano Lett. 5, 403–407.

Karlinsey, J. M., Monahan, J., Marchiarullo, D. J., Ferrance, J. P. and Landers, J. P. (2005). Pressure injection on a valved microdevice for electrophoretic analysis of submicroliter samples. Anal. Chem. 77, 3637–3643.

Kartalov, E. P. and Quake, S. R. (2004). Microfluidic device reads up to four consecutive base pairs in DNA sequencing-by-synthesis. Nucleic Acids Res. 32, 2873–2879.

Keith, J. M., Adams, P., Bryant, D., Cochran, D. A. E., Lala, G. H. and Mitchelson, K. R. (2004a). Algorithms for sequencing aided by mutagenesis. Bioinformatics 20, 2401–2410.

Keith, J. M., Cochran, D. A. E., Lala, G. H., Adams, P., Bryant, D. and Mitchelson, K. R. (2004b). Revealing hidden genomic sequence. Nucleic Acids Res. 32, e35.

Keith, J. M., Hawkes, D. B., Adams, P., Bryant, D. E., Carter, J. C., Cochran, D. A. E. and Mitchelson, K. R. (2007). Sequencing aided by mutagenesis facilitates *de novo* sequencing of megabase DNA fragments by short length reads. In: K. R. Mitchelson (Ed), New High Throughput Technologies for DNA Sequencing and Genomics (pp. 301–326). Elsevier, Amsterdam.

Kowalchuk, G. A., Austin, J. J., Gooding, P. S. and Stephen, J. R. (2007). Valid recovery of nucleic acid sequence information from high contamination risk samples – ancient DNA and environmental DNA. In: K. R. Mitchelson (Ed), New High Throughput Technologies for DNA Sequencing and Genomics (pp. 357–371). Elsevier, Amsterdam.

Krause, J., Dear, P. H., Pollack, J. L., Slatkin, M., Spriggs, H., Barnes, I., Lister, A. M., Ebersberger, I., Pääbo, S. and Hofreiter, M. (2006). Multiplex amplification of the mammoth mitochondrial genome and the evolution of *Elephantidae*. Nature 439, 724–727.

Krishnan, M., Namasivayam, V., Lin, R., Pal, R. and Burns, M. A. (2001). Microfabricated reaction and separation systems. Curr. Opin. Biotechnol. 12, 92–98.

Kumar, S. and Fuller, C. W. (2007). Advances in dye-nucleotide conjugate chemistry for DNA sequencing. In: K. R. Mitchelson (Ed), New High Throughput Technologies for DNA Sequencing and Genomics (pp. 119–149). Elsevier, Amsterdam.

Kumar, S., Sood, A., Wegener, J., Finn, P. J., Nampalli, S., Nelson, J. R., Sekher, A., Mitsis, P., Macklin, J. and Fuller, C. W. (2005). Terminal phosphate labeled nucleotides: synthesis, applications, and linker effect on incorporation by DNA polymerases. Nucleosides Nucleotides Nucleic Acids 24(5–7), 401–408.

Lee, J. W. and Meller, A. (2007). Rapid DNA sequencing by direct nanoscale reading of nucleotide bases on individual DNA chains. In: K. R. Mitchelson (Ed), New High Throughput Technologies for DNA Sequencing and Genomics (pp. 245–263). Elsevier, Amsterdam.

Levene, M. J., Korlach, J., Turner, S. W., Foquet, M., Craighead, H. G. and Webb, W. W. (2003). Zero-mode waveguides for single-molecule analysis at high concentrations. Science 299, 682–686.

Li, M., Diehl, F., Dressman, D., Vogelstein, B. and Kinzler, K. W. (2006). BEAMing up for detection and quantification of rare sequence variants. Nat. Methods 3, 95–97.

Liu, Y., Shin, S., Zeng, X., Zhan, M., Gonzales, R., Mueller, F. J., Schwartz, C. M., Xue, H., Li, H., Baker, S. C., Chudin, E., Barker, D. L., McDaniel, T. K., Oeser, S., Loring, J. F., Mattson, M. P. and Rao, M. S. (2006). Genome wide profiling of human embryonic stem cells (hESCs), their derivatives and embryonal carcinoma cells to develop base profiles of U.S. Federal government approved hESC lines. BMC Dev. Biol. 6(1), 20.

Margulies, E. H., Chen, C. W. and Green, E. D. (2006). Differences between pair-wise and multi-sequence alignment methods affect vertebrate genome comparisons. Trends Genet. 22, 187–193.

Margulies, E. H., Vinson, J. P., NISC Comparative Sequencing Program; Miller, W., Jaffe, D. B., Lindblad-Tok, K., Chang, J. L., Green, E. D., Lander, E. S., Mullikin, J. C. and Clamp, M. (2005b). An initial strategy for the systematic identification of functional elements in the human genome by low-redundancy comparative sequencing. Proc. Natl. Acad. Sci. USA, 102, 4795–4800.

Margulies, M., Egholm, M., Altman, W., Attiya, S., Bader, J., Bemben, L., Berka, J., Braverman, M., Chen, Y., Chen, Z., Dewell, S., Du, L., Fierro, J., Gomes, X., Godwin, B., He, W., Helgesen, S., Ho, C., Irzyk, G., Jando, S., Alenquer, M., Jarvie, T., Jirage, K., Kim, J., Knight, J., Lanza, J., Leamon, J., Lefkowitz, S., Lei, J., Li, K., Lohman, H., Lu, V. B., Makhijani, K., McDade, M., McKenna, E., Myers, M., Nickerson, E., Nobile, J., Plant, R., Puc, B., Ronan, M., Roth, G., Sarkis, G., Simons, J., Simpson, J., Srinivasan, M., Tartaro, K., Tomasz, A., Vogt, K., Volkmer, G., Wang, S., Wang, Y., Weiner, M., Yu, P., Begley, R. and Rothberg, J. (2005a). Genome sequencing in microfabricated high-density picolitre reactors. Nature 437, 376–380.

Margulies, M., Jarvie, T. P., Knight, J. R. and Simons, J. F. (2007). The 454 Life Sciences picoliter sequencing system. In: K. R. Mitchelson (Ed), New High Throughput Technologies for DNA Sequencing and Genomics (pp. 151–186). Elsevier, Amsterdam.

Marziali, A. and Akeson, M. (2001). New DNA sequencing methods. Annu. Rev. Biomed. Eng. 3, 195–223.

Mathé, J., Aksimentiev, A., Nelson, D. R., Schulten, K. and Meller, A. (2005). Orientation discrimination of single-stranded DNA inside the alpha-hemolysin membrane channel. Proc. Natl. Acad. Sci. USA 102, 12,377–12,382.

Mattick, J. S. and Makunin, I. V. (2006). Non-coding RNA. Hum. Mol. Genet. 15(1), R17–R29.

McGrath, A. (2007). Genome sequencing and assembly. In: K. R. Mitchelson (Ed), New High Throughput Technologies for DNA Sequencing and Genomics (pp. 327–355). Elsevier, Amsterdam.

Medini, D., Donati, C., Tettelin, H., Masignani, V. and Rappuoli, R. (2005). The microbial pan-genome. Curr. Opin. Genet. Dev. 15, 589–594.

Meller, A., Mathé, J. and Eid, J. (2005). Ultra-high-throughput opti-nanopore DNA readout platform. PCT/US05/28566.

Meng, Q., Kim, D. H., Bai, X., Bi, L., Turro, N. J. and Ju, J. (2006). Design and synthesis of a photocleavable fluorescent nucleotide 3′-*O*-allyl-dGTP-PC-Bodipy-FL-510 as a reversible terminator for DNA sequencing-by-synthesis. J. Org. Chem. 71(8), 3248–3252.

Metzker, M. L. (2005). Emerging technologies in DNA sequencing. Genome Res. 15, 1767–1776.

Meyers, B. C., Souret, F. F., Lu, C. and Green, P. J. (2006). Sweating the small stuff: microRNA discovery in plants. Curr. Opin. Biotechnol. 17(2), 139–146.

Mineno, J., Okamoto, S., Ando, T., Sato, M., Chono, H., Izu, H., Takayama, M., Asada, K., Mirochnitchenko, O., Inouye, M. and Kato, I. (2006). The expression profile of microRNAs in mouse embryos. Nucleic Acids Res. 34, 1765–1771.

Mitchelson, K. R. (2003). The use of capillary electrophoresis for DNA polymorphism analysis. Mol. Biotechnol. 24, 41–68.

Mitchelson, K. R. (2005). DNA sequencing. In: P. Worsfold, A. Townshend and C. Poole (Eds), Encyclopedia of Analytical Science 2e (vol. 2, pp. 286–292). Elsevier, Oxford.

Mitra, R. D., Shendure, J., Olejnik, J., Krzmanska-Olejnik, E. and Church, G. M. (2003). Fluorescent *in situ* sequencing on polymerase colonies. Anal. Biochem. 320, 55–65.

Nakane, J., Wiggin, M. and Marziali, A. (2004). A nanosensor for transmembrane capture and identification of single nucleic acid molecules. Biophys. J. 87, 615–621.

Nakano, M., Nobuta, K., Vemaraju, K., Tej, S. S., Skogen, J. W. and Meyers, B. C. (2006). Plant MPSS databases: signature-based transcriptional resources for analyses of mRNA and small RNA. Nucleic Acids Res. 34(Database Issue), D731–D735.

Noonan, J. P., Hofreiter, M., Smith, D., Priest, J. R., Rohland, N., Rabeder, G., Krause, J., Detter, J. C., Pääbo, S. and Rubin, E. M. (2005). Genomic sequencing of Pleistocene cave bears. Science 309, 597–599.

Nyren, P., Pettersson, B. and Uhlen, M. (1993). Solid phase DNA minisequencing by an enzymatic luminometric inorganic pyrophosphate detection assay. Anal. Biochem. 208, 171–175.

Oudes, A. J., Roach, J. C., Walashek, L. S., Eichner, L. J., True, L. D., Vessella, R. L. and Liu, A. Y. (2005). Application of Affymetrix array and massively parallel signature sequencing for identification of genes involved in prostate cancer progression. BMC Cancer 5, 86.

Pääbo, S., Poinar, H., Serre, D., Jaenicke-Despres, V., Hebler, J., Rohland, N., Kuch, M., Krause, J., Vigilant, L. and Hofreiter, M. (2004). Genetic analyses from ancient DNA. Annu. Rev. Genet. 38, 645–679.

Paegel, B. M., Blazej, R. G. and Mathies, R. A. (2003). Microfluidic devices for DNA sequencing: sample preparation and electrophoretic analysis. Curr. Opin. Biotechnol. 14, 42–50.

Paegel, B. M., Emrich, C. A., Wedemayer, G. J., Scherer, J. R. and Mathies, R. A. (2002a). High throughput DNA sequencing with a microfabricated 96-lane capillary array electrophoresis bioprocessor. Proc. Natl. Acad. Sci. USA 99, 574–579.

Paegel, B. M., Yeung, S. H. I. and Mathies, R. A. (2002b). Microchip bioprocessor for integrated nano-volume sample purification and DNA sequencing. Anal. Chem. 74, 5092–5098.

Pal, R., Yang, M., Lin, R., Johnson, B. N., Srivastava, N., Razzacki, S. Z., Chomistek, K. J., Heldsinger, D. C., Haque, R. M., Uqaz, V. M., Thwar, P. K., Chen, Z., Alfano, K., Yim, M. B., Krishnan, M., Fuller, A. O., Larson, R. G., Burke, D. T. and Burns, M. A. (2005). An integrated microfluidic device for influenza and other genetic analyses. Lab. Chip 5, 1024–1032.

Peterson, D. G., Schulze, S. R., Sciara, E. B., Lee, S. A., Bowers, J. E., Nagel, A., Jiang, N., Tibbits, D. C., Wessler, S. R. and Paterson, A. H. (2002). Integration of Cot analysis, DNA cloning, and high-throughput sequencing facilitates genome characterization and gene discovery. Genome Res. 12, 795–807.

Plasterk, R. H. (2006). MicroRNAs in animal development. Cell 124(5), 877–881.

Poinar, H. N., Schwarz, C., Qi, J., Shapiro, B., Macphee, R. D., Buigues, B., Tikhonov, A., Huson, D. H., Tomsho, L. P., Auch, A., Rampp, M., Miller, W. and Schuster, S. C. (2006). Metagenomics to paleogenomics: large-scale sequencing of mammoth DNA. Science 311, 392–394.

Pourmand, N., Karhanek, M., Persson, H. H., Webb, C. D., Lee, T. H., Zahradnikova, A. and Davis, R. W. (2006). Direct electrical detection of DNA synthesis. Proc. Natl. Acad. Sci. USA 103, 6466–6470.

Reed, J., Singer, E., Kresbach, G. and Schwartz, D. C. (1998). A quantitative study of optical mapping surfaces by atomic force microscopy and restriction endonuclease digestion assays. Anal. Biochem. 259, 80–88.

Reinartz, J., Bruyns, E., Lin, J. Z., Burcham, T., Brenner, S., Bowen, B., Kramer, M. and Woychik, R. (2002). Massively parallel signature sequencing (MPSS) as a tool for in-depth quantitative gene expression profiling in all organisms. Brief Funct. Genomic Proteomic 1(1), 95–104.

Remington, K. A., Heidelberg, K. and Venter, J. C. (2005). Taking metagenomic studies in context. Trends Microbiol. 13, 404.

Reslewic, S., Zhou, S., Place, M., Zhang, Y., Briska, A., Goldstein, S., Churas, C., Runnheim, R., Forrest, D., Lim, A., Lapidus, A., Han, C. S., Roberts, G. P. and Schwartz, D. C. (2005). Whole-genome shotgun optical mapping of *Rhodospirillum rubrum*. Appl. Environ. Microbiol. 71, 5511–5522.

Rhee, M. and Burns, M. A. (2006). Nanopore sequencing technology: research trends and applications. Trends Biotechnol. Oct 18.

Ronaghi, M. (2001). Pyrosequencing sheds light on DNA sequencing. Genome Res 11, 3–11.

Sanger, F., Air, G. M., Barrell, B. G., Brown, N. L., Coulson, A. R., Fiddes, C. A., Hutchison, C. A., Slocombe, P. M. and Smith, M. (1977a). Nucleotide sequence of bacteriophage phi X174 DNA. Nature 265, 687–695.

Sanger, F., Coulson, A. R., Barrell, B. G., Smith, A. J. and Roe, B. A. (1980). Cloning in single-stranded bacteriophage as an aid to rapid DNA sequencing. J. Mol. Biol. 143, 161–178.

Sanger, F., Coulson, A. R., Hong, G. F., Hill, D. F. and Petersen, G. B. (1982). Nucleotide sequence of bacteriophage lambda DNA. J. Mol. Biol. 162, 729–773.

Sanger, F., Nicklen, S. and Coulson, A. R. (1977b). DNA sequencing with chain termination inhibitors. Proc. Natl. Acad. Sci. USA 74, 5463–5467.

Schena, M., Shanlon, D., Davis, R. W. and Brown, P. O. (1995). Quantitative monitoring of gene expression patterns with a complementary DNA microarray. Science 270, 467–470.

Schumacher, A., Kapranov, P., Kaminsky, Z., Flanagan, J., Assadzadah, A., Yau, P., Virtanen, C., Winegarden, N., Cheng, J., Gingeras, T. and Petronis, A. (2006). Microarray-based DNA methylation profiling: technology and applications. Nucleic Acids Res. 34, 528–542.

Selzer, R. R., Richmond, T. A., Pofahl, N. J., Green, R. D., Eis, P. S., Nair, P., Brothman, A. R. and Stallings, R. L. (2005). Analysis of chromosome breakpoints in neuroblastoma at sub-kilobase resolution using fine-tiling oligonucleotide array CGH. Genes Chromosome Res. 44, 305–319.

Shah, T., de Villiers, E., Nene, V., Hass, B., Taracha, E., Gardiner, M. J., Sansom, C., Pelle, R. and Bishop, R. (2006). Using the transcriptome to annotate the genome revisited: application of massively parallel signature sequencing (MPSS). Gene 366, 104–108.

Shendure, J., Mitra, R. D., Varma, C. and Church, G. M. (2004). Advanced sequencing technologies: methods and goals. Nat. Rev. Genet. 5, 335–344.

Shendure, J., Porreca, G. J., Reppas, N. B., Ln, X., McCutcheon, J. P., Rosenbaum, A. M., Wang, M. D., Zhang, K., Mitra, R. D. and Church, G. M. (2005). Accurate multiplex polony sequencing of an evolved bacterial genome. Science 309, 1728–1732.

Shi, Y. N., Simpson, P. C., Scherer, J. R., Wexler, D., Skibola, C., Smith, M. T. and Mathies, R. A. (1999). Radial capillary array electrophoresis microplate and scanner for high-performance nucleic acid analysis. Anal. Chem. 71, 5354–5361.

Skelley, A. M., Scherer, J. R., Aubrey, A. D., Grover, W. H., Ivester, R. H. C., Ehrenfreund, P., Grunthaner, F. J., Bada, J. L. and Mathies, R. A. (2005). Development and evaluation of a microdevice for amino acid biomarker detection and analysis on Mars. Proc. Natl. Acad. Sci. USA 102, 1041–1046.

Smith, H. O. and Birnstiel, M. L. (1976). A simple method for DNA restriction site mapping. Nucleic Acids Res. 3, 2387–2398.

Sulston, J., Du, Z., Thomas, K., Wilson, R., Hillier, L., Staden, R., Halloran, N., Green, P., Thierry-Mieg, J. and Qiu, L. *et al.* (1992). The *C. elegans* genome sequencing project: a beginning. Nature 356, 37–41.

Thomas, R. K., Nickerson, E., Simons, J. F., Jänne, P. A., Tengs, T., Yuza, Y., Garraway, L. A., LaFramboise, T., Lee, J. C., Shah, K., O'Neill, K., Sasaki, F., Lindeman, N., Wong, K.-K., Borras, A. M., Gutmann, E. J., Dragnev, K. H., DeBiasi, R., Chen, T.-H., Glatt, K. A., Greulich, H., Desany, B., Lubeski, C. K., Brockman, W., Alvarez, P., Hutchison, S. K., Leamon, J. H., Ronan, M. T., Turenchalk, G. S., Egholm, M., Sellers, W. R., Rothberg, J. M. and Meyerson, M. (2006). Sensitive mutation detection in heterogeneous cancer specimens by massively parallel picoliter reactor sequencing. Nat. Med. 12, 852–855; Erratum 12, 1220.

Venter, J. C., Remington, K., Heidelberg, J. F., Halpern, A. L., Rusch, D., Eisen, J. A., Wu, D. Y., Paulsen, I., Nelson, K. E., Nelson, W., Fouts, D. E., Levy, S., Knap, A. H., Lomas, M. W., Nealson, K., White, O., Peterson, J., Hoffman, J., Parsons, R., Baden-Tillson, H., Pfannkoch, C., Rogers, Y. H. and Smith, H. O. (2004). Environmental genome shotgun sequencing of the Sargasso Sea. Science 304, 66–74.

Viasnoff, V., Meller, A. and Isambert, H. (2006). DNA nanomechanical switches under folding kinetics control. Nano Lett. 6, 101–104.

Vos, P., Hoger, R., Bleeker, M., Reijans, M., van der Lee, T., Hornes, M., Frijters, A., Pot, J., Peleman, J. and Kuiper, M. (1995). AFLP: a new technique for DNA fingerprinting. Nucleic Acids Res. 23, 4407–4414.

Washietl, S., Hofacker, I. L., Lukasser, M., Huttenhofer, A. and Stadler, P. F. (2005). Mapping of conserved RNA secondary structures predicts thousands of functional non-coding RNAs in the human genome. Nat. Biotechnol. 23, 1383–1390.

Wilson, G. M., Flibottes, S., Missirlis, P. I., Marra, M. A., Jones, S., Thornton, K., Clark, A. G. and Holt, R. A. (2006). Identification by full-coverage array CGH of human DNA copy number increases relative to chimpanzee and gorilla. Genome Res. 16, 173–181.

Wolfe, J. L., Kawate, T., Sarracino, D. A., Zillmann, M., Olson, J., Stanton, V. P., Jr. and Verdine, G. L. (2002). A genotyping strategy based on incorporation and cleavage of chemically modified nucleotides. Proc. Natl. Acad. Sci. USA 99, 11,073–11,078.

Wolfe, J. L., Wang, B. H., Kawate, T. and Stanton, V. P., Jr. (2003). Sequence-specific dinucleotide cleavage promoted by synergistic interactions between neighboring modified nucleotides in DNA. J. Am. Chem. Soc. 125, 10,500–10,501.

Xiong, Q. and Cheng, J. (2007). Chip capillary electrophoresis and total genetic analysis systems. In: K. R. Mitchelson (Ed), New High Throughput Technologies for DNA Sequencing and Genomics (pp. 45–95). Elsevier, Amsterdam.

Yura, T., Mori, H., Nagai, H., Nagata, T., Ishihama, A., Fujita, N., Isono, K., Mizobuchi, K. and Nakata, A. (1992). Systematic sequencing of the *Escherichia coli* genome: analysis of the 0–2.4 min region. Nucleic Acids Res. 20, 3305–3308.

Zhang, K., Martiny, A. C., Reppas, N. B., Barry, K. W., Malek, J., Chisholm, S. W. and Church, G. M. (2006a). Sequencing genomes from single cells by polymerase cloning. Nat. Biotechnol. 24, 680–686.

Zhang, K., Zhu, J., Shendure, J., Porreca, G. J., Aach, J. D., Mitra, R. D. and Church, G. M. (2006b). Long-range polony haplotyping of individual human chromosome molecules. Nat. Genet. 38(3), 382–387.

Zhou, S., Herschleb, J. and Schwartz, D. C. (2007). A single molecule system for whole genome analysis. In: K. R. Mitchelson (Ed), New High Throughput Technologies for DNA Sequencing and Genomics (pp. 265–300). Elsevier, Amsterdam.

Zhu, J., Shendure, J., Mitra, R. D. and Church, G. M. (2003). Single molecule profiling of alternative pre-mRNA splicing. Science 301, 836–838.

Chapter 2

Chip Capillary Electrophoresis and Total Genetic Analysis Systems

Qiang Xiong[1] and Jing Cheng[1,2]

[1]*Capitalbio Corporation: National Engineering Research Center for Beijing Biochip Technology, 18 Life Science Parkway, Changping District, Beijing 102206, China*
[2]*Medical Systems Biology Research Center, Tsinghua University School of Medicine, Beijing 100084, China*

Contents

PERSPECTIVES IN BIOANALYSIS, VOLUME 2
ISSN 1871-0069 DOI: 10.1016/S1871-0069(06)02002-7

Abstract

The utilization of new sequencing techniques based on capillary array electrophoresis (CAE) has had a great impact on the progress of the Human Genome Project (HGP), and finally led to its successful completion at much lower costs than initially anticipated for the project (Collins *et al.*, 2003). Similarly, chip-based capillary electrophoresis, the technological extension of capillary electrophoresis (CE), is a rapidly emerging technology, which has caused revolution in analytical chemistry. In fact, there has been an explosion of interest in the development of chip-based CE ever since the initial concept "micro-total analysis systems (μ-TAS)" or "lab-on-a-chip" was introduced by Manz and Harrison (Manz *et al.*, 1990; Harrison *et al.*, 1993). With great efforts from leading scientists, this new-born technology has matured rapidly. It has several advantages over conventional methods, such as reduced analysis time, high efficiency, low sample consumption, the potential for integration and automation, disposability, portability and so on. All these features make chip-based CE an attractive technology for the next generation of CE instrumentation. For example, a binary mixture could be successfully resolved in 0.8 ms using chip-based CE separation using a field strength of 53 kV/cm, with an analysis time of several orders of magnitudes less than conventional CE (Jacobson *et al.*, 1998). Other groups have applied chip-based CE for separating FITC-labeled amino acids, and the plate heights obtained could be down to 0.3 μm, which demonstrated the high efficiency of this separation technology (Effenhauser *et al.*, 1993). More recently, a microfabricated 384-lane CAE device has been developed and used for highly parallel genetic analysis, showing great promise as a means for ultra high-throughput bioanalysis (Emrich *et al.*, 2002; Paegel *et al.*, 2003). With the maturation of these technologies, companies have initiated industrialization of chip-based CE products. Several commercial chip products are currently available, such as the 5100 Automated Lab-on-a-chip Platform from Agilent Co., Ltd., and the Labchip 90 Electrophoresis System from Caliper Co., Ltd. This chapter will give an overview of the chip-based CE technology, including microchip design, fabrication and detection, surface modification and applications.

1. INTRODUCTION

1.1. Various chip-based capillary electrophoresis systems

Chip-based CE is based on the microfabrication techniques developed in the semiconductor industry, with microchannels of a desired pattern for the particular application fabricated using photolithography or micromolding processes. These techniques are used to produce both the microchannel separation device and attendant structures, and to integrate electrodes, detection apparatus and other electro-mechanical control devices needed for the operation of the microchannel device. Manufacture will be discussed in detail in Section 3. For a typical electrophoretic separation experiment on a microchip several steps are involved. First, all channels should be filled with running buffer or sieving matrix. After the sample is loaded into a sample reservoir, it is transferred electrokinetically into an injector region, forming a sharp band. Then, a high voltage is applied and the sharp sample band is forced into the separation channel, and it begins to separate according to the properties of the separating environment and media. When each of the resolving species reaches the detection point at the end of the separation channel, signals are produced and recorded to form an electropherogram. The above-mentioned steps are typical for capillary zone electrophoresis (CZE), which is the first electrophoretic format transferred from conventional CE onto microchips. In this format, all samples are electrophoresed in free solution, and separated into zones according to their charge-to-mass ratio. There are other formats besides CZE for chip-based CE, formats such as capillary gel electrophoresis (CGE), isoelectric focusing (IEF), micellar electrokinetic capillary chromatography (MECC), isotachophoresis (ITP), capillary electrochromatography (CEC), etc. A brief introduction to microchips using these different electrophoretic formats are given below.

CGE is a most widely used format for chip-based CE. Microchannels are filled with sieving matrix, which is cross-linked gel or entangled polymer solution, instead of a free solution. The analytes electrophoresed in this media are separated according to their size. DNA fragment sizing, genotyping and sequencing have been demonstrated with chip-based CGE (Effenhauser *et al.*, 1994; Schmalzing *et al.*, 1997; Shi and Anderson, 2003).

IEF is a kind of electrophoresis in a pH gradient set up between a cathode and anode, with the cathode at a higher pH than the anode. During IEF electrophoresis, amphoteric samples (such as proteins) are separated according to their isoelectric point (pI) values, and at the end of the electrophoresis, each species migrate and concentrate at their isoelectric points. For chip-based IEF electrophoresis, which utilizes single-point detection system, forces are needed to mobilize the focused sample band to the detection point. The most common mobilization methods are chemical, hydrodynamic and electro-osmotic flow (EOF)-driven mobilization, among which EOF-driven mobilization proves most suitable for miniaturized systems because of its high speed and low instrumentation requirements (Hofmann *et al.*, 1999). Typically, systems use EOF to drive samples to a single-point for detection. However other systems such as whole column imaging detection have also been developed, such that the sample band mobilization step is eliminated (Mao and Pawliszyn, 1999a, 1999b).

MECC is another kind of electrophoresis that has been successfully transferred into microchip and is widely utilized. MECC is an operational mode of CE developed by Terabe *et al.* (1985) to address the general problem of CE of an inability to separate uncharged species. A surfactant such as sodium dodecyl sulfate (SDS) is added to the running buffer in sufficient concentration to form micelles. In a typical run, the micelles move much more slowly toward the cathode than does the running buffer solution (driven by EOF). The partitioning of solute between the micelles and the running buffer provides a separation mechanism similar to that of liquid chromatography. Separation of various samples such as neutral coumarin dyes (Moore *et al.*, 1995), amino acids (Rodriguez *et al.*, 1999) and explosives (Wallenborg and Bailey, 2000), has been successfully demonstrated using microchip MECC, showing enhanced separating efficiency and decreased analysis time compared to conventional MECC.

For ITP, two different buffer systems are used to create zones into which the analytes separate. For example, to separate anions a leading electrolyte is chosen whose anionic component has a higher mobility than that of the analytes. It will migrate faster than the analyte in an electric field. Similarly, a trailing electrolyte is also chosen with an anionic component that migrates slower than the analytes. When the electric field is applied the anions start to migrate, and the analytes will separate into zones determined by their mobilities, with the fastest analyte moving behind the leading electrolyte. A demonstration of ITP on microchip was provided by Walker and Morris (1998), where they performed ITP separations of herbicides paraquat and diquat on a glass microchip, with an on-chip detection using normal Raman spectroscopy.

CEC is a hybrid separation method that couples the high separation efficiency of CZE with high-performance liquid chromatography (HPLC), and uses an electric field rather than hydraulic pressure to propel the mobile phase through the stationary phase. With this combination, samples with similar electrophoretic mobilities can be separated, even when they cannot be separated by CZE alone. An additional benefit of CEC compared to HPLC is the fact that the flow profile in a pressure-driven system is parabolic, whereas in an electrokinetically driven system it is plug like and therefore much more efficient. There are three types of CEC according to the formation of the stationary phase, open tubular CEC (OTCEC), packed column CEC and monolith column CEC. Transfer of each such technique to the microchip format has been successfully demonstrated by Jacobson *et al.* (1994a), Ceriotti *et al.* (2002) and He *et al.* (1998).

2. CHIP DESIGN AND FLUID MANIPULATION

2.1. Chip design

The design of microchips has undergone significant development, and it is the sophistication of these designed structures that exemplify the differences between microchip devices and conventional capillary devices. The first and simplest chip designs consisted of microchannels laid-out in cross geometries, with a straight separation channel intersected by a second channel for sample injection. Four reservoirs were positioned at the end of each channel, two for sample and background introduction, and the other two serving as waste reservoirs (see Figure 1A). With the maturation of computer-aided design (CAD)

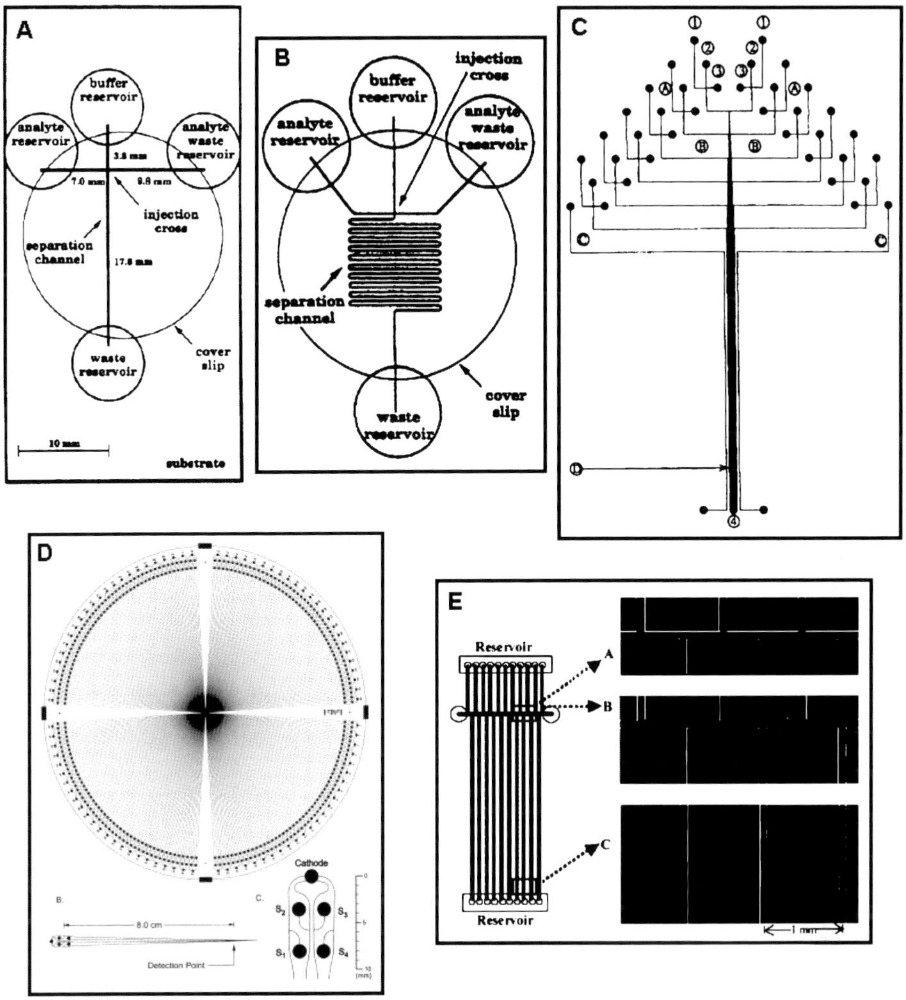

Fig. 1. Various designs for chip-based CE. (A) Microchip with a cross design, reprinted from Jacobson *et al.* (1994b) with permission; (B) microchip with serpentine turns in separation channel, reprinted from Jacobson *et al.* (1994c) with permission; (C) microchip with 12 channels, reprinted from Woolley *et al.* (1997) with permission; (D) microchip with 384-lane radial capillary array, reprinted from Emrich *et al.* (2002) with permission; and (E) microchip design for 2-D electrophoresis, reprinted from Li *et al.* (2004) with permission.

tools and microfabrication techniques more complex microchip designs have been developed, with features such as longer separation channels, serpentine turns, tapered turns, multichannels and capillary arrays for high-throughput analysis, two-dimensional electrophoresis, and others (see Figure 1B–E). With various improvements in microchip designs to overcome particular limitations, electrophoretic analysis with higher efficiency and throughput can be achieved, which greatly enhances the uses of chip-based CE for different types of analytical functions.

2.2. Fluid manipulation

Sample injection is a major concern of fluid manipulation on microchip. So far, three different injection methodologies have been established with chip-based CE: unpinched injection (Wang *et al.*, 1999a, 1999b; Martin *et al.*, 2000), pinched injection (Woolley *et al.*, 1998; Evans, 1997) and gated injection (Liu *et al.*, 2000).

Unpinched injections are performed using a single power supply. First, a high voltage is applied between the sample and the waste reservoirs for a short time. The sample is electrokinetically introduced directly into the separation channel. After the injection is complete, a high voltage is applied to the buffer and the waste reservoirs, which initiates the separation. However, in this unpinched mode, in which no push-back voltages are applied, irreproducible and larger-than-normal sample bands may result (Figure 2A).

Pinched injections are now the most widely used mode for sample injection on microchip. During the pinched injection, samples are injected electrokinetically into a volume-defined injector (cross or double-T shaped) with a pinching voltages applied at both the buffer and the waste reservoirs to define the sample band size. With this pinching voltage, short injection plugs with reproducibility

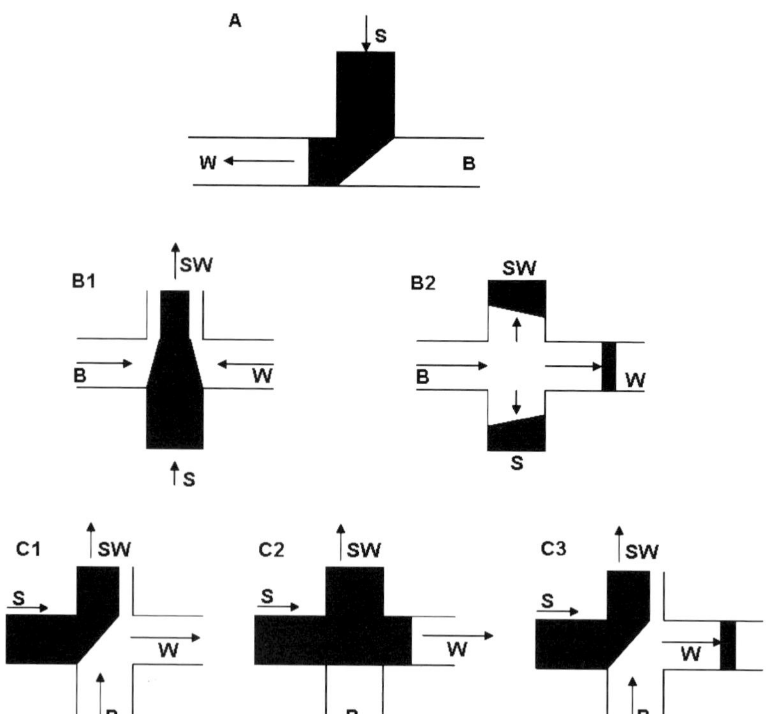

Fig. 2. Schematic views for thee types of injection methodologies. (A) Unpinched injection; (B) pinched injection; and (C) gated injection. Annotations: B, buffer reservoir; S, sample reservoir; SW, sample waste reservoir; and W, waste reservoir.

of peak heights better than 4.1% could be achieved (Evans, 1997). After the injection is complete, a high voltage is applied between the buffer and the waste reservoirs to initiate the separation, with a lower voltage applied at the injection channel to prevent the samples remaining in the injection channel from leaking into the separation channel (Figure 2B). With the pinched injection mode, the electrokinetic bias of the sample can be eliminated.

Gated injection is another widely utilized methodology for sample injection. With the gated injection scheme, the sample is introduced at the head of the separation channel, and buffer is placed in one of the side reservoirs. A high voltage is applied to the buffer reservoir and a fraction of that high voltage is applied to the sample reservoir with the sample waste and waste reservoirs grounded. This results in a flow of sample toward the sample waste reservoirs and a separation flow from the buffer to waste reservoir. Owing to the control of voltage, no sample will mix into the separation channel. To inject a sample plug into the separation channel, the high voltage applied to the buffer reservoir is removed for a short-time period (normally several seconds). Later, the separation step is initiated by resuming the high voltage to the buffer reservoir. This gated injection mode is very convenient to control the amount of sample plug injected into the separation channel. However, electrokinetic bias among samples cannot be avoided with gated injection mode (Figure 2C).

There are also other concerns with fluid manipulation on microchip, including the mixing of sample with labeling reagent or buffer (Kutter *et al.*, 1997), and methods for redirecting the sample after separation for collection of the separated fractions of the analyte (Effenhauser *et al.*, 1995). Successful fluid manipulation could be achieved only if chip design and voltage control are thoroughly studied and understood.

3. MATERIALS AND FABRICATION

3.1. Materials

Silicon is excluded as a material for fabricating CE microchips because of its semiconductivity proves problematic when high voltage is applied (Kopp *et al.*, 1997). Instead, various glass substrates are widely used, from inexpensive soda lime glass through high quality quartz. These substrates are chosen because of their good optical properties, well-understood surface characteristics, high efficiency in dissipating heat and well-developed microfabrication techniques transferred from semiconductor industry (Jacobson *et al.*, 1994d; Fan and Harrison, 1994). However, there are some disadvantages of using glass substrates as microchip materials that hinder the ongoing production and commercialization of glass chip-based CE devices, including the high cost of substrate materials, the many steps and harmful wet chemistry involved in microetching and limitations in geometrical designs available due to the isotropicity of the etching process. Researchers and industry however have explored the use of wide variety of polymer materials for fabricating microchips instead of glass,

Table 1. The properties of common fabrication materials used for microchips

Material property	Si (single crystal)	Glass	SiO_2	PDMS	PMMA	PC
Dielectric strength ($\times 10^6$ V/cm)	3	5–10	2–3	2.1	0.17–0.19	0.39
Coefficient of thermal expansion ($\times 10^{-6}$ C^{-1})	2.6	0.55	0.55	310	55	70.2
Thermal conductivity at 300 K (W/cm/K)	1.57	0.011	0.014	0.0018	0.002	0.002
>70% optical transmittance (nm)	>700	>350	>350	400–700	400–700	400–700
Maximum processing temperature (°C)	1415	550–600	1700	~150	~100	~100
Bulk resistivity (μΩ cm)	2.3×10^{11}	$>10^{10}$	$>10^{10}$	$>10^{20}$	$>10^{20}$	$>10^{20}$
Temperature coefficient of resistance (10^{-3} K^{-1})	−70	—	—	—	—	—
Water contancy angle (advancing)	110°	20–35°	~30°	~110°	60–75°	78°
Glass temperature T_g (°C)	—	—	—	—	106	150

Source: Modified from Lagally and Mathies (2004) and Becker and Gärtner (2000) with permission.

including standard polymer materials. Some properties of these different materials are listed in Table 1.

These materials include polymers such as polyamide (PA), polybutyleneterephthalate (PBT), polycarbonate (PC), polyethylene (PE), polymethylmethacrylate (PMMA), polyoxymethylene (POM), polypropylene (PP), polyphenylene ether (PPE), polystyrene (PS), etc. To date, PMMA and PC are the most popular polymer materials for microfabrication employing hot embossing and injection molding techniques. Poly(dimethylsiloxane) (PDMS) is another very widely used polymer material for fabricating microfluidic devices because of extreme ease of fabrication and uniformity. A cycloolefin copolymer

(COC) is currently being tested for fabricating microchips, which shows very high promise for applications in chemical engineering and molecular biotechnology due to its high chemical stability and optically transparent properties.

3.2. Fabrication

3.2.1. Fabrication procedures for glass materials

Structures on glass substrates are usually generated using standard photolithography techniques (Woolley and Mathies, 1994). Briefly, a glass substrate is coated with a photoresist film. Then, transferring the channel pattern to the film is conducted by exposure to UV radiation through a patterning mask. The exposed portions of the film are dissolved, the remaining parts of the film are hardened by heating, and these then serve as a sacrificial layer for chemical etching. Wet etching using hydrofluoric acid is the most popular way for chemical etching of glass substrates. Finally, the etched substrate was thermally bonded to the top glass plate, which has access holes drilled into it. Figure 3 illustrates the procedure.

3.2.2. Fabrication procedures for polymer materials

Two main methods are utilized for microfabricating polymeric microchips according to the properties of polymer materials: replication technologies and

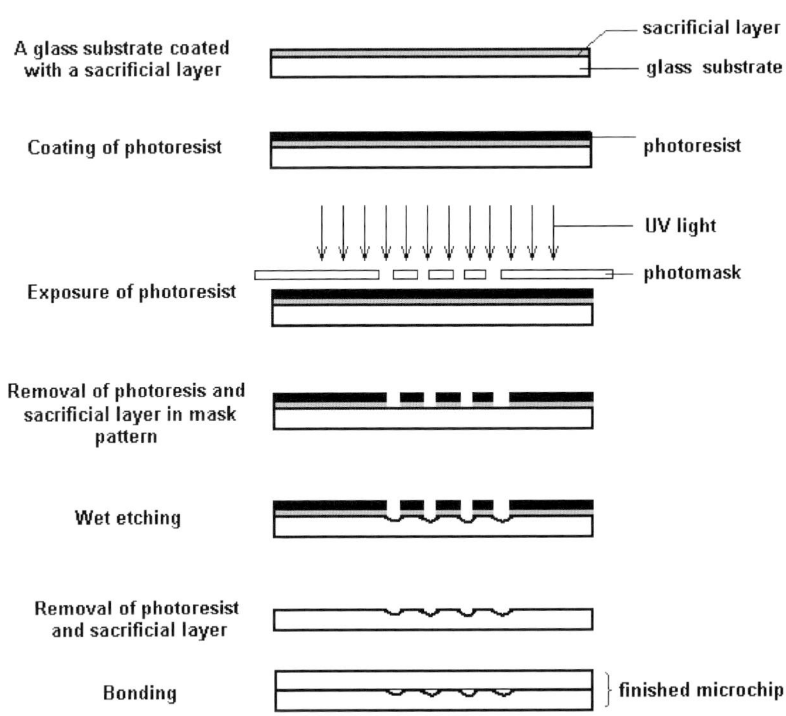

Fig. 3. Procedures for photolithographic fabrication with glass materials.

direct technologies. After the desired patterns are fabricated, a sealing process is needed to form the enclosed microchannels.

3.2.2.1. Replication technologies. The principles behind replication technologies are already well known in the macroworld fabrication. For example, injection molding represents a standard technology for macroscopic polymer component manufacturing. Thus, there is considerable interest to establish a low-cost manufacturing process in the fabrication of microchips using these technologies. However, there are several processes used with replication technologies that give concern for transfer to the microworld. First, undercuts (i.e. structures in the polymer with overhanging edges) cannot be fabricated since the master needs to be removed from the molded structures. Second, the surface quality of the master is a key factor for the lifetime of the mold tools and a limitation on the achievable aspect ratios. Typically, surface roughness values of better than 100 nm root mean square (RMS) are necessary for a good and reliable microstructure replication. Third, interactions between master materials and the polymer substrate should be prevented, as release agents are often used in the demolding step.

The key to the replication technologies is the fabrication of master molds. Generally, three methods are utilized for master fabrication, including micromachining methods (Martynova *et al.*, 1997), electroplating methods (Ehrfeld and Munchmeyer, 1991; Hesch *et al.*, 1995) and silicon micromachining methods (Jansen *et al.*, 1996). After master molds have been fabricated, several methods can be applied for the replication step, such as hot embossing, injection molding and casting.

Hot embossing is currently the most widely used replication process for the fabrication of microchannels onto microfluidic devices. A master with micropatterns is mounted in an embossing system together with a planar polymer substrate, both of which are heated in a vacuum chamber to a temperature just above the glass transition temperature of the polymer substrate. The vacuum is necessary to avoid bubble formation due to the entrapping of air in small cavities. Then, the master mold is brought into contact with the substrate and embossed with a controlled force for some period of time (typically several minutes). Later, the master-substrate is cooled to below T_g with force still applied. Diagrams of a hot embossing machine and a view of a microchannels fabricated using hot embossing are shown in Figure 4A and B.

Injection molding is another widely applied technology, with ability to form almost any geometry from a large variety of thermoplastic materials (McCormick *et al.*, 1997; Piotter *et al.*, 1997). Compact discs (CD) are well-known examples for injection molded microstructured products. However, special care should be taken for microstructures with high aspect ratios. An elevated temperature is reached before injection molding. For amorphous thermoplastics (e.g. PMMA, PC), these temperatures are above their T_g. For semi-crystalline thermoplastics (e.g. POM, PA), crystallite melting points are often chosen. Prior to demolding, both the polymer substrate and the molding tools with mold inserts have to be cooled to a demolding temperature determined by the material and the specific patterned microstructures. Other peripheral

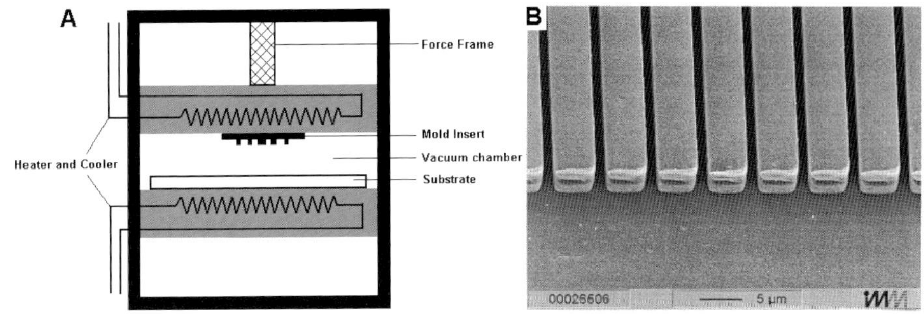

Fig. 4. (A) Schematic diagram for a hot embossing machine. (B) Microchannel structures created on PMMA by hot embossing. Reprinted with permission from Fan and Harrison (1994).

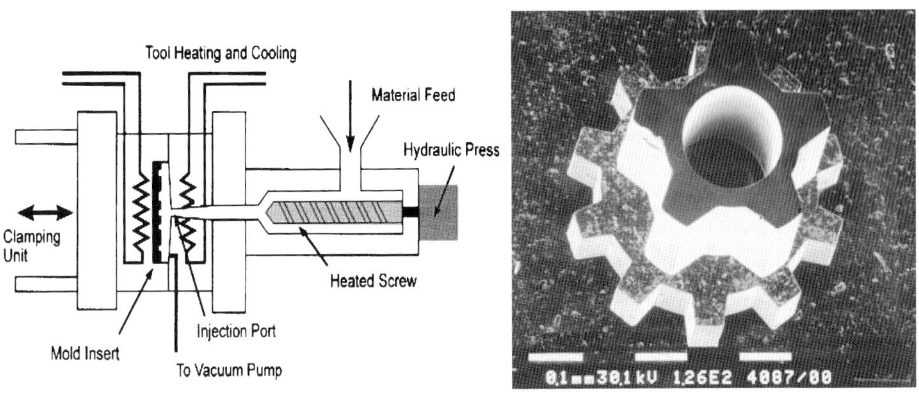

Fig. 5. (A) Schematic diagram for an injection molding machine, reprinted from Becker and Gärtner (2000) with permission. (B) Microgearwheel structures manufactured by injection molding, reprinted from Piotter *et al.* (1997) with permission.

equipment including a special vacuum unit is also needed for evacuation of the mold cavity in the molding tool, as well as a temperature control unit. Diagrams of an injection molding machine and a view of microchannels fabricated using injection molding are shown in Figure 5A and B.

For silicone-based elastomers, a casting process is the easiest and most widely used way of fast fabricating microfluidic devices (Qin *et al.*, 1998; Effenhauser *et al.*, 1997). This type of fabrication is extremely suitable for PDMS elastomers. During the casting process, a mixture of elastomer prepolymer and curing agent are cast against a master with a negative relief, degassed under vacuum and then heated to initiate the polymerization (typically 60~90°C and lasting 1~2 h for PDMS). After the elastomer is cured, the replica is then peeled from the master, and sealed with another planar sheet, forming an enclosed microchip.

Two classes of technologies, soft lithography and rapid prototyping are the most common methods employed for creating a master. Soft lithography is a suite of non-photolithographic methods for replicating patterns. An elastomeric structure with patterns embedded as a bas-relief on the surface acts as the pattern transfer agent.

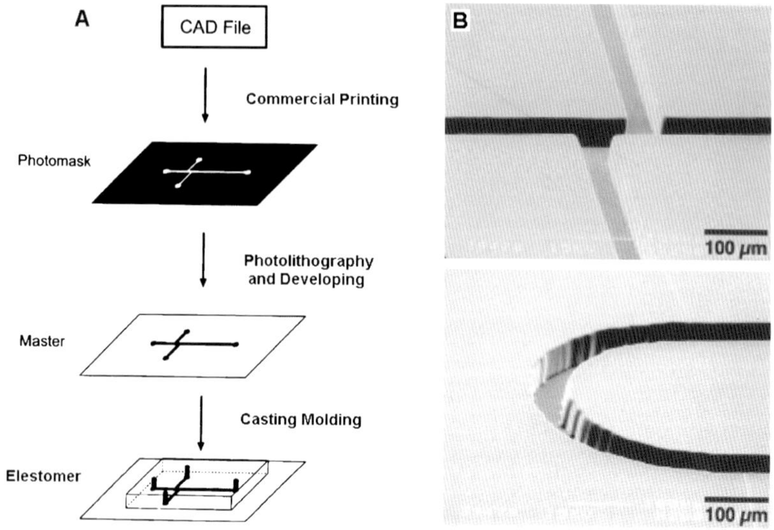

Fig. 6. (A) Scheme for rapid prototyping and casting molding. (B) Microstructure formed on PDMS fabricated by casting molding. Reprinted from Duffy *et al.* (1998a, 1998c) with permission.

The soft lithography methods do not require routine access to a clean room for replication resolution down to <1 μm. Rapid creation of a master prototype (rapid prototyping) begins with creation of a design for a device in a CAD program. Then, the design is printed on to a transparency using a high-resolution image setter. This transparency later serves as the photomask in contact photolithography to produce a positive pattern of photoresist (e.g. Su-8). PDMS is cast against the master made of patterned photoresist, to form a negative relief (Figure 6A).

3.2.2.2. Direct technologies. Besides the above-mentioned techniques using replication processes to produce polymer devices from a single mold, there are several other techniques that allow individual micromachining of each single device. An obvious advantage of these techniques over replication techniques is that no master is needed, especially for manufacturing a single device or a few devices. A widely used technology for the fabrication of microfluidic devices is laser ablation (Roberts *et al.*, 1997; Pethig *et al.*, 1998). In this process, the energy of a laser pulse is used to break chemical bonds in a polymer molecule and to remove the decomposed fragments from the ablation region. With this technology, a wide range of polymeric materials including PMMA, PS, PC and others can be structured. An example of microstructure fabricated by laser ablation is shown in Figure 7A. Optical lithography refers to directly patterning a photosensitive polymer to form microchannels using lithography technologies. Thick photoresists such as Su-8 are applied for the fabrication materials using this technology (Guerin *et al.*, 1997). A view of microchannel fabricated by optical lithography is shown in Figure 7B.

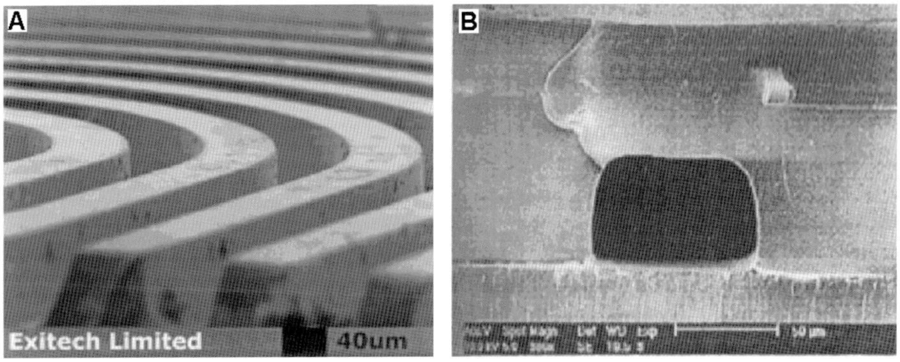

Fig. 7. (A) Microstructures fabricated using laser ablation. Reprinted from Becker and Gärtner (2000) with permission. (B) Microchannels formed in Su-8 photoresist fabricated using optical lithography. Reprinted from Fielden *et al.* (1998) with permission.

Stereolithography is another direct technology that allows actual three-dimensional microfabrication. In this method, a photocuring liquid polymer is exposed to a focused laser light. The polymer cures and forms a solid at the focal point. By moving focal point relative to the polymer, a three-dimensional structure can be constructed.

3.2.2.3. Sealing. Sealing is the last step for manufacturing microchips. For glass substrates, thermal diffusion bonding is the most often applied to enclose open channels (Woolley *et al.*, 1996). Other methods such as chemical-activated bonding and adhesive annealing are also used (Wang *et al.*, 1997). For polymeric substrates such as PMMA, PC, sealing with heat and pressure is normally utilized (Paulus *et al.*, 1998). Sealing with heat and pressure is different from hot embossing in that the temperature is below T_g of the polymer materials for sealing and above T_g for embossing. Care has to be taken not to damage the microstructures, thus this method is advisable mainly for designs with comparatively small structured areas, in comparison to the whole chip surfaces. Lamination is a well-known sealing process used for macroscale fabrication, and it has been transferred into microchip sealing (Roberts *et al.*, 1997). However, the adhesive sometimes tends to block the microchannels, and an inhomogeneous interface can be created, which may lead to band distortion during electrophoretic separation. A further two technologies, laser welding and ultrasonic welding have also been applied for sealing polymeric microchips which avoid the problems of adhesive.

4. DETECTION

The more fluidic functions are integrated on microchips, the more the necessity for high-performance detection is needed. Thus, the final success of a μ-TAS is increasingly determined by the ability of researchers and engineers to realize detection methods that utilize the advantages of reduced diffusion lengths and confined geometries, while also solving the challenges imposed by such miniaturization. There are some requirements for microfluidic detection compared to those of conventional analytical systems. First of all, higher sensitivity is required, because of extremely small injection volume (typically in range of picoliters), as well as the minute detection cell size. Second, faster response times are also necessary, as the time for a sample band to pass the detector is much less than for conventional electrophoresis. Third, special structures are often required to match microchips with sophisticated detector designs. Fourth, the implementation of parallel detection is needed for high-throughput microchips. Last, portability and low cost are other two major requirements for successful microchip commercialization. Three major principles for analyte detection can be described: optical detection (laser-induced Fluorescence, absorbance detection, chemiluminescence detection and others), electrochemical detection (amperometric detection, conductivity detection and potentiometric detection) and mass detection (mass spectrometry). The following section is meant to give a brief introduction to the most widely utilized detection methods employed in chip-based CE.

4.1. Optical detection

4.1.1. Laser-induced fluorescence detection

Among the several above-mentioned detection methods, laser-induced fluorescence (LIF) is so far the most popular detection scheme and the first choice for detection of microchip separation because of its high sensitivity. Generally, detection limits down to 10^{-9}–10^{-12} mol/L can be obtained using LIF detection. With photon counting implemented in LIF detection, even single molecule detection can be achieved (Fister *et al.*, 1998). Both non-confocal and confocal detection systems are used for LIF detection of microchip separation (Jiang *et al.*, 2000). Confocal detection systems (Figure 8) have higher signal/noise (S/N) ratio than non-confocal detection systems, although a more complicated device setup is required. A typical confocal LIF detection involve the following steps: a coherent, collimated laser beam is reflected by a dichroic beam splitter into a high numerical aperture objective, which focuses the laser beam on inside the microchannel to excite fluorescently labeled analytes. The fluorescence emitted from the microchannel is collected by the same objective, and passes back through the dichroic beam splitter. In this same direction, the dichroic beam splitter reflects the laser while passes through the fluorescent analyte solution, which is then focused by another lens onto the entrance of a spatial filter (pinhole). The name "confocal" refers to the fact that this lens is confocal to the objective, that is to say, only light emitted from the focus of the objective can pass through the spatial filter, which will significantly increase the S/N ratio. Then, the fluorescence is directed to detectors, such as photomultiplier tubes (PMTs) or charge-coupled devices (CCDs), and an output signal from these devices will finally be recorded by a computer. There are some drawbacks of LIF detection however, such as the requirement of high-cost and cumbersome

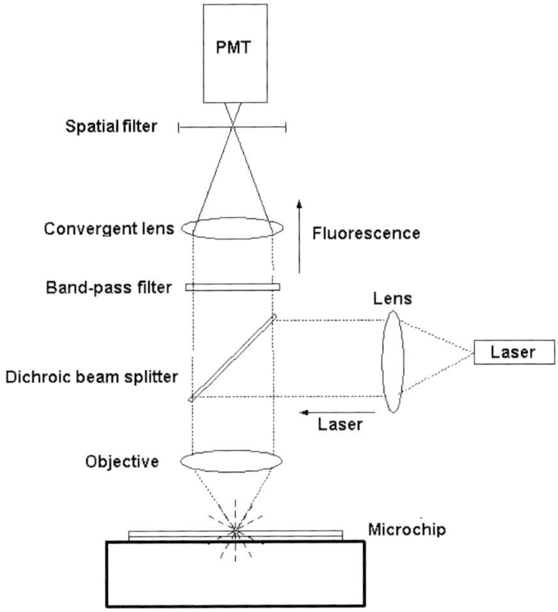

Fig. 8. Schematic view of LIF with confocal detection system.

laser instrument, optical alignment and the need for fluorescently labeled analyte. Therefore, great efforts have been made to ameliorate these costs and to develop various LIF detection systems with features such as device miniaturization and integration with the other microdevice components.

Integrated optics are an example of such a detection system, which aims at miniaturization and integration by utilizing light-emitting diodes (LED) in place of expensive and cumbersome external laser sources. As LEDs with higher output power and shorter wavelengths become available, more and more miniaturized detection systems have been developed using LEDs as light sources (Chabinyc *et al.*, 2001). Other groups have advanced further in developing integrated optics. For example, a blue GaN semiconductor LED, a CDs optical filter and a silicon photodiode were integrated into the same substrate by Chediak *et al.* (2004) to form an integrated optical system, with which the microfluidics component can be disposed after use, while the detection optics can be reused. Another example is from Webster's group (Webster *et al.*, 2001), who recently reported a combination of detector with optical interference filter constructed on the separation device, which has the advantage of the elimination of external optics and alignment procedures. Recently, another technique named liquid core waveguide has also been used in LIF detection. In liquid core waveguide mode, a special material (e.g., Teflon AF) functions as a liquid core waveguide. Excitation light impinges perpendicularly onto the capillary axis and passes through it, without being axially transmitted along the liquid core waveguide, while the fluorescence emitted by the dissolved analyte is transmitted along the waveguide and is then collected by an optical fiber positioned adjacent to the capillary outlet (Wang *et al.*, 2001).

4.1.2. Absorbance detection

Absorbance detection is quite popular in conventional CE, however it is not commonly used for chip-based detection because of the short optical path length in microchannels compromises the sensitivity. Nevertheless, various absorbance detectors have been developed for microchip format (Petsul *et al.*, 2001; Lu and Collins, 2001). In particular, absorbance detectors based on linear photodiode arrays have been developed to image the entire separation channel, which allows the direct visualization of the dynamics of entire IEF (Mao and Pawliszyn, 1999b) or electrophoretic separation processes (Nakanishi *et al.*, 2001). Other groups have devoted their efforts to increase the sensitivity of absorbance detection by increasing optical path length, by devices such as the use of multireflection cells (Salimi-Moosavi *et al.*, 2000) or by use of a thermal lens approach (Sato *et al.*, 1999).

4.1.3. Chemiluminescence detection

The term "chemiluminescence (CL)" was first coined by Eilhardt Weidemann in 1888, referring to the emission of light from a chemical reaction. Using CL as a detection method for chip-based CE has the following advantages. First, high detection sensitivity can be achieved comparable with that achieved with LIF detection. Second, a wide linear range of responding signals will be beneficial for quantitation of the analyte. Third, using CL detection eliminates the need for light sources, which are often expensive and cumbersome in LIF detection. Mangru and Harrison (1998) first demonstrated a CL system to monitor horseradish peroxidase (HRP) and fluorescein-conjugated HRP for microchip electrophoresis. Another CL detection system with still higher sensitivity was developed by Liu *et al.* (2003a) for microchip CE fabricated in PDMS.

Electrochemiluminescence (ECL) is also sometimes called electro-generated CL and it is a form of CL in which the light emitting chemiluminescent reaction is preceded by an electrochemical reaction. With ECL, the advantages of CL are retained, but the electrochemical reaction allows the time and position of the light emitting reaction to be controlled. Furthermore, an additionally beneficial aspect of ECL is that better detection limits may be achieved by rapid electrochemical recycling of reagents generating a rapid release of a chemiluminescent signal (Arora *et al.*, 2001).

4.2. Electrochemical detection

Although LIF detection is the most common detection scheme for microchip separation systems employed so far, its shortcomings are also obvious. Most compounds are not naturally fluorescent, and thus fluorescent labeling or a derivatization step is inevitable for LIF detection, which may artificially alter the separation properties of the analytes. Furthermore, the high cost and large size of the instrumental set up of LIF detection are sometimes incompatible with the concept of μ-TAS, especially with the applications when portability and disposability are necessary, such as point-of-care or *in-situ* analysis. In contrast, electrochemical (EC) detection is an alternative detection mode that is ideally suited to miniaturization and thus suited to chip-based CE analysis. With photolithographic techniques that have already been used for constructing microchips, microelectrodes can be fabricated directly onto the microchip device, leading to a fully integrated system. Furthermore, miniaturization of electrodes does not compromise their sensitivity, unlike absorbance detection which is strictly governed by the length of optical path. Typical limits of detection for microchip EC detection are in the nanomolar range. Generally there are three modes of EC detection: amperometry, conductimetry and potentiometry. The following section will give a brief introduction to these modes of EC detection.

4.2.1. Amperometric detection

Amperometric detection is the most extensively reported EC detection method for chip-based CE, which may be considered in terms of electrolysis at a fixed point along a flowing stream. The stream here is a sequence of analyte zones separated with varying degrees of resolution. These zones pass into a detection cell, where a planar electrode is held at a fixed potential. If the potential is greater (more positive for oxidation or more negative for reduction) than that required for the electrolysis of the analyte, a measurable charge passes from electrode to analyte (or *vice versa*). The resulting current is directly proportional to the concentration of solute passing through the cell (Figure 9). The electrode may be thought of as a chemical reagent. The more positive its potential, the stronger an oxidizing agent it becomes; alternatively when the potential is made more negative, it becomes a stronger reducing agent. In either case, as the concentration of solute rises and falls in passing through the thin-layer cell, the electrolysis current proportionately follows these changes. This current, as a function of time, is

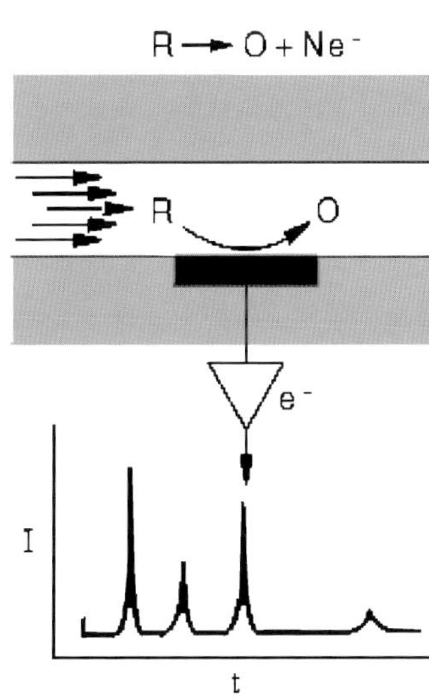

Fig. 9. Schematic diagram of amperometric detection. R and O refer to the reduced and oxidized states of the analyte, respectively.

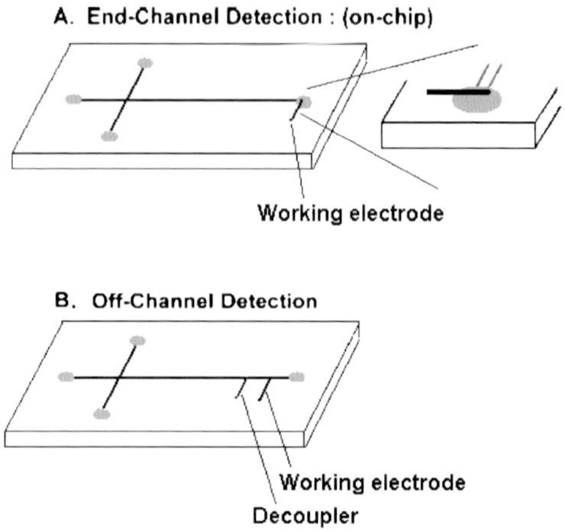

Fig. 10. Two configurations of working electrodes that isolate signal detectors from the high separation voltages.

amplified and sent to a recorder to yield a chromatogram. When utilizing amperometic detection in microchip electrophoresis, an issue of great importance should be addressed, that is to isolate the high separation voltage from the detector. Generally, two approaches have been developed for this purpose, termed end-channel detection and off-channel detection (Figure 10).

For end-channel detection mode, the working electrode is positioned tens of microns from the exit of the separation channel. This distance allows sufficient decoupling of the separation voltage from the detector. Amperometric detection in the end-channel mode was first reported for chip-based CE by Woolley *et al.* (1998). With their microchip electrophoresis system, neurotransmitters were successfully separated with good resolution, and attomole detection sensitivity was achieved. However, with their system, the separation voltage is grounded within the detection reservoir and the remaining separation field may cause potential shifts at the working electrode, therefore, it is necessary to perform a hydrodynamic voltammogram for each given analyte. Another drawback of end-channel detection is band broadening because the sample diffuses when passing from the exit of separation channel to the working electrode. Mathies and co-workers have developed a sheath-flow-supported scheme for end-channel detection to overcome part of this disadvantage (Ertl *et al.*, 2004). In their scheme, two sheath-flow channels were placed at a 30° angle relative to the separation channel, and were joined to it just before the end of the separation channel. A constant gravity-driven flow passed through the sheath-flow channels into the detection reservoir. These flows increased the velocity of analyte in the detection reservoir, which minimized the band broadening. Another way to eliminate band broadening is to employ off-channel amperometric detection. In this mode, the working electrode is placed directly within the separation channel, the analytes migrate over the electrode while still confined to the channel. However, a decoupler is usually needed for isolating the separation voltage from the amperometric detector (Osbourn and Lunte, 2003; Wu *et al.*, 2003; Lai *et al.*, 2004).

4.2.2. Conductimetric detection

Conductimetric detection is another EC detection method that is now gaining more popularity for use with chip-based CE. Conductimetry measures the differences in the conductivity of the bulk solution compared to that of the analyte zones, and can be considered a universal detection method because the analyte is detected without the need of fluorophore, chromophore or electroactive functional group detection. A typical conductimetry system employs two electrodes that are either in direct contact with the background electrolyte (contact mode) (Masár *et al.*, 2004) or are external and capacitively coupled to the solution (contactless mode) (Lichtenherg *et al.*, 2002; Laugere *et al.*, 2003). For measuring the conductivity, an alternating current (AC) potential is applied between the two electrodes, and when the ions of the background electrolyte are displaced by the analyte in the passing zone, a change in conductivity occurs, and the concentration of the analyte can then be correlated with the deviation of conductivity from the baseline. A major consideration for conductimetric detection is that the conductivity of the background solution must be different from that of the analyte. In addition, the electrolyte must be carefully considered as a highly conductive may also result in too high a background signal, which could possibly interfere with the detection of the analyte zones.

4.2.3. Potentiometric detection

Potentiometric detection is a technique used to measure the potential that arises upon a membrane between two solutions with different ionic activities, under conditions of on-current flow. When analyte solution flows through a semipermeable, ion-selective membrane of an ion-selective electrode, a potential difference between the activity of the external and that of the internal solutions of the electrode is created. This potential is measured against the fixed potential of a reference electrode and can be correlated logarithmically to the concentration of the analyte. However, potentiometry is a difficult technique to apply to the detection of separation-based systems, because it is difficult to create an ion-selective membrane, which must be semipermeable to multiple ionic species that are to be analyzed, but not highly permeable to the background buffer ions. Consequently, there are few papers that report use of potentiometric detection for microfluidic systems (Tantra and Manz, 2000; Ferrigno *et al.*, 2004).

4.3. Mass spectrometry

Mass spectrometry is a powerful alternative to optical and electrochemical detection techniques that can be used to identify unknown compounds molecularly, to quantify known compounds, and to elucidate the structure and chemical properties of molecules. Detection of compounds can be accomplished with very minute quantities (as little as 10^{-12} g, or 10^{-15} m for a compound of mass 1000 Da), which means that compounds can be identified at very low concentrations (one part in 10^{12}) in chemically complex mixtures. Thus, examples of the application of mass spectrometric detection for low concentration and complex mixtures are increasing – is it increasingly being applied for the detection of modifications in the methylation state and other modifications of nucleic acids in biological systems (see this volume, Ehrich *et al.*, 2007). It is also very suitable for the study of proteomics with complex mixtures of proteins, or for small modifications in the abundance of modified species of proteins.

To date, most work on coupling mass spectrometry with microchip separation systems is focused on the design of the interfacing between the systems. Electrospray ionization is the most frequently used method of ionizing chemical or biological compounds at the interface between separation microsystems and mass spectrometer. An emitter is needed for this method, which delivers the sample liquid to be ionized by creating a strong electric field between the emitter and the mass spectrometer. The two most important properties of the emitter are – a low liquid flow rate that results in high ionization efficiency, and the sharpness of the emitter that leads to the creation of high electric fields. Many reports have described efficient sample introduction from microchip into mass spectrometer using electrospray ionization. For example, Schilling *et al.* (2004) have reported a new on-chip electrospray ionization nozzle that can be used as an interface for coupling microfluidic devices with mass spectrometric detection. The nozzle was micromilled in a polymer foil (PMMA, 750-μm thick) with three different inner nozzle diameters, and two different apex angles (Figure 11). The analysis of the tetrapeptide MRFA was readily achieved using this interface nozzle as stable electrospray conditions could be generated between the chip ionization system and the mass spectrometer.

Another example is provided by Tachibana *et al.* (2003), who have developed a robust and simple interface for microchip elecrophoresis-MS. A spray nozzle was connected to the exit of the separation channel of the microchip by use of a polyether ether ketone screw without glue, allowing easy replacement. Using this instrumental set up a few basic drugs were separated by microchip electrophoresis and the separation efficiency was improved by using high-viscosity separation matrix and a spray nozzle with small core size (20 μm).

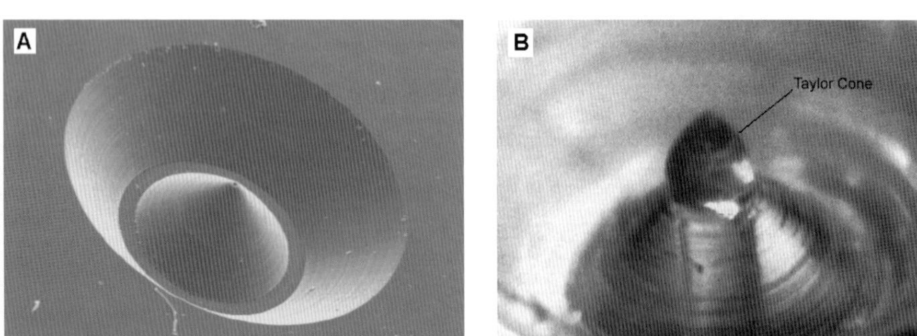

Fig. 11. (A) SEM picture of a micromachined ESI nozzle with an inner diameter of the tip hole of 40 μm and an apex angle of 90°. (B) Taylor cone at the nozzle tip with fluoropolymer coating. Reprinted from Schilling *et al.* (2004) with permission.

5. SURFACE MODIFICATION

It is well known that for conventional CE, the performance as well as the reproducibility of electrophoretic separations can considerably be improved by appropriate surface modification. Internal capillary coatings are typically applied to control the EOF and to suppress interactions between the analyte and the capillary wall, especially for separations involving macromolecules such as protein and DNA. Chip-based CE can be considered as a highly miniaturized version of CE, and with the reduction of the size of channels the properties of channel walls become more important than for conventional capillaries. Surface modification is a key factor for a successful demonstration of chip-based CE, as well as for other microfluidic device applications. A number of obvious advantages can result from choice of the appropriate surface modification, such as the change from hydrophobicity to hydrophilicity that enables use of polar liquids, and the generation or elimination of EOF effects that can facilitate fluid manipulation.

There are two major categories for surface modification of microchannel: dynamic coatings (physical-adsorbed coatings) and permanent coatings. Dissolved surface adsorptive compounds are usually used for dynamic coating of the channel. By rinsing the microchannel with a solution of the modifier prior to separation or by addition of the modifier to the electrolyte, microchannel can be dynamically coated with modifier, which then remains during electrophoresis. In permanent coatings, the coating materials are covalently bounded to functional groups on the microchannel surface and are thus immobilized and insoluble to the electrolyte. However, the achievement of internal surface modification for chip-based CE devices is particularly challenging, because chips are formed from diverse materials with various different inherent surface properties. Here, we focus only on the surface modification of the materials most widely used for microchips, such as glass, PMMA and PDMS.

5.1. Dynamic coating

Dynamic coating is the easiest way to achieve surface modification and is widely used for the control the EOF in microchip separations. Dynamic coating can be accomplished by adding selected surface-active compounds like polymers or surfactants to the running buffer and modifying the surface, and then applying a rinsing step to remove excess compound immediately prior to the separation step. Depending on the charge of the modifier compounds adsorbed to the microchannel walls, the EOF can be suppressed, enhanced or even reversed. For example, poly(dimethyl acrylamide) (PDMA), hydroxyethylcellulose (HEC), hydroxypropylmethylcellulose (HPMC) and hydroxypropylcellulose (HPC) are frequently used as dynamic coatings to diminish EOF and serve simultaneously as a sieving matrix for DNA fragments sizing (Bean and Lookhart, 1998; Albarghouthi *et al.*, 2003). To reverse the EOF, cationic detergents such as didodecyldimethylammonium bromide polycations, and cetyltrimethylammonium (CATB) can be employed (Landers *et al.*, 1992; Legaz and Pedrosa, 1996; Ding and Fritz, 1997; Melanson *et al.*, 2000).

5.1.1. Dynamic coating for glass/quartz substrates

For the dynamic coating of glass/quartz microchip substrates, the methods employed for conventional fused-silica capillary can be easily transferred to these devices. The dynamic coating of glass microchannels with PDMA is a well-known example, where 5% PDMA was used as a sieving matrix as well as dynamic wall coating in a 384-lane microchip for ultra high-throughput genetic analysis (Emrich *et al.*, 2002). In addition, various cellulose derivatives such as HEC, HPMC and HPC are also very popular buffer additives for producing dynamic coatings on glass chip surfaces used for DNA amplification and separation (Tian and Landers, 2002). Another non-conventional approach to dynamic coatings of glass substrates is the use of gold nanoparticles provided by Pumera *et al.* (2001). Here, the surface was first coated with a layer of poly(diallyldimethylammonium chloride) (PDADMAC) to support the adsorption of citrate-stabilized gold nanoparticles, which were subsequently collected on that surface. With the presence of gold nanoparticles on the channel surface, the resolution of aminophenol isomers was greatly increased due to selective interactions of the different solutes with this modified surface.

5.1.2. Dynamic coating for PMMA substrates

PMMA is an inexpensive polymer with good optical properties in visible light, which has been widely employed for fabricating microchips. However, its surface is rather hydrophobic and it exhibits moderate electro-osmotic mobility in aqueous solution, which is most likely due to non-esterified carboxyl groups. Several dynamic coating compounds have been tested for their ability to modify surface of PMMA microchips. For example, Xu *et al.* (2002) have reported separation of DNA using HPMC as a sieving buffer matrix on a PMMA microchip. To further improve analyte resolution, polyhydroxyl additives such as mannitol, glucose and glycerol have been employed as well. Another demonstration of dynamic coating is provided by Dang *et al.* (2003), who aimed at reducing analyte adsorption of microchannel walls by modifing PMMA surfaces using several low-molecular-weight compounds (amines, SDS, CTAB) and some hydrophilic neutral polymers (PEG, HEC, HPMC, MC). Oligosaccharide ladders could be successfully separated in surface-modified PMMA microchips.

5.1.3. Dynamic coating for PDMS substrates

PDMS is another popular material for making microchips because of its optical transparency and utility for fabrication. However the surface of PDMS is also hydrophobic similar to PMMA, making it difficult to fill the PDMS microchannels with aqueous electrolyte, and also because of its hydrophobicity, significant physical adsorption of proteins on PDMS surface has also been reported. Dou *et al.* (2002) utilized MES to modify surfaces of PDMS microchips, where the addition of 2-morpholinoethanesulfonic acid (MES) to the electrolyte buffer improved the separation efficiencies of analytes, such as arginine, glucose and methionine-glycine. A dynamic coating composed from multiple layers of positively and negatively charged polymers was applied to PDMS chips by Liu *et al.* (2000). Polyelectrolyte multilayers were created by exposing the channel wall to alternating solutions of positively (polybrene) and negatively (dextransulfate) charged polyelectrolytes. The dynamic coating exhibited a stable and nearly pH-independent EOF in the range of 5–10.

5.2. Permanent coatings

Though much more laborious to apply than dynamic coating, permanent coating is considered as the most effective way to achieve surface modification because of its better stability and performance.

5.2.1. *Permanent coating for glass/quartz substrates*

The surface of glass substrate contains silanol groups, similar to fused silica, although at a lower density. Therefore, the first choice to permanently modify a glass surface is to undertake chemical reactions with its silanol groups. A permanent coating for glass/quartz microchips with linear polyacrylamide (LPA) proposed by Hjertén (1985) is now the most widely applied method. Briefly, microchannels are first flushed with NaOH, and then filled with γ-methacryloxypropyltrime-thoxysilane in diluted acetic acid and/or acetonitrile for 1 h. In a second step, an aqueous solution of acrylamide with ammonium persulfate (APS) and *N*,*N*,*N*′,*N*′-tetramethylenediamine (TEMED) is then pumped into the channels and polymerize at room temperature. Lastly, the channels are flushed with water and dried by vacuum. Glass microchips with this permanent coating have been widely used for genetic analysis.

5.2.2. *Permanent coating for PMMA substrates*

Permanent coating for PMMA substrates is employed to generate amine-terminated surfaces that could be utilized for immobilization of enzymes and dsDNA (Henry *et al.*, 2000). Further reaction of this surface with *n*-octadecane-1-isocyanate will generate an octadecane-chain-terminated surface, which could be applied to reversed-phase CEC separations of DNA ladder (Soper *et al.*, 2002). Another modification is to apply a pulsed UV excimer Laser (KrF, 248 nm) at sub-ablation fluency (Johnson *et al.*, 2001). Using low laser power, the surface chemistry of PMMA could be altered, producing carboxylic groups at the surface without changing the physical morphology.

5.2.3. *Permanent coating for PDMS substrates*

Treatments of the PDMS surface with oxygen plasma, UV light or corona discharge will replace methyl (Si-CH_3) groups with hydroxyl (Si-OH) groups, thus changing its hydrophobic surface properties from hydrophobicity to hydrophilicity (Hillborg and Gedde, 1998; Efimenko *et al.*, 2002). However, oxidized PDMS surfaces are observed to exhibit dynamic surface properties due to recovery of hydrophobicity after oxidation, probably because of diffusion of low-molecular-mass PDMS chains in the polymer bulk onto the surface, or diffusion of the oxidized PDMS chains into the polymer bulk. Another approach namely radiation-induced graft polymerization can be employed in place of modification by exposure to the energy sources described above. If the polymer surface has no chemically reactive functional groups, irradiation would be needed to generate free radicals on the surface, which then act as sites for graft polymerization. Hu *et al.* (2002) demonstrated this radiation-induced graft polymerization process, by modifying the surface of PDMS with acrylic acid (AA), acrylamide (AM), dimethyl acrylamide (DMA), 2-hydroxyethylacrylate (HEA) and poly(ethylene glycol)mono methoxylacrylate (PEGA). The DMA and PEGA grafted microchannels were then selected for electrophoresis of two peptides, and both modified surfaces exhibited little adsorption of the peptides on the channel walls.

6. APPLICATIONS

As the technologies required for chip-based CE have matured, there have been a wide application of this platform for the analysis of small molecules such as amino acids (Jacobson *et al.*, 1998), dyes (Effenhauser *et al.*, 1993) and explosives (Wallenborg and Bailey, 2000), and also for macro molecules such as polysaccharides (Emrich *et al.*, 2002), proteins (Li *et al.*, 2004) and DNA (Dou *et al.*, 2002). It has also been readily integrated with other functional miniaturized instrument components to realize the concept of "lab-on-a-chip" (Manz *et al.*, 1990). However, here we only focus on the applications relevant to nucleic acid analysis by chip-based CE, and on integrated lab-on-a-chip devices (genetic micro total analysis systems, g-μTAS).

6.1. Nucleic acid analyses

Conventional CE offers many advantages over slab-gel electrophoresis in terms of higher resolution, shorter separation and lower sample consumption. The employment of high-throughput CAE sequencing methods also allowed the HGP to be successfully completed in advance of projected time. Chip-based CE is now widely considered as a viable alternative to conventional CE and will undoubtedly have a great impact on analytical science, especially for the analysis of nucleic acids. CGE with various different sieving matrices is always employed for the analysis of nucleic acids because of the nearly identical charge-to-mass ratios of nucleic acids regardless of chain length. Therefore, we will first give a brief introduction to sieving matrices utilized for nucleic acids separation. Then according to the resolution required, several specific applications including fragments sizing, genotyping and sequencing are discussed.

6.1.1. Sieving matrices

Although cross-linked LPA is widely used in slab-gel electrophoresis and was also employed in early CE development, there are several deleterious factors that hinder its use in chip-based CE systems. Several major factors are that both gel breakage and gas bubble formation can occur because of matrix shrinkage during *in situ* polymerization. In addition, the high field strengths and alkaline pH used for DNA separation exacerbate gel hydrolysis and degradation and result in short lifetimes and low reproducibility. Hence, for chip-based capillary electrophoretic analysis of nucleic acids, cross-linked gels have been replaced by non-cross-linked polymer solutions (Heiger *et al.*, 1990) and other novel types of matrix.

The main polymer solutions used as sieving matrix for nucleic acids analysis are listed in Table 2, along with their respective viscosities. These physical characteristics are a key factor in the efficiency of the matrix performance. Among the above-listed polymer solutions, the non-cross-linked, highly hydrophilic LPA solution provides the highest resolution, unsurpassed by other linear or branched polymers. Unfortunately, LPA has no self-coating ability, thus it must be used in pre-coated separation channels, whereas some of the other polymer solutions including PDMA, PVP, PEO and cellulose derivates (see Section 5), have self-coating properties and thus are also widely used for DNA separation.

Viscosity is an important factor in matrix loading due to the loading pressure limitation of the bonded microchip. For most glass chips, up to 200 psi pressure can be tolerated, while most plastic microchips can only tolerate 50 psi. However, it is generally true that higher resolution can be achieved using more concentrated polymer solution, which usually results in higher viscosity. To address this issue, several temperature-dependent viscosity-adjustable (thermo-responsive) polymer solutions have been developed for DNA separation (Albarghouthi *et al.*, 2001; Doherty *et al.*, 2004; Barron, 2004). These matrices contain two viscosity zones between loading and separating stages, allowing rapid loading with a lower viscosity and rehabilitating good separations with a higher viscosity. One type of thermo-responsive polymer solutions is the thermo-associating polymer solution, of which the viscosities increase with elevated temperature.

A good example is the low-molar-mass triblock copolymer of the poly(ethylene oxide)-poly(propylene oxide)-poly(ethylene oxide) ($PEO_{99}PPO_{69}PEO_{99}$). At 5°C, a 25% w/v $PEO_{99}PPO_{69}PEO_{99}$ solution has a low viscosity of 50 cP, allowing for easy injection loading. When at 20°C, the viscosity increases sharply to 250 cP making it suitable for DNA fragments sizing. Another type of thermo-responsive polymer is the thermo-thinning polymer solution that has temperature-response characteristics opposite to that of thermo-associating polymers. For example, the copolymer of *N,N*-dimethylacrylamide (DMA) and *N,N*-diethylacrylamide (DEA) has a low viscosity of 10 cP at 70°C suitable for matrix loading, whereas at 40°C this matrix permits separation of DNA to be achieved at single-base resolution (Albarghouthi *et al.*, 2001).

Large DNA molecules are conventionally separated by pulsed-field gel electrophoresis (PFGE), which is extremely time consuming, with a typical running

Table 2. Polymer solutions used as sieving matrices for nucleic analysis

Polymer	Molar mass (kDa)	Concentration (wt%)	Buffer	Temperature (°C)	Viscosity (cP)
LPA	9000	2	50 mM Tris-TAPS, 2 mM EDTA, 7 M urea	25	27,400
PDMA	200	6.5	100 mM TAPS, 8 M urea, pH 8	30	1200
PDMA	98	6.5	100 mM TAPS, 8 M urea, pH 8	30	75
PEG	35	6	100 mM TAPS, 6.6 M urea		10,000
HEC	97	2	89 mM Tris-borate, 2 mM EDTA, 6 M urea, 10% formamide	25	5000
PEO	8000	1.5	89 mM Tris-borate, 2 mM EDTA	Ambient	1200
	600	1.4	3.5 M urea, pH 8.2		
PVP	1000	4.5	89 mM Tris-borate, 2 mM EDTA, pH 8.0	20	27
HPMC-4000		2	50 mM Tris-borate, 2.5 mM EDTA, pH 8.3		4390
HPMC-50		50	89 mM Tris-borate, 2 mM EDTA, pH 8.3, polyhydroxy	25	40
HPMC-5		5	100 mM Tris-borate, 2 mM EDTA, pH 8.0, mannitol	20	5.7
MC-4000		2	50 mM Tris-borate, 2.5 mM EDTA, pH 8.3	20	4390
MC-8000		2	50 mM Tris-borate, 2.5 mM EDTA, pH 8.3	20	7980
HPC-11000		2	50 mM Tris-borate, 2.5 mM EDTA, pH 8.3	20	11,000

Source: Reprinted with permission from Xu and Baba (2004).

time more than 10 h. A different idea for separating large DNA molecules involves fabricated nanostructures in the chip channel instead of using polymer solution-based sieving matrices. Entropic-based separation is a good example of this process, where separation is based on artificially fabricated nanofluidic channels with narrow constrictions that create mobility differences for separating long DNA molecules (>5 kbp) without using polymer solutions or pulsed fields (Han and Craighead, 2000). The internal conformational entropy is one of the dominant properties of long flexible macromolecules, such as DNA, which is directly proportional to the molecular contour length. The entropic trapping

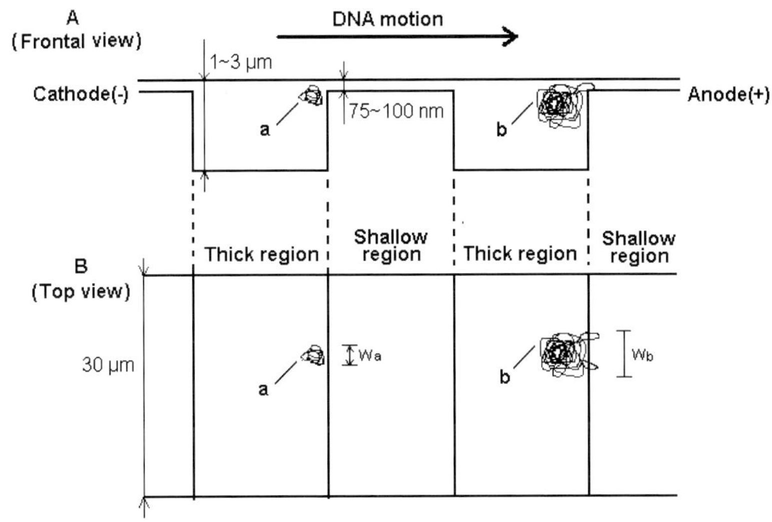

Fig. 12. Schematic view of entropic-based DNA fragment separation, where (A) illustrates the separation of smaller DNA molecules; and (B) illustrates the separation of larger DNA molecules. Reprinted from Han and Craighead (2000) with permission.

(ET) effect will occur when the radius of gyration (R_g) of DNA molecule is comparable to the mean pore size of separation gel. In entropic-based separations, DNA molecules are migrated by electrophoresis along a channel composed of a periodic array of alternate deep and shallow regions. In the deep regions the DNA molecules can form spherical equilibrium shapes, because the dimensions are much larger than the R_g of the DNA. In contrast, the height of shallow regions is much less than the R_g of DNA, thus the DNA becomes trapped at the entrance to the shallow regions and has to be deformed from its equilibrium shape to fit into the constriction, which can be used to determine the apparent mobility of DNA molecules. Longer DNA molecules will escape more easily from this ET than short molecules, because long molecules will have a larger contact area in the thin slit, resulting in a faster electrophoretic mobility. A schematic view of entropic-based separation is shown in Figure 12.

6.1.2. DNA fragment sizing

Fragments sizing requires the least resolution, and is thus the easiest achievable application of nucleic acid analysis. Various nucleic acids such as short oligonucleotides, polymerase chain reaction (PCR) products, restriction fragments and ribosomal RNA have been sized using chip-based CE, with some representative electropherograms shown in Figure 13.

6.1.3. Genotyping

Applications of chip-based CE for genotyping include analysis of genetic polymorphisms such as single-nucleotide polymorphism (SNP), single-stranded

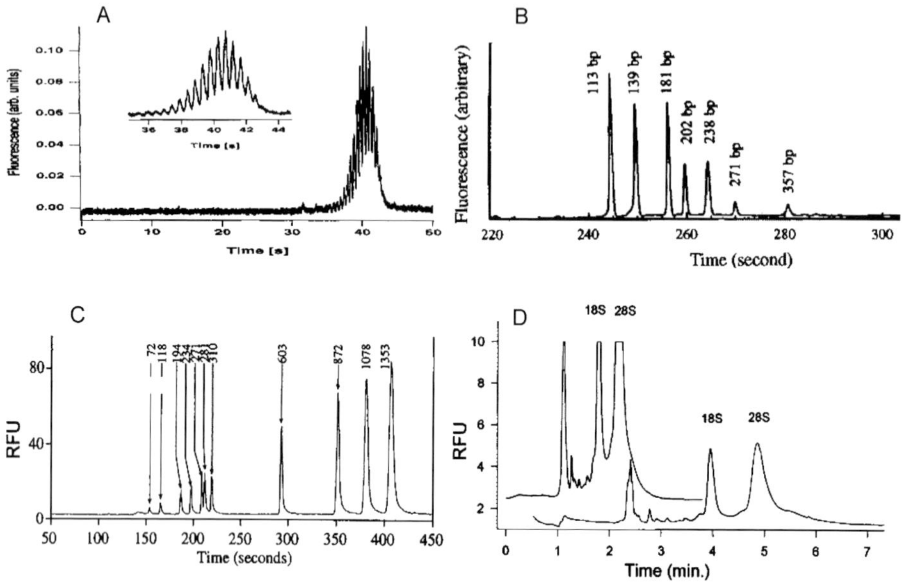

Fig. 13. Representative electropherograms of various nucleic acids fragments separated by chip-based CE. (A) Short oligonucleotide ladders, reprinted from Effenhauser and Mathies (1994) with permission; (B) PCR products, reprinted from Cheng *et al.* (1998c) with permission; (C) ΦX174 *Hae III* fragments, reprinted from Woolley and Mathies (1994), with permission; and (D) total RNA, reprinted from Ogura *et al.* (1998) with permission.

conformation polymorphism (SSCP), short-tandem repeat (STR) and so on. In these applications, efficient and sensitive separation of DNA molecules according to their size or conformation is required.

SNPs are the most abundant type of genetic variation in mammalian genomes. There are estimated to be about 3 million SNPs within an individual, and SNPs are widely used as markers for gene mapping and for genetic polymorphism analysis. SNP genotyping promises to reveal some of the genetic reasons why some people are more susceptible to diseases such as cancer or diabetes, and what predisposes others to suffer adverse reactions to drugs. As a result, SNP genotyping is a booming market with annual expenditure on SNP research predicted to grow from US$158 million in 2001 to more than $1.2 billion in 2005 (Melton, 2003). Most SNP scanning technologies utilize hybridization-based methods, however prior knowledge of the sequence of interest as well as the use of large amounts of synthetic DNA probes are often required (Park *et al.*, 2002; Wang *et al.*, 2003). Instead, high-throughput electrophoresis-based methods such as temperature gradient CE (TGCE) can also be applied for the discovery and mapping of SNPs. Liu *et al.* (2003b) have used chip-based TGCE for fast screening of SNPs, and some representative electropherograms are shown in Figure 14.

SSCP is an example of an alternative method for mutation detection based on electrophoretic DNA separations. SSCP analysis involves an electrophoretic separation of single-stranded nucleic acids based on differences in sequence.

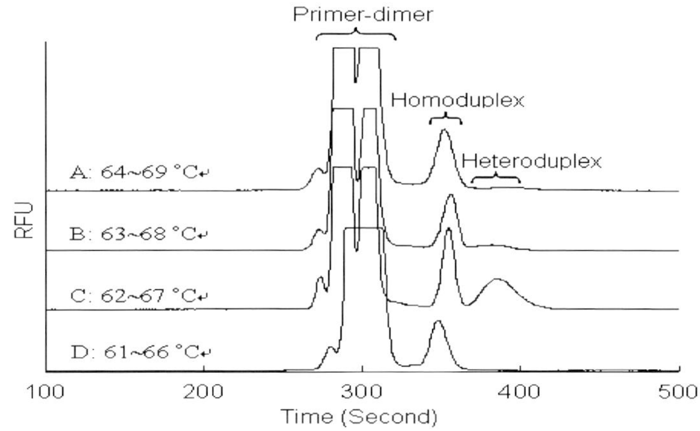

Fig. 14. Chip-based temperature gradient capillary electrophoretic separation of DNA fragments with two SNPs. Reprinted from Liu *et al.* (2003b) with permission.

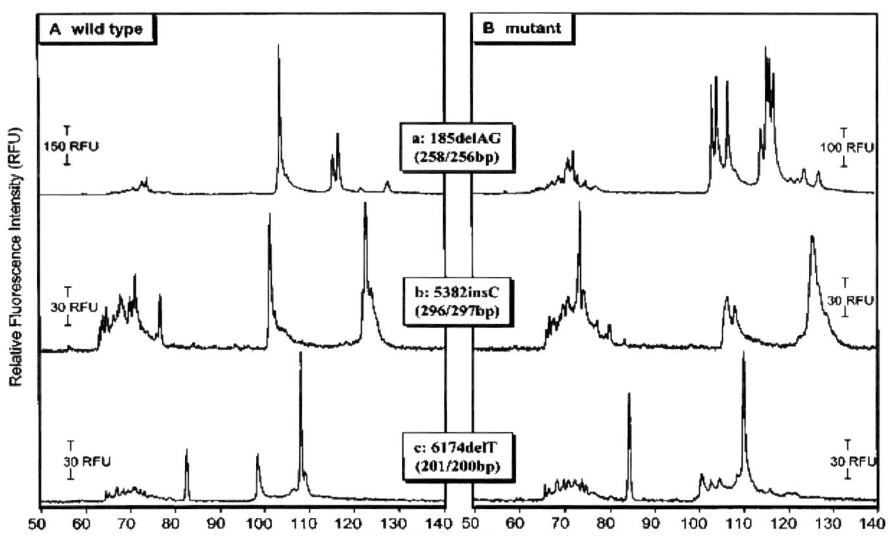

Fig. 15. Fast SSCP analysis of DNA fragments by chip-based CE. Reprinted from Tian *et al.* (2000) with permission.

Single base mutation may be detected because single base mutation may alter or disrupt the secondary conformation of the single-stranded nucleic acids sufficiently, resulting in an electrophoretic mobility shift. An example of mutation detection by SSCP analysis is provided by Tian Tian *et al.* (2000), who used both conventional and chip-based CE for detection of common mutations in BRCA1 and BRCA2, representative electropherograms are shown in Figure 15. Simple tandem repeats (STRs) are short stretches of repetitive DNA sequence that are distributed throughout the genome, typically, with each STR locus consisting of 7–20 repeats of specific 2- to 7-base sequences, resulting in another

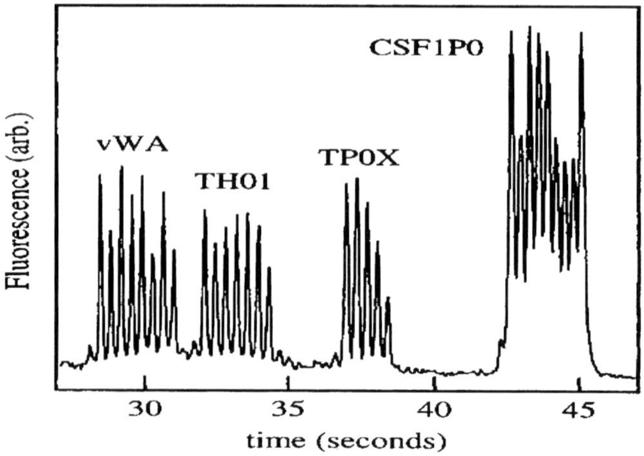

Fig. 16. Electropherogram of the four-locus CTTv allelic sizing standard. Reprinted from Schmalzing *et al.* (1997) with full permission.

common measurable genetic variation. STR analysis has developed as an important tool in forensic analysis, gene mapping and discovery, paternity testing and clinical diagnostics (Edwards *et al.*, 1991; Hammond *et al.*, 1994; Carey and Mitnik, 2002). Among various advances in STR typing technologies, the most popular is a PCR-based method with the product sizes analyzed by CE.

Resolution of 2–3 bp over several hundred bp is required to achieve unambiguous resolution of the different STR alleles. Schmalzing *et al.* (1997) demonstrated single-channel microchip STR analysis using a 2.6 cm separation length and a replaceable LPA matrix to analyze both fluorescently labeled CTTv PCR samples and STRs (shown in Figure 16).

7. DNA SEQUENCING

DNA sequencing has become an area of intense interest as one of the most important scientific accomplishments in human history is the completion of the human genome sequence in 2003 by an international collaborative Human Genome Program, coinciding with the 50th anniversary of the discovery of the DNA double-helix structure. The present state-of-the-art method for DNA sequencing still relies on an advanced form of the Nobel Prize-winning Sanger dideoxy chain termination reaction (Sanger *et al.*, 1977). Briefly, the genome is fragmented into small pieces, and the Sanger reaction (a controlled interruption of the enzymatic replication of an ssDNA template by a DNA polymerase) is used to produce a ladder of template-complementary DNA fragments that differ in length by one base and bear unique fluorescent labels, according to their terminal base. Electrophoretic separation of DNA fragments with single-base resolution is then applied, detecting the base-specific labels and reading the base sequence of each fragment into their original order. Until 1999, the majority of all DNA sequence had been generated by a manual slab-gel electrophoresis process. Subsequently, CE, because of its speed and ease of process automation became a realistic option for increasing the throughput of any sequencing project. It is because of this successful transfer from slab-gel-based to capillary-array-based instrumentation that such a tremendous progress in data accumulation by HGP was achieved (Collins *et al.*, 2003). In the February 15, 2001 issue of *Nature*, the International HGP consortium published the draft sequence and initial analysis of the human genome, only several years after this transfer of genomic sequencing to array-CE.

Initially cross-linked polyacrylamide was employed in CGE, but this separation medium created practical difficulties for the preparation, operation and indeed the shelf life of capillaries filled with such materials. Karger and colleagues (Heiger *et al.*, 1990) soon developed a replacement for it with linear LPA in 1990. Among many materials employed for DNA sequencing, LPA shows the best performance: a 2% w/w solution of high molecular weight (M_r) LPA-producing single base-length fragment separation for more than 1000 bases in 80 min, which permits a read accuracy of approximately 97% was reported by Karger's lab in 1996 (Carrilho *et al.*, 1996). The same group continued their efforts of improving the performance of LPA by additional fine-tuning of the LPA composition, as well as optimization of electric field strength, run temperature and dye chemistry. The sequencing of 1000 bases in less than 1 h with a base calling accuracy of 98–99% was achieved using the optimized 2.5% LPA matrix, consisting of a mixture of 2% w/w high M_r LPA and 0.5% w/w low M_r LPA operated at 60°C and 200 V/cm (Salas-Solano *et al.*, 1998).

LPA is also not the sole choice of sieving matrix for DNA sequencing. Yeung and colleagues (Fung *et al.*, 1998) introduced the use of replaceable linear poly(ethylene oxide) (PEO) for DNA analysis, and Kim and Yeung (1997) reported the separation of a single-color sequencing ladder up to 1000 bases (resolution of raw data = 0.5 at base number 966) in a mixture of 1.5% high M_r PEO (8 MDa) and 1.4% low M_r PEO (600 kDa) at 75 V/cm over a separation capillary distance of 70 cm.Yeung's group also explored the use of polyvinyl pyrrolidone (PVP) for DNA sequencing, which is a self-coating material. Single-color sequencing with 7% commercially available PVP (1 MDa) showed reasonable separation up to 350 bases. Separation could be extended to 530 bases when a 5% solution of high M_r PVP extracted from the 1 M Da material was used. The separation was performed at 150 V/cm and room temperature with an effective column length of 50 cm, and the 530 bases eluted in 83 min (Gao and Yeung, 1998). Poly(dimethylacrylamide) (PDMA) is another self-coating material for DNA sequencing. In four-color sequencing runs, approximately 600 bases could be analyzed with a final resolution of 0.59 and a total run time of 125 min.The run conditions were 6.5% w/v PDMA at 160 V/cm and 42°C with an effective separation distance of 40 cm (Madabhushi, 1998).

Although CE with polymer solutions provides both better read-lengths and faster analysis speed (more than 1000 bases in less than 1 h) compared to that of slab-gel electrophoresis (routinely 600–700 bases requiring up to 10 h), unfortunately, this was not sufficient to compete with the throughput capable on the parallel lane slab-gel instruments. To turn it into parallel equipment with the required DNA sequencing throughput, CAE was introduced in 1992, combining the high efficiency of conventional CE and the parallel feature of slab-gel electrophoresis (Huang *et al.*, 1992). Several different commercial capillary array sequencers are now available and this powerful instrument soon became the major workhorse for *de novo* genomic DNA sequencing. The two most powerful commercially available capillary array sequencers of that era were the ABI PRISM 3700 DNA Sequencer from PE Biosystems (Applied Biosystems) and the MegaBACE 1000 from Molecular Dynamics (GE Healthcare, Amersham-Pharmacia Biotech), both of which could analyze 96 samples per run. The MegaBACE 1000 DNA sequencer was based on the system developed by Mathies and colleagues (Huang *et al.*, 1992). It had automated sample and separation-matrix loading with a total turnaround time per sequencing run of less than 2 h between subsequent injections. This sequencer had confocal detection consisting of a microscope objective for focusing the laser light inside the capillaries and at the same time for collecting the emitted light from excited fluorophore tags at the center of the fluid column. A scanning system was used to collect the signal from all capillaries. With the MegaBACE 1000 Long Read Matrix, which contains LPA, the average read length exceeds 600 bases. The ABI PRISM 3700 DNA Sequencer from PE Biosystems was based on the developments of Kambara and Takahashi (1993) Dovichi (1997) employed post-column detection with liquid sheath flow. The capillary bundle was aligned inside a quartz cuvette. A buffer solution was pumped through the cell, along the dead space between the capillaries and the walls of the cuvette, where the liquid sheath flowing on the outside drags down the DNA zones eluting from each of the columns, tapering them to a small diameter without mixing. A laser beam crosses all flow streams and excites the fluorescent molecules. Light collection is made at 90° from the laser plane, and fluorescent light is imaged onto a cooled CCD camera for detection. Later iterations of these two systems employed 384 capillaries, yet despite the increased throughput these Sanger sequencing machines are now seen as limited in the present era of mammalian genome sequencing (with multigigabase data generation) and comparative genomics. Thus, the limitations of both throughput and sensitivity of these sequencing instruments have become critical as this technology was initially developed some 15 yr previously and require significant amounts of costly consumables (separation matrix, fluorescent sequencing chemicals, etc). Many researchers and companies are making concerted efforts to develop more powerful sequencers. These efforts include the development of integrated microchip bioprocessors that continue to use nanolitre amounts of Sanger sequencing chemistries (see Sections 7.1 and 7.3). Several of these other technology developments for DNA sequencing are also detailed in this volume, including, nanopore systems for single molecule analysis (Lee and Meller, 2007), and integrated solid-phase sequencing by synthesis systems that avoid use of electrophoretic fragment analysis characteristic of Sanger sequencing (Margulies *et al.*, 2007; Edwards *et al.*, 2007; Hebert and Braslavsky, 2007).

Another major development that led to integrated systems in which all steps of DNA sequencing analysis could be performed automatically was devised by Tan and Yeung (1997). They developed an integrated sequencer that included thermal cycling, product purification, in-line loading and CAE separation.

A dye-labeled terminator cycle-sequencing reaction is performed in a fused-silica capillary. The sequencing ladder was directly injected into a size-exclusion chromatography column at 95°C for purification. Online injection into the capillary was accomplished at a junction. The system was subsequently improved to process eight samples simultaneously (Tan and Yeung, 1998). The raw data allowed base calling up to 460 bases with an accuracy of 98%, and the design was scalable to 96 capillary arrays. Another general problem with most capillary array instruments was the need to extend read length, as it is generally difficult to achieve sequencing to at least >800 bases with >99% accuracy. Endo *et al.* (1999) proposed electric field strength gradients, with a typical duty cycle of an initial voltage ramp (up to 220 V/cm), for accelerating short fragments, followed by a plateau, a voltage decrement and finally a constant, lower voltage of 90–130 V/cm for separation of longer DNA fragments. These electric field strength gradients, coupled to column temperatures of 60°C, allow extension of the reading length up to 800 bases and similar measures are employed in current commercial 384-lane capillary instruments.

7.1. MicroChip DNA sequencing

However, the goals for the future of genomic research are not satisfied with currently developed CAE technology, and new advances in ultrahigh-throughput sequencing technologies will continue to be demanded for a significant period of time. The re-sequencing of a spectrum human genomes from different ethnic groups is only one project among the almost 100 genomes of other model and domestic organisms currently being sequenced, and a major concern is that the high cost of Sanger sequencing has become a significant obstacle to the continued genome sequencing of other mammals and to future progress in genomic science. To stimulate further technology development, an ambitious goal was set by the National Human Genome Research Institute (NHGRI) to develop tools capable of sequencing a human individual's genome for under $1000, in anticipation of a future genomic medicine. Here, chip-based CE appears promising as one emerging technology potentially capable of interim cost performance, while delivering long sequence reads typical of CE. Sequencing on chip-based CE offers many advantages over conventional CAE methods. A typical run time on microchip is in ranges of seconds to minutes, an order of magnitude less than that required for conventional methods. Furthermore, using lithography and MEMS technologies, microchips with high lane density and compactness can be easily fabricated, allowing massive parallel analysis and increasing throughput yet further. Another advantage provided by microchip format is low sample consumption that reduces the cost, typically picoliters of sample consumed for each run, compared with nanoliters required for conventional CE and microliters for slab-gel electrophoresis.

DNA sequencing with microchips was first demonstrated in 1995 by Mathies's group (Woolley and Mathies, 1995), who achieved a single-base resolution read length of 150–200 bases in 10–15 min with a separation channel distance of 3.5 cm using glass microchip with a denaturing LPA sieving matrix. Later efforts have been made to improve the read length, as well as the throughput (multichannel microchip). Schmalzing *et al.* (1998, 1999) evaluated the relationship of sequencing read length with separation length and applied voltages in microdevices. Employing a single lane microchip, they separated up to 400 bases in 14 min at 200 V/cm by using a 4% solution of LPA. In a second approach, Liu *et al.* (1999) after optimizing some experimental parameters (extension of the separation channel length to 7 cm, use of optimized LPA, increase of the electrophoresis channel depth to 50 μm, and the use of low-fluorescence background borofloat glass wafers) reported sequencing read lengths of over 500 bases in 20 min with a accuracy of 99.4%. In a separate research development, the optimization of DNA sequencing was performed on a chip device containing a 150 μm twin-T injector and an 11.5-cm long separation channel (Salas-Solano *et al.*, 2000). Using a separation matrix composed of 3% w/w 10 MDa plus 1% w/w 50 kDa LPA, an elevated temperature (50°C) and 200 V/cm field, high-speed DNA sequencing of 580 bases was achieved in 18 min with a base-calling accuracy of 98.5%. This read length extended to 640 bases at 98.5% accuracy by reducing the electric field strength to 125 V/cm and with an increased analysis time of 30 min. Backhouse *et al.* (2000) reported an increased sequencing read length of 640 bases with 98% accuracy by using a 50 cm long microchannel at room temperature and 200 V/cm, utilizing POP-6^{TM} as sieving matrix, showing that the performance of the microchip was identical to a fused-silica capillary with similar cross-sectional area.

Although problems still remain in the performance of microfabricated sequencers, significant advances have been made since Heller quoted that

> "for high-resolution applications such as DNA sequencing, the use of miniaturized separation devices in matrix-based electrophoresis remains an illusion". (Heller, 2000)

This was because base-call 600 or 800 bases, a separation length of at least a minimum of 14 cm, or as much as 30 cm, is usually required, a path length thought incompatible with microchips. An initial effort toward high-throughput DNA sequencing with long path multichannel microdevices was made by Ehrlich and colleagues (Koutny *et al.*, 2000), which with some 32 identical, separate 40-cm long channels, each with a 15 μm twin-T injector were fabricated on a 50-cm long and 25-cm wide glass chip, was hardly a microdevice. The separation achieved however was excellent, with average read lengths of up to 800 bases at an accuracy of 98%, and with a separation time of 80 min.

Mathies' group took up this challenge and devised long functional channel lengths within a small planar area by using tapered turns to prevent fragment separation distortions due to different path lengths across a turning channel (Paegel *et al.*, 2002, 2003). Signal detection in their microfabricated 96-lane radical CAE processor was by a 4-color rotary confocal fluorescence scanner, and the tapered turns extended the separation path length to 15.9 cm on a compact 150-mm diameter wafer. They obtained an average read length of 430 bases with a quality of PHRED >20. The sequencing output of one lane that exhibits average performance of this device is shown in Figure 17.

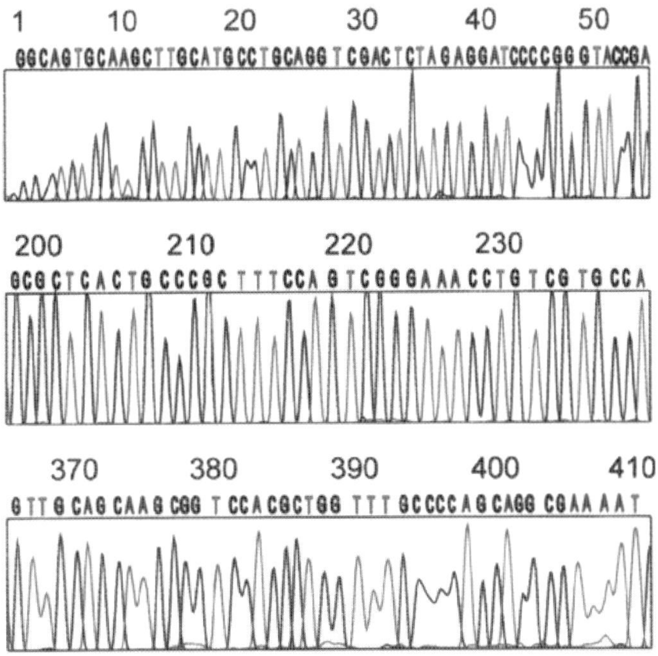

Fig. 17. The processed sequencing output display from a lane that exhibits the average performance of the system. The three panels display the typical quality of data obtained at the start, middle and the end of the run. Reprinted from Paegel *et al.* (2002) with permission.

7.2. Total genetic analysis systems

An intense area of development involves the creation of total genetic analysis systems (G-μTAS), formed by the integration of chip-based CE with other functional microfabricated components for processes such as sample preparation and other enzymatic bioreactions (PCR, restriction, labeling, etc.), and also with assay detection (such as fluorescence or CL) and potentially with other detection methods (such as microarray hybridization technologies). There are several advantages of the integration of such microanalysis systems. First, scaling down can increase the sensitivity. Suppose a single gene of interest is to be analyzed within the volume of a typical mammalian cell, where the concentration is of the order of 10^{-12} M. However, if it is diluted into a more conventional analysis volume of 10 μL, the concentration would drop to $<10^{-18}$ M, which makes it impossible to detect even with the most sensitive systems. Second, integration can eliminate external contaminations since the analytes need no longer be open to the environment. Third, reduction to a microscale makes it possible for batch fabrication to process many systems in parallel, as well as to fabricate many parallel systems on a single wafer, which can serve a unique role in clinical and research settings, since parallelism using conventional technologies is often prohibitively difficult. Finally, automation, portability and disposability can also be realized in an integrated microsystem, which are all key factors for successful commercialization and industrialization. Typically, conventional genetic analysis consists of three separate tasks: sample preparation, bioreactions and detection. For sample preparation, several time-consuming steps are involved, including cell culturing, nucleic acids extracting and purification. Then, bioreactions such as PCR amplification and enzyme digestion are carried out. Finally, samples of interest are detected via electrophoresis (Harrison *et al.*, 1993) or hybridization. Prior to realizing a total nucleic acids system, successful transfer of the above-mentioned tasks onto microchip format must be carried out separately. Then, these functional parts must be integrated by sophisticated fluid manipulation using microfabricated pumps and valves.

7.2.1. Sample preparation on microchips

Sample preparation is essential for all bioanalysis and is a step with the most diverse procedures. Therefore, various techniques have been developed for different purposes, such as cell trapping, or selection from an undesirable context via dielectrophoresis (DEP) or filtration, then cell lysis using ultrasonics or electric fields, and finally sample concentration or dilution. Some different representative techniques that have been enabled on microchip devices are illustrated in Figure 18.

7.2.2. Bioreactions on microchips

PCR amplification was first introduced by Saiki *et al.* (1985), and has been widely used for bioanalysis, because it can amplify trace amounts of nucleic acids to a detectable level, although concerns for the fidelity of the representative products are described in an accompanying chapter (Kowalchuk *et al.*, 2007). A brief introduction to the PCR amplification process is as follows. Template DNA molecules are denatured to form two complementary single strands at an elevated temperature (about 95°C). Temperature is then lowered for the annealing step: primers specifically bind to the complementary sequences of the DNA templates (usually 50–65°C). Finally, the temperature is raised to allow polymerase catalyzed DNA extension: the template is typically replicated by a thermostable DNA polymerase at about 72°C. Theoretically, the number of molecules generated by a PCR amplification starting with a single molecule after n cycles is $(1+p)^n$, where p is the probability for a molecule to duplicate, and is usually close to 1 for a good reaction system. A schematic description of PCR amplification is shown in Figure 19.

Microchip-based PCR amplification has many advantages over its conventional counterpart. For example, the costs mostly due to the price of the enzyme will be significantly reduced as the volume scales are down. Additionally, miniaturized systems present less inertial mass to temperature changes, and thus rates for heating or cooling are drastically increased, giving rise to more rapid template amplification. Nowadays, there are two main types of PCR amplification on microchip format: well-based amplification and continuous flow-through-based amplification. Microwell-based PCR amplification was first introduced by Northrup *et al.* (1993) and by Wilding *et al.* (1994). The device

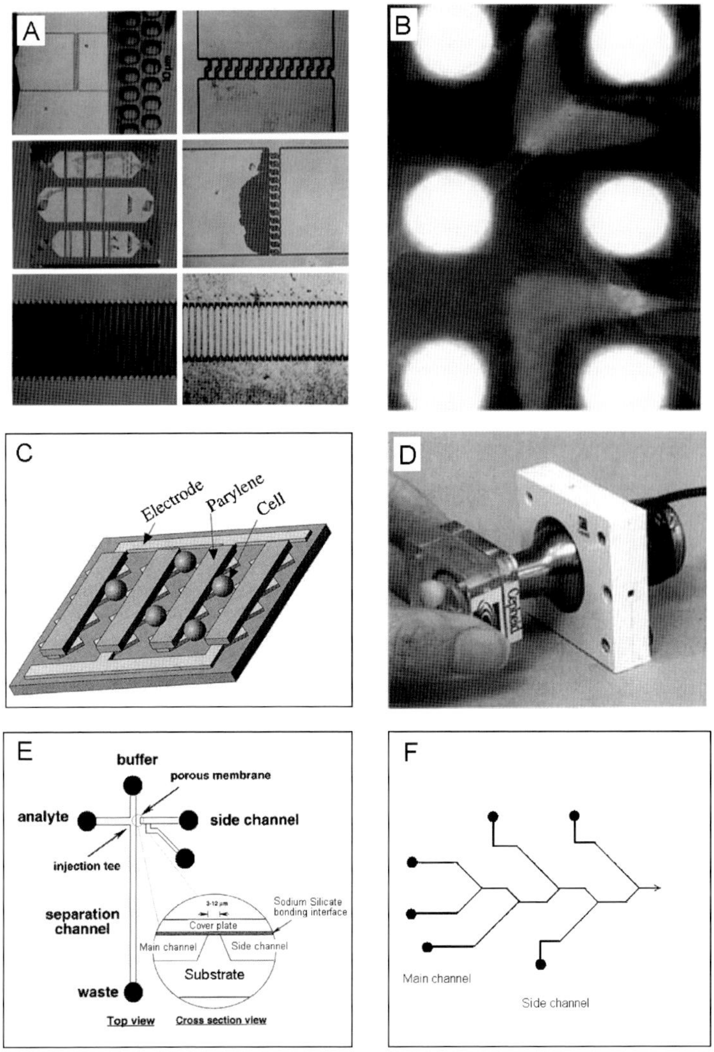

Fig. 18. Various types of preparative microchip created for DNA sample preparation. (A) Microfabricated filters for cell selecting, reprinted from Cheng *et al.* (1998a) with permission. (B) Separation of *Escherichia coli* (white) from human blood cells (red) by DEP on a microfabricated electronic microchip, reprinted from Cheng *et al.* (1998b) with permission. (C) An electronic cell lysis device, reprinted from Lee and Tai (1999) with permission. (D) An ultrasonic cell lysis device, reprinted from Belgrader *et al.* (1999) with permission. (E) Sample concentration using porous membrane structures, reprinted from Khandurina *et al.* (1999) with permission. (F) Microchannels for sample dilution with low error introduction, reprinted from Cheng *et al.* (1998d) with permission.

consisted of a silicon chip with a microwell in which the sample was loaded. The entire chip was heated and cooled to provide the adequate thermo-cycling conditions (Figure 20A). For continuous flow-through-based amplification, the transition times to change temperature depend only on the sample pumping rate

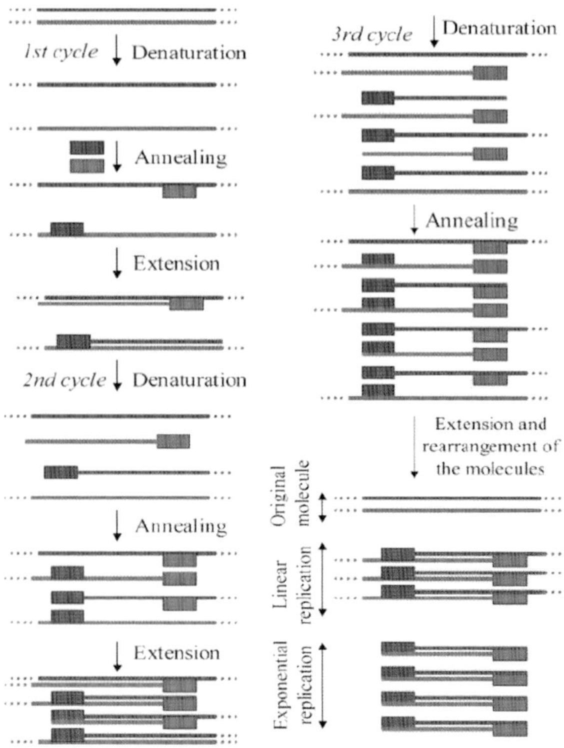

Fig. 19. Schematic illustration of PCR amplification.

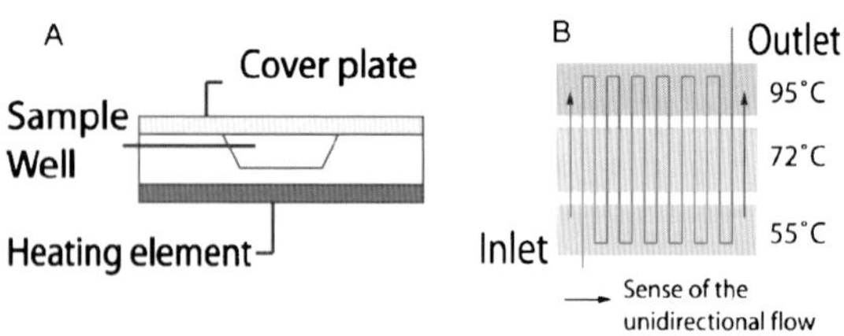

Fig. 20. Schematic view of (A) well-based PCR amplification on a microchip; and (B) continuous flow-through-based PCR amplification on a microchip.

and the time the sample needs to reach temperature equilibrium. It was first introduced by Nakano *et al.* (1994) using capillary-based systems, and the first continuous flow PCR chip was presented by Kopp *et al.* (1998), who fabricated a serpentine channel that passed through three heating zones maintained at constant temperature by copper blocks (Figure 20B).

7.2.3. System integration

To integrate various functional parts to form a miniaturized system, fluid control and manipulation must be realized by using micropumps and microvalves.

To create micropumps on chips, non-mechanical, mechanical–electrical and mechanical forces, such as electro-osmotic pumps (Takamura *et al.*, 2003) and bubble-driven pumps (Tsai and Lin, 2002) have been used. Various types of microvalves have also been microfabricated. For example, Harmon *et al.* (2003) introduced an actuator, which was based on a thermo-responsive hydrogel that shrank or swelled with fluid from a separate reservoir and thereby displaced a PDMS membrane to actuate fluid in the microchannel underneath. More types of microvalves such as check valves, diverter valves and micropipettes were provided by Hasselbrink *et al.* (2002), who employed moving micropistons formed *in situ* by laser polymerization. Some successful demonstrations of total genetic analysis systems are detailed below.

Cheng *et al.* (1998a, 1998c) introduced the first total genetic analysis system that incorporated sample preparation, bioreactions and detection. First, *Echerichia coli* were separated from blood cells using DEP on a silicon chip. An array of 25 individually addressable microelectrodes were microfabricated in two patterns (square-wall and checkerboard); both patterns functioned well (Figure 21A and B). The separated bacteria were retained above the electrodes after the blood cells had been washed off. Then, the isolated prokaryotic cells were lysed electronically by applying a series of high-voltage pulse to release a full size spectrum of nucleic acids. Next, protein was digested by introducing proteinase K. Finally, the released RNA and DNA were transferred onto another electronic microchip, where electronically enhanced hybridization of DNA or RNA was performed for detection of the initial bacterial cells (Figure 21C).

Another development of an integrated miniaturized system for DNA analysis was also introduced Burns *et al.* (1998), in which almost all functional parts were integrated onto one single microchip, this included a nanoliter liquid injector, a sample mixing and positioning system, a temperature-controlled

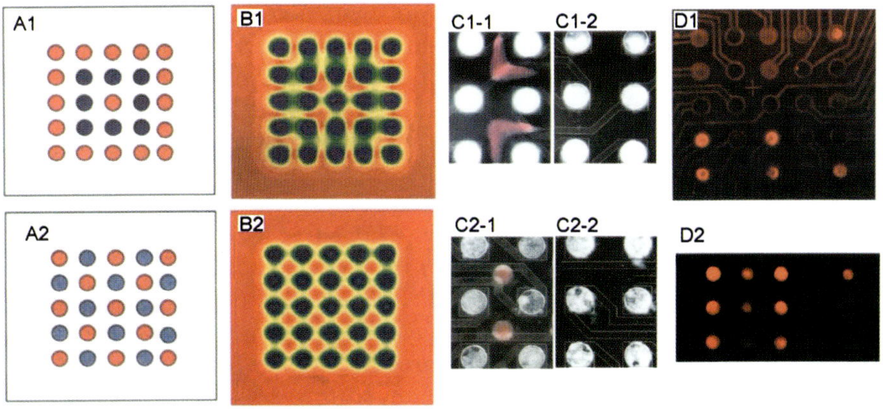

Fig. 21. The first complete total genetic analysis system. A1, A2: square-wall and checkerboard design of electrode array. B1, B2: simulated electronic field generated from corresponding types of electrodes. C1, C2: cells separation by means of DEP using the two electrodes designs. D1: hybridization results of DNA generated by electronic cell lysis. D2: hybridization results of RNA generated by electronic cell lysis. Reprinted from Cheng *et al.* (1998a, 1998c) with permission.

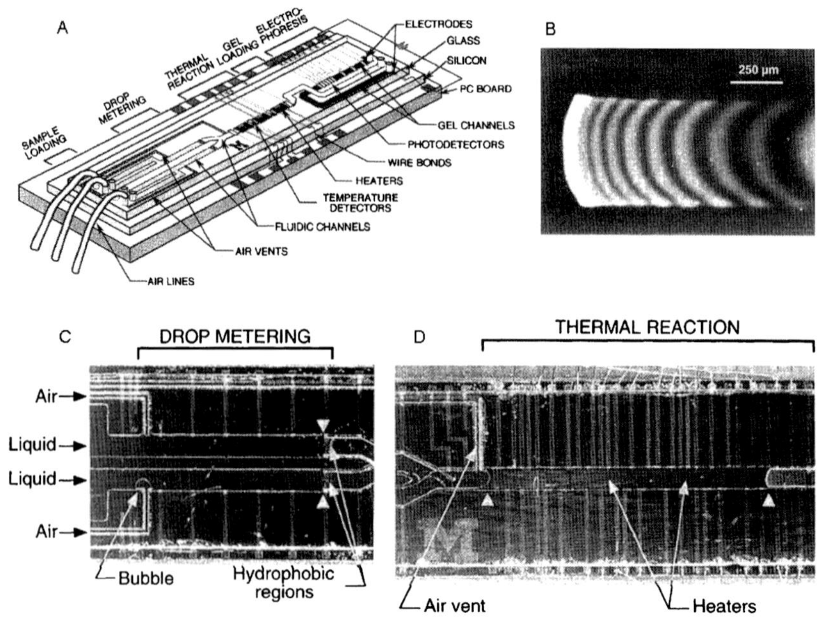

Fig. 22. (A) Overview of Burns's integrated device for DNA analysis. (B) Optical micrograph of a 50 bp DNA ladder separated via on-chip gel electrophoresis, where the separation path length is about 1 mm. (C) Micrograph of the drop-metering region. (D) Micrograph of reaction region. Reprinted from Burns *et al.* (1998) with permission.

reaction chamber, an electrophoretic separation system and a fluorescence detector (Figure 22). Liquid samples were injected via capillary action and were metered by means of an air vent and a hydrophobic patch positioned sequentially along the microchannels. Liquid aliquots with a volume of 120 nL could be defined precisely. The aliquots in different microchannels were then mixed together and moved forward the reaction chamber, where integrated metal heaters and sensors were positioned to facilitate PCR amplification, or enzymatic digestion of the sample. After processing in the bioreactor, the products were subjected to an on-chip gel electrophoresis. *In situ* polymerized polyacrylamide was used as sieving matrix, with an efficient effective separation path length of several millimeters. Diode photodetectors, as well as a quarter-wavelength interference filter, were microfabricated directly beneath the separation channel to facilitate fluorescence detection.

More recently, a fully integrated miniaturized system created by Liu *et al.* (2004) from the Motorola Corporation provided a *tour de force* demonstration of a self-contained DNA analysis. This miniaturized system consisted of micofluidic mixers, valves, pumps, channels chambers heaters and a microarray of DNA sensors. Sample preparation (including magetic bead-based cell capture, cell preconcentration, cell purification and cell lysis), PCR amplification, DNA hybridization and electrochemical detection were each performed on chip with no external pressure source, fluid storage, mechanical pumps or valves required for fluid manipulation (Figure 23).

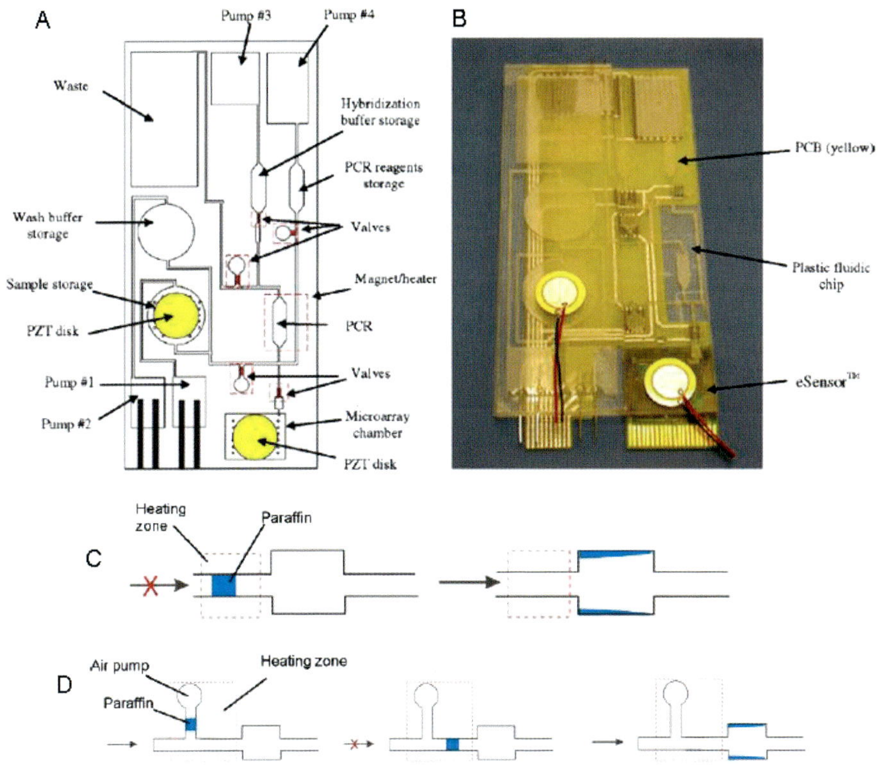

Fig. 23. (A) Design layout and (B) photograph of the fully integrated microsystem for DNA analysis. (C) Design of the close–open and (D) The open–close–open valves employed for fluid manipulation. Reprinted from Liu *et al.* (2004) with permission.

A biological sample requiring analysis (such as blood solution) of a solution containing immunomagnetic capture beads was first loaded into the sample storage chamber, and other solutions including washing buffer, PCR reagents and hybridization buffer were separately loaded into other storage chambers. Cavitation microstreaming generated by vibrating bubbles was used to mix the sample and other reagents in various chambers. Two types of micropumps, electrochemical pumps (driven by the generation of gas by electrolysis of water to move a piston liquid) and thermo-pneumatic air pumps (driven by the expansion of gas via heating to move a piston liquid) as well as paraffin microvalves were employed for fluid manipulation. The detection of pathogenic bacteria from approximately one milliliter of whole blood sample by employing SNP analysis (directly on the diluted blood sample) was successfully demonstrated with this fully integrated microsystem.

7.3. DNA sequencing lab-on-a-chip

As the above section demonstrates, the progressive development of microfabrication techniques and determination of suitable compatible materials is

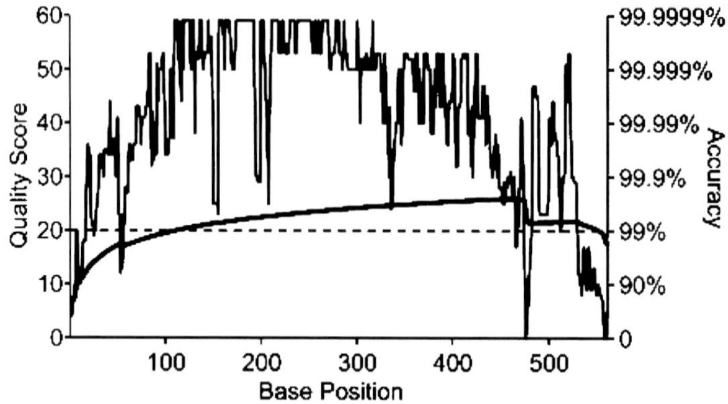

Fig. 24. The base-call accuracies and sequence read length predicted by PHRED. The percent accuracy is related to the PHRED quality score by the relationship 100 × (1 − P_e), where P_e, the probability that the base call is incorrect, is equal to $1/10^{Q/10}$. A one-in-a-hundred error rate is indicated by the dashed line. The gray line plots the PHRED quality scores at each base position, and the black line charts predicted the read accuracy at each base position, 100 × $(\text{base}_i - \Sigma P_{ei})/\text{base}_i$. Reprinted from Blazej *et al.* (2006). Copyright (2006), reprinted with permission from National Academy of Sciences, USA.

allowing the realization of fully integrated microsystems, which are capable of integrated sample manipulation, separation and analysis. The Mathies' group have now recently reported an integrated "lab-on-a-chip" DNA sequencing system (Blazej *et al.*, 2006). It involved the construction and application of an efficient, nanoliter-scale microfabricated bioprocessor that combined all three Sanger sequencing processes of thermal cycling, sample purification and CE into a single analytical device. The design had several advanced and unique features for miniaturization and integration. It used both electrophoretic and pneumatic forces for sample movement and for improved sample transfer through holes into channels. It also used a hybrid glass–PDMS assembly that was necessary for parallel processing, and the multilayer construction also enabled a much greater design complexity and permitted the exchange of materials across fluidic and pneumatic lines. The wafer-scale device was constructed to form a single microfabricated instrument with pneumatic valves and pumps, different 250-nL reactor chambers, affinity-capture purification chambers, and high-performance CE channels. This device was shown to be capable of undertaking complete Sanger sequencing from only 1 fmol of DNA template with reads of >500 bp. The base-call accuracies and sequence read length predicted by PHRED of a typical sequence are shown in Figure 24. Further advances in lab-on-a-chip electrophoretic sequencing technologies are soon to be expected.

7.3.1. Alternative DNA sequencing technologies

A number of non-Sanger DNA sequencing technologies have recently begun to emerge as alternative contenders for the $1000 per genome prize. None of these technologies use electrophoresis or fragment sizing. Several of these technologies involve detection of short DNA extension process events (typically a single nucleotide addition at each cycle) catalyzed simultaneously on different DNA templates bound at numerous fixed locations (100,000 to 1,000,000 features) arrayed on a planar "chip" surface. Each extension reaction at each feature is detected by imaging fluorescent dye nucleotides incorporated by each unique template (Edwards *et al.*, 2007; Kumar and Fuller, 2007; Hebert and Braslavsky, 2007) or by CL detection of a linked enzymatic process (pyrosequencing) (Margulies *et al.*, 2007). These technologies are characterized by their reliance on precision microengineering, microfluidics and ultra-precision optics and ultra-sensitive imaging devices. Although rapid progress is being made different aspects of this approach, in the length of sequence read, in the number of arrayed templates, and in efficient data generation with limited amounts of sequencing chemicals, these sequencing technologies are characterized by short reads (30–100+ bp), which at present cannot compete for data utility with the high quality of Sanger sequence reads obtained even on high-throughput microfabricated CE devices. Readers are directed to chapters by Margulies *et al.* (2007) on pyrosequencing, Edwards *et al.* (2007) and Kumar and Fuller (2007) on dye-nucleotide extension cyclic sequencing and to Hebert and Braslavsky (2007) on single molecule cyclic sequencing for fuller accounts of the development of these exciting new sequencing technologies.

REFERENCES

Albarghouthi, M. N., Buchholz, B. A., Doherty, E. A. S., Bogdan, F. M., Zhou, H. H. and Barron, A. E. (2001). Impact of polymer hydrophobicity on the properties and performance of DNA sequencing matrices for capillary electrophoresis. Electrophoresis 22, 737–747.

Albarghouthi, M. N., Stein, T. M. and Barron, A. E. (2003). Poly-*N*-hydroxyethylacrylamide as a novel adsorbed coating for protein separation by capillary electrophoresis. Electrophoresis 24, 1166–1175.

Arora, A., Eijkel, J. C. T., Morf, W. E. and Manz, A. (2001). A wireless electrochemiluminescence detector applied to direct and indirect detection for electrophoresis on a microfabricated glass device. Anal. Chem. 73, 3282–3288.

Backhouse, C., Caamano, M., Oaks, F., Nordman, E., Carrillo, A., Johnson, B. and Bay, S. (2000). DNA sequencing in a monolithic microchannel device. Electrophoresis 21, 150–156.

Barron, A. E. (2004). Sparsely cross-linked "nanogel" matrixes as fluid, mechanically stabilized polymer networks for high-throughput microchannel DNA sequencing. Anal. Chem. 76, 5249–5256.

Bean, S. R. and Lookhart, G. L. (1998). Faster capillary electrophoresis separation of wheat proteins through modifications to buffer composition and sample handling. Electrophoresis 19, 3190–3198.

Becker, H. and Gärtner, C. (2000). Polymer microfabrication methods for microfluidic analytical applications. Electrophoresis 21, 12–26.

Belgrader, P., Hansfor, D., Kovacs, G. T. A., Venkateswaran, K., Mariella, R., Milanovich, F., Nasarabadi, S., Okuzumi, M., Pourahmadi, F. and Northrup, M. A. (1999). A minisonicator to rapidly disrupt bacterial spores for DNA analysis. Anal. Chem. 71, 4232–4236.

Blazej, R. G., Kumaresan, P. and Mathies, R. A. (2006). Microfabricated bioprocessor for integrated nanoliter-scale Sanger DNA sequencing. Proc. Natl. Acad. Sci. USA 103, 7240–7245.

Burns, M. A., Johnson, B. N., Brahmasandra, S. N., Handique, K., Webster, J. R., Krishnan, M., Sammarco, T. S., Man, P. M., Jones, D., Heldsinger, D., Mastrangelo, C. H. and Burke, D. T. (1998). An integrated nanoliter DNA analysis device. Science 282, 484–487.

Carey, L. and Mitnik, L. (2002). Trends in DNA forensic analysis. Electrophoresis 23, 1386–1397.

Carrilho, E., Ruiz-Martinez, M. C., Berka, J., Smirnov, I., Goetzinger, W., Miller, A. W., Brady, D. and Karger, B. L. (1996). Rapid DNA sequencing of more than 1000 bases per run by capillary electrophoresis using replaceable linear polyacrylamide solutions. Anal. Chem. 68, 3305–3313.

Ceriotti, L., de Rooij, N. F. and Verpoorte, E. (2002). An integrated fritless column for on-chip capillary electrochromatography with conventional stationary phases. Anal. Chem. 74, 639–647.

Chabinyc, M. L., Chiu, D. T., McDonald, J. C., Stroock, A. D., Christian, J. F., Karger, A. M. and Whitesides, G. M. (2001). An integrated fluorescence detection system in poly(dimethylsiloxane) for microfluidic application. Anal. Chem. 73, 4491–4498.

Chediak, J., Luo, Z., Seo, J., Cheung, N., Lee, L. P. and Sands, T. D. (2004). Heterogeneous integration of CdS filters with GaN LEDs for fluorescence detection microsystems. Sensors Actuators A 111, 1–7.

Cheng, J., Kricka, L. J., Sheldon, E. L., Wilding, P. (1998a). Sample preparation in microstructured devices. In: A. Manz and H. Becker (Eds), Microsystem Technology in Chemistry and Life Sciences. Topics in Current Chemistry, (vol. 194, pp. 215–231). Springer, Berlin.

Cheng, J., Sheldon, E. L., Wu, L., Uribe, A., Gerrue, L. O., Carrino, J., Heller, M. J. and O'Connell, J. P. (1998b). Preparation and hybridization analysis of DNA/RNA from *E. coli* on microfabricated bioelectronic chips. Nature 16, 541–546.

Cheng, J., Waters, L. C., Fortina, P., Hvichia, G., Jacobson, S. C., Ramsey, J. M., Kricka, L. J. and Wilding, P. (1998c). Degenerate oligonucleotide primed-polymerase chain reaction and capillary electrophoretic analysis of human DNA on microchip-based devices. Anal. Biochem. 257, 101–106.

Cheng, S. B., Skinner, C. D. and Harrison, D. J. (1998d). Integrated serial dilution on a microchip for immunoassay sample treatment and flow injection analysis. In: D. J. Harrison and A. Van den berg (Eds), Micro-Total Analysis Systems'98, (pp. 157–160). Kluwer Publishing, Dordrecht.

Collins, F. S., Morgan, M. and Patrinos, A. (2003). The human genome project: lessons from large-scale biology. Science 300, 286–290.

Dang, F., Zhang, L., Hagiwara, H., Mishina, Y. and Baba, Y. (2003). Ultrafast analysis of oligosaccharides on microchip with light-emitting diode confocal fluorescence detection. Electrophoresis 24, 714–721.

Ding, W. L. and Fritz, J. S. (1997). Separation of basic proteins and peptides by capillary electrophoresis using a cationic surfactant. J. High Resol. Chromatogr. 20, 575–580.

Doherty, E. A. S., Kan, C. W., Paegel, B. M., Yeung, S. H. I., Cao, S. T., Mathies, R. A. and Barron, A. E. (2004). Sparsely cross-linked "nanogel" matrixes as fluid, mechanically stabilized polymer networks for high-throughput microchannel DNA sequencing. Anal. Chem. 76, 5249–5256.

Dou, Y. H., Bao, N., Xu, J. J. and Chen, H. Y. (2002). A dynamic modified microfluidic poly(dimethylsiloxane) chip with electrochemical detection for biological analysis. Electrophoresis 23, 3558–3566.

Dovichi, N. J. (1997). DNA sequencing by capillary electrophoresis. Electrophoresis 18, 2393–2399.

Duffy, D. C., McDonald, J. C., Schueller, O. J. A. and Whitesides, G. M. (1998). Rapid prototyping of microfluidic systems in poly(dimethylsiloxane). Anal. Chem. 70, 4974–4984.

Edwards, A., Civitello, A., Hammond, H. A. and Caskey, C. T. (1991). DNA typing and genetic mapping with trimeric and tetrameric tandem repeats. Am. J. Hum. Genet. 49, 746–756.

Edwards, J., Dae, H. K. and Ju, J-Y. (2007). An integrated system for DNA sequencing by synthesis. In: K. R. Mitchelson (Ed), New High Throughput Technologies for DNA Sequencing and Genomics (pp. 187–205). Elsevier, Amsterdam.

Effenhauser, C. S., Bruin, G. J., Paulus, A. and Ehrat, M. (1997). Integrated capillary electrophoresis on flexible silicone microdevices: analysis of DNA restriction fragments and detection of single DNA molecule on microchips. Anal. Chem. 69, 3451–3457.

Effenhauser, C. S., Manz, A. and Widmer, H. M. (1993). Glass chips for high-speed capillary electrophoresis separations with submicrometer plate heights. Anal. Chem. 65, 2637–2642.

Effenhauser, C. S., Manz, A. and Wodmer, H. M. (1995). Manipulation of sample fractions on a capillary electrophoresis chip. Anal. Chem. 67, 2284–2287.

Effenhauser, C. S., Paulus, A., Manz, A. and Widmer, H. M. (1994). High-speed separation of antisense oligonucleotides on a micromachined capillary electrophoresis device. Anal. Chem. 66, 2949–2953.

Efimenko, K., Wallace, W. E. and Genzer, J. (2002). Surface modification of Sylgard-184 poly(dimethylsiloxane) networks by ultraviolet and ultraviolet/ozone treatment. J. Colloid Interf. Sci. 254, 306–315.

Ehrfeld, W. and Munchmeyer, D. (1991). Three-dimensional microfabrication using synchrotron radiation. Nucl. Instrum. Methods Phys. Res. A 303, 523–532.

Ehrich, M., Hillenkamp, F. and van den Boom, D. (2007). Comparative sequence analysis by MALDI-TOF mass spectrometry – Utilizing the known to discover the new. In: K. R. Mitchelson (Ed), New High Throughput Technologies for DNA Sequencing and Genomics (pp. 97–117). Elsevier, Amsterdam.

Emrich, C. A., Tian, H., Medintz, I. L. and Mathies, R. A. (2002). Microfabricated 384-lane capillary array electrophoresis bioanalyzer for ultrahigh-throughput genetic analysis. Anal. Chem. 74, 5076–5083.

Endo, Y., Yoshida, C. and Baba, Y. (1999). DNA sequencing by capillary array electrophoresis with an electric field strength gradient. Biochem. Biophys. Methods 41, 133–141.

Ertl, P., Emrich, C. A., Singhal, P. and Mathies, R. A. (2004). Capillary electrophoresis chips with a sheath-flow supported electrochemical detection system. Anal. Chem. 76, 3749–3755.

Evans, C. E. (1997). Direct on-line injection in capillary electrophoresis. Anal. Chem. 69, 2952–2954.

Fan, Z. H. and Harrison, D. J. (1994). Micromachining of capillary electrophoresis injectors and separators on glass chips and evaluation of flow at capillary intersections. Anal. Chem. 66, 177–184.

Ferrigno, R., Lee, J. N., Jang, X. Y. and Whitesides, G. M. (2004). Potentiometric titrations in a poly(dimethylsiloxane)-based microfluidic device. Anal. Chem. 76, 2273–2280.

Fielden, P.R., Baldock, S.J., Goddard, N.J., Pickering, L.W., Prest, J.E., Snook, R.D., Treves-Brown, B.J., Vaireanu, D.I. (1998). Proceedings of Micro-TAS' 98, Banff, Canada, pp. 323–326.

Fister, J. C., Jacobson, S. C., Davis, L. M. and Ramsey, J. M. (1998). Counting single chromophore molecules for ultrasensitive analysis and separations on microchip devices. Anal. Chem. 70, 431–437.

Fung, E. N., Pang, H. M. and Yeung, E. S. (1998). Fast DNA separations using poly (ethylene oxide) in non-denaturing medium with temperature programming. J. Chromatogr. A. 806, 157–164.

Gao, Q. and Yeung, E. S. (1998). A Matrix for DNA separation: genotyping and sequencing using poly(vinylpyrrolidone) solution in uncoated capillaries. Anal. Chem. 70, 1382–1388.

Guerin, L. J., Bossel, M., Demierre, M., Calmes, S. and Renaud, P. (1997). Proceedings of 1997 IEEE International Conference on Solid-State Sensors Actuators. Chicago, IL, pp. 1419–1422.

Hammond, H. A., Jin, L., Zhong, Y., Caskey, C. T. and Chakraborty, R. (1994). Evaluation of 13 short tandem repeat loci for use in personal identification applications. Am. J. Hum. Genet. 55, 175–189.
Han, J. and Craighead, H. G. (2000). Separation of long DNA molecules in a microfabricated entropic trap array. Science 288, 1026–1029.
Harrison, D. J., Fluri, K., Seiler, Z., Fan, Z., Effenhauser, C. S. and Manz, A. (1993). Micromachining a miniaturized capillary electrophoresis-based chemical analysis system on a chip. Science 261, 895–897.
Harmon, M. E., Tang, M. and Frand, C. W. (2003). A microfluidic actuator based on thermoresponsive hydrogels. Polymer 44, 4547–4556.
Hasselbrink, E. F., Shepodd, T. J. and Rehm, J. E. (2002). High-pressure microfluidic control in lab-on-a-chip devices using mobile polymer monoliths. Anal. Chem. 74, 4913–4918.
He, B., Tait, N. and Regnier, F. (1998). Fabrication of nanocolumns for liquid chromatography. Anal. Chem. 70, 3790–3797.
Hebert, B. and Braslavsky, I. (2007). Single-molecule fluorescence microscopy and its applications to single-molecule sequencing by cyclic synthesis. In: K. R. Mitchelson (Ed), New High Throughput Technologies for DNA Sequencing and Genomics (pp. 207–244). Elsevier, Amsterdam.
Heiger, D. N., Cohen, A. S. and Karger, B. L. (1990). Separation of DNA restriction fragments by high performance capillary electrophoresis with low and zero crosslinked polyacrylamide using continuous and pulsed electric fields. J. Chromatogr. 516, 33–48.
Heller, C. (2000). Influence of electric field strength and capillary dimensions on the separation of DNA. Electrophoresis 21, 593–602.
Henry, A. C., Tutt, T. J., Galloway, M., Davidson, Y. Y., McWhorter, C. S., Soper, S. A. and McCarley, R. L. (2000). Surface modification of poly(methyl methacrylate) used in the fabrication of microanalytical devices. Anal. Chem. 72, 5331–5337.
Hesch, K., Amold, J., Dasbach, U., Ehrfeld, W. and Löwe, H. (1995). Combination of excimer laser micromachining and replication process suited for large scale production. Appl. Surf. Sci. 86, 251–258.
Hillborg, H. and Gedde, U. W. (1998). Hydrophobicity recovery of polydimethylsiloxane after exposure to corona discharge. Polymer 39, 1991–1998.
Hjertén, S. (1985). High performance electrophoresis. Elimination of electroendosmosis and solute adsorption. J. Chromatogr. 347, 191–198.
Hofmann, O., Che, D., Cruickshank, K. A. and Muller, U. R. (1999). Adaptation of capillary isoelectric focusing to microchannels on a glass chip. Anal. Chem. 71, 678–686.
Hu, S. W., Ren, X. Q., Bachman, M., Sims, C. E., Li, G. P. and Allbritton, N. (2002). Surface modification of poly(dimethylsiloxane) microfluidic devices by ultraviolet polymer grafting. Anal. Chem. 74, 4117–4123.
Huang, X. C., Quesada, M. A. and Mathies, R. A. (1992). Capillary array electrophoresis using laser-excited confocal fluorescence detection. Anal. Chem. 64, 967–972.
Jacobson, S. C., Culbertson, C. T., Daler, J. E. and Ramsey, J. M. (1998). Microchip structures for submillisecond electrophoresis. Anal. Chem. 70, 3476–3480.
Jacobson, S. C., Hergenröder, R., Koutny, L. B., Moore, A. W. and Ramsey, J. M. (1994a). Effects of injection schemes and column geometry on the performance of microchip electrophoresis devices. Anal. Chem. 66, 1107–1118.
Jacobson, S. C., Hergenröder, R., Koutny, L. B. and Ramsey, J. M. (1994b). High-speed separations on a microchip. Anal. Chem. 66, 1114–1118.
Jacobson, S. C., Hergenröder, R., Koutny, L. B. and Ramsey, J. M. (1994c). Open channel electrochromatography on a microchip. Anal. Chem. 66, 2369–2373.
Jacobson, S. C., Hergenröder, R., Moore, A. W. and Ramsey, J. M. (1994d). Precolumn reactions with electrophoretic analysis integrated on a microchip. Anal. Chem. 66, 4127–4132.
Jansen, H., Gardeniers, H., de Boer, M., Elwenspoek, M. and Fluitman, J. (1996). A survey on the reactive ion etching of silicon in microtechnology. J. Micromech. Microeng. 6, 14–28.

Jiang, G. F., Attiya, S., Ocvirk, G., Lee, W. E. and Harrison, D. J. (2000). Red diode laser induced fluorescence detection with a confocal microscope on a microchip for capillary electrophoresis. Biosens. Bioelectron. 14, 861–868.

Johnson, T. J., Ross, G., Gaitan, M. and Locascio, L. E. (2001). Laser modification of preformed polymer microchannels: application to reduce band broadening around turns subject to electrokinetic flow. Anal. Chem. 73, 3656–3661.

Kambara, H. and Takahashi, S. (1993). Multiple-sheath flow capillary array DNA analyzer. Nature 361, 565–566.

Khandurina, J., Jacobson, S. C., Waters, L. C., Foote, R. S. and Ramsey, J. M. (1999). Microfabricated porous membrane structure for sample concentration and electrophoretic analysis. Anal. Chem. 71, 1815–1819.

Kim, Y. and Yeung, E. S. (1997). Separation of DNA sequencing fragments up to 1000 bases by using poly(ethylene oxide)-filled capillary electrophoresis. J. Chromatogr. A 781, 315–325.

Kopp, M. U., Crabtree, H. J. and Manz, A. (1997). Developments in technology and applications of microsystems. Curr. Opin. Chem. Biol. 1, 410–419.

Kopp, M. U., Mellp, A. J. and Manz, A. (1998). Chemical amplification: continuous-flow PCR on a chip. Science 280, 1046–1048.

Koutny, L., Schmalzing, D., Salas-Solano, O., El-Difrawy, S., Adourian, A., Buonocore, S., Abbey, K., McEwan, P. and Ehrlich, P. M. (2000). Eight hundred-base sequencing in a microfabricated eectrophoretic device. Anal. Chem. 72, 3388–3391.

Kowalchuk, G. A., Austin, J. J., Gooding, P. S. and Stephen, J. R. (2007). Valid recovery of nucleic acid sequence information from high contamination risk samples – ancient DNA and environmental DNA. In: K. R. Mitchelson (Ed), New High Throughput Technologies for DNA Sequencing and Genomics (pp. 357–371). Elsevier, Amsterdam.

Kumar, S. and Fuller, C. W. (2007). Advances in dye-nucleotide conjugate chemistry for DNA sequencing. In: K. R. Mitchelson (Ed), New High Throughput Technologies for DNA Sequencing and Genomics (pp. 119–149). Elsevier, Amsterdam.

Kutter, J. P., Jacobson, S. C. and Ramsey, J. M. (1997). Integrated microchip device with electrokinetically controlled solvent mixing for isocratic and gradient elution in micellar electrokinetic chromatography. Anal. Chem. 69, 5165–5171.

Lagally, E. T. and Mathies, R. A. (2004). Integrated genetic analysis microsystems. J. Phys. D. Appl. Phys. 37, R245–R261.

Lai, C. C. J., Chen, C. H. and Ko, F. H. (2004). In-channel dual-electrode amperometric detection in electrophoretic chips with a palladium film decoupler. J. Chromatogr. A 1023, 143–150.

Landers, J. P., Oda, R. P., Madden, B. J. and Spelsberg, T. C. (1992). High-performance capillary electrophoresis of glycoproteins: the use of modifiers of electroosmotic flow for analysis of microheterogeneity. Anal. Chem. 205, 115–124.

Laugere, F., Guijt, R. M., Bastemeijer, J., van der Steen, G., Berthold, A., Baltussen, E., Sarro, P., van Dedem, G. W. K., Vellekoop, M. and Bossche, A. (2003). On-chip contactless four-electrode conductivity detection for capillary electrophoresis devices. Anal. Chem. 75, 306–312.

Lee, J. W. and Meller, A. (2007). Rapid DNA sequencing by direct nanoscale reading of nucleotide bases on individual DNA chains. In: K. R. Mitchelson (Ed.), New High Throughput Technologies for DNA Sequencing and Genomics (pp. 245–263). Elsevier, Amsterdam.

Lee, S. W. and Tai, Y. C. (1999). A micro cell lysis device. Sensors Actuators A 73, 74–79.

Legaz, M. E. and Pedrosa, M. M. (1996). Effect of polyamines on the separation of ovalbumin glycoforms by capillary electrophoresis. J. Chromatogr. A 719, 159–170.

Li, Y., Buch, J. S., Rosenherger, F., DeVoe, D. L. and Lee, C. S. (2004). Integration of isoelectric focusing with parallel sodium dodecyl sulfate gel electrophoresis for multidimensional protein separations in a plastic microfluidic network. Anal. Chem. 76, 742–748.

Lichtenherg, J., de Rooij, N. F. and Verpoorte, E. (2002). A microchip electrophoresis system with integrated in-plane electodes for contactless conductivity detection. Electrophoresis 23, 3769–3780.

Liu, B. F., Ozaki, M., Utsumi, Y., Hattori, T. and Terabe, S. (2003a). Chemiluminescence detection for a microchip capillary electrophoresis system fabricated in poly(dimethylsiloxane). Anal. Chem. 75, 36–41.
Liu, P., Xing, W. L., Liang, D., Huang, G. L., Zhou, Y. X. and Cheng, J. (2003b). Fast screening of single-nucleotide polymorphisms using chip-based temperature gradient capillary electrophoresis. Anal. Lett. 36, 2823–2834.
Liu, R. H., Yang, J., Lenigk, R., Bonanno, J. and Grodzinski, P. (2004). Self-contained, fully integrated biochip for sample preparation, polymerase chain reaction amplification, and DNA microarray detection. Anal. Chem. 76, 1824–1831.
Liu, S., Shi, Y., Ja, W. W. and Mathies, R. A. (1999). Optimization of high-speed DNA sequencing on microfabricated capillary electrophoresis channels. Anal. Chem. 71, 566–573.
Liu, Y., Fanguy, J. C., Bledsoe, J. M. and Henry, C. S. (2000). Dynamic coating using polyelectrolyte multilayers for chemical control of electroosmotic flow in capillary electrophoresis microchips. Anal. Chem. 72, 5939–5944.
Lu, Q. and Collins, G. E. (2001). Microchip separations of transition metal ions via LED absorbance detection of their PAR complex. Analyst 126, 429–432.
Madabhushi, R. S. (1998). Separation of 4-color DNA sequencing extension products in noncovalently coated capillaries using low viscosity polymer solutions. Electrophoresis 19, 224–230.
Mangru, S. D. and Harrison, D. J. (1998). Chemiluminescence detection in integrated post-separation reactors for microhip-based capillary electrophoresis and affinity electrophoresis. Electrophoresis 19, 2301–2307.
Manz, A., Graber, N. and Widmer, H. M. (1990). Miniaturized total analysis systems: a novel concept for chemical sensors. Sens. Actuators B. 1, 244–248.
Mao, Q. L. and Pawliszyn, J. (1999a). Capillary isoelectric focusing with whole column imaging detection for analysis of proteins and peptides. J. Biochem. Biophys. Methods 39, 93–110.
Mao, Q. L. and Pawliszyn, J. (1999b). Demonstration of isoelectric focusing on an etched quartz chip with UV absorption imaging detection. Analyst 124, 637–641.
Margulies, M., Jarvie, T. P., Knight, J. R. and Simons, J. F. (2007). The 454 life sciences picoliter sequencing system. In: K. R. Mitchelson (Ed), New High Throughput Technologies for DNA Sequencing and Genomics (pp. 151–186). Elsevier, Amsterdam.
Martin, R. S., Gawron, A. J. and Lunte, S. M. (2000). Dual-electrode electrochemical detection for poly(dimethylsiloxane)-fabricated capillary electrophoresis microchips. Anal. Chem. 72, 3196–3202.
Martynova, L., Locascio, L. E., Gaitan, M., Kramer, G. W., Christensen, R. G. and MacCrehan, W. A. (1997). Fabrication of plastic microfluid channels by imprinting methods. Anal. Chem. 69, 4783–4789.
Masár, M., Danková, M., Ölvecká, E., Stachurová, A., Kaniansky, D. and Stanislawski, B. (2004). Sample pretreatment on a column-coupling chip. J. Chromatogr. A 1026, 31–39.
McCormick, R. M., Belson, R. J., Alonso-Amigo, M. G., Benvegnu, D. J. and Hooper, H. H. (1997). Microchannel electrophoretic separations of DNA in injection-molded plastic substrates. Anal. Chem. 69, 2626–2630.
Melanson, J. E., Baryla, N. E. and Lucy, C. A. (2000). Double-chained surfactant for semipermanent wall coating in capillary electrophoresis. Anal. Chem. 72, 4110–4114.
Melton, L. (2003). On the trail of SNPs. Nature 42, 917–923.
Moore, A. W., Jacobson, S. C. and Ramsey, J. M. (1995). Microchip separations of neutral species via micellar electrokinetic capillary chromatography. Anal. Chem. 67, 4184–4189.
Nakanishi, H., Nishimoto, T., Arai, A., Abe, H., Kanai, M., Fujiyama, Y. and Yoshida, T. (2001). Fabrication of quartz microchips with optical slit and development of a linear imaging UV detector for microchip electrophoresis systems. Electrophoresis 22, 230–234.
Nakano, H., Matsuda, K., Yohda, M., Nagamune, T., Endo, I. and Yamane, T. (1994). High-speed polymerase chain-reaction in constant flow. Biosci. Biotechnol. Biochem. 58, 349–352.

Northrup, M. A., Ching, M. T., White, R. M. and Watson, R. T. (1993). DNA amplification with a microfabricated reaction chamber. Transducers. Proceedings of Transducers '93. The Seventh International Conference on Solid-State Sensors and Actuators. (pp. 924–926). Pacifico, Yokohama, Japan, June 7–10, 1993.

Ogura, M., Agata, Y., Watanabe, K., McCormick, R., Hamaguchi, Y., Aso, Y. and Mitsuhashi, M. (1998). RNA chip: quality assessment of RNA by microchannel linear gel electrophoresis in injection-molded plastic chips. Clin. Chem. 44, 2249–2255.

Osbourn, D. M. and Lunte, G. E. (2003). On-column electrochemical detection for microchip capillary electrophoresis. Anal. Chem. 75, 2710–2714.

Paegel, B. M., Blazej, R. G. and Mathies, R. A. (2003). Microfluidic devices for DNA sequencing: sample preparation and electrophoretic analysis. Curr. Opin. Biotechnol. 14, 42–50.

Paegel, B. M., Emrich, C. A., Wedemayer, G. J., Scherer, J. R. and Mathies, R. A. (2002). High throughput DNA sequencing with a microfabricated 96-lane capillary array electrophoresis bioprocessor. Proc. Natl. Acad. Sci. USA 99, 574–579.

Park, S. J., Taton, T. A. and Mirkin, C. A. (2002). Array-based electrical detection of DNA with nanoparticle probes. Science 295, 1503–1506.

Paulus, A., Williams, S. J., Sassi, A. P., Kao, P. K., Tan, H. and Hooper, H. H. (1998). Integrated capillary electrophoresis using glass and plastic chips for multiplexed DNA analysis. Proceedings of SPIE Microfluidic Devices and Systems, Santa, Clara, CA 3515, pp. 94–103.

Pethig, R., Burt, J. P. H., Parton, A., Rizvi, N., Talary, M. S. and Tame, J. A. (1998). Development of biofactory-on-a-chip technology using excimer laser micromachining. J. Micromech. Microeng. 8, 57–63.

Petsul, P. H., Greenway, G. M. and Haswell, S. J. (2001). The development of an on-chip micro-flow injection analysis of nitrate with a cadmium redactor. Anal. Chim. Acta 428, 155–161.

Piotter, V., Hanemann, T., Ruprecht, R. and Hausselt, J. (1997). Injection molding and related techniques for fabrication of microstructures. Microsystem Technol 3, 129–133.

Pumera, M., Wang, J., Grushka, E. and Polsky, R. (2001). Gold nanoparticle-enhanced microchip capillary electrophoresis. Anal. Chem. 73, 5625–5628.

Qin, D., Xia, Y., Rogers, J. A., Jackman, R. J., Zhao, X. M. and Whitesides, G. M. (1998). In: A. Manz and H. Becker (Eds), Micro System Technology in Chemistry and Life Science (pp. 2–20). Springer, Heidelberg.

Roberts, M. A., Rossier, J. S., Bercier, P. and Girault, H. (1997). UV laser machined polymer substrates for the development of microdiagnostic systems. Anal. Chem. 69, 2035–2042.

Rodriguez, I., Lee, H. K. and Li, S. F. (1999). Microchannel electrophoretic separation of biogenic amines by micellar electrokinetic chromatography. Electrophoresis 20, 118–126.

Salas-Solano, O., Carrilho, E., Kotler, L., Miller, A. W., Goetzinger, W., Sosic, Z. and Karger, B. L. (1998). Routine DNA sequencing of 1000 bases in less than one hour by capillary electrophoresis with replaceable linear polyacrylamide solutions. Anal. Chem. 70, 3996–4003.

Salas-Solano, O., Schmalzing, D., Koutny, L., Adourian, B. S., Matsudaira, A. and Ehrlich, P. (2000). Optimization of high-performance DNA sequencing on short microfabricated electrophoretic devices. Anal. Chem. 72, 3129–3137.

Saiki, R. K., Scharf, S., Faloona, F., Mullis, K., Horn, G. and Erlich, H. (1985). Enzymatic amplification of betaglobin genomic sequences and restriction site analysis for diagnosis of sickle cell anemia. Science 230, 1350–1354.

Salimi-Moosavi, H., Jiang, Y., Lester, L., McKinnon, G. and Harrison, D. J. (2000). A multireflection cell for enhanced absorbance detection in microchip-based capillary electrophoresis devices. Electrophoresis 21, 1291–1299.

Sanger, F., Nicklen, S. and Coulson, A. R. (1977). DNA sequencing with chain termination inhibitors. Proc. Natl. Acad. Sci. USA 74, 5463–5467.

Sato, K., Kawanishi, H., Tokeshi, M., Kitamori, T. and Sawada, T. (1999). Sub-zeptomole detection in a microfabricated glass channel by thermal-lens microscopy. Anal. Sci. 15, 525–529.

Schilling, M., Nigge, W., Rudzinski, A., Neyer, A. and Hergenroder, R. (2004). A new on-chip ESI nozzle for coupling of MS with microfluidic devices. Lab on a Chip 4, 220–224.
Schmalzing, D., Adourian, M., Koutny, L., Ziaugra, L., Matsudaira, P. and Ehrlich, D. (1998). DNA sequencing on microfabricated electrophoretic devices. Anal. Chem. 70, 2303–2310.
Schmalzing, D., Koutny, L., Adourian, A., Belgrader, P., Matsudaira, P. and Ehrlich, D. (1997). DNA typing in thirty seconds with a microfabricated device. Proc. Natl. Acad. Sci. USA 94, 10,273–10,278.
Schmalzing, D., Tsao, N., Koutny, L., Chisholm, D., Srivastava, A., Adourian, A., Linton, L., McEwan, P., Matsudaira, P. and Ehrlich, D. (1999). Toward real-world sequencing by microdevice electrophoresis. Genome Res. 9, 853–858.
Shi, Y. and Anderson, R. (2003). High-resolution single-stranded DNA analysis on 4.5 cm plastic electrophoresis microchannels. Electrophoresis 24, 3371–3377.
Soper, S. A., Henry, A. C., Vaidya, B., Galloway, M., Wabuyele, M. and McCarley, R. L. (2002). Surface modification of polymer-based microfluidic devices. Anal. Chim. Acta 470, 87–99.
Tachibana, Y., Otsuka, K., Terabe, S., Arai, A., Suzuki, K. and Nakamura, S. (2003). Robust and simple interface for microchip electrophoresis-mass spectrometry. J. Chromatogr. A 1011, 181–192.
Takamura, Y., Onoda, H., Inokuchi, H., Adachi, S., Oki, A. and Horiike, Y. (2003). Low-voltage electroosmosis pump for stand-alone microfluidics devices. Electrophoresis 24, 185–192.
Tan, H. D. and Yeung, E. S. (1997). Integrated on-line system for DNA sequencing by capillary electrophoresis: from template to called bases. Anal. Chem. 69, 644–674.
Tan, H. D. and Yeung, E. S. (1998). Automation and integration of multiplexed on-line sample preparation with capillary electrophoresis for high-throughput DNA sequencing. Anal. Chem. 70, 4044–4053.
Tantra, R. and Manz, A. (2000). Integrated potentiometric detector for use in chip-based flow cells. Anal. Chem. 72, 2875–2878.
Terabe, S., Otsuka, K. and Ando, T. (1985). Electrokinetic chromatography with micellar solution and open-tubular capillary. Anal. Chem. 57, 834–841.
Tian, H., Jaquins-Gerstl, A., Munro, N., Trucco, M., Brody, L. C. and Landers, J. P. (2000). Single-strand conformation polymorphism analysis by capillary and microchip electrophoresis: a fast, simple method for detection of common mutations in BRCA1 and BRCA2. Genomics 63, 24–34.
Tian, H. J. and Landers, J. P. (2002). Hydroxyethylcellulose as an effective polymer network for DNA analysis in uncoated glass microchips: optimization and application to mutation detection cia heteroduplex analysis. Anal. Biochem. 309, 212–223.
Tsai, J. H. and Lin, L. (2002). Active microfluidic mixer and gas bubble filter driven by thermal bubble micropump. Sensors Actuators A 97–98, 665–671.
Walker, P. A., III. and Morris, M. D. (1998). Isotachophoretic separations on a microchip-normal Raman spectroscopy detection. Anal. Chem. 70, 3766–3769.
Wallenborg, S. R. and Bailey, C. G. (2000). Separation and detection of explosives on a microchip using micellar electrokinetic chromatography and indirect laser-induced fluorescence. Anal. Chem. 72, 1872–1878.
Wang, H. Y., Foote, R. S., Jacobson, S. C., Schneibel, J. H. and Ramsey, J. M. (1997). Low temperature bonding for microfabrication of chemical analysis devices. Sensors Actuators B 45, 199–207.
Wang, J., Tian, B. and Sahlin, E. (1999a). Integrated electrophoresis chips/amperometric detection with sputtered gold working electrodes. Anal. Chem. 71, 3901–3904.
Wang, J., Tian, B. and Sahlin, E. (1999b). Micromachined electrophoresis chips with thick-film electrochemical detectors. Anal. Chem. 71, 5436–5440.
Wang, S. L., Huang, X. J. and Fang, Z. L. (2001). A miniaturized liquid core waveguide-capillary electrophoresis system with flow injection sample introduction and fluorometric detection using light emitting diodes. Anal. Chem. 73, 4545–4549.

Wang, Y., Vaidya, B., Farquar, H. D., Stryjewski, W., Hammer, R. P., McCarley, R. L., Soper, S. A., Cheng, Y. W. and Barany, F. (2003). Microarrays assembled in microfluidic chips fabricated from poly(methyl methacrylate) for the detection of low-abundant DNA mutations. Anal. Chem. 75, 1130–1140.

Webster, J. R., Burns, M. A., Burke, D. T. and Mastrangelo, C. H. (2001). Monolithic capillary electrophoresis device with integrated fluorescence detector. Anal. Chem. 73, 1622–1626.

Wilding, P., Shoffner, M. A. and Kricka, L. J. (1994). PCR in silicon microstructure. Clin. Chem. 40, 1815–1818.

Woolley, A. T., Hadley, D., Landre, P., Demello, A. J., Mathies, R. A. and Northrup, M. A. (1996). Functional integration of PCR amplification and capillary electrophoresis in a microfabricated DNA analysis device. Anal. Chem. 68, 4081–4086.

Woolley, A. T. and Mathies, R. A. (1994). Ultra-high-speed DNA fragment separations using microfabricated capillary array electrophoresis chips. Proc. Natl. Acad. Sci. USA 91, 11348–11352.

Woolley, A. T., Lao, K., Glazer, A. N. and Mathies, R. A. (1998). Capillary electrophoresis chips with integrated electrochemical detection. Anal. Chem. 70, 684–688.

Woolley, A. T. and Mathies, R. A. (1995). Ultra-high-speed DNA sequencing using capillary electrophoresis chips. Anal. Chem. 67, 3676–3680.

Woolley, A. T., Sensabaugh, G. F. and Mathies, R. A. (1997). High-speed DNA genotyping using microfabricated capillary array electrophoresis chips. Anal. Chem. 69, 2181–2186.

Wu, C. C., Wu, R. G., Huang, J. G., Lin, Y. C. and Chang, H. C. (2003). Three-electrode electrochemical detector and platinum film decoupler integrated with a capillary electrophoresis microchip for amperometric detection. Anal. Chem. 75, 947–952.

Xu, F. and Baba, Y. (2004). Polymer solutions and entropic-based systems for double-stranded DNA capillary electrophoresis and microchip electrophoresis. Electrophoresis 25, 2332–2345.

Xu, F., Jabasini, M. and Baba, Y. (2002). DNA separation by microchip electrophoresis using low-viscosity hydroxypropylmethylcellulose-50 enhanced by polyhydroxy compounds. Electrophoresis 23, 3608–3614.

Chapter 3

Comparative Sequence Analysis by MALDI-TOF Mass Spectrometry – Utilizing the Known to Discover the New

Mathias Ehrich,[1] Franz Hillenkamp[2] and Dirk van den Boom[1]

[1]*Sequenom Corporation, 3595 John Hopkins Court, San Diego, CA 92121, USA*
[2]*Institute for Medical Physics and Biophysics, University of Münster, Robert-Koch-Str. 31, Münster D-48149, Germany*

Contents

Abstract

This chapter introduces the use of matrix-assisted laser desorption/ionization (MALDI) time-of-flight (TOF) mass spectrometry for high-throughput comparative sequence analysis. We first briefly define the concept of comparative sequence analysis in the context of current genomic and genetic research. The chapter then summarizes the basic principles and challenges of MALDI-TOF mass spectrometry (MS) of nucleic acids and introduces the concept of base-specific cleavage analyzed by MALDI-TOF MS for discovery of sequence changes.

PERSPECTIVES IN BIOANALYSIS, VOLUME 2
ISSN 1871-0069 DOI: 10.1016/S1871-0069(06)02003-9

We provide an overview of how the concept applies to some of the predominant research questions in genomics and genetics, such as discovery of new mutations, analysis of cytosine methylation or rapid identification and characterization of pathogenic organisms. Finally, we provide an outlook of further developments, which will significantly expand the capability and performance of the introduced methods and concept.

1. THE CONCEPT OF COMPARATIVE SEQUENCING

Large-scale sequencing efforts during the past decade have allowed us to obtain the nucleotide sequence of entire genomes. The most important factors for the rapid progress in DNA sequencing have been the invention of powerful DNA sequencing methods like Sanger sequencing, but also significant advances in bioinformatics that finally allow the efficient use of all the sequence data to assemble genome sequences. Today several eukaryotic and prokaryotic organisms have been fully sequenced and an almost complete sequence of the human genome is available. Sequencing technologies are still under development and we can expect to see further milestones in their development over the next years. Especially massive parallel sequencing technologies such as those developed by 454 Life Science Corporation (Margulies *et al.*, 2005), Solexa (Bennett *et al.*, 2005) or Helicos (Kartalov and Quake, 2004) have already shown remarkable success in the construction of full genome sequences, mostly for microbial or viral genomes. With these developments, we can expect to have many more full-genome sequences of organisms available in the near future.

Discovering novel genomic sequences is always a scientifically valuable task and it provides a necessary foundation for further research. Once the genomic sequence is available it creates the opportunity to systematically study differences from this genetic blueprint. Sequence variations are used to explore genotype–phenotype correlations and to elucidate the genetic pathways that contribute to the physiology and pathophysiology of an organism. One major focus of current international research projects is to investigate and catalogue genetic variation. The most abundant genetic markers, e.g., microsatellites and single-nucleotide polymorphisms (SNPs), have found widespread use in academic as well as commercial areas.

1.1. Population genotyping

A variety of methods have been developed in the last two decades to investigate genetic variations on a population-wide level. These include methods for targeted *de novo* discovery of polymorphisms and mutations as well as their large-scale genotyping in various sample populations (as required in disease association studies and pharmacogenomics studies). Among the developed methods, mass spectrometry has set a mark as a versatile nucleic acid analysis technology that offers both, highest analytical accuracy and throughput (Gut, 2004; Jurinke *et al.*, 2004; Tost and Gut, 2005). This chapter focuses on the use of matrix-assisted laser desorption/ionization (MALDI) time-of-flight (TOF) mass spectrometry (MS) for high-throughput discovery of sequence variants and comparative sequence analysis.

The following sections will introduce the underlying principles of MALDI-TOF MS of nucleic acids and the assay formats used in comparative sequence analysis. We also hope that they make the advantages of mass spectrometry more evident and spark further interest in reading the original publications, for which we can only provide brief summaries here.

2. MALDI-TOF MS-BASED NUCLEIC ACID ANALYSIS

The method of MALDI-TOF MS was developed in the late 1980s by Karas and Hillenkamp (Karas and Hillenkamp, 1988). Thus, for the first time both MALDI-TOF MS as well as electrospray ionization (ESI) MS (Fenn *et al.*, 1989) enabled the mass spectrometric analysis of large biomolecules such as proteins and nucleic acids. These technologies are nowadays the cornerstones of routine proteomics and genomics research.

2.1. Sample preparation

With MALDI-TOF MS, the analysis of proteins and nucleic acids is accomplished by embedding the analyte in a crystalline structure of organic molecules, referred to as matrix. This matrix later serves as the "launching" material for mass spectrometric analysis. The matrix-analyte co-crystal is volatilized with laser bursts. Usually, in UV MALDI, nitrogen lasers with a wavelength of 337 nm are used. The type of matrix and the laser wavelength are aligned such that during this volatilization the matrix molecules absorb the laser energy. During the last decade, for example, the matrix of choice for nucleic acid analysis by UV MALDI has been 3-hydroxy-picolinic acid. Introduction of energy into the crystal structure leads to a micro explosion, which generates a particle cloud. Analyte molecules are desorbed into the gas-phase along with the matrix molecules. Because the matrix absorbs the energy, analyte molecules remain intact and can be analyzed as intact molecules. The volatilization process is accompanied by gas-phase proton transfer reactions, which generate both analyte and matrix ions. An electric field of approximately 20 kV is used to accelerate the ions to nearly uniform kinetic energy. The ions then travel through a field-free drift region (usually 1 m length) and separate by their mass-to-charge ratio. The ions finally reach a detector, which allows the measurement of their TOF. This TOF is directly proportional to the mass-to-charge ratio. Because the MALDI process generates predominantly singly charged ions, measured signals directly represent the molecular mass of the analyte.

What sparked the interest of researchers to analyze nucleic acids by MS? The desire to establish mass spectrometric methods as a means to analyze nucleic acids was related to deficiencies in other methods commonly used: they were (and still are) based on indirect detection methods requiring a fluorescent or radioactive reporter and some of them used and still use fairly time-consuming separation methods such as gel or capillary electrophoresis (at least in DNA sequencing). In contrast, the molecular mass of the analyte is an intrinsic molecular property. It allows direct and highly accurate characterization of the underlying nucleic acid reaction product. The MALDI process is extremely fast and therefore enables high-throughput applications. Both accuracy and speed are tremendously important to set performance standards in genomic/genetic research and molecular diagnostics. Furthermore, the availability of molecular mass information also significantly improves the assay development process. Unanticipated reaction byproducts can be easily characterized on the basis of their mass, which significantly improves the ability to trouble-shoot undesired enzyme properties and enzymatic reactions, and eventually allows their optimization.

3. THE BASE-SPECIFIC CLEAVAGE ASSAY

In recent years, several new biochemical concepts have been applied to circumvent some of the existing issues in measuring larger DNA products with MALDI-TOF MS. Each of them use base-specific cleavage of amplification products as a means to analyze the amplified sequence for potential sequence changes (Hahner *et al.*, 1997; Elso *et al.*, 2002; Hartmer *et al.*, 2003; Krebs *et al.*, 2003; Stanssens *et al.*, 2004). Complete base-specific cleavage generates a set of short oligonucleotides from the amplification product, where the distance between the cleavage sites determines the length of the oligonucleotides. The products, thus, most often fall into a mass range preferable for current MALDI-TOF MS. Methods such as base-specific cleavage alleviate the major issues limiting the read length when classical Sanger sequencing was combined with MALDI-TOF MS analysis and enables sequence determination by MS. Conceptually, these approaches resemble earlier methods for RNA sequencing and also methods for protein identification by peptide mapping.

The identification of sequence changes using base-specific cleavage requires a different analysis approach than Sanger sequencing. In its current format, base-specific cleavage is used as a re-sequencing method and is not applied to *de novo* sequencing. Thus, we assume that a reference sequence is available for each target region. This reference sequence can be used to generate *in silico* cleavage and mass signal patterns. If a sample carries a sequence change in the amplified target region, then this sequence change will have an impact on the mass signal pattern: the sequence change can introduce a new cleavage site leading to two shorter cleavage products; it can remove a cleavage site leading to a longer cleavage product; if no cleavage-base is affected, it can shift the mass of an existing cleavage product higher or lower. Hence, a comparison of experimental mass signal patterns with *in silico* patterns can be employed to identify additional and missing mass signals, which then can be used to interpret the discovered sequence change. Each cleavage product can consist of a combination of the three non-cleavage nucleotides and a single-nucleotide residue of the cleavage nucleotide. It is thus fairly straightforward to calculate for each additional mass signal, which combination of A, C, G and T results in the measured molecular mass. Once the composition is identified, an algorithm can determine which sequence change can account for the observed mass signal changes. Usually this requires the integration of multiple base-specific cleavage reactions. For further reading, please refer to the original articles describing the concept (Pomerantz *et al.*, 1993; Bocker, 2003; Stanssens *et al.*, 2004).

Figure 1 exemplifies the concept. In this theoretical case, the reference sequence is cleaved at every T. The *in silico* cleavage of the reference sequence creates cleavage products. To demonstrate the detection of a sequence change, the reference nucleotide subjected to a substitution is indicated in bold. The cleavage products are then converted into compomers by simply counting the occurrence of each nucleotide in a cleavage product. Using the four nucleotide masses, we can then calculate the mass of each compomer (representing a

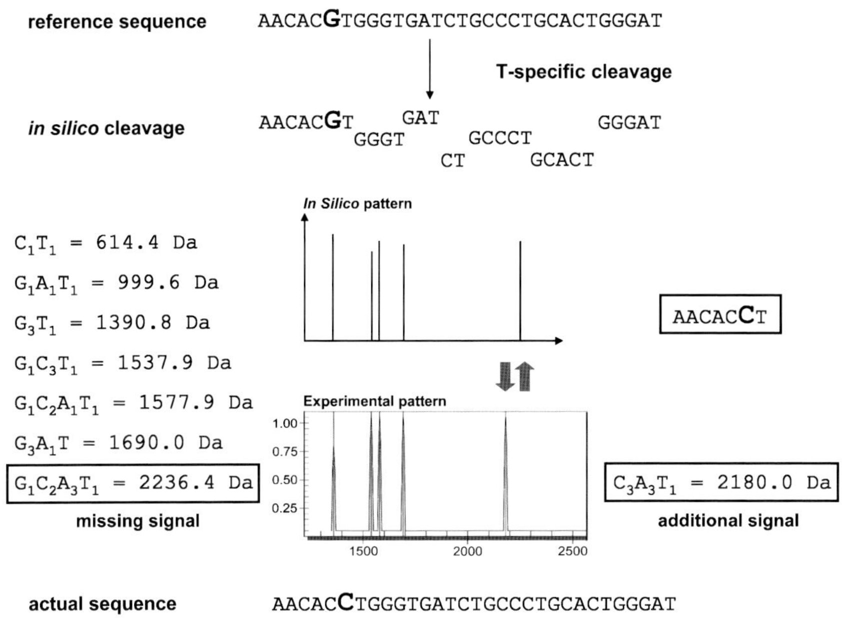

Fig. 1. The scheme illustrates the concept of base-specific cleavage and analysis of cleavage products by MALDI-TOF MS for discovery and identification of sequence changes. A reference sequence is cleaved *in silico* at every T. The resulting T-specific cleavage products are converted into compomers by simply summing the occurrence of each nucleotide in a cleavage product. Each compomer can be assigned a molecular mass. The list of compomers yields a list of masses, which represent the *in silico* mass signal pattern (*in silico* pattern). To discover/identify the sequence changes, the target region is amplified from a sample and subjected to the base-specific cleavage process. The resulting experimental mass spectrum is then compared to the *in silico* mass spectrum derived from the reference sequence. In the depicted case the experimental mass spectrum shows a new mass signal (additional signal) at 2180 Da and the mass signal at 2236.4 Da (predicted from the reference sequence) is missing. These mass signal changes can be interpreted by first finding the compositional explanation of the additional mass signal. An algorithm will search for a combination of the four nucleotides, which yields the detected mass. In the depicted case, we know that the compomer can only contain one T, because we performed a T-specific cleavage. A compositional explanation for a mass of 2180 Da is given by C2A3T1. We also know from the missing signal at 2236.4 Da, which represents the first cleavage product of the target sequence (compomer G1C2A3T1), that there must be a sequence change in this region of the target sequence. We can conclude from these two observations, that the target region in the sample had a G/C sequence change at position 6 of the target sequence studied.

cleavage product). The list of masses then represents our *in silico* mass spectrum (the upper mass spectrum in Figure 1). The lower mass spectrum represents analysis of the same genomic region from a real sample. Comparison of the mass signals shows that the mass signal representing the compomer G1C2A3T1 (mass signal at 2236.4 Da) predicted from the reference sequence is missing. We also find an additional mass signal with the mass of 2180 Da. For the next step, we can search for a compositional explanation of this new mass signal. From the cleavage reaction we know that the composition can only contain one T and need to find the combination of CxAyGzT1, which can account for a mass of 2180 Da. Searching the compositional space under this constraint will yield C3A3T1 as a solution. We further see that the mass signals representing other cleavage products of the genomic region are present. This allows us to restrict potential sequence changes (causing the change in mass signals) to a specific cleavage product and to link it with the sequence region representing the first 7 nucleotides of the target region. In fact, the "missing" compomer (G1C2A3T1) can easily be transformed in the compomer derived of the additional mass signal (C3A3T1) by a G/C substitution. Because there is only one G within the first 7 nucleotides, we can directly pinpoint the G/C substitution and assign an "actual" sequence fully explaining the observed mass signal pattern. Supplementary cleavage reactions could be used to verify this result.

3.1. Methods for base-specific cleavage

Several methods have been developed to obtain base-specific cleavage (Hahner *et al.*, 1997; Shchepinov *et al.*, 2001; Elso *et al.*, 2002; von Wintzingerode *et al.*, 2002; Hartmer *et al.*, 2003). Among them are enzymatic as well as chemical means of cleaving DNA and RNA. The most prominent methods use the following process. The target region is amplified by conventional PCR using primers tagged with a T7 promoter sequence. This allows subsequent generation of a single-stranded RNA transcript from either the forward or reverse direction. See Figure 2 for an illustration of the process flow). The RNA transcript can then be cleaved to completion with a base-specific RNAse, such as RNAse T1, which yields for example G-specific cleavage (Hahner *et al.*, 2000; Hartmer *et al.*, 2003; Krebs *et al.*, 2003). A variation to this concept uses a mutant T7 RNA polymerase capable of incorporating dNTPs (Stanssens *et al.*, 2004). The selective incorporation of a dNTP in the RNA transcript allows the use of less-specific RNAse. RNAse A, for example, cleaves at every C and U residue of a transcript. This would degrade the transcript into very small cleavage products and most of the sequence information would be lost. However, if rCTP is fully replaced by dCTP during the transcription, RNAse A is rendered specific for U (T) cleavage. Similarly, the use of dTTP during transcription renders RNAse A cleavage C-specific. The combination of C- and U-specific cleavage from forward and reverse direction in four separate cleavage reactions allows cleavage after virtually all four bases. This provides the most comprehensive scan of a target region for potential polymorphisms or mutations. The process has been applied successfully to target regions ranging in size from 150 bp, up to 1 kb.The detection rates depend upon amplicon length and sequence context. In a 500 bp amplicon, on average 95% of all single-base sequence changes can be detected when solely heterozygous sequence changes are considered and about 85% of these can be mapped unambiguously to a nucleotide position in the amplicon. For haploid organisms or for the detection of homozygous sequence changes, the detection rates increase to 99% of all possible single base changes because the combination of additional and missing mass signals can be used. Figure 2 displays the results of a simulation of SNP discovery rates for a 4 Mb sequence region surrounding the ApoE gene region (Lai *et al.*, 1998).

3.2. MassCLEAVE

Figure 3 depicts a schema of the current base-specific cleavage process, called MassCLEAVE™, as implemented on the MassARRAY® system. The process is homogeneous and does not require any intermediate purification steps. The reagents for each reaction step are simply added to the well of a microtiterplate. After completion of the cleavage reaction, ion-exchange resin is added to each well in order to remove salts from the phosphate backbone of the nucleic acid cleavage products. These are then transferred onto a miniaturized chip array and analyzed by MALDI-TOF MS. Spectra interpretation and identification of polymorphisms is performed by algorithms as described in Bocker (2003).

4. APPLICATIONS FOR COMPARATIVE SEQUENCING

This section will highlight the applications for which the concept of base-specific cleavage analyzed by MALDI-TOF MS has provided proof-of-concept. We will briefly summarize the application and refer to the original publications for further reading.

4.1. Signature sequence identification/pathogen identification

One of the first implementations of base-specific cleavage and MALDI-TOF MS for comparative sequence analysis is related to microbial analysis. Here, the progress in DNA sequencing has allowed the detailed study of genome plasticity by sequencing multiple, unrelated bacterial strains of the same species. The generated sequence information allowed identification of candidate genes or signature sequences that can be used as molecular typing tools for the characterization of microbial isolates and the study of population genetics of the particular organism. The most widely accepted marker region is the partial small subunit ribosomal RNA sequence (16S rRNA). It has been studied extensively and a public sequence database (e.g. http://rdp.cme.msu.edu./) presents the current structure of the microbial kingdom. This allows for sequencing-based 16S rDNA typing to clearly impact on the discovery of new bacterial species (Woese, 1987).

A more recent approach able to assess the population structure of bacterial isolates was proposed in 1998. Multi-locus sequence typing (MLST) explores the genomic relatedness at the inter- and intraspecies level by sequence analysis

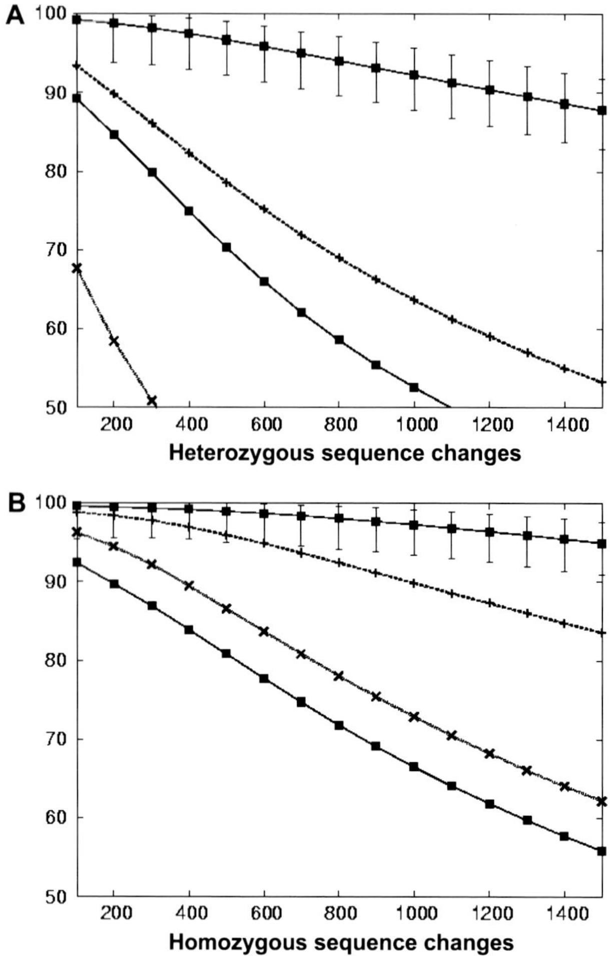

Fig. 2. This figure depicts simulation results for the discovery of single-nucleotide polymorphisms using the MassCLEAVE process. The simulation is based on a 4 Mb sequence region surrounding the ApoE gene. The region was randomly divided into amplicons of varying length using a windowed approach. For each amplicon, we simulated all possible single base substitutions and single base insertions and deletions and determined if the introduced sequence change would be detectable. The graphs display the percentage of detectable single base sequence changes (*y*-axis) with increasing amplicon length (from 100 bp to 1.5 kbp). Panel A provides results for sequence changes detected as heterozygous changes. In this case, the detection is purely based on additional mass signals. Signal intensity changes were not taken into account to interpret sequence changes. The continuous line with squares represents the fraction of detectable SNPs with at least one additional mass signal. The dashed line represents detectable SNPs with at least 2 two additional mass signals. The dotted line represents SNPs detected with at least 3 additional mass signals. The second continuous line represents the fraction of SNPs that can be localized exactly (given the presence of potential sequence repeats and stretches of the same nucleotide). It can be concluded from these simulations that about 95% of all possible SNPs can be detected at an amplicon length of 500 bp when the sequence change is present as a heterozygous sequence change. Panel B shows the same simulation assuming homozygous sequence changes as they would occur

of housekeeping genes (Maiden *et al.*, 1998; Urwin and Maiden, 2003). Sample sequences of multiple marker regions are matched to references in a public database (http://www.mlst.net/). Each sample is defined by its combination of the matching results. As of today the MLST database covers 18 different species. Genotypic identification and characterization of microbial organisms using 16S rDNA analysis or MLST are nowadays key elements of modern epidemiological surveillance and monitoring.

The initial concept of base-specific cleavage was the transformation of the sequence variability of these marker regions into mass signal patterns, which allow for unambiguous identification of the bacterial species from a reference set of sequences. Figure 4 illustrates this concept. The upper panel of the figure shows an overlay of sequences in the 16S rRNA gene of three different *Mycobacteria* strains. The hypervariable regions are indicated with a box. The lower panel depicts the corresponding mass signal patterns generated through PCR amplification of the region and base-specific cleavage at every T. Signals allowing the discrimination between the three *Mycobacteria* are indicated with arrows.

The concept has been successfully reduced to practice for 16S rDNA signature sequence-based identification of environmental *Bordetella* strains (von Wintzingerode *et al.*, 2002; Hartmer *et al.*, 2003) and of *Mycobacteria* (Lefmann *et al.*, 2004) with 24 isolates unequivocally identified in the study. These experiments serve as a general model for microbial or viral genotypic fingerprints that can be utilized to differentiate between samples and extract corresponding best-matching reference sequences from reference databases. The advantage MALDI-TOF MS offers in this respect is the speed of analysis, the degree of automation and the opportunity for standardization. All of these are important attributes required to cope with new threats from infectious microorganisms, such as agents of bioterrorism, emerging pathogens and their resistance to antibiotic treatment, and also to allow high-level quality control of pharmaceutical products.

4.2. SNP discovery and mutation detection

The concept of base-specific cleavage has also been implemented as a tool for high-throughput SNP discovery and mutation detection. The MassCLEAVE process, as depicted in Figure 3, has been used to discover new SNPs in regions of the CETP gene. This initial study was designed to validate the concept on a

Fig. 2. (*Continued*)

when haploid organisms, such as bacteria, are sequenced or when the SNP is highly frequent. The description of the lines is the same as in panel A. At the same amplicon lengths (500 bp) the detection rates increase from around 95% to around 99% of all possible SNPs. This increased detection rate is based on the fact that additional and missing signals provide information for possible sequence changes. Similarly, the number of SNPs detected by multiple mass signals is much higher compared to heterozygous sequence changes and the localization rate is increased.

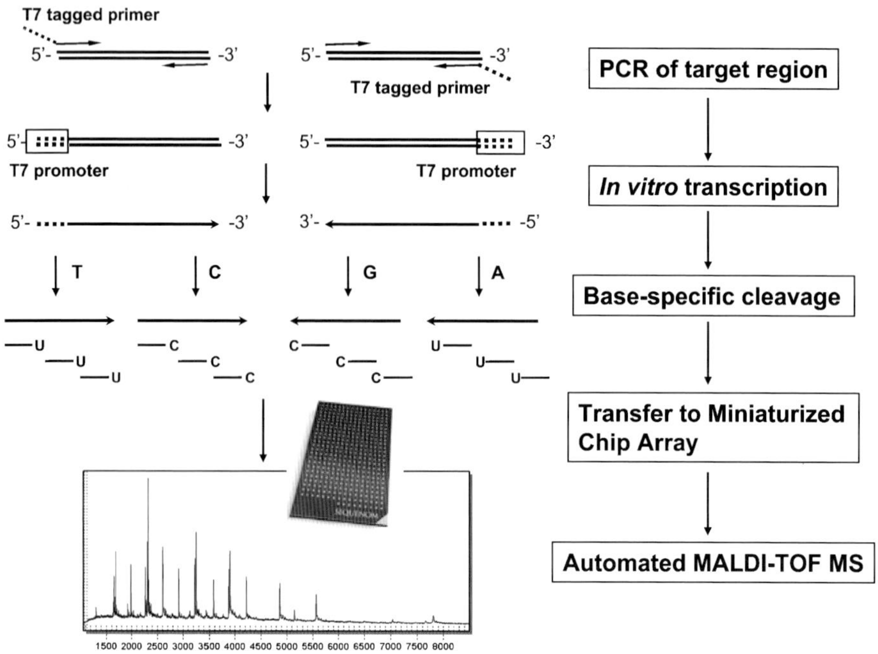

Fig. 3. Shown is the current process flow of the MassCLEAVE concept. The target region is amplified using PCR primers carrying a T7 promoter tag either at the 5′ end of the forward or the reverse PCR primer. Subsequent to PCR, unincorporated nucleotides are degraded with shrimp alkaline phosphatase. The PCR product is then transcribed into a single-stranded RNA molecule (either from the forward or reverse strand). During this transcription either rCTP or rUTP is fully replaced by their deoxyNTP counterparts. The resulting transcript has a mosaic structure comprised of three rNTPs and one dNTP (either dCTP or dTTP). This renders the transcript insensitive to cleavage at either U (when UTP is replaced by dTTP) or C (when CTP is replaced by dCTP). The RNA transcript is then subjected to complete cleavage by RNAse A. This RNAse normally cleaves at every C and U in an RNA molecule, but is rendered base-specific by the incorporation of respective dNTPs. The combination of forward and reverse transcription with the two possible cleavages (either C-specific or U-specific) virtually allows for cleavage at all four nucleotides. The cleavage products are conditioned for mass spectrometric analysis by the addition of ion-exchange resin and finally are transferred onto miniaturized chip-arrays for fully automated MALDI-TOF MS analysis and interpretation. The assay format is homogenous and can be accomplished with simple add-on steps, where the next reaction cocktail is added in the same reaction vessel.

region with several known SNPs and served as a model system for discovery of sequence variations in human disease candidate genes (Stanssens *et al.*, 2004).

A second study served as a model system for high-throughput mutation detection in microbial genomes. Here, base-specific cleavage was employed to identify DNA sequence changes that occurred in *Escherichia coli* K-21 MG1655 during laboratory adaptive evolution to new optimal growth phenotypes (Honisch *et al.*, 2004). To identify mutations underlying a 40-day adaptive laboratory evolution on glycerol, 4.4% (202 kb) of the *E. coli* K12 MG1655 genome was re-sequenced in

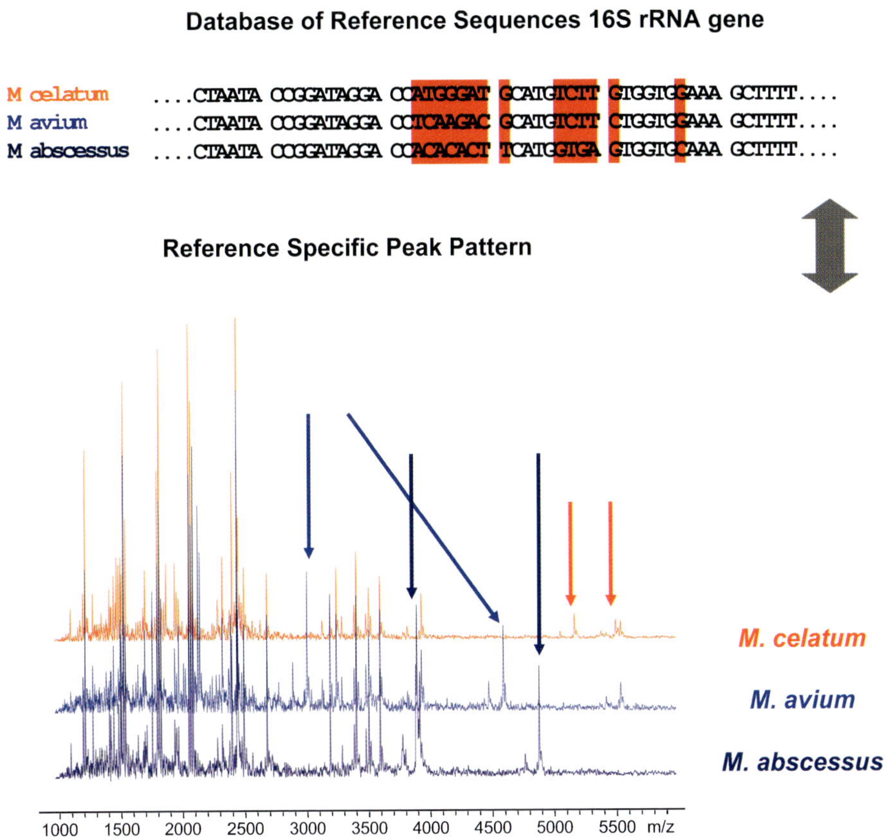

Fig. 4. Shown is an example of the use of base-specific cleavage for pathogen identification. The three depicted DNA sequences represent a hypervariable region in the 16S rRNA gene. The sequence differences between each of three different mycobacteria (*M. celatum*, *M. avium* and *M. abscessus*) are marked with red boxes. Shown below the reference sequences are three experimental base-specific cleavage spectra representing each of the *Mycobacteria* strain. The mass signals allowing discrimination and unambiguous identification of the strain are marked with colors matching the reference strain.

several clones picked at the end of the evolutionary process. The screen focused on 125 genes involved in glycerol metabolism, transcription factors, σ factors and mutator genes. Including all clones analyzed, the screen covered 1.54 Mb and was completed in 13.5 h. This is among the fastest of currently available re-sequencing technologies that allow parallel screening of multiple samples (compared with 2.8 Mb/day for a 384-capillary instrument) (Kling, 2003).

Figure 5 illustrates the simplicity and strength of mutation detection by base-specific cleavage. It shows an overlay of mass spectrometric traces of four individuals covering a region that harbors multiple sequence changes. The trace marked "Ref" provides the results of a sample with wild-type sequence, samples S1–S3 carry different mutations, which can be identified by the new mass signals marked with an arrow. The method allows the determination of haplotypes in

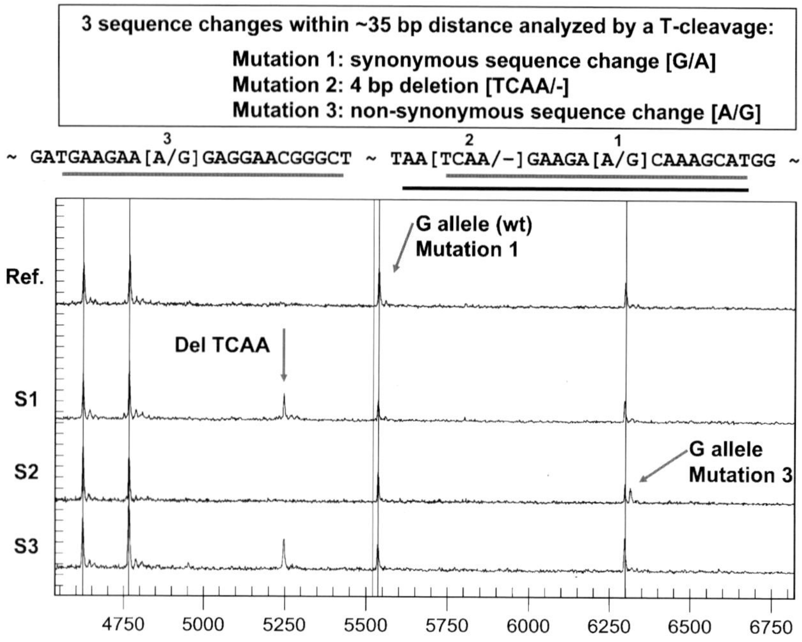

Fig. 5. Displayed are examples for detection and interpretation of complex sequence changes using base-specific cleavage and MALDI-TOF MS. The analyzed sequence region carried three different mutations (two A/G substitutions at different positions and a 4bp deletion. The mass spectra represent results of the sequence region generated by T-specific cleavage of the amplification product. Four mass spectra are shown: Ref. represents the mass spectrum for a reference sample with wild-type sequence; S1–S3 are derived from individuals harboring different mutations. Dotted lines mark mass signals predicted from the wild-type reference sequence. As can be seen, the occurrence of new mass signals (marked with arrows) easily identifies mutations in the sample. Samples S1 and S3 carry the 4 bp deletion (TCAA), sample S2 carries the mutant G allele for mutation 3. This experiment also demonstrates one of the strength of mass spectrometric analysis. For sequence changes in close proximity, the described method will be able to determine the haplotype structure. To illustrate this, the T-cleavage products harboring the mutations are marked with gray/black lines. Mutations 1 and 2 are residing in the cleavage product. The combination of the 4 bp deletion with the wild-type A allele will differ in mass from a combination of the 4 bp deletion with the mutant G allele by 16 Da. Using the molecular mass information, an algorithm can therefore easily deconvolute compound heterozygotes and determine if a patient carries two mutations on the same chromosome and hence has an intact copy of the gene.

compound heterozygotes as exemplified by mutations 1 and 2. In a T-specific cleavage, the combination of mutations 1 and 2 will generate different cleavage products and hence different mass signals. Mutation 2 (a 4 bp deletion) will change the mass by roughly 1200 Da (the equivalent of 4 bp), Mutation 1 will change the mass by 16 Da (mass difference between G/A). It is therefore straightforward to deconvolute which combination of mutations an allele carries. This can be of high medical relevance, as is, for example, the case in hemoglobinopathies.

Recently, the concept of re-sequencing by base-specific cleavage and MALDI-TOF MS has been enriched by a new aspect that may help to excel the technology for targeted re-sequencing of exonic/coding regions (Ehrich *et al.*, 2005a). Most of the exonic regions are relatively short in length and their analysis does not require long read length. It is therefore desirable to be able to analyze these regions in a multiplexed fashion. Base-specific cleavage has multiple features that enable multiplexed analysis. The most interesting part is the analysis process. Base-specific cleavage fractionalizes a target sequence region or amplification product in multiple independent cleavage products, which in their entirety represent the analyzed region. The process unties the cleavage products from their sequence context and generates a defined experimental mass signal pattern where each mass signal represents at least one fragment evolved from the target sequence. This experimental pattern is then compared to an *in silico* reference pattern derived from a reference sequence. As described earlier, differences between the expected and the observed mass signal pattern are interpreted to identify sequence changes by collecting all unexpected additional signals and calculation of their nucleotide compositions. This process only involves part of the reference sequence, more specifically the affected cleavage product, and not the complete sequence. Hence, this process can be performed on multiple independent sequences in parallel with the contingency that the rendered cleavage products have to allow mapping of their origin (meaning to which of the analyzed multiple sequence regions they belong) in order to assign the location of a sequence change to a particular target region of the multiplex.

The biochemistry of base-specific cleavage supports multiplexing in a very straightforward fashion. Multiple target regions can be amplified in a multiplex-PCR with promoter-tagged PCR primers. This allows for parallel RNA transcription and RNase cleavage, and simultaneous detection by MALDI-TOF MS. To explore the capabilities and limitations of multiplexed re-sequencing, an extensive simulation has been performed on sequences from the human genome. This simulation demonstrated that multiplexing does not significantly reduce detection rates for sequence variations, provided that the overall nucleotide count of the multiplexed regions does not exceed the nucleotide count usually used for uniplex re-sequencing (also see, Figure 2, for the simulation results on uniplexed re-sequencing). The ability to multiplex can exhibit significant improvements for targeted re-sequencing, because it reduces the associated costs (reagent cost and process flow improvements) by the factor of simultaneously analyzed regions. In addition to exon re-sequencing, this feature may prove extremely valuable to applications such as HLA typing or simultaneous screening of multiple regions for mutations conferring drug resistance.

4.3. Methylation detection

Base-specific cleavage and MALDI-TOF MS has recently also been implemented as an effective means for high-throughput quantitative analysis of DNA methylation. The analysis of covalent modifications of nucleotides in genomic DNA, e.g., cytosine methylation, has gained significant momentum when it became apparent that genetic information is not only stored in the arrangement of four nucleotide bases, but also in the covalent modification of selected bases. As indicated above, methylation of cytosine is the most common modification in mammals. The covalent addition of methyl groups to cytosine is catalyzed by the DNA methyltransferase (DNMT) enzyme family (Pradhan *et al.*, 1999; Szyf and Detich, 2001). More specifically, DNA methyltransferases target cytosines in CpG dinucleotides. Such CpG dinucleotides are generally underrepresented in the human genome and are concentrated in distinct areas called CpG islands. A large proportion of these CpG islands are found in the promoter regions of genes. The conversion of cytosine into 5-methylcytosine in promoter-associated CpG islands causes changes in chromatin structure, usually resulting in transcriptional silencing of the genes controlled by this promoter region.

While the study of the role of methylation changes in the context of cell biology is a rapidly expanding research field, the importance of methylation is already highlighted through its connection to mammalian development, imprinting, X-chromosome inactivation (Li, 2002), suppression of parasitic DNA (Walsh *et al.*, 1998) and cancer etiology (Costello *et al.*, 2000; Costello and Plass, 2001; Feinberg, 2001; Jones and Baylin, 2002). DNA methylation analysis has received particular attention in cancer research, because several studies have shown that changes in the methylation status of nucleosomal DNA promise to be a powerful marker for the early detection of neoplastic events (Tsou *et al.*, 2002; Etzioni *et al.*, 2003; Laird, 2003).

Several methods have been developed to analyze DNA methylation. Because common techniques for amplification of DNA cannot preserve the cytosine methylation profile (the presence of methylated cytosine cannot be mirrored in the PCR product), most methods rely on bisulfate treatment of genomic DNA prior to amplification (Clark *et al.*, 1994). Bisulfite treatment converts all non-methylated cytosine into uracil, whereas all methylated cytosine are inert. This treatment translates the methylation mark into a sequence change. Amplification products of the target region therefore carry the methylation pattern as a pattern of C/T sequence changes. Correspondingly, sequence analysis of PCR products from bisulfate treatment can be used to conclude on the methylation status of CpGs in a genomic region.

We have shown in earlier sections how the mass signal changes of a base-specific cleavage patterns can be used to identify sequence changes. The process of base-specific cleavage lends itself also for the analysis of DNA methylation. The mass signal pattern is affected by methylation through the C/T sequence changes introduced into the bisulfate-treated genomic DNA. The exact change depends on the type of cleavage reaction: methylation can introduce new cleavage sites resulting in new, shorter products (and hence new mass signals with lower mass); methylation can lead to a replacement of an existing cleavage site with a non-cleavable nucleotide and therefore connect two existing fragments together resulting in a new, longer product (new mass signals with higher mass); and methylation can generate a sequence change in an existing cleavage product that does not affect cleavage, but generates a mass shift. To illustrate this, consider bisulfate treatment of a fully unmethylated genomic region.

All cytosines will be converted into uracils, and correspondingly there will no existing cleavage sites in a C-specific cleavage of the forward strand. Figure 2 illustrates the cleavage reactions. However, each methylated cytosine remains a cytosine and it will therefore generate a cleavage site and corresponding cleavage products/mass signals, which allow discovery of the methylation and its identification. Cleavage reactions, in which a methylation event generates a new cleavage site, are perfectly suited for sensitive discovery of methylation.

The third option, in which methylation does not affect a cleavage site but leads to mass shifts, is the most interesting when quantitation of the relative amount of methylated DNA is desired. The T-reverse cleavage, for example, always cleaves outside of a CpG dinucleotide. In this reaction methylation events will be represented as mass shifts of 16 Da (the mass difference between G/A, which represents the reverse strand complement of the C/T changes). If multiple CpG sites are enclosed in one cleavage product, the mass will shift in multiples of 16 Da. The coexistence of mass signals representing methylated DNA and unmethylated DNA in close proximity (in terms of mass difference) enables the estimation of the relative amount of methylated DNA by the use of the peak area ratios.

To experimentally verify the concept, base-specific cleavage and MALDI-TOF MS have been used to successfully reconstruct the methylation pattern of IGF2/H19, a well-described genomic region commonly hemi-methylated (Ehrich *et al.*, 2005b). It has then been used successfully for large-scale quantitative profiling of methylation patterns in lung cancer. In this study, the methylation status of 47 promoter regions (1426 CpG sites in total) was assessed quantitatively in 48 lung cancer tissue samples and compared to their normal adjacent lung tissue to identify differentially methylated CpGs that allow accurate classification of samples (Ehrich *et al.*, 2005b). Using the technology described here, this study was completed within a single day (measurement and analysis time).

What are the advantages of the described approach over other technologies? Conventionally, DNA methylation is either analyzed by PCR sequencing or by assays such as methylation-specific PCR (MSP), which target individual CpGs (Herman *et al.*, 1996). PCR sequencing delivers very limited quantitative information and usually has to be supplemented by sequencing from clones to verify the results. To minimize statistical effects, a large number of clones need to be sequenced. This makes the approach very cumbersome and expensive. Assays targeting individual CpGs can provide quantitative information with sufficient precision, but are limited in that they only provide information on the targeted CpG. To obtain a reasonable overview of the methylation status of a genomic region, multitudes of assays have to be developed and optimized, which again is a cumbersome task. Base-specific cleavage and MALDI-TOF MS allow the study of DNA methylation in target regions between 200 and 600 bp in length. A single cleavage reaction can generate quantitative methylation data for up to 70 CpGs. Combined with the high-throughput of MALDI-TOF MS acquisition and analysis, this allows efficient screening of genomic regions for differentially methylated CpGs in large numbers of samples.

5. SUMMARY

Presently, PCR/primer extension and MALDI-TOF MS detection is widely accepted as the method of choice for fine-mapping in SNP genotyping and association studies. We have shown in this section how the concept of base-specific cleavage and mass spectrometric analysis can be used in comparative sequence analysis in general, and in the rapidly developing fields of mutation detection, pathogen identification and DNA methylation analysis in particular. The assay concept significantly expands the application portfolio of MALDI-TOF MS because it allows the analysis of larger target regions. In essence, this enables the use of MS in applications currently dominated by capillary electrophoresis separation/fluorescent detection.

What are the major differentiators of this method versus the gold standard Sanger sequencing and versus newer technologies such as 454 Life Science's pyrosequencer GS20, and array-based sequencing by hybridization assays? One of the most striking features of the described concept is the collateral security. The discovery and identification of sequence changes is, in most cases, based on multiple "observations", or so called "witnesses". This means the combination of mass spectra provides for multiple mass signals (witnesses) and mass signal changes, which contribute to the identification process. This can simplify the process of automated data interpretation, and appropriate tools provided, enables very high accuracy. A conventional sequencing approach requires the analysis of two colors overlaying each other at the mutation position. This is a much more challenging task, because the number of observations is low. A further tremendous advantage is the opportunity to derive quantitative information from the mass spectra. We have exemplified this for DNA methylation analysis, but this can also be translated into analysis of mixed viral populations, as for example required in treatment monitoring of HIV patients.

6. OUTLOOK

We have shown in previous sections of this contribution how MALDI-TOF MS can be applied for high-throughput comparative sequence analysis and discovery of new sequence variants. As outlined earlier, several other methods have been recently developed for whole genome sequencing and *de novo* sequencing of individual samples. Our approach provides a method to rapidly analyze large numbers of target regions in large cohorts of different populations, as required for example in epidemiologic studies. The strength of the approach has its foundation in the accuracy and degree of automation, both enabled through the use of an intrinsic molecular property of the analyte, the molecular mass and the multitude of observations used to identify a sequence change. These are also key features required for transition of DNA sequencing into molecular diagnostics. We therefore anticipate that MALDI-TOF MS of nucleic acids will play an increasingly important role in the implementation of further molecular diagnostic tests.

6.1. Improvements in instrumentation and processing

Which further opportunities does this concept have to increase performance and competitiveness in comparative sequence analysis? Improvements in performance can be achieved through changes in the instrumentation, the biochemistry and continuous evolution of the software algorithms applied for mass spectral interpretation. One of the more simple advancements with a direct impact on throughput would be the use of faster lasers. Most of the MALDI-TOF MS instruments currently applied in nucleic acid analysis use nitrogen lasers with a repetition rate of 20 Hz. X/Y movement of sample stage, spectra acquisition (usually 10–20 laser shots are acquired and summed) and real-time interpretation of the spectrum, e.g., the calling of a genotype can currently be achieved in about 1.5 s, with the majority of the time consumption related to data acquisition. The use of much faster lasers (e.g., 200 Hz nitrogen lasers) therefore would increase the sequencing throughput by 2.5-fold (and of course would then also improve depreciation of instrument cost significantly).

Despite increases in throughput, researchers are usually also interested in the opportunity to reduce reagent cost associated with sample processing. The base-specific cleavage assay, as introduced earlier in this chapter, offers considerable cost-saving opportunities. In the long run, these can best be achieved through advances in microfluidics and sample processing in microfluidic devices. A simple calculation reveals the cost saving potential. The PCR is currently performed in 384-well microtiter plate (MTP) format with 5 μl volumes. Some 2 μl volumes of the PCR reaction are used for an individual transcription and cleavage reaction, which usually is performed in 7 μl total volume. The sample is then diluted and the cleavage products are conditioned for MALDI-TOF MS using ion-exchange resin. The final sample volume for each reaction well is usually around 30 μl. Then, only 15 nl of sample are dispensed on a miniaturized chip array for automated MALDI-TOF MS analysis. This means that only about 1/2,000 of the sample is actually required for the final analysis step and therefore cost savings through miniaturization of sample processing are tremendous. The proof-of-principle for analysis is already established in the current format.

Finally, we should also take into account those performance improvements that relate to the applications listed in previous sections of this chapter. The main limitation in an application such as re-sequencing is related to loss of information: some cleavage products may be too small to be detected (their masses fall below the low mass cut-off around 1200 Da), some cleavage products may be too long to be detected (their masses fall above the high mass cut-off, where the sensitivity of axial DE-MALDI-TOF MS is limited) and multiple cleavage products may overlap in mass so that they cannot be resolved and changes in the mass signals remain obscured.

6.2. High-resolution mass spectrometers and isotopically depleted nucleotides

A MALDI-TOF instrument with much higher resolution (M_r 10,000 instead of the M_r 800 currently achieved) and higher sensitivity in the high mass range

(8000+ Da) will be helpful to improve identification and characterization of sequence changes. These performance characteristics can, for example, be achieved with recently introduced orthogonal-extraction MALDI-TOF mass spectrometers (also called O-TOF) (Berkenkamp *et al.*, 2003; Loboda *et al.*, 2003). The higher mass resolution reduces spectral overlap of mass signals (with the restriction of course that the isotopic envelope inherent to "natural" nucleic acid building blocks is still limiting) and enables processing of spectra with much higher peak densities. This is particularly interesting, if instrumental improvement in mass resolving power is coupled with the use of isotopically depleted nucleotides during the transcription and cleavage process. In simulations using the same data set as employed for the results displayed in Figure 2, the combination of O-TOF and isotopically depleted nucleotides for base-specific cleavage increased the discovery rates of heterozygous sequence changes (all possible single base sequence changes simulated) from an average 95% (at 500 bp amplicon length) to an average 98% (Sebastian Bocker, University of Bielefeld, personal communication). Early experiments to improve performance of MALDI-TOF MS analysis of nucleic acids with isotopically depleted nucleotides have already been published (Abdi *et al.*, 2001, 2002; Tang *et al.*, 2002). An additional benefit of the use of orthogonal TOF mass spectrometers may be the achievable throughput. They can be equipped with 1 kHz lasers, which will allow further improvements and data acquisition speed.

While the reduction of overlapping mass signals (by improved mass resolution and use of isotopically depleted nucleotides) helps significantly in avoiding ambiguities in the identification and characterization of sequence changes, it cannot resolve issues related to the reconstruction of regions where significant information is lost in the low mass range. This issue, however, can be tackled by modification of the biochemistry. The base-specific cleavage process described above uses complete cleavage. Due to the complete cleavage, we isolate the cleavage products from their original position within the amplification product. Hence, we lose the information about how the cleavage products were ordered. Furthermore, the size distribution of cleavage products is dependent on the sequence context and distribution of cleavage sites. In some cases, many of the cleavage products will fall under our low-mass cut-off.

6.3. Use of cleavable/non-cleavable nucleotide mixtures

The use of partial cleavage would help to decrease the above issues if one can control the distribution of partial cleavage products. Obtaining partial cleavage by limited endonucleolytic digestion (through limited amount of enzyme or limited incubation time) is surely possible, but very sensitive to experimental variations. We therefore propose the use of a mixture of cleavable and non-cleavable nucleotides during the RNA transcription/RNase cleavage process. The ratio of cleavable and non-cleavable nucleotides can be adjusted to obtain the desired ratio of completely cleaved products, first partial cleavage, second partial cleavage and so on, while the cleavage process itself (the RNase action) is still a complete cleavage and therefore more robust. The proposed approach has been modeled recently and initial results have been published (Bocker, 2004).

ACKNOWLEDGEMENTS

We thank Christiane Honisch, Niels Storm, Yong Chen and Bryan Coon for their contributions to the technology. Special thanks go to Stefan Berkenkamp for discussions on improvements on MALDI-TOF MS instrumentation and Sebastian Boecker, University of Bielefeld, for sharing early results of his simulations on SNP Discovery rates.

REFERENCES

Abdi, F., Bradbury, E. M., Doggett, N. and Chen, X. (2001). Rapid characterization of DNA oligomers and genotyping of single nucleotide polymorphism using nucleotide-specific mass tags. Nucleic Acids Res. 29(13), e61.

Abdi, F. A., Mundt, M., Doggett, N., Bradbury, E. M. and Chen, X. (2002). Validation of DNA sequences using mass spectrometry coupled with nucleoside mass tagging. Genome Res. 12(7), 1135–1141.

Bennett, S. T., Barnes, C., Cox, A., Davies, L. and Brown, C. (2005). Toward the $1,000 human genome. Pharmacogenomics 6(4), 373–382.

Berkenkamp, S., Hillenkamp, F. and van den Boom, D., (2003). Orthogonal ion extraction MALDI-TOF with a single-quadrupole. In: Features and Applications for DNA/RNA Analysis. International Conference of the American Society of Mass Spectrometry (ASMS), Montreal, Canada, Extended Abstract.

Bocker, S. (2003). SNP and mutation discovery using base-specific cleavage and MALDI-TOF mass spectrometry. Bioinformatics 19(Suppl 1), I44–I53.

Bocker, S. (2004). Sequencing from compomers: using mass spectrometry for DNA *de novo* sequencing of 200+ nt. J. Comput. Biol. 11(6), 1110–1134.

Clark, S. J., Harrison, J., Paul, C. L. and Frommer, M. (1994). High sensitivity mapping of methylated cytosines. Nucleic Acids Res. 22(15), 2990–2997.

Costello, J. F., Fruhwald, M. C., Smiraglia, D. J., Rush, L. J., Robertson, G. P., Gao, X., Wright, F. A., Feramisco, J. D., Peltomaki, P., Lang, J. C., Schuller, D. E., Yu, L., Bloomfield, C. D., Caligiuri, M. A., Yates, A., Nishikawa, R., Su Huang, H., Petrelli, N. J., Zhang, X., O'Dorisio, M. S., Held, W. A., Cavenee, W. K. and Plass, C. (2000). Aberrant CpG-island methylation has non-random and tumour-type-specific patterns. Nat. Genet. 24(2), 132–138.

Costello, J. F. and Plass, C. (2001). Methylation matters. J. Med. Genet. 38(5), 285–303.

Ehrich, M., Bocker, S. and van den Boom, D. (2005a). Multiplexed discovery of sequence polymorphisms using base-specific cleavage and MALDI-TOF MS. Nucleic Acids Res. 33(4), e38.

Ehrich, M., Nelson, M. R., Stanssens, P., Zabeau, M., Liloglou, T., Xinarianos, G., Cantor, C. R., Field, J. K. and van den Boom, D. (2005b). Quantitative high-throughput analysis of DNA methylation patterns by base-specific cleavage and mass spectrometry. Proc. Natl. Acad. Sci. USA 102(44), 15785–15790.

Elso, C., Toohey, B., Reid, G. E., Poetter, K., Simpson, R. J. and Foote, S. J. (2002). Mutation detection using mass spectrometric separation of tiny oligonucleotide fragments. Genome Res. 12(9), 1428–1433.

Etzioni, R., Urban, N., Ramsey, S., McIntosh, M., Schwartz, S., Reid, B. *et al.* (2003). The case for early detection. Nat. Rev. Cancer 3(4), 243–252.

Feinberg, A. P. (2001). Cancer epigenetics takes center stage. Proc. Natl. Acad. Sci. USA 98(2), 392–394.

Fenn, J. B., Mann, M., Meng, C. K., Wong, S. F. and Whitehouse, C. M. (1989). Electrospray ionization for mass spectrometry of large biomolecules. Science 246, 246.

Gut, I. G. (2004). DNA analysis by MALDI-TOF mass spectrometry. Hum. Mutat. 23(5), 437–441.

Hahner, S., Ludemann, H. C., Kirpekar, F., Nordhoff, E., Roepstorff, P., Galla, H. J. and Hillenkamp, F. (1997). Matrix-assisted laser desorption/ionization mass spectrometry (MALDI) of endonuclease digests of RNA. Nucleic Acids Res. 25(10), 1957–1964.

Hahner, S., Schneider, A., Ingendoh, A. and Mosner, J. (2000). Analysis of short tandem repeat polymorphisms by electrospray ion trap mass spectrometry. Nucleic Acids Res. 28, 18.

Hartmer, R., Storm, N., Boecker, S., Rodi, C. P., Hillenkamp, F., Jurinke, C. *et al.* (2003). RNase T1 mediated base-specific cleavage and MALDI-TOF MS for high-throughput comparative sequence analysis. Nucleic Acids Res. 31(9), e47.

Herman, J. G., Graff, J. R., Myohanen, S., Nelkin, B. D. and Baylin, S. B. (1996). Methylation-specific PCR: a novel PCR assay for methylation status of CpG islands. Proc. Natl. Acad. Sci. USA 93(18), 9821–9826.

Honisch, C., Raghunathan, A., Cantor, C. R., Palsson, B. O. and van den Boom, D. (2004). High-throughput mutation detection underlying adaptive evolution of *Escherichia coli*-K12. Genome Res. 14(12), 2495–2502.

Jones, P. A. and Baylin, S. B. (2002). The fundamental role of epigenetic events in cancer. Nat. Rev. Genet. 3(6), 415–428.

Jurinke, C., Oeth, P. and Van Den Boom, D. (2004). MALDI-TOF mass spectrometry: a versatile tool for high-performance DNA analysis. Mol. Biotechnol. 26(2), 147–164.

Karas, M. and Hillenkamp, F. (1988). Laser desorption ionization of proteins with molecular masses exceeding 10,000 Daltons. Anal. Chem. 60(20), 2299–2301.

Kartalov, E. P. and Quake, S. R. (2004). Microfluidic device reads up to four consecutive base pairs in DNA sequencing-by-synthesis. Nucleic Acids Res. 32(9), 2873–2879.

Kling, J. (2003). Ultrafast DNA sequencing. Nat. Biotechnol. 21(12), 1425–1427.

Krebs, S., Medugorac, I., Seichter, D. and Forster, M. (2003). RNaseCut: a MALDI mass spectrometry-based method for SNP discovery. Nucleic Acids Res. 31(7), e37.

Lai, E., Riley, J., Purvis, I. and Roses, A. (1998). A 4-Mb high-density single nucleotide polymorphism-based map around human APOE. Genomics 54(1), 31–38.

Laird, P. W. (2003). The power and the promise of DNA methylation markers. Nat. Rev. Cancer 3(4), 253–266.

Lefmann, M., Honisch, C., Bocker, S., Storm, N., von Wintzingerode, F., Schlotelburg, C., Moter, A., van den Boom, D. and Gobel, U. B. (2004). Novel mass spectrometry-based tool for genotypic identification of mycobacteria. J. Clin. Microbiol. 42(1), 339–346.

Li, E. (2002). Chromatin modification and epigenetic reprogramming in mammalian development. Nat. Rev. Genet. 3(9), 662–673.

Loboda, A. V., Ackloo, S. and Chernushevich, I. V. (2003). A high-performance matrix-assisted laser desorption/ionization orthogonal time-of-flight mass spectrometer with collisional cooling. Rapid Commun. Mass Spectrom. 17(22), 2508–2516.

Maiden, M. C., Bygraves, J. A., Feil, E., Morelli, G., Russell, J. E., Urwin, R., Zhang, Q., Zhou, J., Zurth, K., Caugant, D. A., Feavers, I. M., Achtman, M. and Spratt, B. G. (1998). Multilocus sequence typing: a portable approach to the identification of clones within populations of pathogenic microorganisms. Proc. Natl. Acad. Sci. USA 95(6), 3140–3145.

Margulies, M., Egholm, M., Altman, W. E., Attiya, S., Bader, J. S., Bemben, L. A., Berka, J., Braverman, M. S., Chen, Y. J., Chen, Z., Dewell, S. B., Du, L., Fierro, J. M., Gomes, X. V., Godwin, B. C., He, W., Helgesen, S., Ho, C. H., Irzyk, G. P., Jando, S. C., Alenquer, M. L., Jarvie, T. P., Jirage, K. B., Kim, J. B., Knight, J. R., Lanza, J. R., Leamon, J. H., Lefkowitz, S. M., Lei, M., Li, J., Lohman, K. L., Lu, H., Makhijani, V. B., McDade, K. E., McKenna, M. P., Myers, E. W., Nickerson, E., Nobile, J. R., Plant, R., Puc, B. P., Ronan, M. T., Roth, G. T., Sarkis, G. J., Simons, J. F., Simpson, J. W., Srinivasan, M., Tartaro, K. R., Tomasz, A., Vogt, K. A., Volkmer, G. A., Wang, S. H., Wang, Y., Weiner, M. P., Yu, P., Begley, R. F. and Rothberg, J. M. (2005). Genome sequencing in microfabricated high-density picolitre reactors. Nature 437(7057), 376–380.

Pomerantz, S. C., Kowalak, J. A. and McCloskey, J. A. (1993). Determination of oligonucleotide composition from mass spectrometrically measured molecular weight. J. Am. Soc. Mass Spectrom. 4, 204–209.

Pradhan, S., Bacolla, A., Wells, R. D. and Roberts, R. J. (1999). Recombinant human DNA (cytosine-5) methyltransferase. I. Expression, purification, and comparison of *de novo* and maintenance methylation. J. Biol. Chem. 274(46), 33002–33010.

Shchepinov, M. S., Denissenko, M. F., Smylie, K. J., Worl, R. J., Leppin, A. L., Cantor, C. R. and Rodi, C. P. (2001). Matrix-induced fragmentation of P3′-N5′ phosphoramidate-containing DNA: high-throughput MALDI-TOF analysis of genomic sequence polymorphisms. Nucleic Acids Res. 29(18), 3864–3872.

Stanssens, P., Zabeau, M., Meersseman, G., Remes, G., Gansemans, Y., Storm, N., Hartmer, R., Honisch, C., Rodi, C. P., Bocker, S. and van den Boom, D. (2004). High-throughput MALDI-TOF discovery of genomic sequence polymorphisms. Genome Res. 14(1), 126–133.

Szyf, M. and Detich, N. (2001). Regulation of the DNA methylation machinery and its role in cellular transformation. Prog. Nucleic Acid Res. Mol. Biol. 69, 47–79.

Tang, K., Shahgholi, M., Garcia, B. A., Heaney, P. J., Cantor, C. R., Scott, L. G. and Williamson, J. R. (2002). Improvement in the apparent mass resolution of oligonucleotides by using 12C/14N-enriched samples. Anal. Chem. 74(1), 226–231.

Tost, J. and Gut, I. G. (2005). Genotyping single nucleotide polymorphisms by MALDI mass spectrometry in clinical applications. Clin. Biochem. 38(4), 335–350.

Tsou, J. A., Hagen, J. A., Carpenter, C. L. and Laird-Offringa, I. A. (2002). DNA methylation analysis: a powerful new tool for lung cancer diagnosis. Oncogene 21(35), 5450–5461.

Urwin, R. and Maiden, M. C. (2003). Multi-locus sequence typing: a tool for global epidemiology. Trends Microbiol. 11(10), 479–487.

von Wintzingerode, F., Bocker, S., Schlotelburg, C., Chiu, N. H., Storm, N., Jurinke, C., Cantor, C. R., Gobel, U. B. and van den Boom, D. (2002). Base-specific fragmentation of amplified 16S rRNA genes analyzed by mass spectrometry: a tool for rapid bacterial identification. Proc. Natl. Acad. Sci. USA 99(10), 7039–7044.

Walsh, C. P., Chaillet, J. R. and Bestor, T. H. (1998). Transcription of IAP endogenous retroviruses is constrained by cytosine methylation. Nat. Genet. 20(2), 116–117.

Woese, C. R. (1987). Bacterial evolution. Microbiol. Rev. 51(2), 221–271.

Chapter 4

Advances in Dye-Nucleotide Conjugate Chemistry for DNA Sequencing

Shiv Kumar and Carl W. Fuller

GE Healthcare, 800 Centennial Avenue, Piscataway, NJ 08855, USA

Contents

Abstract

In the last decade, the efficiency of DNA sequencing has increased dramatically, making it possible to sequence the entire human genome and genomes of many other organisms. This advance featured the development of sensitive fluorescence resonance energy transfer (FRET)-based dye terminators, DNA polymerases that incorporate these terminators very efficiently and high-throughput capillary electrophoresis instruments. Further improvements may be obtained using improved FRET dyes and charge-modified nucleotides for "direct-loading" of sequencing reaction products. The chemistry of FRET-based dye-nucleotide terminators, charge-modified nucleotides (negative as well as positive) and terminal phosphate-labeled nucleotides for DNA sequencing is discussed in this review.

1. INTRODUCTION

Successful systems for DNA sequencing were first described in 1977 by Maxam and Gilbert (1977), and independently by Frederick Sanger and colleagues

PERSPECTIVES IN BIOANALYSIS, VOLUME 2
ISSN 1871-0069 DOI: 10.1016/S1871-0069(06)02004-0

working at Cambridge University (Sanger *et al.*, 1977). The Maxam–Gilbert sequencing method is based on the cleavage of DNA fragments by nucleotide-specific chemical reagents. The DNA fragment to be sequenced is first end-labeled with ^{32}P or ^{35}S at the 5′-end and then four or more different chemical cleavage reactions are carried out, resulting in the cleavage of the DNA at a particular base or bases (G, A, T or C). The cleaved fragments are then loaded on the sequencing gel in four separate lanes and separated by size. The sequence information is deduced from the pattern of band sizes beginning with the labeling site. In contrast, the Sanger or chain termination method relies on the template-directed synthesis of a new, labeled DNA strand by a DNA polymerase using a DNA template and four deoxynucleoside-5′-triphosphates (dNTPs) and a dideoxynucleoside-5′-triphosphate (ddNTP) that acts as a chain terminator. The synthesis is initiated where the primer is annealed to the template. Synthesis continues on each template molecule until a ddNTP is incorporated. Four separate reactions are performed, each with a different ddNTP (G, A, T or C). This results in a population of molecules with a common 5′-end, and 3′-ends with the same ddNTP base of different sizes depending on the site at which ddNTP was incorporated. In the original description of the method, these were radioactively tagged using either radiolabeled dNTPs or labeled primer. The reaction products are then separated by size in four separate lanes of a high-resolution denaturing polyacrylamide gel and detected by autoradiography, exactly like the procedure used for Maxam–Gilbert sequencing.

For detection of sequencing fragments, one of the components, the primer, a dNTP or the terminator (ddNTP), must be labeled. Fluorescent sequencing has become the preferred method because of ease of use and for safety reasons. The Sanger sequencing method also rapidly became the method of choice because it does not require distinct complex chemical manipulations, and it is easily reproducible. In current practice, there are two dideoxy sequencing methods, dye-primer and dye-terminator. In the case of primer sequencing, the fluorescent dye is attached at the 5′-end of the primer. Four separate extension–termination reactions are carried out. Each reaction contains four dNTPs and a single ddNTP. Four distinct dye primers corresponding to each ddNTP are used, each labeled with different fluorescent dye. After the extension reaction, the four reaction products are pooled and the labeled oligonucleotide fragments are separated by gel electrophoresis and the sequence is automatically determined from the size and color information collected by the instrument. For dye-terminator sequencing, the primer is unlabeled and a distinct fluorescent dye is attached to each of the ddNTPs. Because the dyes are spectrally resolvable from each other, they can be mixed together and sequencing reactions run in a single tube. After the extension reaction the products are separated and read essentially the same way as for dye-primer sequencing. The dye-terminator sequencing method has superseded dye-primer method largely because it is simple to run, it is universally adaptable to different sequencing template and primer combinations, and it is free of artifacts caused by ddNTP-independent polymerase stops, as the detectible chain termination depends upon terminator incorporation.

2. FLUORESCENT DNA SEQUENCING

2.1. Single dye-labeled primers and terminators

The fluorescent dyes most commonly used for DNA sequencing are xanthine dyes (fluoresceins and rhodamines). These dyes are mostly excited by an Argon-ion laser line at 488 nm, which is used in most of the common DNA sequencers such as the *ABI PRISM®373* and *ABI PRISM®377* (Applied BioSystems Corp.), and capillary sequencers *MegaBACE™ 1000* and *MegaBACE™ 4000* (Amersham-GE Healthcare) and *ABI PRISM® 3100 and 3700*. The other classes of dyes used for sequencing are Cyanine dyes (Cy5.0 and Cy5.5) which were used on *ALFexpress™* and *SEQ 4 × 4™* (Amersham Biosciences now GE Healthcare), *Clipper™* or *Long-Read Tower™* (Visible Genetics now Bayer Diagnostics). Besides these reagents, Bodipy dyes (Metzker *et al.*, 1996) and IR dyes (*IRDye700™* and *IRDye800™*) have also been used (Middendorf *et al.*, 1992; Roemer *et al.*, 2000). The structures of these various dyes are shown in Figure 1. In 1986, Smith and colleagues reported the first fluorescent automated DNA sequencing using four different fluorescent dye-labeled primers (Smith *et al.*, 1986). For dye-primer sequencing, the dyes are usually attached at the 5′-end of the sequencing primer. In contrast, for dye-terminator sequencing, the dyes are attached at the C-5 position of pyrimidines (ddC and ddU) and C-7 position of 7-deazapurines (7-deaza-ddA and 7-deaza-ddG) via a propargylamino linker. Fluorescently dye-labeled ddNTPs were first introduced by Prober and colleagues working at Dupont (Prober *et al.*, 1987). The fluorescent dyes used were four different succinyl fluoresceins. Similar terminators were later developed by Bergot and colleagues at Applied Biosystems using four different rhodamine dyes in place of fluoresceins (Bergot *et al.*, 1994). The structures of these dye-terminators are given in Figure 2.

These dye-terminators were used with *Taq* DNA polymerase, and dITP was used in place of dGTP to relieve the compression artifacts upon electrophoresis. A second set of fluorescein terminators was also developed for use with T7 DNA polymerase, and α-thio dNTPs were used in place of regular dNTPs to relieve compression artifacts (Lee *et al.*, 1992). There were advantages to both systems. While the use of T7 DNA polymerase and terminators generated better quality peak patterns, sequencing with *Taq* terminators was more sensitive because amplification by cycle sequencing could be used with lower amounts of template.

DNA sequence bands generated by dye-labeled terminators do not all have the same intensity. In fact, the intensities vary predictably in a sequence-dependent fashion. The type of dye used, the charge on the dye, the linker arm length between the dye and the nucleotide, and the use of Manganese and the DNA polymerase employed all have effects on the uniformity of band intensities and on the quality of sequence data. We have studied the effect of linker between the nucleotide and dye. We have observed that a linker with more than 10 atoms gave more uniform bands than shorter arm linkers (Kumar *et al.*, 2005a). More even bands have also been observed when these terminators are used with manganese in the reaction mixture (Fuller *et al.*, 1999). When used

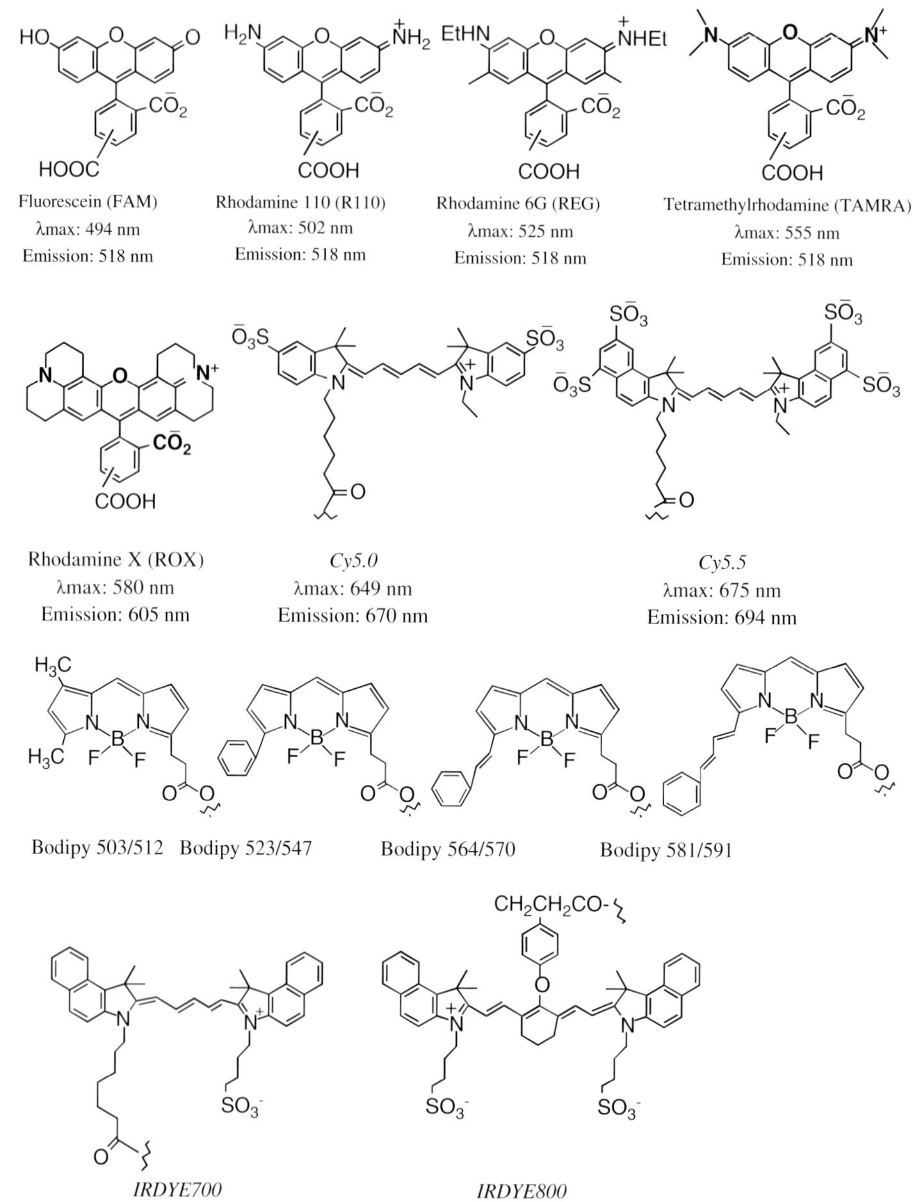

Fig. 1. Structures of fluorescent dyes used for DNA sequencing.

with *Thermo Sequenase*™ DNA polymerase, single color Cyanine dye-labeled terminators also produce uniform bands similar to dye-primer sequencing methods (Kumar *et al.*, 1999; Duthie *et al.*, 2002). A preferred set of Cyanine dye terminators includes two pyrimidines with 17-atom linkers, and two purines with 10-atom linkers between the dye and the nucleotide (Figure 3).

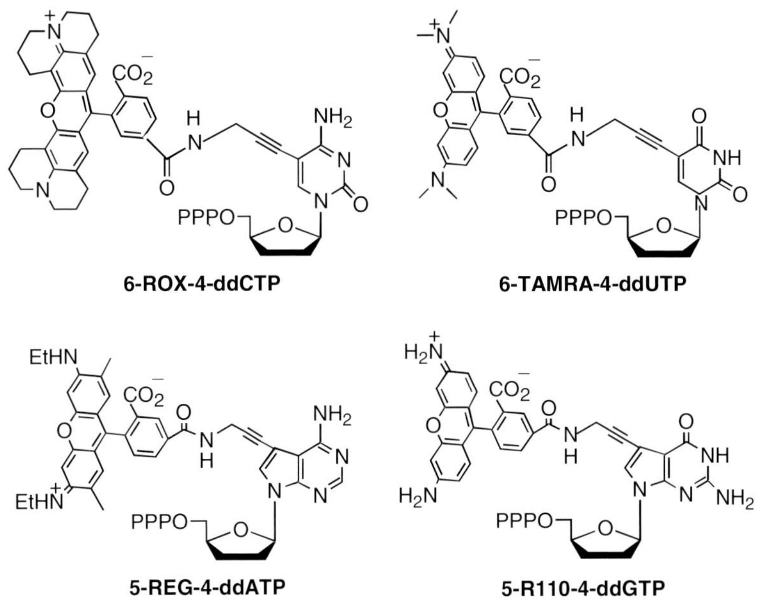

Fig. 2. Structures of rhodamine terminators first developed by Applied Biosystems.

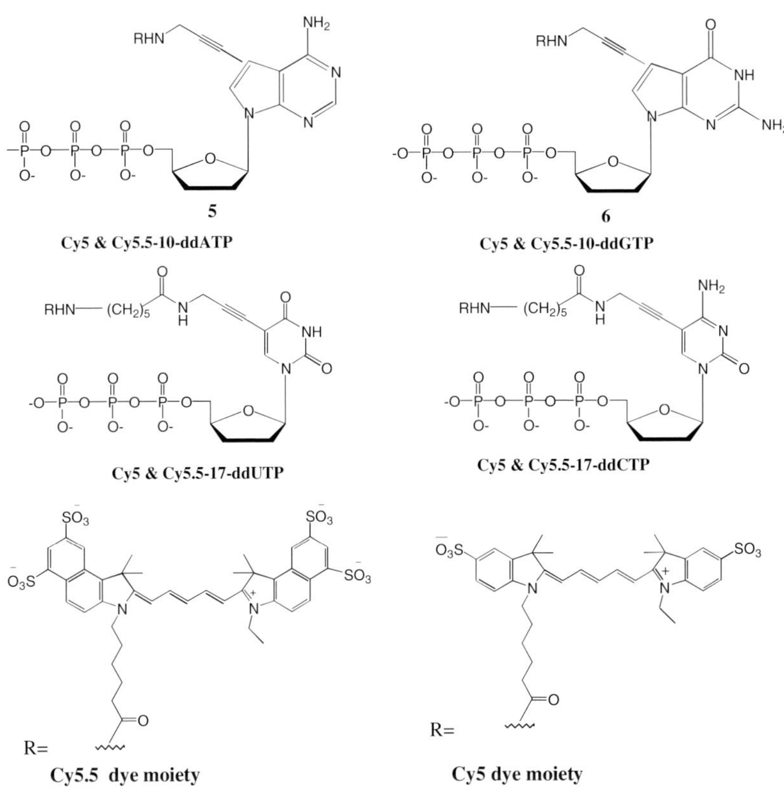

Fig. 3. Structures of Cy5.0 and Cy5.5 dye-labeled terminators.

The intensities of bands generated using *Taq* DNA polymerase and the original rhodamine terminators vary by more than 10-fold. The development of *Thermo Sequenase* DNA polymerase, with a simple phenylalanine (F) to tyrosine (Y) amino acid change in *Taq* DNA polymerase reduces discrimination against ddNTP incorporation by more than 1,000-fold (Tabor and Richardson, 1995). A DNA polymerase with this F to Y mutation generates more uniform band intensities with unlabeled ddNTPs, but little improvement is seen when using the original rhodamine dye-labeled ddNTPs (Reeve and Fuller, 1995; Vander Horn *et al.*, 1997). Uniform band intensities have been achieved using Cyanine dye terminators (Duthie *et al.*, 2002), dichlororhodamine terminators (Rosenblum *et al.*, 1997) or energy transfer (ET) dye-labeled terminators (Lee *et al.*, 1997; Kumar *et al.*, 2004c).

2.2. Fluorescence resonance energy transfer (FRET) based primers and terminators

The rhodamine dyes (R110, REG, TAMRA and ROX) that are currently the most commonly used dyes for DNA sequencing are usually excited by an argon-ion laser at 488 nm.These dyes absorb and emit light optimally at different wavelengths. The ideal dye set should have high molecular absorbance at a common excitation wavelength and exhibit strong, well-separated fluorescence emission with high quantum yield. This criterion is not met by the rhodamine dye set as the absorbance of TAMRA and ROX are only 1/8 and 1/10th respectively of that of fluorescein at 488 nm (Figure 4).

In order to compensate these differences in dye absorbance, one must either use more template DNA in the reactions involving TAMRA and ROX dye-labeled primers, or use more TAMRA and ROX dye-labeled terminators in the dye-terminator sequencing reactions, or use more than one laser to excite all the

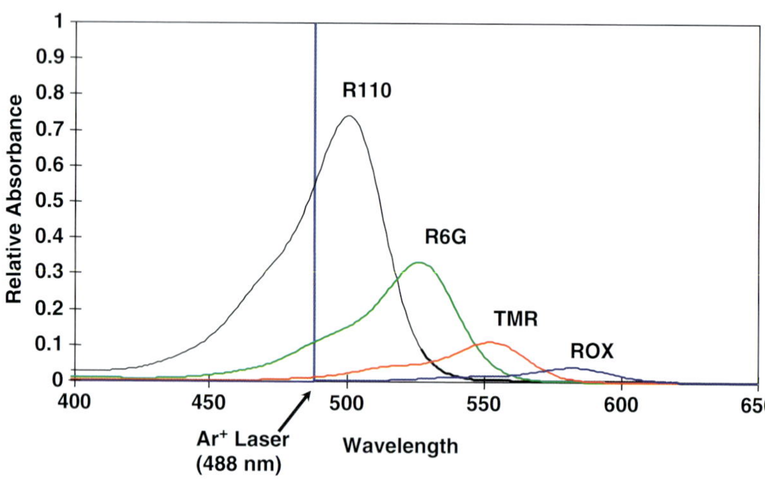

Fig. 4. Relative absorbance of single rhodamine dye terminators at 488 nm.

dyes maximally at their excitation wavelengths. These approaches all add complexity and expense to the sequencing apparatus and associated method. Mathies and his co-workers (Ju *et al.*, 1995a, b) exploited FRET or ET to overcome the constraints imposed by the use of single laser with four dyes of different absorption wavelengths. FRET is the physical process by which energy is transferred non-radiatively from an excited chromophore (donor) to another chromophore (acceptor) by means of long-range dipole–dipole coupling. The energy transfer efficiency and the ratio of acceptor-to-donor emission can be influenced by small changes in the distance between the donor and acceptor, since Föster established that the efficiency of FRET is proportional to the inverse sixth power of the distance between the two chromophores (Förster, 1965).

Mathies and his research group first devised primers in which fluorescein (as a donor dye) was attached at the 5′-end of the oligonucleotide, and the acceptor rhodamine dye (R110, R6G, TAMRA or ROX) was attached some bases away from the 5′-end through a modified Thymidine (T*) nucleoside base which carries a primary amino group for the covalent coupling of the dye (Ju *et al.*, 1995a, b).

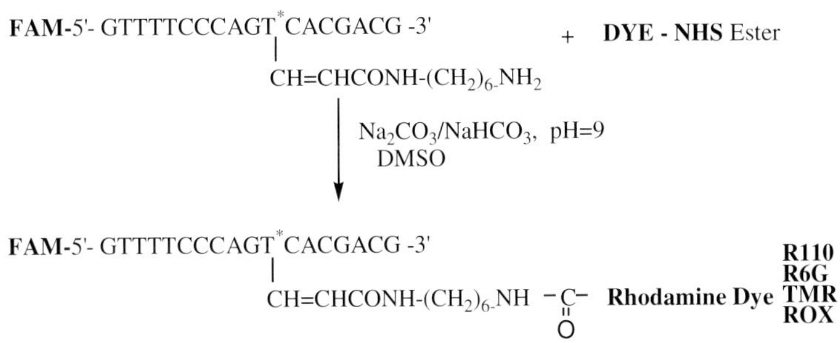

While Mathies spaced the dyes only up to 10 nucleotides apart, experimentation with longer oligo dT primers (Vander Horn *et al.*, unpublished results) showed that optimum distance for efficient ET was 8–10 bases (Figure 5). The synthesis of ET primers is straightforward as most of the synthesis is carried out on a DNA synthesizer, and only the conjugation of the acceptor dye is carried out post synthesis. Several primers set for sequencing are available from Amersham Biosciences (now GE Healthcare) as *DYEnamic*™ ET primers. Mathies' research group also developed ET primers, based on using a "universal spacer" (S) approach and a universal ET cassette labeling reagent (Figure 6) for thiol coupling (Ju *et al.*, 1996; Hung *et al.*, 1996, 1997; Berti *et al.*, 2001).

3. ENERGY TRANSFER DYE TERMINATORS

The approaches described above are good for producing ET primers, but are not suitable for generating ET terminators, since attaching a ddNTP to an

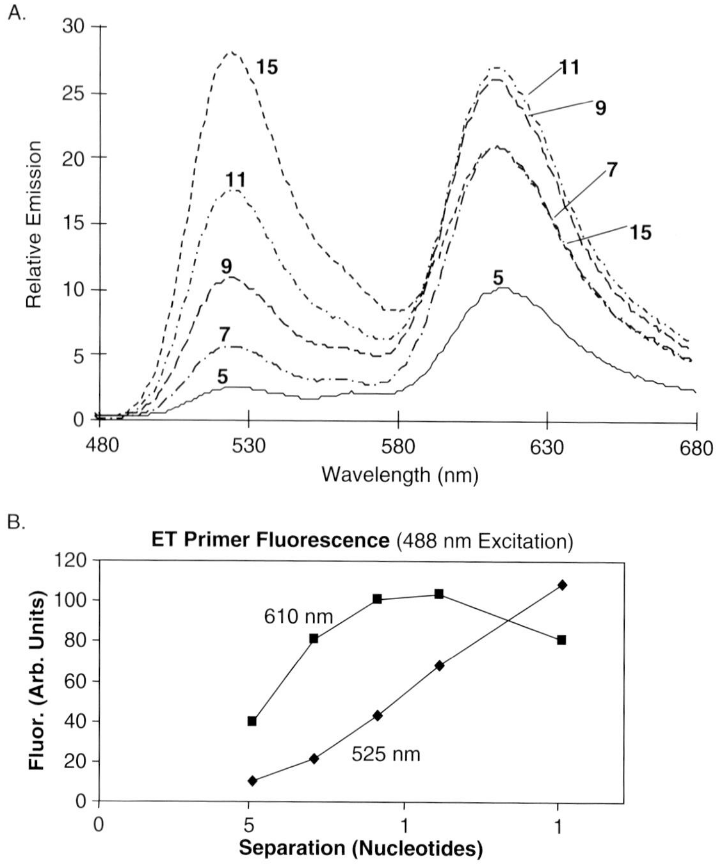

Fig. 5. The effect of spacing between the donor and acceptor dyes on fluorescence emission spectra. The oligonucleotides sequence 5′-FAM-TTTTTTTTTTTTTTTGTAAAA CGACGGCCAGT was made with FAM at the 5′ end. One of the T nucleotides at position 5, 7, 9, 11 or 15 was an amino-modified nucleotide to which X-Rhodamine (ROX) was attached. (A) Fluorescence emission spectra normalized for O.D. at 490 nm.(B) Emission as a function of separation distance between the dyes. Donor dye quenching is strongest at the shortest separation, but acceptor dye emission diminishes with separation either shorter or longer than the optimum of approximately 10 nucleotides.

oligonucleotide primer is neither simple nor straightforward. Our approach for the synthesis of ET terminators relied on the use of amino acids as suitable linkers. Initial attempts at using lysine and cysteine as possible linkers failed to produce efficient ET, apparently due to the quenching of the donor dye. Since aromatic amino acids (such as phenylalanine) may also be modified to contain two amino groups to attach the donor and acceptor dyes along with a carboxylic acid group to attach an amino-modified ddNTP, we undertook this approach for the generation of ET cassettes and terminators. A number of different aromatic, trifunctional amino acids were tried as linkers (Figure 7), and based on the ET efficiency and sequencing results, 4-propargylamino-phenylalanine

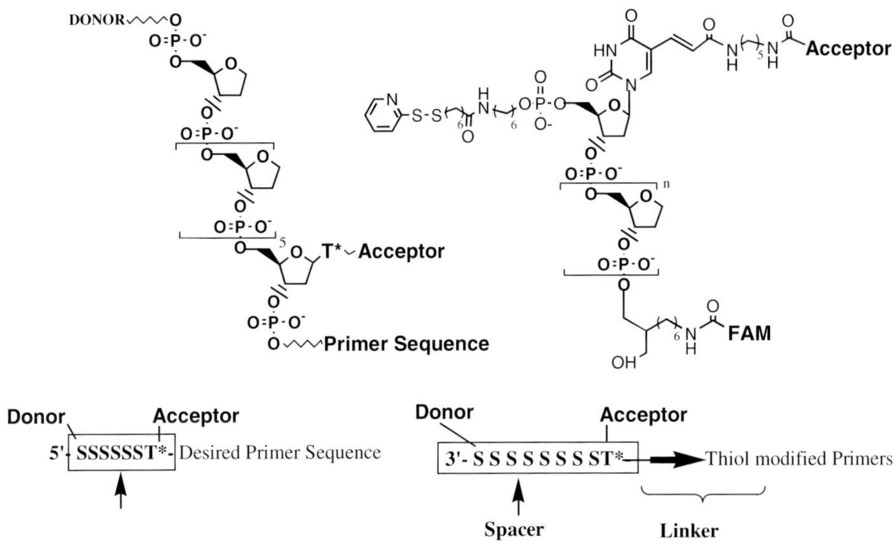

Fig. 6. Universal ET cassette labeled primer synthesis approach and thiol coupling in disulfide exchange reaction approach for ET primer synthesis.

was selected as the suitable linker for generating ET terminators (Rao *et al.*, 2001; Nampalli *et al.*, 2000, 2001, 2002; Kumar *et al.*, 2001b, 2004c).

Another synthetic approach employed to produce ET cassettes and terminators is shown diagrammatically in Figure 8. The commercially available *t*-Boc-L-4-iodophenylalanine was reacted with TFA-protected propargylamine in the presence of $(Ph_3P)_4Pd(0)/CuI/Et_3N/DMF$ to produce the linker suitable to attach the donor and acceptor dyes at the orthogonally protected amino groups. The attachment of the common donor dye (fluorescein) to one of the amino groups, and acceptor dyes (R110, REG, TAMRA and ROX) to the other amino groups provided four ET cassettes in high yield. These ET cassettes are suitable for attachment to amino-modified primers or ddNTPs after activation of the carboxylic acid group. The conjugation of ddNTP to the ET cassette was carried out by *in situ* activation of the carboxylic group to the NHS ester followed by addition of an aqueous solution of amino-modified ddNTP (Nampalli *et al.*, 2000). Out of the possible 64 combinations of ET terminators, a number of terminators were synthesized and tested in DNA sequencing reactions with a variety of DNA templates and *Thermo Sequenase* DNA polymerase.

Based on the sequencing band uniformity, read lengths and an absence of sequencing artifacts such as dye-induced mobility problems, an optimized four-color ET terminator set of FAM-R110-11-ddGTP, FAM-REG-11-ddUTP, FAM-TMR-11-ddATP and FAM-ROX-11-ddCTP (Figure 9A) was selected, which is now available from GE Healthcare as *DYEnamic* ET terminators. The ET terminators are brighter than the single dye-labeled terminators, particularly in the case of the TAMRA- and ROX-labeled terminators. The FAM-ROX-labeled ET terminator is more than 18-fold brighter than the single ROX-labeled terminator (Figure 9B).

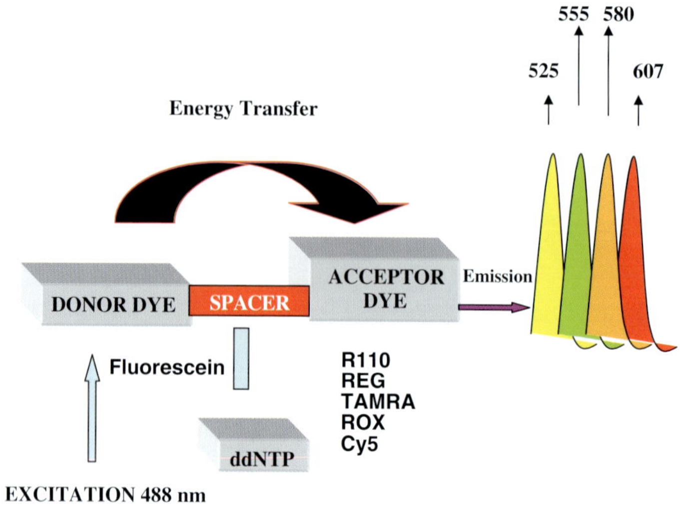

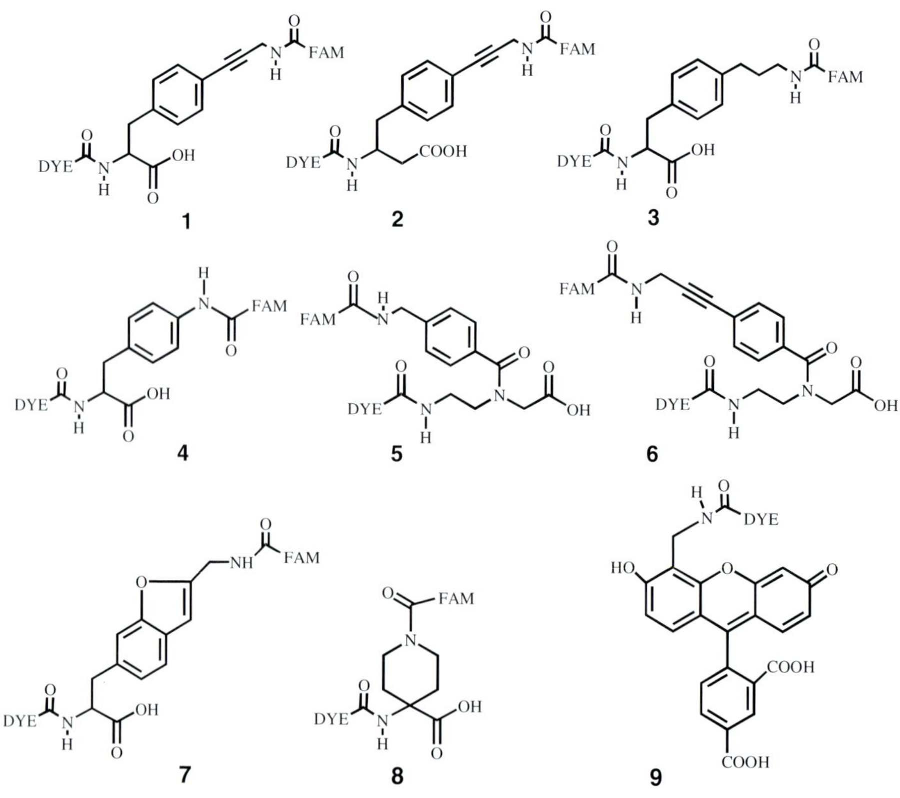

Fig. 7. Trifunctional linkers for the synthesis of ET dyes and terminators.

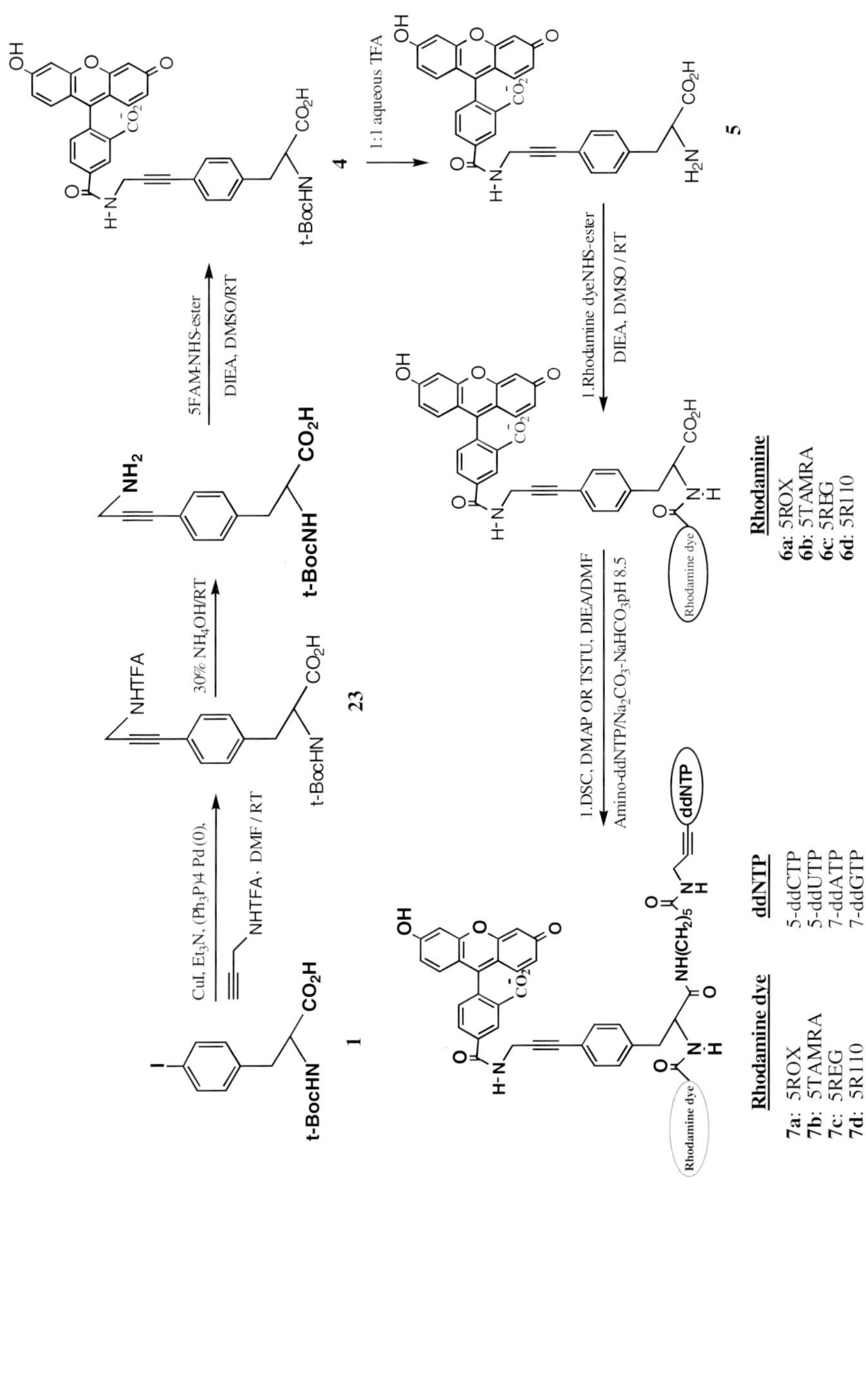

Fig. 8. Synthetic scheme for generation of FRET cassettes and terminators.

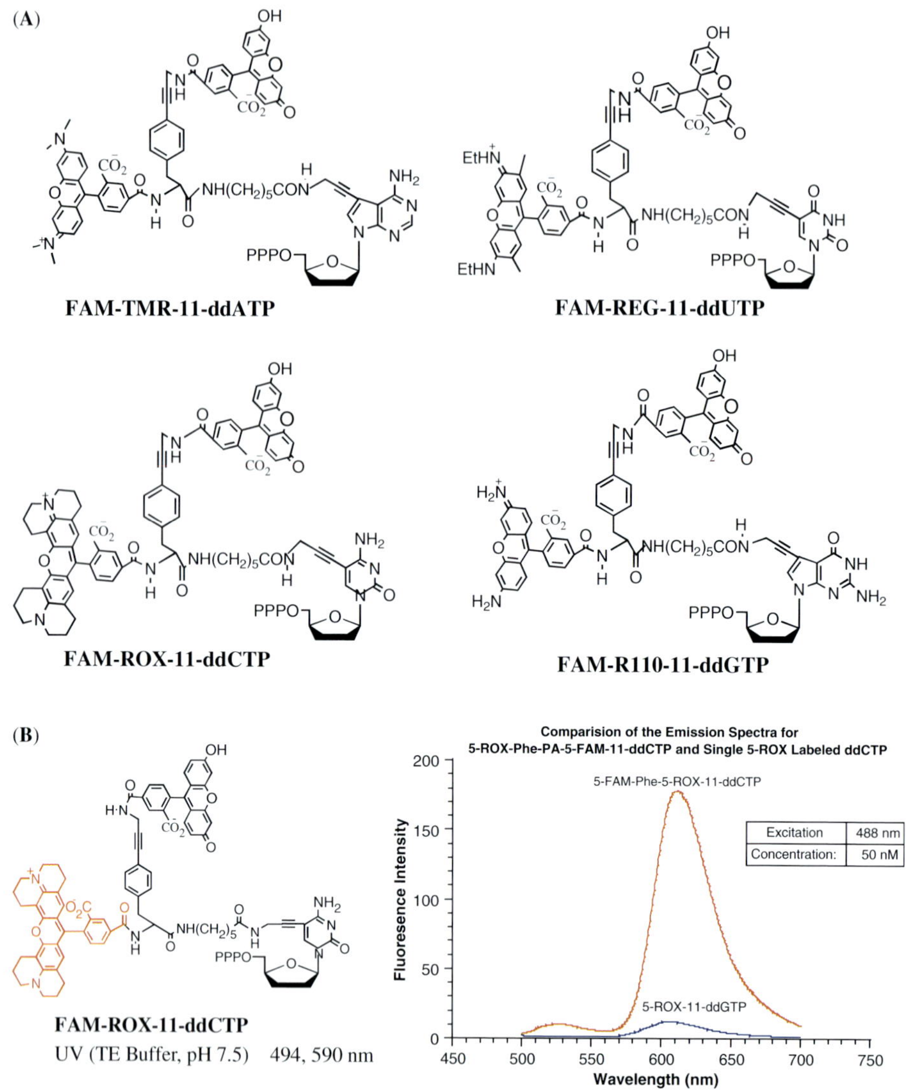

Fig. 9. (A) A commercially available four-color set of ET terminators. (B) Comparisons of brightness of single dye-labeled (ROX-11-ddCTP) and FRET-based (FAM-ROX-11-ddCTP).

Lee and colleagues working at Applied Biosystems used a different approach, which is based on the use of 4-aminomethyl benzoic acid as the linker (Lee *et al.*, 1999). The donor dye (4′-aminomethyl-fluorescein-5(6)-carboxylic acid) is attached to the carboxylic acid of the linker, and the amino group of the linker is attached to the acceptor dye (dichloro-rhodamines). The 5(6)-carboxylic group of the donor dye is used to attach the primer or amino-modified ddNTP. These terminators are available from Applied Biosystems as *Big Dye* terminators. The general structure of the Big Dye terminators is given below.

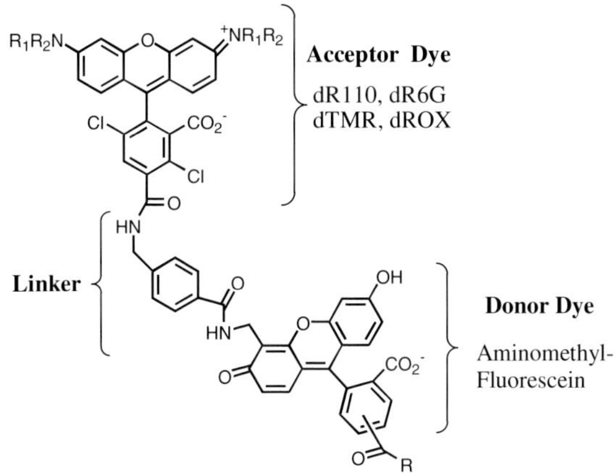

R= Primer Sequence or a Dideoxynucleotide (Terminator)

In an effort to develop yet better and brighter ET cassettes and terminators which have higher ET efficiency than the commercially available *DYEnamic* ET terminators, we have undertaken the synthesis of cassettes and terminators derived from other trifunctional amino acid-based linkers (Figure 10). The ET efficiencies of terminators derived from *p*-amino-phenylalanine (pAPhe) and piperidinyl-1,1-amino carboxylic acid (Pip) were found to be higher than that of the terminators derived from 4-aminopropargyl-L-phenylalanine (Kumar, 2002; Kumar *et al.*, 2004a). Thus, the ET efficiency of FAM-pAPhe-ROX-11-ddCTP (36%) and FAM-Pip-Rox-11-ddCTP (40%) is better than the FAM-Phe-ROX-11-ddCTP (19%).

Another set of terminators which are much brighter than any commercially available terminators and provide uniform band intensities are derived from 4′(5′)-aminomethyl fluorescein-5(6)-carboxylic acid, which serves as both donor dye and linker. This terminator set has two single dye-labeled terminators (R110 and REG derived) and two ET terminators (TAMRA and ROX). Two single dye-labeled terminators were selected because they absorb strongly at the laser wavelength and are already quite as bright as the ET terminators, although the electrophoretic mobility differences between them require software

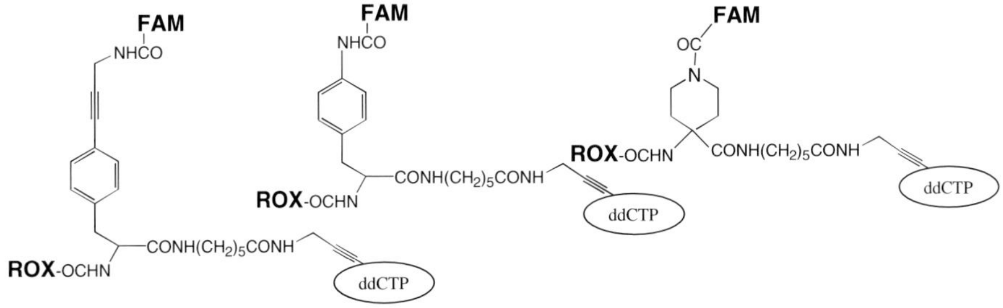

Fig. 10. Percentage energy transfer (PET).

correction (Kumar *et al.*, 2004b). The structures of this terminator set are given in Figure 11.

The comparison of uniformity of sequencing bands generated using *DYEnamic* ET terminators (GE Healthcare), BigDye v.2 and BigDye v.3 terminators (Applied Biosystems), and these new terminators is shown in Figure 12. It is clear that this new terminator set provides the best uniformity of sequencing bands compared to other terminator sequencing sets. These *DYEmamic* ET terminators are incorporated into the nascent DNA with a uniformity and efficiency superior to other dye terminator sets, and this feature in combination with the matched fluorescent output of the dye set results in superior uniformity of sequencing bands.

The sequencing data using these terminators and the MegaBACE 1000 DNA analysis system are shown in Figure 13.

In conclusion, the peak height variance using different dye primers, terminators and DNA polymerases is summarized in Figure 14. The values reported are the variance of the normalized peak heights. Sequencing reactions using dye-primers and T7 DNA polymerase had the best peak height variation with a value of 0.07. Peak height variation for the Cy5.5 terminators was found to be considerably better (~0.3) than the values reported for the four-color dye terminators currently used in the majority of high-throughput sequencing facilities.

3.1. Charged terminators for "direct-load" DNA sequencing

During the thermal cycling of the DNA sequencing reactions, some of the dye-labeled ddNTPs (terminators) are not used up, and another small fraction of the dye terminators degrades to generate dye-labeled by-products (e.g. dye-labeled mono- and diphosphate nucleotides). All of these migrate in the sequencing electrophoresis media among the labeled sequencing fragments, and these contaminants must be removed to avoid interfering with the interpretation of the full range of sequence. If the sequencing reaction products are directly loaded on slab gels or capillary sequencers, these by-products appear as dye-blobs on the electropherogram. Removal of these labeled, interfering compounds is generally addressed by ethanol precipitation of the sequencing products, or by size fractionation by passing the reaction products through a gel filtration column. However, these techniques are variable, and extreme care must be taken during ethanol precipitation to ensure the complete removal of unreacted and degraded nucleotides. Further, if conditions are not properly optimized, it is possible to remove the shorter Sanger sequencing fragments during ethanol precipitation, destroying the initial portion of the sequence result. In a high-throughput sequencing environment, this process is not only laborious and time consuming but also expensive. If these purification step(s) could be either eliminated or simplified, it would make the dye-terminator sequencing process more efficient, reliable and robust.

We hypothesized that this problem could be resolved if the dideoxynucleotide terminators could be changed to either have increased *negative charges* so that the labeled by-products move faster in the sequencing gel than the actual

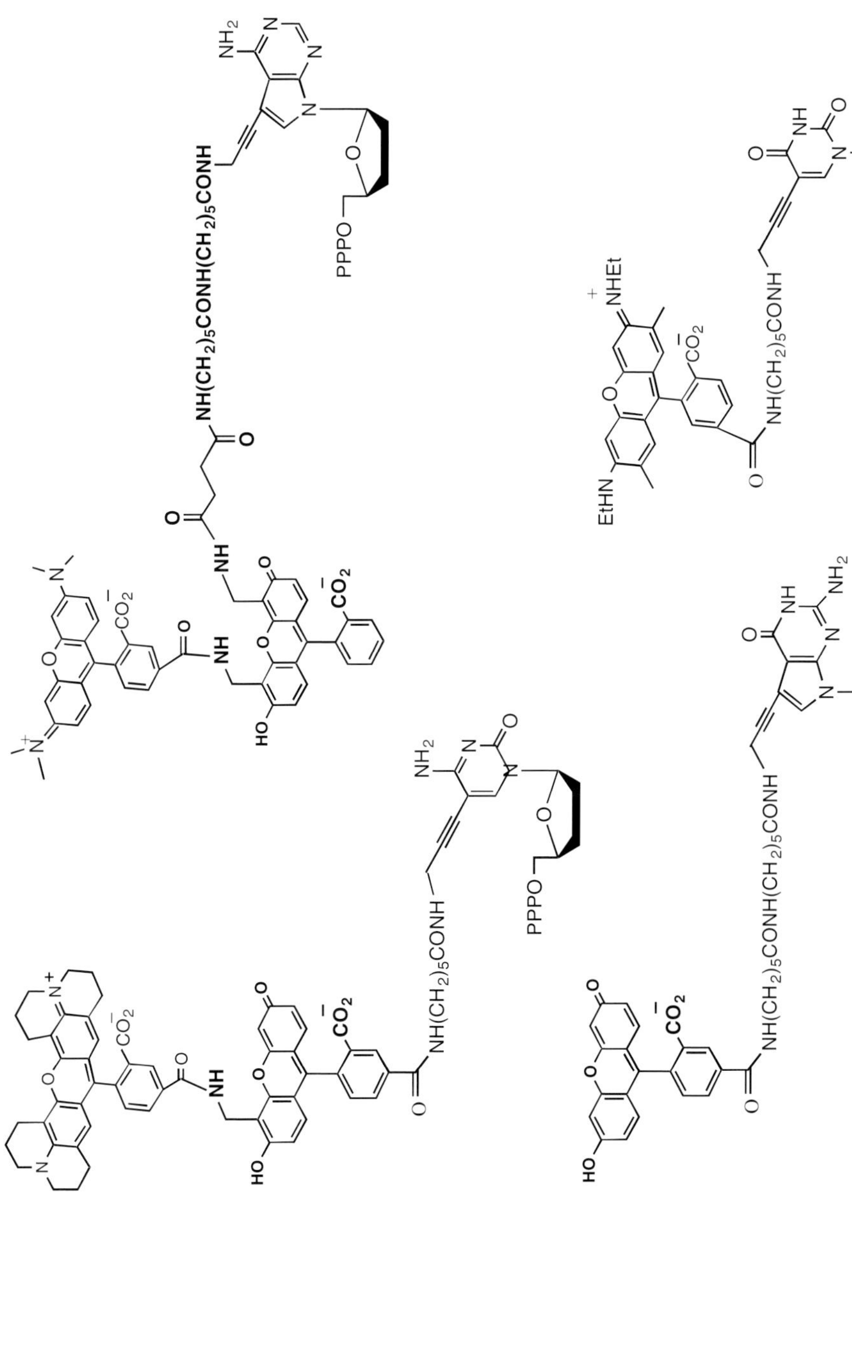

Fig. 11. Structures of newly developed two single dye terminators and two ET terminators for high-throughput DNA sequencing.

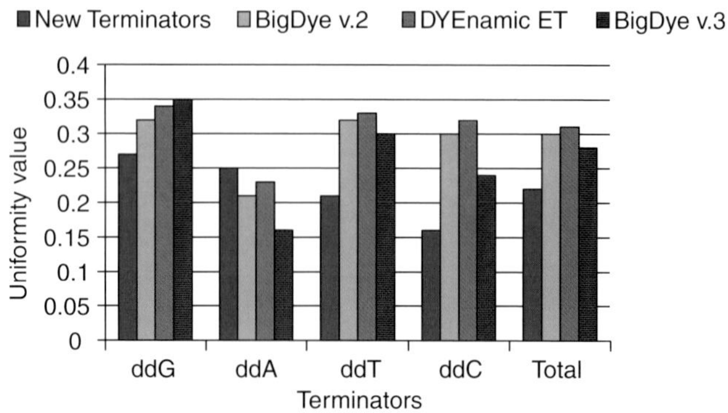

Fig. 12. Uniformity of sequencing bands using new terminators and their comparison with the existing commercially available terminators (*DYEnamic* ET terminators and *Big Dye* v.2 and v.3 terminators).

Sanger sequencing fragments, or have enough *positive charge* so that labeled by-products do not move in the gel, or move backward. To test these hypotheses, we have synthesized three series of novel dye-labeled ddNTPs containing linkers with either negative charges or positive charges or a neutral moiety that becomes positively charged under sequencing gel conditions (Kumar *et al.*, 2001a). These linkers connect the nucleobase and fluorescent label. The synthesis of these charged terminators, their activity with DNA polymerases (Davis *et al.*, 2005) and sequencing potential for direct-load sequencing applications are described below.

3.2. Negatively charged terminators

Initially, dye-labeled terminators with one, two or three negative charges were synthesized to determine how many charges are required to alter the electrophoretic mobility of the unreacted nucleotides and their breakdown products to a point where they migrate faster than the desired sequence information. The negatively charged amino acid (α-sulfo-β-alanine) was found to be ideal linker as no protection/deprotection was required, and a number of these amino acids can be easily linked together using routine peptide chemistry. A series of fluorescein-labeled 2′,3′-dideoxycytidine-5′-triphosphates with various number of α-sulfo-β-alanine moieties in the linker arm were synthesized. It was found that a total of three negative charges (two on α-sulfo-β-alanine moieties and one on fluorescein) are required (Figure 15) to cause the unreacted nucleotides and their degradation products to migrate ahead of the sequence fragment information (Finn *et al.*, 2002).

After determining that three negative charges are required to directly load the sequencing reactions, a four-color set of (−3) charged single dye-labeled terminators and ET dye-terminators was synthesized (Finn *et al.*, 2002).

These terminators were tested with *Thermo Sequenase* DNA polymerase and with other mutant DNA polymerases for their incorporation efficiency.

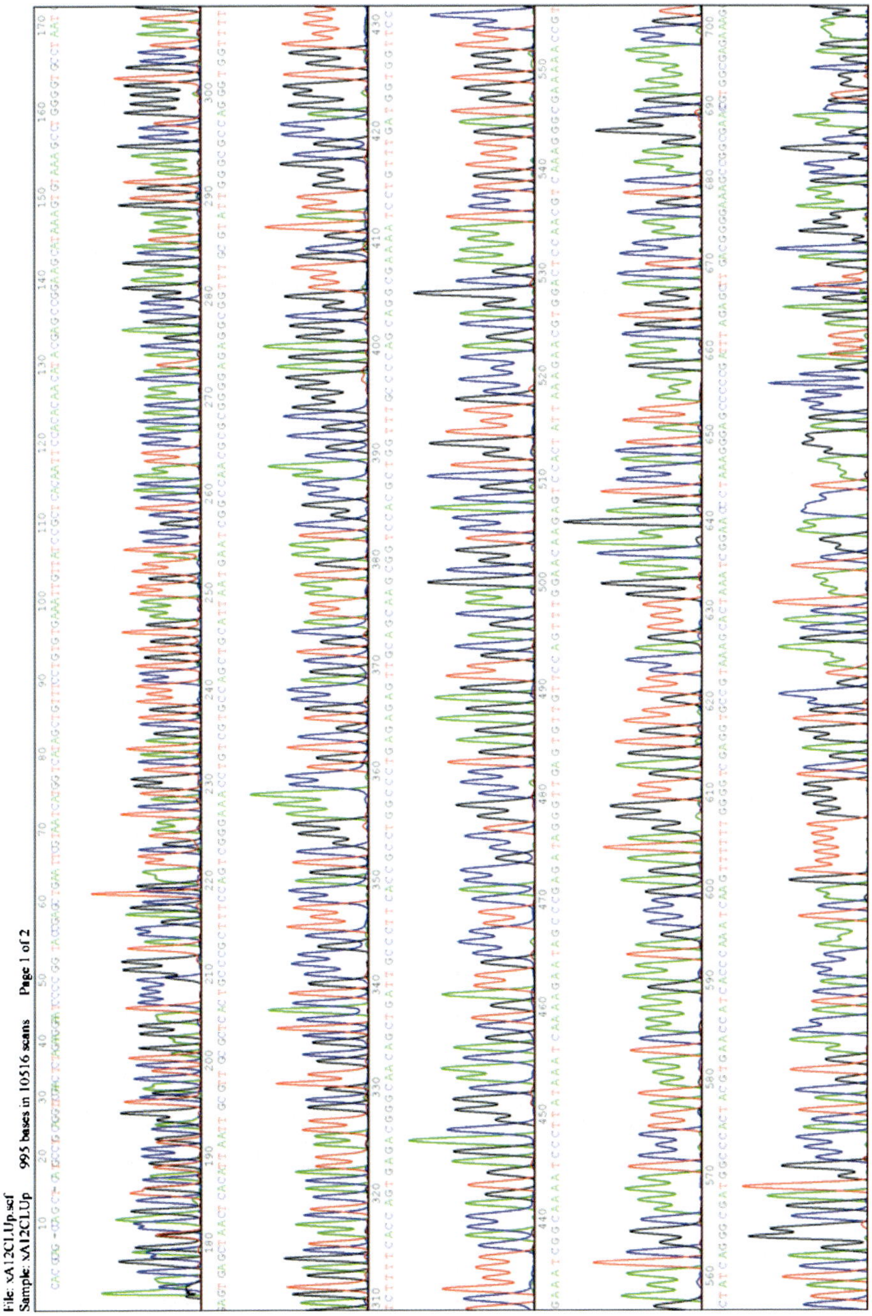

Fig. 13. Sequencing data from MegaBACE 1000 DNA analysis system using dye terminators.

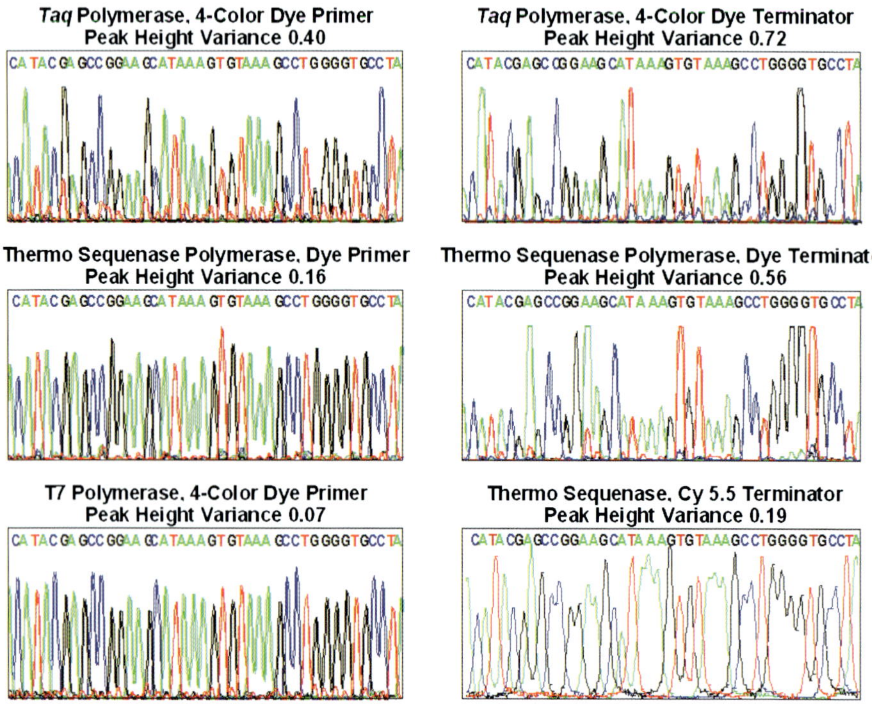

Fig. 14. Visual representation of the peak height variation observed when using different dye terminators and DNA polymerases.

An interesting correlation between the number of charges and reactivity ratio was observed. Increased negative charge resulted in decreased incorporation efficiency. Although, a mutant *Taq* DNA polymerase (E681R) showed less discrimination against the negatively charged nucleotides than Thermo Sequenase II DNA polymerase, their reactivity was still relatively poor (10:1; ddNTP:dNTP) in comparison with *DYEnamic* ET terminators (200:1). Increasing the linker arm length between the charged moiety and nucleobase partially compensates for this effect, improving the incorporation reactivity to about 30:1 (Finn *et al.*, 2002) (Figure 16). A poorly incorporated terminator is obviously not ideal for DNA sequencing, because large quantities of dye-labeled terminator are then required to generate the desired read lengths. In spite of this limitation, we have successfully used these terminators for direct-load sequencing reactions with read lengths in excess of 600 bases (Finn *et al.*, 2002).

3.3. Lysine-derived charge terminators

The synthesis of terminators with oligo-lysine as linker was based on the assumption that the amino group of lysine would acquire positive charge at the pH of electrophoretic separation. The ε-amino group of lysine has a pK_a of

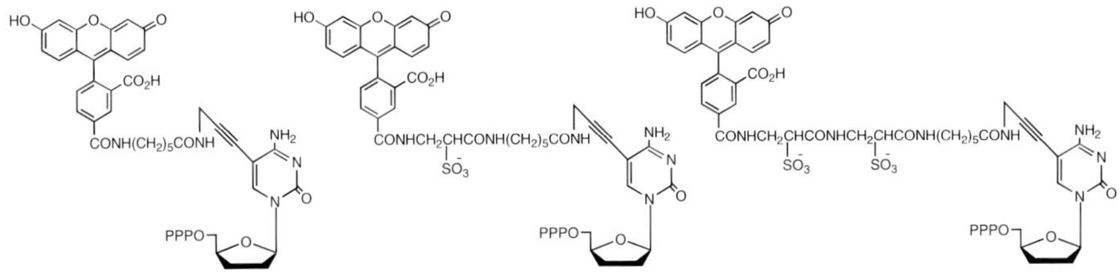

FAM(-1)-11-ddCTP **FAM(-2)-11-ddCTP** **FAM(-3)-11-ddCTP**

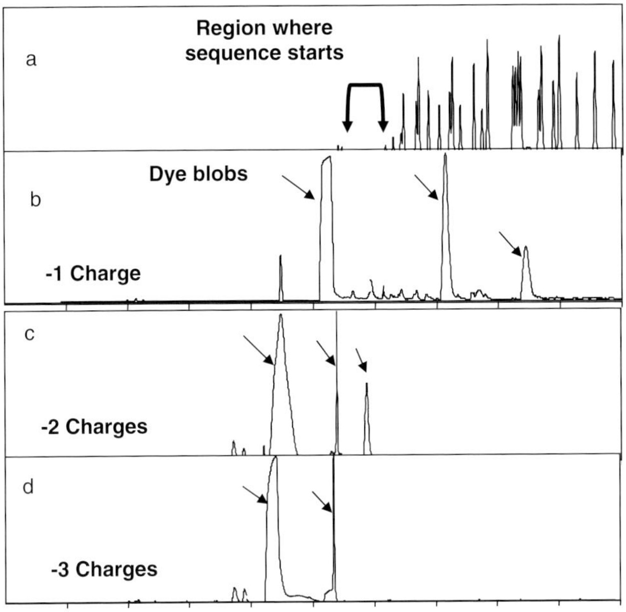

Fig. 15. Electropherogram using −1, −2 and −3 charge modified ddCTP. The arrow in the panel indicates dye-blobs. It is clear from this figure that -3 charges are required to move dye-blobs ahead of the starting sequence.

approximately 10, so it should be protonated at the pH of electrophoretic separation, typically pH 7.8–8.5. Lysines were selected as a charge carrier linker because of the ease of synthesis of oligo-lysines and compatibility of the protecting groups of lysines with nucleotide chemistry.

Initial electrophoresis experiments demonstrated that five lysine moieties were sufficient to remove fluorescent dye artifacts from a sequencing gel using rhodamine-labeled dideoxynucleotides and six lysine moieties using fluorescein-labeled dideoxynucleotides (fluorescein has (−1) negative charge). The synthesis of these terminators is simple, as properly protected (trifluoroacetamido) oligo-lysine is reacted first with the fluorescent dye in high yield, followed by activation of the carboxylic group and then by reaction with suitably linker arm attached dideoxynucleoside-5′-triphosphate (Finn *et al.*, 2003).

The synthesis of ET terminators with lysines as charge carrier moieties was carried out by attachment of fluorescein attached 4-propargylamino-phenylalanine (Nampalli *et al.*, 2000) with hexa-trifluoroacetamidolysine. This was

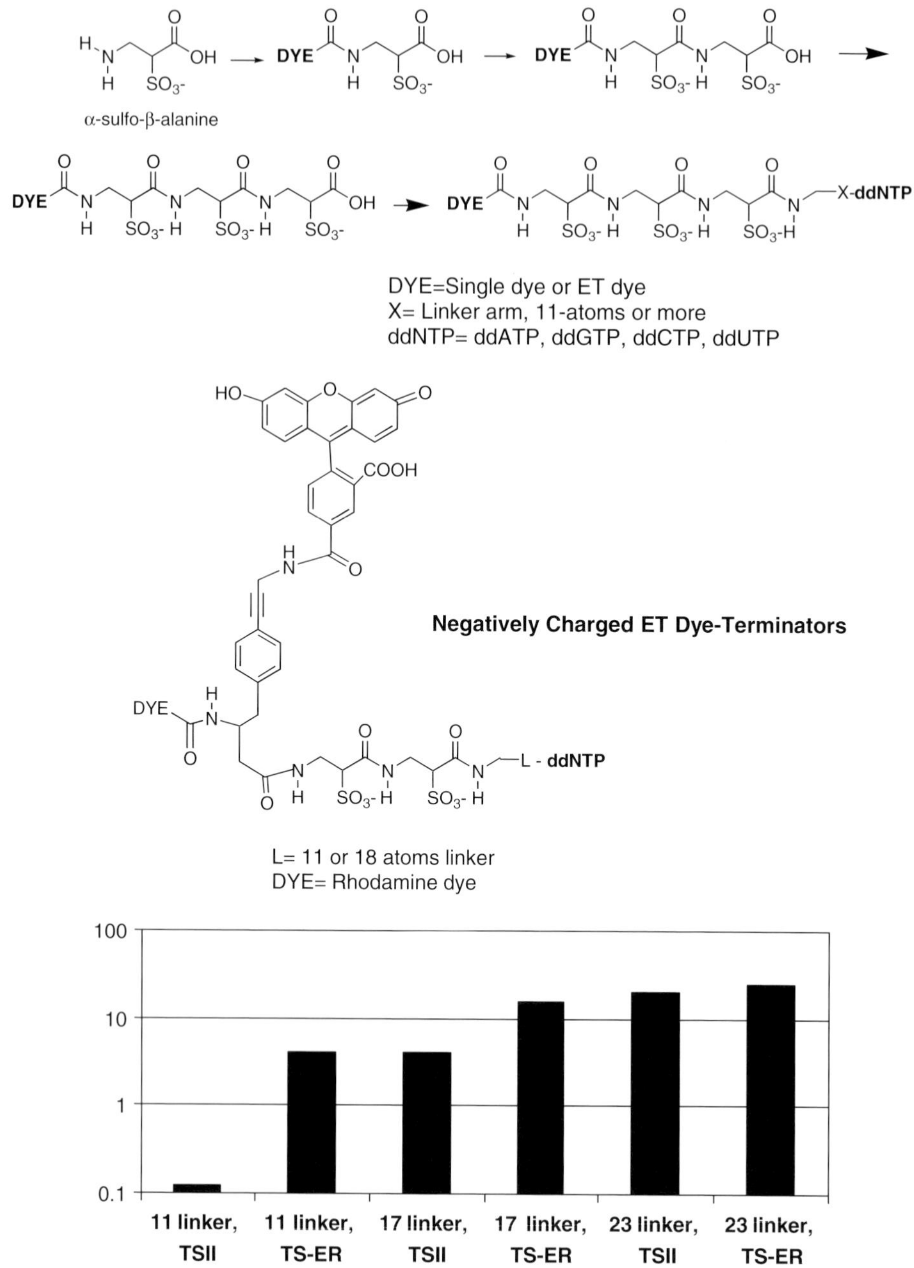

Fig. 16. Structures of negatively charged single dye-labeled and ET terminators and the efficiency of incorporation of (−3) charged and extended linker arm terminators with Thermo Sequenase II DNA polymerase and mutated *Taq* DNA polymerase.

followed by removal of t-Boc group with aqueous acid, and then by attachment of second (acceptor) dye to the amino group.

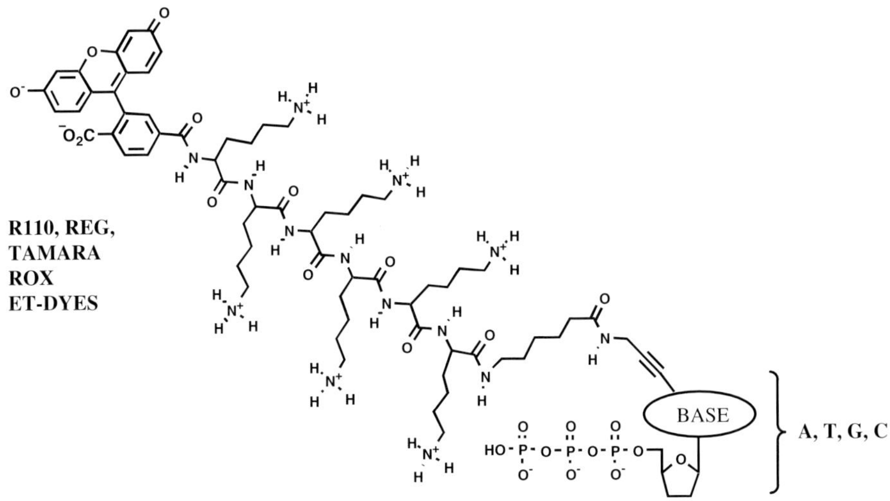

Finally, attachment of the desired dideoxynucleoside-5′-triphosphate was carried out by activation of the peptide carboxylic acid group to provide the desired ET terminator. These terminators were purified by ion-exchange chromatography followed by reverse-phase HPLC (Finn *et al.*, 2003). The general structure of these ET terminators is given below.

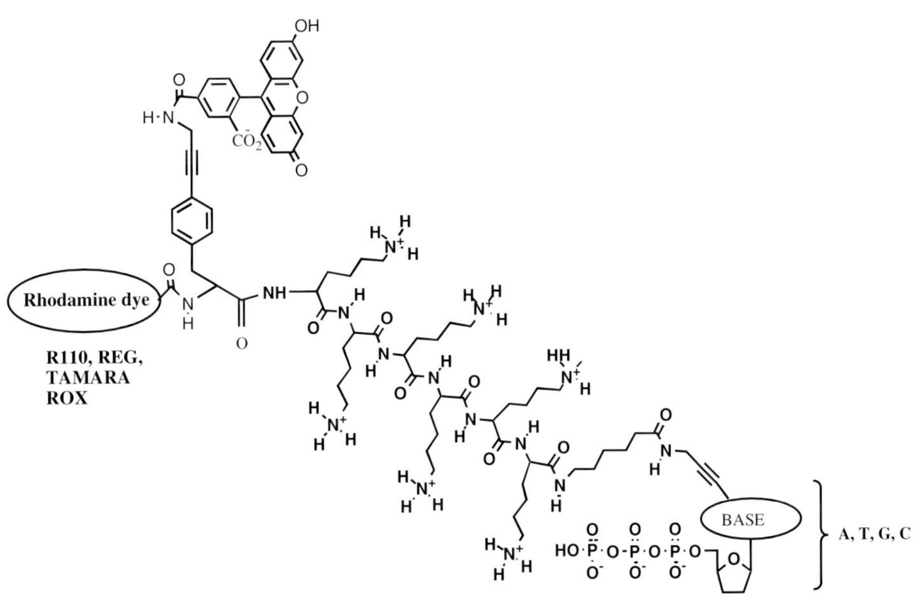

A directly loaded four-color sequencing experiment using an M13 template showed the absence of dye-blobs, and the primary sequence was as expected. However, the results with lysine- modified terminators showed lower electrophoretic resolution (band broadening), in comparison with regular dye-terminators. We felt that this effect may be due to the acid–base equilibrium of the lysine side chains causing heterogeneity of the net charge within a population of oligonucleotide fragments of identical sequence length, which results in the observed band broadening. This was confirmed by the observation that decreasing the pH of the electrophoresis gel loading solution, giving more complete protonation of the linker lysines, could improve the sequence resolution for the earlier (resolving) part of the sequence. However, this peak to peak improvement was not significant beyond 350 bases, possibly due to alkalization during prolonged electrophoresis, restoring lysine charge heterogeneity. To overcome this variability in resolution, we decided to investigate a linker containing a positive charge, such as ε-trimethyllysine, which contains quaternary nitrogen that would have constant charge, independent of the pH at which separation occurs.

3.4. Trimethyllysine-derived terminators

The positively charged trimethyllysine-derived terminators were synthesized in the same manner as described above for the synthesis of lysine-derived terminators. Four trimethyllysine moieties for the single dye-labeled rhodamine terminators, and five for the ET terminators are required for the absence of dye-blobs in the sequencing electropherogram. If rhodamine 110 is used as a donor dye, then four trimethyllysines are sufficient (Finn *et al.*, 2003). The structures of trimethyllysine linker based single dye and ET dye terminators are shown in Figure 17.

The DNA polymerase incorporation of lysine- and trimethyllysine-derived terminators was compared with sulfonated, negatively charged terminators and regular dye-labeled terminators. The positively charged terminators were incorporated much more efficiently than the unmodified dye-labeled terminators with a number of different DNA polymerases. The reactivity of trimethyllysine-derived terminators is 3–4-fold higher than the unmodified dye-terminators (Finn *et al.*, 2003). This is opposite to the behavior of negatively charged terminators whose reactivity is 1/4 to 1/10th that of the uncharged dye-terminators (Finn *et al.*, 2002). The reactivity and comparison of (+5) charged terminators with negatively charged terminators using *Thermo Sequenase* II and mutant *Taq* DNA polymerase (TS-E682R) is shown in Figure 18.

The incorporation efficiency of unmodified dye-labeled terminators was also compared with lysine linker attached terminators and trimethyllysine linker attached terminators using *Thermo Sequenase* II DNA polymerase (Figure 19). It is clear that trimethyllysine linked terminators are more reactive than lysine linked terminators, which in turn are much more reactive than the unmodified dye-labeled terminators.

A four-color sequencing reaction was set up using the optimized concentrations of single dye-labeled terminators shown in Figure 20. The resulting

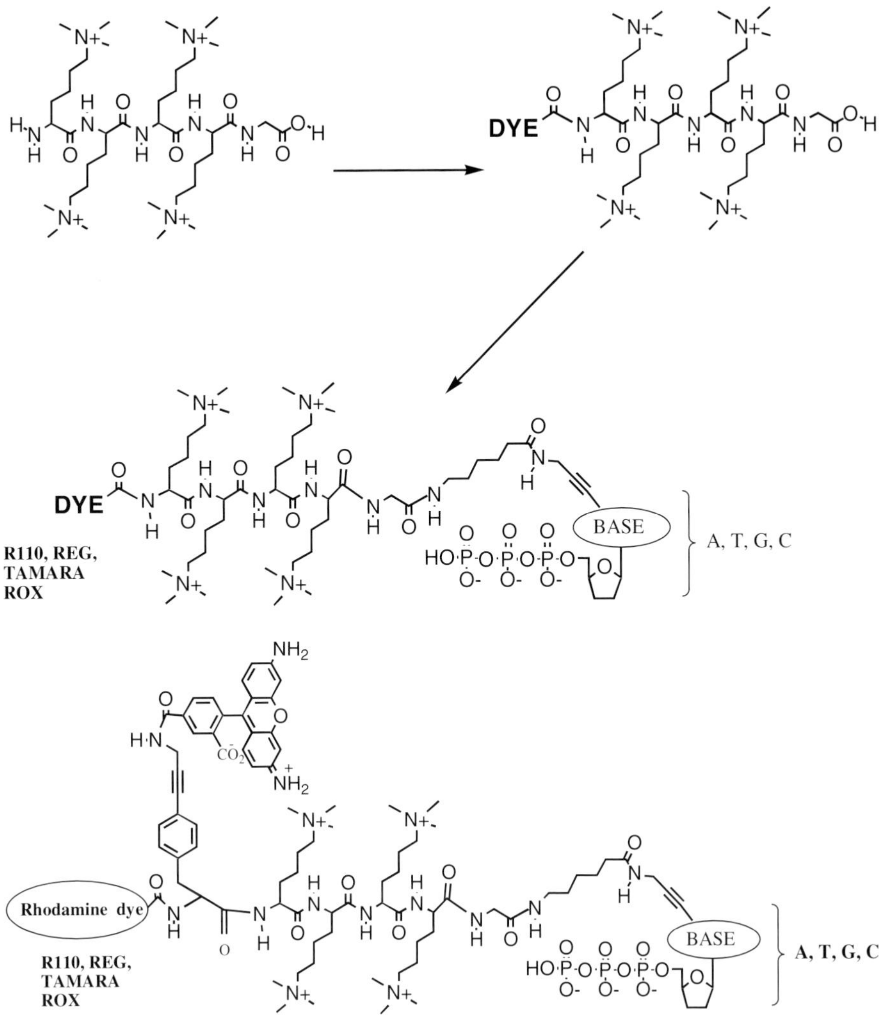

Fig. 17. Structures of single dye-labeled and ET dye-labeled positively charged trimethyllysine-derived terminators.

electropherogram indicated the absence of unreacted nucleotides and their by-products, and the sequence obtained was 100% correct (Finn *et al.*, 2003). The overall resolution of sequencing bands was improved, compared with oligo-lysine linked terminators and overall read length comparable with sequence obtained using regular dye-labeled terminators. The reaction products can be directly loaded onto a capillary sequencer, MegaBACE, after diluting with water. Since samples are electro-kinetically loaded onto the capillary sequencers and ionic strength of the samples affects the loading process, it is important that the samples be diluted to optimize the injection for proper signal strength.

In conclusion, both negatively and positively charged terminators can be used for "direct-load", blob-free DNA sequencing. The negatively charged

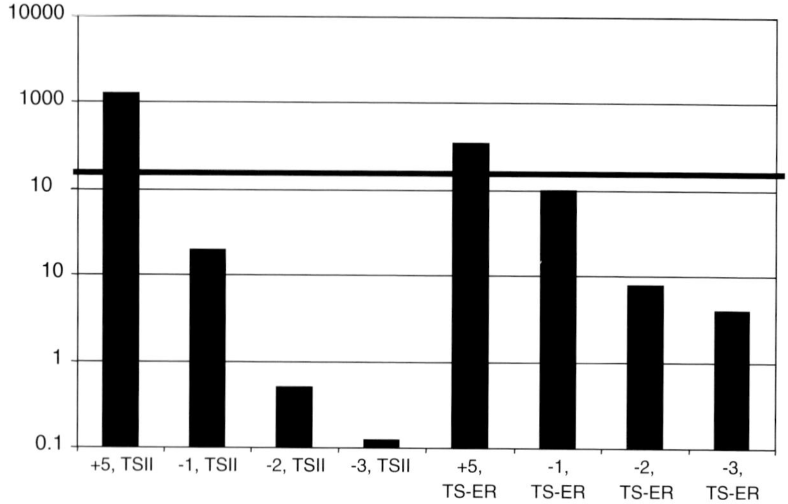

Fig. 18. Reactivity and comparison of (+5) charged terminators with negatively charged terminators. *Thermo Sequenase* II and *Thermo Sequenase* with E681R mutation are used. The thick black line indicates the reactivity of unmodified dye-terminators. It is clear from this figure that positively charged terminators are much more reactive than unmodified dye terminators. Negatively charged terminators are relatively poor substrates for DNA polymerases.

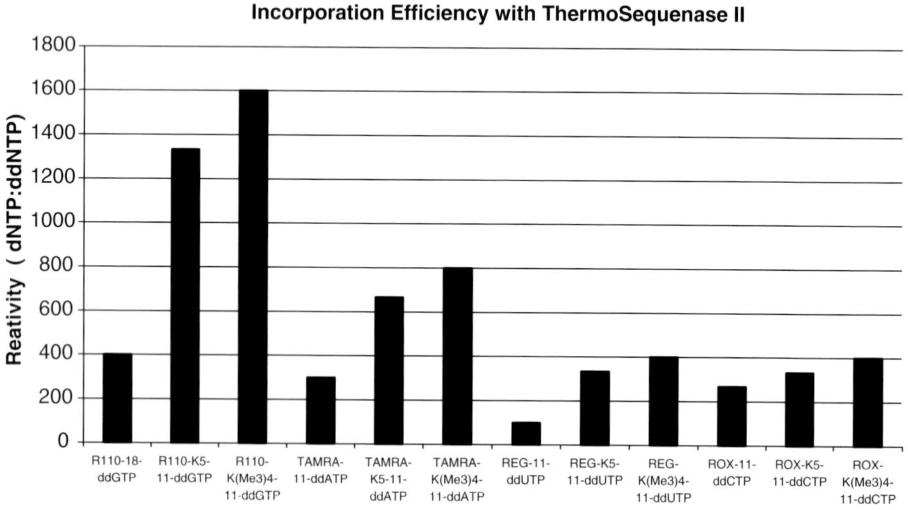

Fig. 19. Comparison of reactivity of unmodified dye-labeled terminators with lysine linked terminators and trimethyllysine linked terminators using *ThermoSequenase* II DNA polymerase.

terminators are relatively poor substrates for DNA polymerases, and positively charged terminators are 3–4 fold better substrates than the regular dye-labeled terminators. The positively charged terminators with oligo-ε-trimethyllysine linker showed better electrophoretic performance than oligo-lysine linker

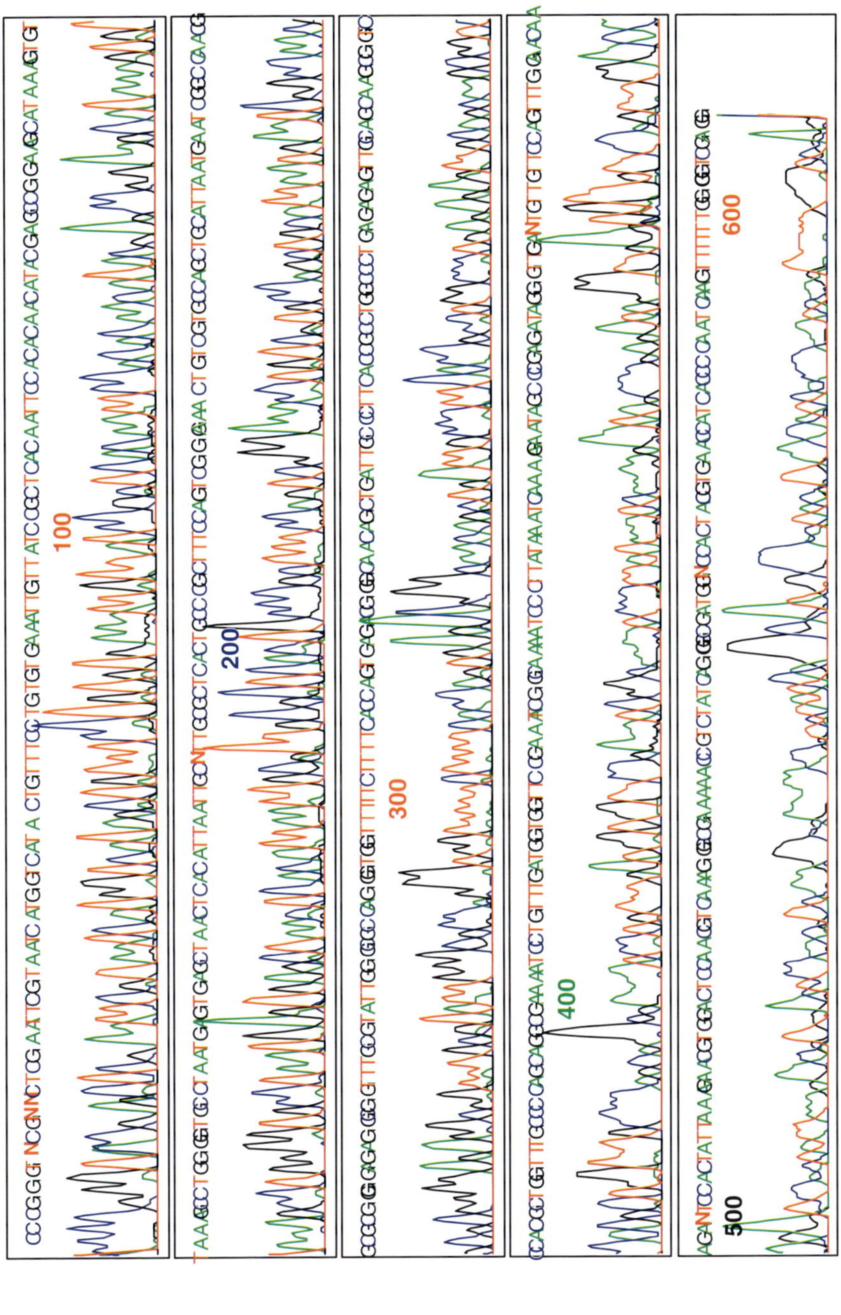

Fig. 20. Direct-Load DNA sequence generated using positively charged oligo-trimethyl lysine modified dideoxynucleoside-5′-triphosphate terminators.

terminators, which yielded broader peaks that limit read lengths to <500 bases. This study also provided useful insight into developing better nucleotide substrates, or inhibitors for DNA polymerases (Finn *et al.*, 2002, 2003).

4. TERMINAL PHOSPHATE-LABELED NUCLEOTIDES

The nucleotides described above are all labeled by a fluorescent dye attached to the base moiety of the nucleoside. When incorporated by a DNA polymerase, these nucleotides contribute the labeled nucleotide base to the growing DNA chain and the incorporated base can be detected by its fluorescence. However, incorporation of base labeled nucleotides results in the modification of newly synthesized DNA structure. There would be great advantage in detecting the addition of a base without modification of the DNA structure in many applications, particularly those that require the product DNA to be continually used as a template for further incorporation. In fact, for several applications, such as real-time sequencing and SNP analysis, it is enough to detect whether a nucleotide has been or has not been added by a DNA polymerase (Ronaghi *et al.*, 1996).

We have developed a new series of nucleoside polyphosphates containing three or more phosphates, with either dye attached directly to the terminal phosphate, or with a linker between the dye and the terminal phosphate. These nucleotides are efficient substrates of DNA polymerases, particularly when more than three phosphates are present at the 5′-position of the nucleoside. We have shown that tetra- (dN4P, deoxynucleoside-5′-tetraphosphate) or higher phosphates are better substrates (up to 50-fold better) for DNA polymerases than the corresponding triphosphates. We have also studied the effect of linker between the dye and terminal phosphate and their incorporation by different DNA polymerases (Sood *et al.*, 2005; Kumar *et al.*, 2005b). In general, the intermediate length linkers (such as hepta-methylene linker), aromatic or positively charged linkers provide better substrate properties for incorporation into DNA with a variety of different DNA polymerases (Kumar *et al.*, 2005b).

Two different types of dyes have been used to attach to the terminal phosphate of the nucleotide (Figure 21). The first type is simple dyes that are initially fluorescent and remain fluorescent throughout the application process (initially some fluorescence quenching may take place depending upon the nucleotide base used). The second type is dyes that are initially colorless, non-fluorescent or weakly fluorescent, and which after enzymatic treatment change their color and become highly fluorescent (e.g. dichlorodimethylacridin-one (DDAO), resorufin, 4-methylcoumarin, alkyl-fluorescein, etc). These latter dyes are useful in designing many kinds of homogenous assays, such as SNP scoring assays, real-time sequencing, allele-specific primer extension assays and general analyte detection assays (Sood *et al.*, 2003; Kumar *et al.*, 2004d). These labeled nucleotides have also been proposed for single molecule sequencing, and Webb and his co-workers (Levene *et al.*, 2003) have recently reported the observation of DNA polymerase activity using zero-mode waveguides that allow single-molecule experiments at high concentrations of labeled nucleotides to be performed.

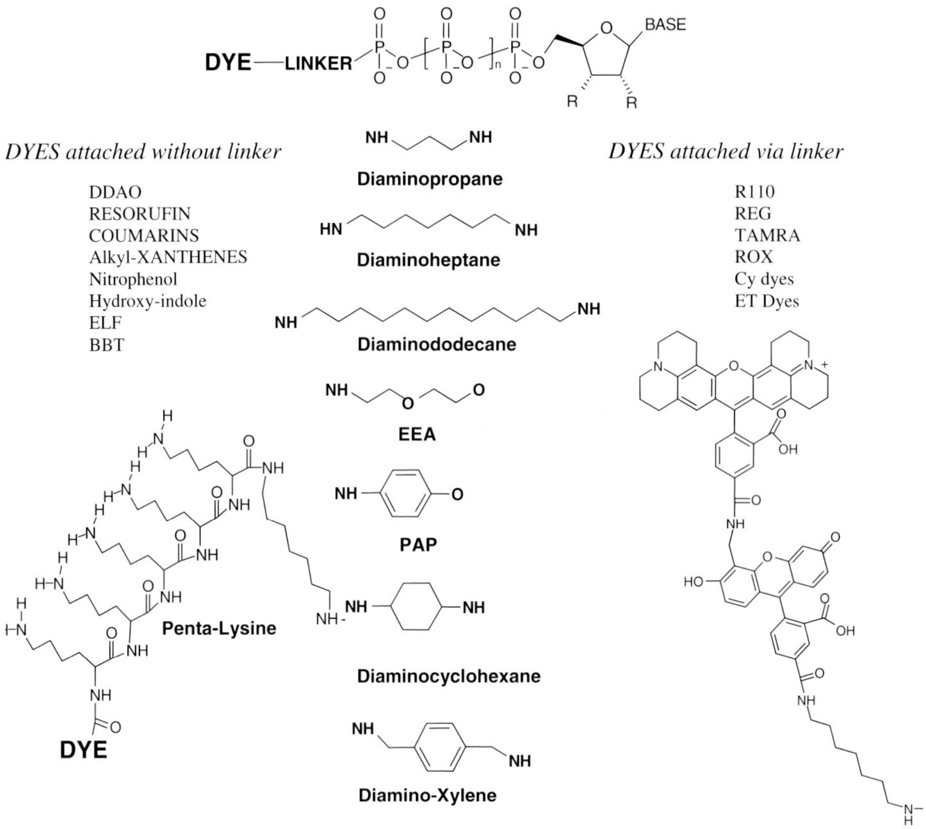

Fig. 21. Structure of terminal phosphate-labeled nucleotides with different linkers and dyes.

The utility of these terminal phosphate-labeled nucleotides in DNA sequencing without the use of any separation matrix is demonstrated in a simple experiment and results are shown in Figure 22. Here, a self-priming oligonucleotide was effectively sequenced by adding nucleoside tetraphosphates (4-methylcoumarin-dN4P) one at a time and detecting any dye released from the nucleotide. If the correct nucleotide is added, the released 4-methylcoumarin-triphosphate (non-fluorescent) is hydrolyzed by alkaline phosphatase present in the reaction mixture to form free 4-methylcoumarin which is highly fluorescent. When an "incorrect" nucleotide was added, little or no product was detected. When a "correct" nucleotide was added, an amount of fluorescent product proportional to the number of "correct" nucleotides in a row was detected. While the experiment we performed used a template immobilized on magnetic beads, one could readily imagine automated instruments capable of carrying out many such sequencing experiments in parallel, resulting in an efficient, high-throughput sequencing device with no requirement for electrophoretic separation (Sood *et al.*, 2003). Many other kinds of homogeneous assays using these dye-labeled nucleotides might also be devised (Nelson *et al.*, 2003a, b; Sood *et al.*, 2004a, b).

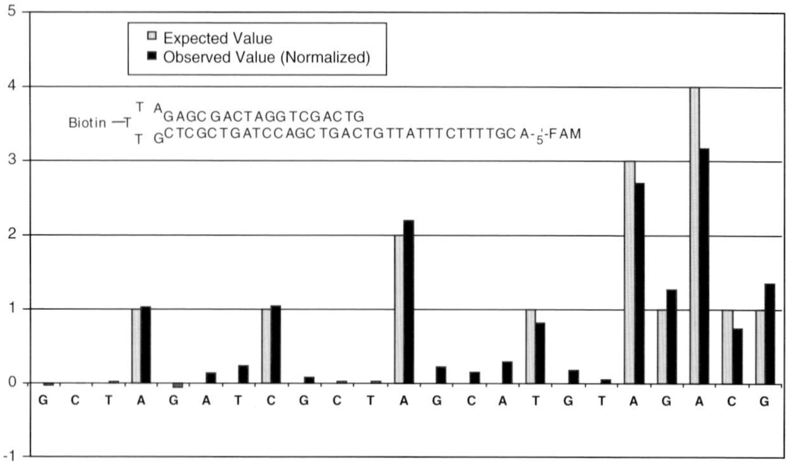

Fig. 22. A self-priming (hairpin) oligonucleotide was effectively sequenced by sequential addition of dye-labeled nucleoside-5′-tetraphosphates. After each addition, production of free dye was assayed spectroscopically. Shown are the relative amounts of product made after each nucleotide addition step. Addition of nucleotides that do not complement the next base in the sequence resulted in little or no free dye formation. When complementary nucleotides were added, an amount of dye proportional to the number of bases added was found. From these results, a correct sequence can be inferred.

5. CONCLUSIONS

DNA sequencing and other kinds of DNA analysis have been carried out using dyes and other tags attached to nucleotides for over 20 years. The modern dye-conjugated nucleotides are among the most complex commercially synthesized compounds ever made, having phosphorylated sugar, nucleobase, linkers and multiple dyes. Each of the parts of these molecules serves a specific purpose, and many of them interact specifically and non-specifically with enzymes, particularly DNA polymerases. Some modifications of the nucleotides can result in diminished (or even enhanced) activity with enzymes, and the enzymes can be modified to compensate if necessary. This complex set of components make DNA sequencing a rich and interesting technology. The lessons learned in developing this technology should have applications in other fields including other analytical fields, synthetic chemistry, enzyme engineering and even medicinal chemistry. As new methods of DNA analysis emerge, one can expect conjugated nucleotides to continue to play an essential part.

REFERENCES

Amersham Biosciences/GE Healthcare, Piscataway, NJ, USA (http://amershambiosciences.com).
Applied Biosystems Inc., Foster City, CA, USA (http://home.appliedbiosystems.com/).
Bergot, B. J., Chakerian, V., Connell, C. R., Eadie, J. S., Fung, S., Hershey, D., Lee, L. G., Menchen, S. M. and Woo, S. L. (1994). Spectrally resolvable rhodamine dyes for nucleic acid sequencing determination. US Patent 5, 366, 860.

Berti, L., Xie, J., Medintz, I. L., Glazer, A. N. and Mathies, R. A. (2001). Energy transfer cassettes for facile labeling of sequencing and PCR primers. Anal. Biochem. 292, 188–197.
Davis, M., Nelson, J., Kumar, S., Finn, P., Nampalli, S. and Flick, P. (2005). Taq DNA polymerase having an amino acid substitution at E681 and homologs thereof exhibiting salt tolerance. European Patent EP 121 0440.
Duthie, R. S., Kalve, I. M., Samols, S. B., Hamilton, S., Livshin, I., Khot, M., Nampalli, S., Kumar, S. and Fuller, C. W. (2002). Novel cyanine dye-labeled dideoxynucleoside triphosphates for DNA sequencing. Bioconj. Chem. 13, 699–706.
Finn, P., Bull, M., Xiao, H., Nelson, J. R., Mamone, J. A., Grossmann, G., Nampalli, S., Flick, P. K., Fuller, C. W. and Kumar, S. (2003). Efficient incorporation of positively charged 2′, 3′-dideoxynucleoside-5′-triphosphates by DNA polymerases and their application to 'direct load' DNA sequencing. Nucleic Acids Res. 31, 4769–4778.
Finn, P., Sun, L., Nampalli, S., Xiao, H., Nelson, J. R., Mamone, J. A., Grossmann, G., Flick, P. K., Fuller, C. W. and Kumar, S. (2002). Synthesis and application of charge-modified dye-labeled dideoxynucleoside-5′-triphosphates to 'direct load' DNA sequencing. Nucleic Acids Res. 30, 2877–2885.
Förster, T. (1965). Modern quantum chemistry. In: O. Sinanoğlu (Ed), Istanbul Lectures, Part III (pp. 93–137). Academic Press, New York.
Fuller, C. W., Mamone, J. A., McArdle, B. F. and Hujer, K. J. (1999). A method, reagent solution and kits for DNA sequencing. PCT WO 99/37810.
Hung, S. C., Ju, J., Mathies, R. A. and Glazer, A. N. (1996). Cyanine dyes with high absorption cross section as donor chromophores in energy transfer primers. Anal. Biochem. 243, 15–27.
Hung, S. C., Mathies, R. A. and Glazer, A. N. (1997). Optimization of spectroscopic and electrophoretic properties of energy transfer primers. Anal. Biochem. 252, 78–88.
Ju, J., Glazer, A. N. and Mathies, R. A. (1996). Cassette labeling for facile construction of energy transfer fluorescent primers. Nucleic Acids Res. 24(6), 1144–1148.
Ju, J., Kheterpal, I., Scherer, J. R., Ruan, C., Fuller, C. W., Glazer, A. N. and Mathies, R. A. (1995a). Design and synthesis of fluorescence energy transfer dye-labeled primers and their application for DNA sequencing and analysis. Anal. Biochem. 231, 131–140.
Ju, J., Ruan, C., Fuller, C. W., Glazer, A. N. and Mathies, R. A. (1995b). Fluorescence energy transfer dye-labeled primers for DNA sequencing and analysis. Proc. Natl. Acad. Sci. USA 92, 4347–4351.
Kumar, S. (2002). Fluorescent dye nucleotide conjugates for DNA sequencing. In: Loakes, D. (Ed.), Modified Nucleotides, Synthesis and Applications Research Signpost (pp. 87–110). Trivandrum, India.
Kumar, S., Chen, C. Y. and Fuller, C. W. (2004a). Synthesis and spectroscopic analysis of energy transfer dye-labeled nucleotides and oligonucleotides for multiplex applications. In: S. G. Pandalai (Ed), Recent Developments in Nucleic Acids Research (pp. 251–274). Transworld Research Network, Trivandrum, India.
Kumar, S., Chen, C. Y. and Rao, T. S. (2004b). Energy transfer dyes, terminators and use thereof. PCT WO 04/029578.
Kumar, S., Finn, P., Nampalli, S., Nelson, J., Flick, P. and Bull, M. (2001a). Charge modified nucleic acid terminators. PCT WO 01/19841.
Kumar, S., Fuller, C. W., Nampalli, S., Khot, M., Livshin, I., Sun, L., Hamilton, S., Samols, S. B., Mamone, J. A., Hujer, K. M., McArdle, B. F., Nelson, J. R. and Duthie, S. (1999). Uniform band intensities in fluorescent dye terminator sequencing. Nucleosides Nucleotides 18, 1101–1103.
Kumar, S., Nampalli, S., Finn, P. J., Rao, T. S., Chen, C. Y., Xiao, H., Nelson, J. R., Mamone, J. A., Flick, P. K. and Fuller, C. W. (2001b). FRET terminators for DNA sequencing: synthesis, enzymatic activity and effect of linker on the energy transfer. Presented at NACON-V, Sheffield, UK, April 8–12, 2001; and at the 13th GSAC Meeting, San Diego (CA), Oct. 25–28, 2001.
Kumar, S., Nampalli, S. and Khot, M. (2004c). Energy transfer dyes. European Patent EP 111 0264.

Kumar, S., Nampalli, S., McArdle, B. F. and Fuller, C. W. (2005a). Dideoxy dye terminators. US Patent 6, 949, 635.
Kumar, S., Sood, A., Nelson, J. R., McDougall, M., Fuller, C. W., Macklin, J. and Mitsis, P. (2004d). Terminal phosphate labeled nucleotides with new linkers. PCT WO 04/072238.
Kumar, S., Sood, A., Wegener, J., Finn, P., Nampalli, S., Nelson, J. R., Sekher, A., Gao, W., McKay, R., Mitsis, P., Macklin, J. and Fuller, C. W. (2005b). Terminal phosphate labeled nucleotides: synthesis, applications and linker effect on incorporation by DNA polymerases. Nucleosides Nucleotides Nucleic Acids 24(5–7), 401–408.
Lee, L. G., Connell, C. R., Woo, S. L., Cheng, R. D., McArdle, B. F., Fuller, C. W., Halloran, N. D. and Wilson, R. K. (1992). DNA sequencing with dye-labeled terminators and T7 DNA polymerase: effect of dyes and dNTPs on incorporation of dye-terminators and probability analysis of termination fragments. Nucleic Acids Res. 20(10), 2471–2483.
Lee, L. G., Spurgeon, S. L., Heiner, C. R., Benson, S. C., Rosenblum, B. B., Menchen, S. M., Graham, R. J., Constantinescu, A., Upadhya, K. G. and Cassel, J. M. (1997). New energy transfer dyes for DNA sequencing. Nucleic Acids Res. 25, 2816–2822.
Lee, L. G., Spurgeon, S. L. and Rosenblum, B., 1999. Energy transfer dyes with enhanced fluorescence. US Patent 5, 863, 727.
Levene, M. J., Korlach, J., Turner, S. W., Foquet, M., Craighead, H. G. and Webb, W. W. (2003). Zero-mode waveguides for single-molecule analysis at high concentrations. Science 299, 682–686.
Maxam, A. M. and Gilbert, W. (1977). A new method for sequencing DNA. Proc. Natl. Acad. Sci. USA 74(2), 560–574.
Metzker, M. L., Lu, J. and Gibbs, R. A. (1996). Electrophoretically uniform fluorescent dyes for automated DNA sequencing. Science 271, 1420–1422.
Middendorf, L. R., Bruce, J. C., Bruce, R. C., Eckles, R. D., Grone, D. L., Roemer, S. C., Sloniker, G. D., Steffens, D. L., Sutter, S. L. and Brumbaugh, J. A. (1992). Continuous, on-line DNA sequencing using a versatile infrared laser scanner/electrophoresis apparatus. Electrophoresis 13(8), 487–494.
Nampalli, S., Khot, M. and Kumar, S. (2000). Fluorescence resonance energy transfer terminators for DNA sequencing. Tetrahedron Lett. 41, 8867–8871.
Nampalli, S., Khot, M., Nelson, J. R., Flick, P. K., Fuller, C. W. and Kumar, S. (2001). Fluorescent resonance energy transfer dye nucleotide terminators: a new synthetic approach for high-throughput DNA sequencing. Nucleosides Nucleotides Nucleic Acids 20(4–7), 361–367.
Nampalli, S., Zhang, W., Rao, T. S., Xiao, H., Kotra, L. P. and Kumar, S. (2002). Unnatural amino acid derived FRET cassettes, terminators and their DNA sequencing potential. Tetrahedron Lett. 43(11), 1999–2003.
Nelson, J., Fuller, C., Sood, A. and Kumar, S. (2003a). Terminal phosphate labeled nucleotides and methods of use. PCT WO 03/020984, US Patent 7,052,839.
Nelson, J., Fuller, C., Sood, A. and Kumar, S. (2003b). Single nucleotide amplification and detection by polymerase. PCT WO 03/020891, US Patent 7,033,762.
Prober, J. M., Trainor, G. L., Dam, R. J., Hobbs, F. W., Robertson, C. W., Zagursky, R. J., Cocuzza, A. J., Jensen, M. A. and Baumeister, K. (1987). A system for rapid DNA sequencing with fluorescent chain-terminating dideoxynucleotides. Science 238, 336–341.
Rao, T. S., Nampalli, S., Lavrenov, K., Zhang, W., Xiao, H., Nelson, J. and Kumar, S. (2001). Four color FRET dye nucleotide terminators for DNA sequencing. Nucleosides Nucleotides Nucleic Acids 20(4–7), 673–676.
Reeve, M. A. and Fuller, C. W. (1995). A novel thermostable polymerase for DNA sequencing. Nature 376, 796–797.
Roemer, S. C., Boveia, V. R., Johnson, C. M. and Olive, D. M. (2000). Sequencing BAC DNA with near infrared fluorescent non-nucleotide terminators. 12th International Genome Sequencing and Analysis Meeting, Miami, USA.
Ronaghi, M., Karamohamed, S., Pettersson, B., Uhlen, M. and Nyren, P. (1996). Real-time DNA sequencing using detection of pyrophosphate release. Anal. Biochem. 242(1), 84–89.

Rosenblum, B. B., Lee, L. G., Spurgeon, S. L., Khan, S. H., Menchen, S. M., Heiner, C. R. and Chen, S. M. (1997). New dye-labeled terminators for improved DNA sequencing patterns. Nucleic Acids Res. 25, 4500–4504.

Sanger, F., Nicklen, S. and Coulson, A. R. (1977). DNA sequencing with chain-terminating inhibitors. Proc. Natl. Acad. Sci. USA 74(12), 5463–5467.

Smith, L. M., Sanders, J. Z., Kaiser, R. J., Hughes, P., Dodd, C., Connell, C. R., Heiner, C., Kent, S. B. and Hood, L. E. (1986). Fluorescence detection in automated DNA sequence analysis. Nature 321, 674–679.

Sood, A., Kumar, S., Nampalli, S., Nelson, J. R., Macklin, J. and Fuller, C. W. (2005). Terminal phosphate labeled nucleotides with improved substrate properties for nucleic acid assays. J. Am. Chem. Soc. 127, 2394–2395.

Sood, A., Kumar, S., Nelson, J. and Fuller, C. (2003). Solid phase sequencing. PCT WO 03/020734.

Sood, A., Kumar, S., Nelson, J. and Fuller, C. (2004a). Analyte detection. PCT WO 04/020603.

Sood, A., Kumar, S., Nelson, J., Fuller, C. and Sekar, A. (2004b). Nucleic acid amplification. PCT WO 04/072304, US Patent 7,125,671.

Tabor, S. and Richardson, C. C. (1995). A single residue in DNA polymerases of the *Escherichia coli* DNA polymerase I family is critical for distinguishing between deoxy- and dideoxyribonucleotides. Proc. Natl. Acad. Sci. USA 92(14), 6339–6343.

Vander Horn, P. B., Davis, M. C., Cunniff, J. J., Ruan, C., McArdle, B. F., Samols, S. B., Szasz, J., Hu, G., Hujer, K. M., Domke, S. T., Brummet, S. R., Moffett, R. B. and Fuller, C. W. (1997). Thermo-sequenase DNA polymerase and *T. acidophilum* pyrophosphatase: new thermostable enzymes for DNA sequencing. Biotechniques 22, 758–765.

Vander Horn, P. B., Holecek, J., Ruan, C. and Fuller, C. W. (unpublished results).

Sequencing by Synthesis Platforms

Chapter 5
The 454 Life Sciences Picoliter Sequencing System

Marcel Margulies, Thomas P. Jarvie, James R. Knight and Jan Fredrik Simons

454 Life Sciences Corporation, 20 Commercial Street, Branford, CT 06405, USA

Contents

Abstract

We describe a novel sequencing system with a 100-fold improvement in throughput over state-of-the-art capillary electrophoresis instruments. The apparatus uses a single $60 \times 60\,\text{mm}^2$ fiber optic slide containing 1,600,000 individual 75 pL wells and routinely generates 25 million bases of sequence in one 4-h run. When applied to bacterial genomes, it achieves consensus accuracies of better than 99.99%, at an average 12-fold depth of coverage. A single library preparation is sufficient to sequence an entire genome. We describe here the components of this system, the sequencing of several bacterial genomes and their *de novo* shotgun assembly. We also discuss a few early applications of this system in comparative genomics and the sequencing of PCR-generated products. Based on current performance and modeling of the existing system, we predict that the system's inherent scalability will lead to two additional orders of magnitude improvement in throughput, possibly enabling routine sequencing of individual human genomes.

PERSPECTIVES IN BIOANALYSIS, VOLUME 2
ISSN 1871-0069 DOI: 10.1016/S1871-0069(06)02005-2

1. INTRODUCTION

The ability to sequence complete genomes has dramatically changed the nature of biomedical research and medicine. It has affected our understanding, diagnosis, prevention and treatment of disease, led to advances in agriculture, and furthered our understanding of evolution. Genomic information on emerging pathogens and drug resistant strains is of great importance to public health policy. Unfortunately, the cost, complexity, and time required to sequence bacterial and eukaryotic genomes currently limit the routine use of genome sequence information in many areas. As important as a significant decrease in the cost of sequencing is the need to develop a complete platform that brings to any research laboratory the capability to perform sequencing of large organisms without a large and expensive infrastructure. There is little doubt that overcoming current obstacles to routine sequencing will have significant scientific, economic and cultural impact.

Early sequencing methods by chemical degradation (Maxam and Gilbert, 1977), although groundbreaking, proved inefficient for sequencing large genomes. Genome sequencing by random insertion of chain-terminators was first conducted by Frederick Sanger (Sanger *et al.*, 1977). This was better suited to whole genomes, and Sanger-based dideoxy nucleotide triphosphate sequencing by random termination of DNA synthesis became widely accepted. This approach required gel electrophoresis of four radio-labeled reactions (one for each dNTP) in order to determine the correct sequence. Visualization of fragments required cumbersome manipulation of gels and unavoidable long film exposures. These methods had short read lengths (50–150 bases) due to the limited ability of denaturing acrylamide gels to resolve DNA fragments just a single nucleotide shorter than the previous fragment in the gel. In addition, electrophoretic anomalies created by secondary structures in ssDNA fragments confounded the elucidation of sequences (Mills and Kramer, 1979).

Much effort throughout the past 20 years has been aimed at automating the process, starting with the development of four dye fluorescent sequencing (Prober *et al.*, 1987). Currently practiced genomic sequencing methods require the cloning of restriction fragments, or mechanically sheared fragments, into bacterial plasmid vectors. This approach has proven unwieldy for large-scale sequencing tasks and the ability routinely and somewhat cost-effectively to sequence bacterial and larger genomes has been limited to a few sequencing centers capable of supporting a sizeable infrastructure that includes large-scale liquid handling automation for fragment cloning and amplification, large-scale colony picking automation and banks of sequencers. Most of the improvements have focused on automating the workflow and on improvements in electrophoresis, resulting in the successful sequencing of larger and larger genomes (Fleischmann *et al.*, 1995; Venter *et al.*, 2001). These efforts, over the last 20 years, have led to more than 2 orders of magnitude increase in the throughput, with a concomitant reduction in cost, of Sanger sequencing (Collins *et al.*, 2003); nevertheless, current estimates still put the cost of sequencing a human genome at a minimum of $10 million (NIH News Release, 2004).

Alternative methods have been described that generate short read lengths but eliminate electrophoresis of fragments to generate sequences (Nyren *et al.*, 1993;

Ronaghi *et al.*, 1996). Additionally, sequencing DNA has been described using mass spectrometry (Jacobson *et al.*, 1991), hybridization (Bains and Smith, 1988) and single-molecule approaches (Jett *et al.*, 1989). Some of these technologies have been in development for years without successful application to whole genomes; most face the formidable molecular challenge of controlling single molecules. As a result, to date no technology has displaced Sanger sequencing and electrophoresis as the main generator of sequencing information.

Recently, a new, highly parallel sequencing system with throughput approximately 100 times that of state-of-the-art capillary electrophoresis instruments was described (Margulies *et al.*, 2005). This approach, developed by the 454 Life Sciences Corporation of Branford, CT, centers on simplifying and parallelizing the many steps necessary to perform genomic analysis. This system is inherently scalable in many dimensions and represents the first in a new generation of very high throughput sequencing instruments. Its power lies in the ability to sequence complete genomes of significant sizes without need for a large infrastructure and at speeds heretofore unachievable. This system relies on a single sample preparation, for any genome, and dramatically reduces the space needed for library creation, amplification, sequencing and data processing.

In this chapter, we discuss this novel technology and its application to whole genome sequencing and the sequencing of polymerase chain reactions (PCR) products. The effectiveness of this approach comes both from the use of an emulsion based method that enables a single person to handle an entire genome as a single sample, irrespective of its size, and the ability to sequence hundreds of thousands of bead-bound DNA templates simultaneously in individual picoliter-sized wells on the surface of a $60 \times 60\,mm^2$ fabricated slide. This system currently generates 80–120 bp of sequence information from each of more than 300,000 DNA templates in a 4-hr instrument run. In a typical run over 25 million bases with a Phred 20 or better quality score (predicted to have an accuracy of 99% or higher) are generated, yielding a throughput of slightly more than 6 million Phred 20 bases per h. While this Phred 20 quality throughput is on the order of 100 times that of capillary sequencing, it is currently at the cost of shorter reads, and lower individual read accuracy: Sanger-based capillary electrophoresis sequencing systems produce up to 700 bp of sequence information from each of 96 DNA samples in 1 h, or 67 kb per h at 99% accuracy (Ogawa *et al.*, 2005).

2. THE 454 LIFE SCIENCES PICOLITER SEQUENCING SYSTEM

2.1. Sample preparation

Using a new approach (Figure 1) that remains unchanged as far as its complexity is concerned, no matter what the genome size, randomly fragmented libraries of DNA fragments are generated by shearing an entire genome using a generic nebulizer. The nebulizer is adjusted to yield DNA fragments in the 50–800 bp range, with a mean fragment size of 375 ± 50 bp. Following nebulization, the fragmented DNA is rendered blunt ended and phosphorylated

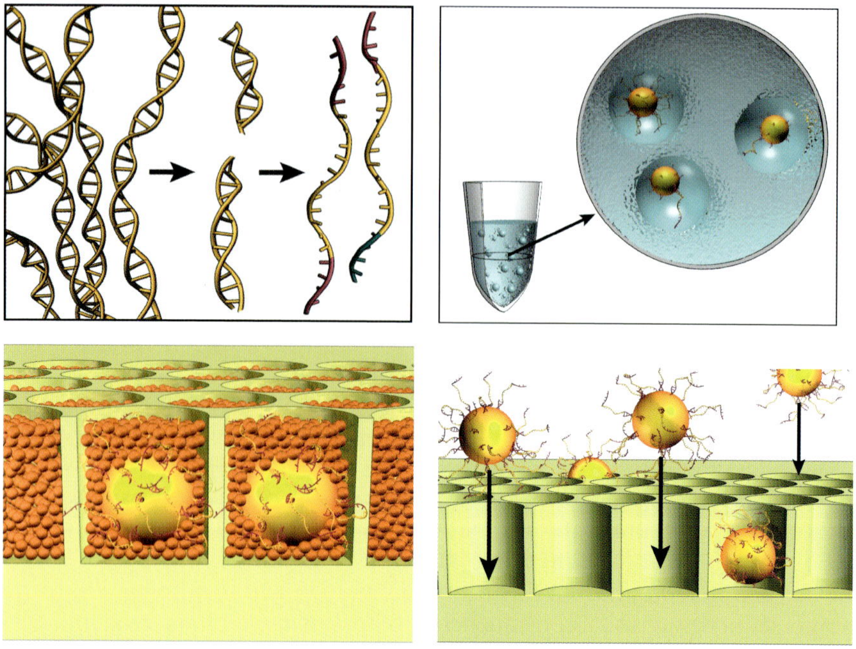

Fig. 1. Sample preparation. Clockwise from top left: (i) genomic DNA is isolated, fragmented, bound to adaptors and separated into single strands; (ii) fragments are bound to beads under conditions which favor at most one fragment per bead, the beads are captured in the droplets of a PCR-reaction-mixture-in-oil emulsion and PCR amplification occurs within each droplet; (iii) the emulsion is broken, the DNA strands are denatured, and beads carrying single-stranded DNA clones are deposited into the wells of a fiber optic slide; (iv) smaller beads carrying immobilized enzymes required for pyrophosfate sequencing are deposited into each well.

through a standard reaction involving T4 DNA polymerase and T4 polynucleotide kinase.

Following fragmentation and polishing of the genomic DNA library, non-phosphorylated oligonucleotide adaptors are added to the each end of the DNA fragments. The 44-base adaptors consist of a 5′ 20 base PCR amplification primer portion, followed by a 20 base sequencing primer portion and a 3′, 4 base, non-palindromic sequencing "key" comprised of one of each deoxyribonucleotide (e.g. AGTC). As explained below, this key sequence is necessary for the software to identify wells containing beads that are performing properly, and to assist in normalizing observed signals. Two classes of adaptors, termed "adaptor A" and "adaptor B", are used in each reaction. The A and B adaptors differ in both nucleotide sequence and the presence of a 5′ biotin tag on the B adaptor. Streptavidin–biotin interactions are used to remove fragments flanked by homozygous adaptor sets (A/A and B/B) and to generate single-stranded library templates (Figure 2).

Fragments carrying a biotinylated B adaptor are bound to Streptavidin beads while unbound material (composed of homozygous A/A adaptor sets, which

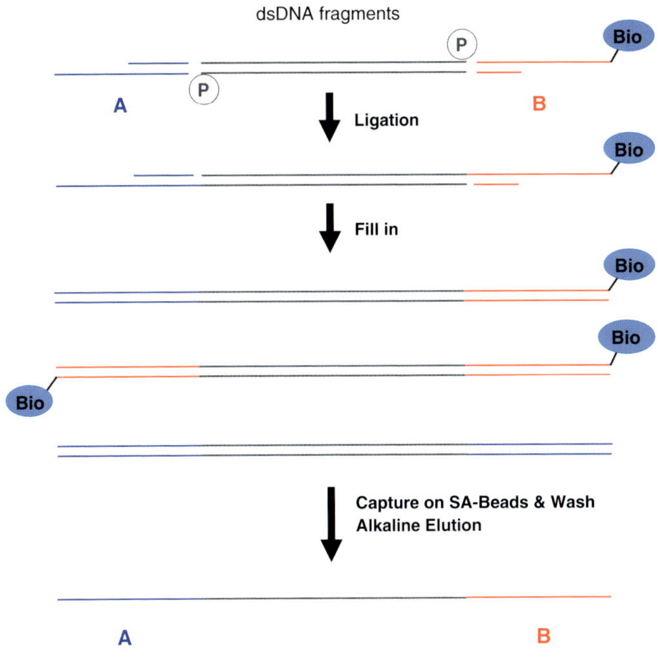

Fig. 2. Selection of AB-adapted library fragments.

lack biotin) is washed away. The immobilized fragments are then alkali-denatured; both strands of the B/B fragments remain immobilized through the biotinylated B adaptor, while A/B fragments are washed free and used in subsequent sequencing steps. Typically, the final library has a concentration that is well in excess of 10^8 molecules/μL.

Next, the template molecules are annealed to complementary primers on DNA-capture beads in a UV-treated laminar flow hood, under conditions that favor capture of one fragment per bead. These capture beads come from a Sepharose HP affinity column; they are NHS activated, after removal from the column, and sieved to yield a population with size distribution between 25 and 36 μm and carry in excess of 10^7 covalently linked capture primers that are complementary to one of the library adaptors. Owing to the limiting amount of template added, most beads are associated with only one template.

The individual DNA-carrying beads are isolated in separate aqueous droplets (on the order of 2×10^6/mL) made through the creation of a PCR-reaction-mixture-in-oil emulsion (Tawfik and Griffiths, 1998; Ghadessy *et al.*, 2001; Dressman *et al.*, 2003). The emulsification process creates a heat-stable water-in-oil emulsion whose individual droplets serve as a matrix for single molecule, clonal, amplification of the individual molecules of the target library in millions of parallel PCR, thereby overcoming the current limitation on amplification. At the conclusion of this process, each bead carries approximately 10^7 clonal copies of an individual sequencing template. Inherently, the volume of the emulsion can be scaled up to contain millions of beads, depending on the size of the genome and the number of fragments needed to cover that genome completely.

A 2 Mbp genome, oversampled 10 times, requires approximately 1.6 mL of emulsion.

After amplification, the emulsion is broken to release the beads; amplified DNA, immobilized on the capture beads, is then rendered single stranded by removal of the secondary strand through incubation in an alkaline melt solution. Up to this point the bead mass is comprised of both beads with amplified, immobilized DNA strands, and null beads with no amplified product. An enrichment process is used to selectively capture beads with sequenceable amounts of template DNA while rejecting the null beads. This ensures optimal usage of the wells in the fabricated slide, and results in a corresponding improvement in sequencing throughput.

Unlike in current sequencing technology, this approach does not require subcloning in bacteria or the handling of individual clones; the templates are handled in bulk within the emulsions, and clonality is achieved by limiting dilution. The result of this all-in-one sample preparation and amplification reaction is the creation of hundreds of thousands of beads, each containing millions of copies of their respective clonal fragment, sized to optimize the sequencing process.

2.2. Sequencing

Sequencing is performed by successive synthesis along the single-stranded DNA, using a process known as "pyrosequencing®" (Hyman, 1988; Nyren *et al.*, 1993; Ronaghi *et al.*, 1996), which was optimized to achieve high efficiency on solid support. Nucleotide incorporation is detected by inorganic pyrophosphate (PP_i) release.

$$\text{DNA} + \text{dNTP} \underset{\text{polymerase}}{\longrightarrow} \text{DNA}_{+1} + \text{PP}_i$$

$$\text{PPi} + \text{APS} \underset{\text{Sulfurylase}}{\longrightarrow} \text{ATP}$$

$$\text{ATP} + \text{Luciferin} \underset{\text{Luciferase}}{\longrightarrow} \text{Oxyluciferin} + \text{photon}$$

In a cascade of enzymatic reactions, visible light is generated in proportion to the number of incorporated nucleotides. The cascade starts with a nucleic acid polymerization reaction that releases inorganic PP_i only when a nucleotide is incorporated by polymerase. Released PP_i is converted to ATP by ATP-sulfurylase, providing energy to luciferase to oxidize luciferin and generate light. To increase the signal-to-noise ratio, the natural dATP is replaced by dATPαS. Since the added nucleotide is known, the sequence of the template can be determined.

The reactions take place in the open wells of a novel fiber optic slide (Figure 3). The slides are obtained by slicing of a fiber optic block that is manufactured by repeated drawing and fusing of optic fibers. At each iteration, the diameters of the individual fibers decrease as they are hexagonally packed into bundles of increasing cross-sectional sizes. Each optic fiber core is about 44 µm in diameter, surrounded by a 2–3 µm cladding; etching of each fiber's core creates wells of

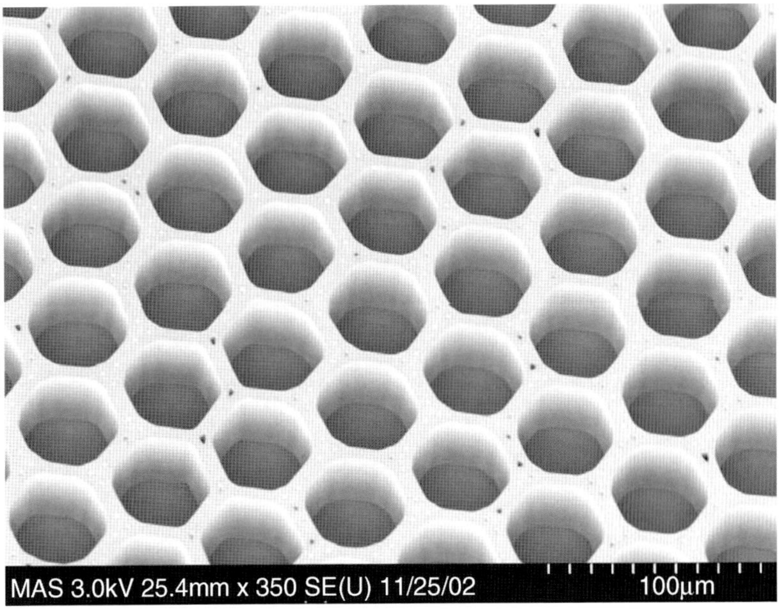

Fig. 3. SEM photograph of portion of a fiber optic slide, showing fiber optic cladding and wells prior to bead deposition.

approximately 55 μm in depth, with a well center-to-center pitch of 50 μm, resulting in a calculated well size of 75 pL and a well density of 480 wells/mm^2.

The fiber optic fabricated substrate, which currently contains approximately 1.6 million wells, is unique in that it both contains an array of wells in which sequencing reactions take place, and allows the capture of emitted photons from the bottom of each individual well. The substrate also has the advantage that it can be fabricated from fibers of any diameter, thus paving the way for further increase in well density, beyond the current one. To avoid having to load beads individually into wells, a new approach to template delivery was developed. The enriched template-carrying beads are simultaneously deposited by centrifugation into the slide's wells in a single step, one bead per well, together with a mixture of smaller beads that carry immobilized sulfurylase and luciferase enzymes necessary to generate light from free pyrophosphate. The DNA-carrying beads (average diameter is ~28 μm) are sized to ensure that no more than one bead fits in most reaction wells. Bead deposition is accomplished with the help of a custom-designed jig (Figure. 4).

The sequencing instrument consists of three major assemblies: a fluidics subsystem, a fiber optic slide cartridge/flow chamber and an imaging subsystem. Reagent inlet lines, a multi-valve manifold, a de-bubbler and a peristaltic pump, actuated by a stepper motor controlled by the computer, form part of the fluidics subsystem (Figure 5).

The individual reagents are connected to the appropriate reagent inlet lines, which allows for reagent delivery into the flow chamber, one reagent at a time, at a pre-programmed flow rate and duration. The fiber optic slide cartridge/flow

Fig. 4. Fiber optic slide in a custom-designed jig, used for bead deposition. After insertion of the slide in the jig, a slurry containing the DNA beads and the enzyme beads is slowly injected in the thin channel above the slide's wells. The jig is then mounted in a centrifuge to force the beads into the wells.

chamber has a 300µm space between the slide's etched side and the flow chamber ceiling. This chamber and the flow regimes are designed such that laminar flow is achieved everywhere in the chamber but at very short distances from the inlet and outlet ports, which are located outside of the imaging area. Laminar flow is critical to the efficient diffusion of reagents into and out of the reaction wells. The flow chamber also includes means for temperature control of the reagents and fiber optic slide, as well as a light-tight housing. The polished (unetched) side of the slide is placed directly in contact with a second fiber optic imaging bundle that is bonded to a CCD sensor, allowing the capture of emitted photons from the bottom of each well.

Sequencing is achieved by the cyclical delivery of sequencing reagents into the flow chamber and washing out of the sequencing reaction by-products. In the flow chamber, reagents flow perpendicularly to the wells and they, as well as the enzymatic reaction products, enter and exit the wells through lateral diffusion; the process does not rely on the individually controlled dispensing of reagents to individual wells. This configuration allows simultaneous extension reactions on template carrying beads within hundreds of thousands of open wells and relies on convective and diffusive transport to control the addition or removal of reagents and by-products. The generated light, transmitted through the base of the fiber optic slide, is detected by a large format CCD (4096 × 4096 pixels). A bead carrying 10 million copies of a template yields approximately 10,000 photons at the CCD sensor, per incorporated nucleotide.

Following the flow of each nucleotide, a wash containing apyrase is used to ensure that nucleotides do not remain in any well prior to the next nucleotide

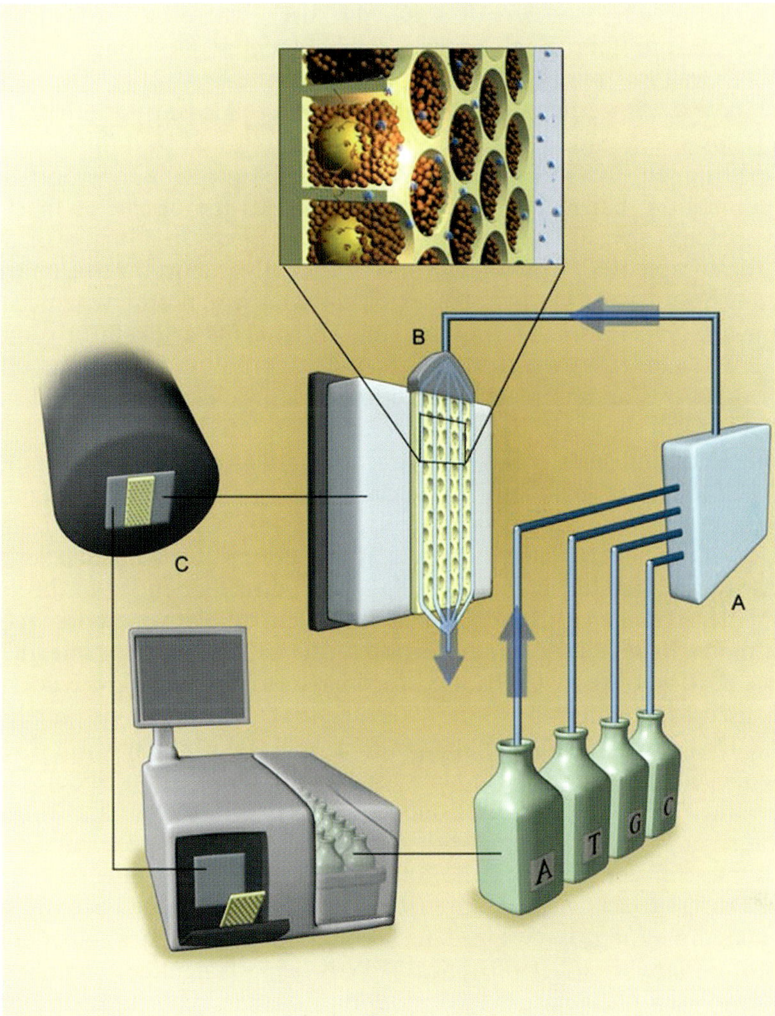

Fig. 5. The sequencing instrument and its three major subsystems: a fluidic assembly (A), a flow chamber that includes the well-containing fiber optic slide (B) and a CCD camera-based imaging assembly (C).

being introduced. The cyclical delivery of sequencing reagents into the fiber optic slide wells and washing of the sequencing reaction by-products from the wells is achieved by a pre-programmed operation of the fluidics system. The program is written in the form of an Interface Control Language (ICL) script, specifying the reagent name (Wash, dATPαS, dCTP, dGTP, dTTP and PP_i standard), flow rate and duration of each script step. Flow rate is set at 4 mL/min for all reagents and the linear velocity within the flow chamber is approximately 1 cm/s. A typical run consists of 42 cycles, organized as follows: a first "kernel" consisting of a PP_i flow (21 s), followed by 14 s of substrate flow, 28 s of apyrase wash and 21 s of substrate flow. That PP_i kernel is then followed by 21 cycles of dNTP flows (dC, dG, dT, dA), where each dNTP cycle is composed of four individual kernels. Each kernel is 84 s long (dNTP,21 s; substrate flow, 14 s; apyrase wash, 28 s; and substrate flow, 21 s). After 21 cycles of dNTP flow, a PP_i kernel is introduced, and then followed by another 21 cycles of dNTP flows. The end of the sequencing run is followed by a third PP_i kernel. Because of the sensitivity of pyrophosphate-based sequencing to various sources of contamination, particular attention was paid to the design of the reagent delivery system to ensure complete washout of reagents in all parts of the system, at the completion of each flow. This is critically important to eliminate the loss of synchronism during template extension that would otherwise occur. The total run time is 244 min. During the run, all reagents are kept at room temperature. The temperature of the flow chamber and flow chamber inlet tubing is controlled at 30°C and all reagents entering the flow chamber are pre-heated to 30°C. Such a run results in an average read length of 100 bp per template; and increasing the number of cycles yields correspondingly longer read lengths.

The optical system includes a CCD camera with a one-to-one imaging fiber bundle (6 μm fibers) and associated electronics; the CCD is cooled to improve sensitivity. The camera currently used is a Spectral Instruments Series 800 camera with a Fairchild Imaging LM485 sensor, directly bonded to the imaging fiber bundle. The camera can be operated in either of two modes: (i) frame transfer mode, in which the center portion of the CCD is used for imaging while the outer portion of the CCD is used for image storage and slow readout (this mode is used for smaller fiber optic slides) or (ii) full frame mode, in which the entire CCD is used for imaging and readout occurs during the wash (i.e. dark) portion of each flow cycle (this mode is used for the $60 \times 60\,mm^2$ slide). The data are read out through four ports, one at each corner of the CCD. Signal integration is set at 28 s per frame, with a frame shift time of approximately 0.25 s in the frame transfer mode; in the full frame mode, signal integration (frame duration) is set at 21 s (wash capture frame) and 63 s (nucleotide capture frame).

The imaging system is designed to accommodate a large number of small wells and the large number of optical signals being generated from individual wells during each nucleotide flow. It does not require the wells to be registered with respect to the imaging system: once mounted, the fiber optic slide's position does not shift, making it possible for the image analysis software to determine the location (in CCD pixel coordinates) of each well, based on light generation during the pyrophosphate solution flow which precedes each sequencing run. In operation, the entire slide is simultaneously imaged by the camera. A single well is imaged by approximately nine 15 μm pixels. For each nucleotide flow, the light intensities collected by the pixels covering a particular well are summed to generate a signal for that particular well at that particular nucleotide flow. Each image captured by the CCD produces 32 MB of data. The images are processed to yield sequence information simultaneously for all wells. In order to perform the necessary signal processing in real time, the control computer relies on an accessory board, hosting a 6 million gate FPGA (Mehta *et al.*, 1993; Fagin *et al.*, 1993). The FPGA essentially performs the role of a high-speed, on-the-fly re-programmable, co-processor. The system includes software that downloads to the FPGA, in the space of milliseconds, the necessary binary modules that encode *in hardware* the algorithms necessary to perform the successive image processing steps. In this manner, image processing itself is completely off-loaded from the computer, leaving to the main computer only the task of downloading appropriate binaries to the FPGA and processing flow signals. This approach enables the instrument to carry out all of the required image processing in *real time*; at the conclusion of a sequencing run, all of the necessary data are available to the on-board computer to execute final signal adjustments and to map the fragments to a specified genome or to perform *de novo* assembly.

2.3. Image processing

Raw signals are background-subtracted, normalized and corrected. The first step in processing data is to perform background subtraction for each acquired image at the pixel level, using an "erosion-dilation" algorithm that automatically determines the local background for each pixel. Then, for each nucleotide flow, the light intensities collected, over the entire duration of the flow by the pixels covering a particular well, are summed to generate a signal for that particular well at that particular flow. The acquired images are corrected to eliminate cross talk between wells due to optical bleed (the fiber optic cladding is not completely opaque and transmits a small fraction of the light generated within a well into an adjacent well) and to diffusion of ATP or PP_i (generated during synthesis) from one well to another further downstream. To perform this correction, the extent of cross talk under low-occupancy conditions was empirically determined and de-convolution matrices were derived to remove from each well's signals, the contribution coming from neighboring wells. In order to account for variability in the number of enzyme-carrying beads in each well and variability in the number of template copies bound to each bead, two types of normalization are carried out: (i) raw signals are first normalized by reference to the pre- and post-sequencing run PP_i standard flows and (ii) these signals are further normalized by reference to the signals measured during incorporation of the first three bases of the known "key" sequence included in each template.

The normalized and corrected signal intensity at each nucleotide flow, for a particular well, indicates the number of nucleotides, if any, that were incorporated. This linearity in signal intensity is observed to remain valid through homopolymers of length at least up to eight (Figure 6). However, in sequencing by synthesis a very small number of templates on each bead lose synchronism (i.e. either get ahead of or fall behind, all other templates in sequence) (Ronaghi, 2001). The effect is mostly due to undegraded or leftover nucleotides in a well (creating "carry forward") or to impaired polymerase activity (creating "incomplete extension"). Typically, carry forward rates of 1–2% and incomplete extension rates of 0.1–0.3% are seen. It is important to correct signals for these effects, because the loss of synchronism is a cumulative error that degrades the quality of sequencing at longer read lengths.

As a result, the impact of carry forward and incomplete extension is felt particularly toward the end of reads as illustrated in Figure 7 which shows the average read accuracy, at the single read level, as a function of base position.

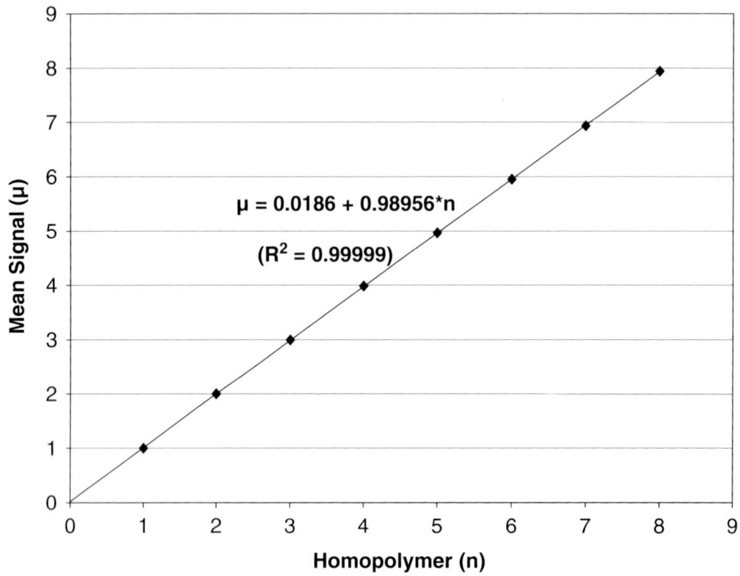

Fig. 6. Average of the flow signals ascribed to various homopolymers for the mapped reads of an *M. genitalium* run.

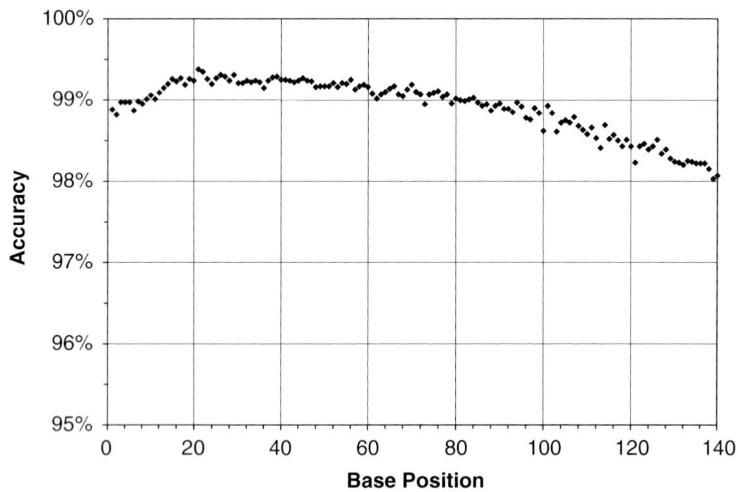

Fig. 7. Average read accuracy, at the single read level, as a function of base position for the mapped reads obtained in sequencing a typical microbial genome.

Based on detailed models of the underlying physical phenomena, one can develop bootstrapping methods that yield the amounts of both carry forward and incomplete extension occurring in individual wells, even when the underlying sequence is unknown. The approach is based on an iterative technique and two-dimensional minimization to achieve a least squares fit between the measured signals and the model's output. Based on these results, it is then

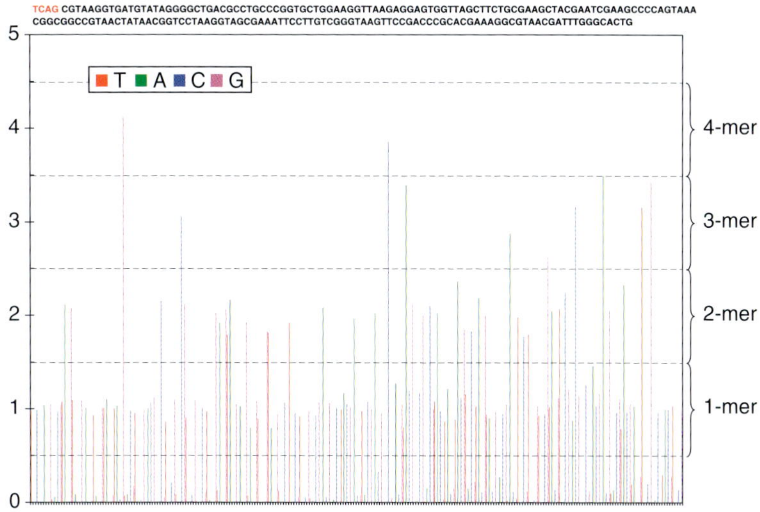

Fig. 8. Flowgram for a 191 base read from a *S. aureus* run. Nucleotides are flowed in the order T, A, C, G. The sequence is shown above the flowgram. The signal value intervals corresponding to the various homopolymers are indicated on the right. The first four bases (in red, above the flowgram) constitute the "key" sequence, used to identify wells containing a DNA-carrying bead.

possible to correct the observed signal traces and accurately base call each fragment.

Figure 8 shows the processed result, illustrating a trace (referred to as a "flowgram") for an error-free 191 bp long read generated in a 63 cycle run of a S. aureus library. Such error free reads are a product of the iterative technique and two-dimensional minimization to achieve a least squares fit between the measured signals and the model's output.

2.4. Sequencing accuracy

The sequencing performance of this system and the effectiveness of the correction algorithms can be verified independently of possible artifacts introduced during sample preparation or amplification by using synthetic templates with difficult-to-sequence stretches of identical bases of increasing length. Using such fragments, a read error rate of approximately 0.3%, at read lengths in excess of 100 bases, is observed. Approximately, half of the errors result from single-base insertion or deletions and the remainder from errors in homopolymer calls.

2.5. Base calling

Before base calling or aligning reads, poor quality reads are filtered out using the measured signals without relying on any information on the genome or template being sequenced. The approach is based on the observation that poor quality reads have a high proportion of indeterminate signals, i.e. signals that do not allow a clear distinction between a flow during which no nucleotide was incorporated and a flow during which one or more nucleotide was incorporated.

Each flow, in each well, results in no incorporation, or incorporation of one, or two, or three, etc. nucleotides. For any sequencing run, a histogram of signal intensities for each of these groups can be compiled (when dealing with a known sequence). As illustrated in Figure 9, the signal strengths of the various groups overlap slightly. Generally, good reads (i.e. those that map to a reference genome with few errors) have most of their signals close to integral values equal to the number of incorporated nucleotides. Those reads in which a substantial number of signals fall in the overlap region between a negative flow (one in which no nucleotide is incorporated) and a positive flow (one in which at least one nucleotides is incorporated) ($0.5 < \text{signal} < 0.7$) are found to be of poor quality (i.e. do not map anywhere in the genome or do so with a large number of errors), mostly because such reads originate from beads that carry copies of two or more templates. This forms the basis for an *a priori* filter that selects high-quality reads: for each read, the number of flows that fall in the overlap region is counted and only those reads whose number of such flows is less than 5% of the total number of flows are selected.

For a read that does not meet this criterion, the read is progressively trimmed by eliminating flows, starting from the end of the read, until the criterion is either satisfied (number of flows in indeterminate region <5% of remaining flows) or the number of flows has been reduced to less than half the original number of flows, at which point the read is considered to have been filtered out of the pool of high-quality reads. This selection process has been verified to be very effective, as most of the remaining reads map to the underlying genome at high accuracy.

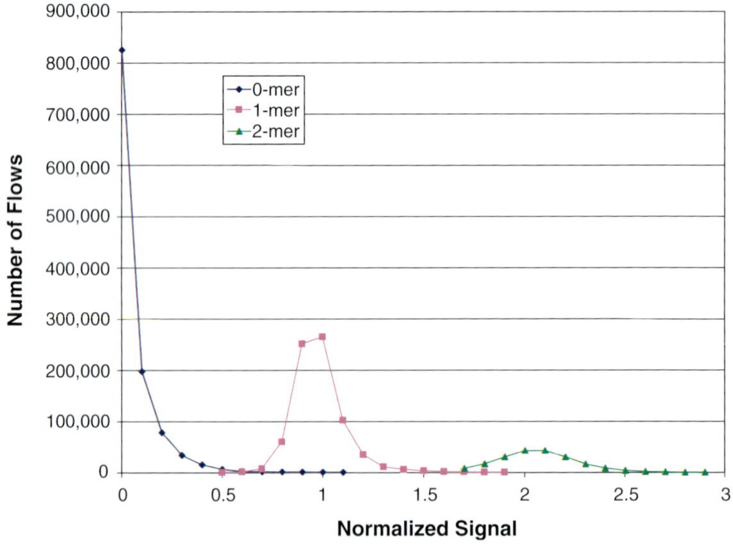

Fig. 9. Typical histogram of signal intensities for negative and positive flows.

When base calling individual reads, in principle, the intensity of an observed signal directly indicates the number of incorporated nucleotides. However, as previously mentioned, signal strengths for the various homopolymers are not fully resolved. Were it not for this, it would be possible to base call unambiguously any given sequence of signals. In pyrophosphate-based sequencing, the two types of direct errors are overcalls (calling one more base than actually present in the genome) or undercalls (calling one less base than actually present in the genome). The identity of a base is not in question since it is determined by the addition of one known nucleotide at a time. Substitution errors (miscalling one base for another) result from the occurrence of two consecutive errors (undercall followed by overcall or vice versa) and are therefore significantly rarer. For libraries, the observed error rates are consistent with the hypothesis that some beads carry copies of more than one template. Most of these reads get filtered out by the selection process described above. Those, however, for which the admixture significantly favors one template, may not be filtered out and contribute heavily to the overall error rate. Based on the observed distributions of signals, it is possible to estimate *ab initio* the quality (or probability of correct base call) of each base of a read, analogous to the Phred score (Ewing *et al.*, 1998) used by current Sanger-based sequencers.

The confidence in (or "quality" of) any particular base call associated with a given signal value is a function of where that signal falls in the distribution of signals, for a given homopolymer length. Based on a large number of runs in which various known genomes were sequenced, and mapping the resulting reads, it was found that negative flows follow a lognormal distribution, while all positive flows are normally distributed with mean and standard deviation proportional to the underlying homopolymer length; furthermore these distributions remain remarkably invariant across different genomes. This observation allows the calculation of a quality score for each individual base called. To estimate a quality score for a particular base call, the probability must be determined that the measured signal originates from a homopolymer of length at least equal to the called length. For instance, if two As are called for a particular signal, the quality score for the second A is given by the probability that the observed signal came from a homopolymer of length two or greater. Since the probability of measuring a signal, given a homopolymer length, was empirically established, Bayes' theorem can be used to determine the probability that a particular homopolymer length produced the observed signal, as follows:

$$P(n|s) = \frac{P(s|n)P(n)}{\sum_j P(s|j)P(j)}$$

where s is the observed signal and n the length of the homopolymer that produced the signal. As described above, the probability $P(s|n)$ of measuring signal s given a homopolymer of length n follows a Gaussian distribution. For a random nucleotide sequence, the probability $P(n)$ of encountering a homopolymer of length n is simply $1/4^n$ (ignoring a multiplicative normalization constant). The quality score assigned to each base called for each fragment can then be reported as a Phred-equivalent using the following transformation:

$$Q = -10\log_{10}[P(\geq n|s)]$$

The validity of this approach was demonstrated by correlating calculated and observed Phred scores, sequencing known genomes other than those used to establish the distribution of signals (see Figure 10).

2.6. Sequence alignment

In pyrophosphate-based sequencing, the base call is sensitive to the threshold values used to convert each signal to the corresponding integer number of bases. A difference in base calls may result from very minor differences in signal strength near a threshold value. To increase accuracy, the system uses algorithms that take advantage of the large oversampling present in a typical sequencing experiment. In this scheme, referred to as "flow-space mapping", fragments are aligned to one another using the signal strengths at each nucleotide flow, rather than individual base calls, to determine optimal alignment.

Given the order in which nucleotides are flowed, a given reference genome implies a known succession of ideal signal values. This ideal flowgram is divided into contiguous, overlapping, sub-flowgrams of a particular length (default length is 24 flows), which are indexed so as to allow very rapid searching (each sub-flowgram starts at a positive flow). To map the query flowgram to the targetthat is divided into sliding sub-flowgrams having the length that was used in the indexing step and the space of indexed ideal sub-flowgrams is searched. A perfect match anchors the query flowgram against the reference genome. The

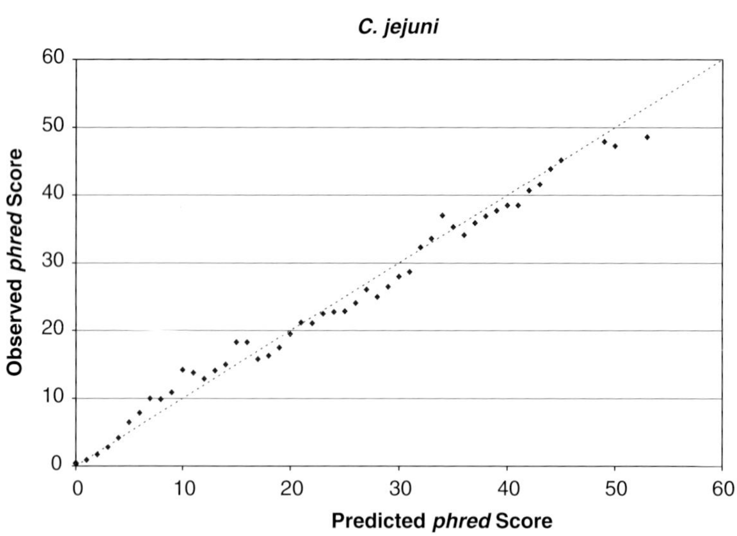

Fig. 10. Correlation between predicted and observed quality scores for a sequencing run of *C. jejuni.*

alignment of the read is then assessed beginning at the 5′ end, moving down the entire length of the read. The longest segment that meets a user-specified total mismatch threshold is selected at which point the alignment is terminated and the read is trimmed. The reads are aligned to the reference at a very low level of stringency in order to detect mutations or other genomic variations. Once such alignments have been performed, all the flow signals from the various reads that correspond to the same location in the target are arithmetically averaged, after which individual base calling is performed.

This procedure is extremely effective in reducing error rates (Figure 11); it is equally applicable whether re-sequencing or consensus base calling a *de novo* assembly. The quality of the consensus base call (without relying on knowledge of the underlying sequence) can be estimated by measuring the absolute value of the average signal's distance from the closest signal threshold for the corresponding homopolymer, and dividing it by the normalized standard deviation of all the signals measured at that particular genome location. This ratio is called the *Z*-score. To enhance the reliability of observed variations, the consensus sequence is filtered by imposing a minimum *Z*-score to give rise to a high-quality consensus sequence. By using a few exactly known sequences the number of errors in the consensus sequence can be determined, in order to estimate the quality of the consensus calls and the correlation between minimum *Z*-score and consensus accuracy. The system reports genome coverage based on regions with consensus sequence accuracy of 99.99% or better, which typically is achieved by selecting a minimum *Z*-score equal to four. Without using *Z*-scores as a cut-off, consensus sequences cover the genome except for repeat

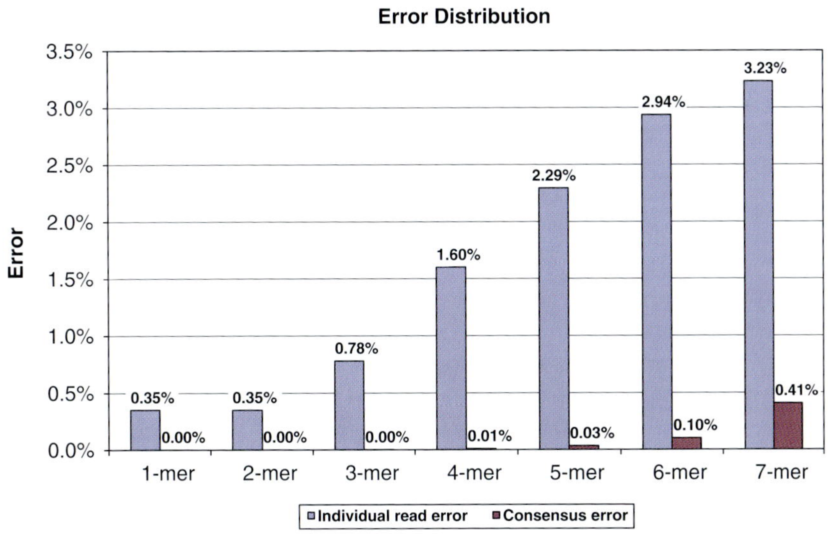

Fig. 11. Error rates in sequencing an *E. coli* library, as a function of homopolymer length. Single-base error rates are referenced to the total number of single-bases sequenced; for homopolymers, the error rate is referenced to the total number of bases sequenced that belong to homopolymers of each length. The error rates are shown for individual reads and after the consensus sequence was formed using all reads, without *Z*-score restriction.

regions that cannot be uniquely resolved with 100 bp reads and very small regions where no reads exist because of sample preparation bias or the unavoidable statistics of random fragmentation (Lander and Waterman, 1988). As bases are selected by selecting higher minimum Z-score values, the consensus accuracy increases substantially, while coverage decreases somewhat because lower oversampling in some genome regions leads to lower Z-scores.

Current Sanger-based sequencers require a depth of coverage at any base of at least 3 in order to achieve a consensus accuracy of 99.99%. To achieve a minimum of three-fold coverage of more than 95% of a typical genome requires approximately sevenfold oversampling. Because of higher individual read errors, this system achieves comparable consensus accuracies, over a similar fraction of a genome, with a local depth of coverage of four or more, requiring approximately 10–12 times oversampling.

3. APPLICATIONS

3.1. *De novo* sequence assembly

Historically, performing assembly of shotgun sequence reads has been based on a comparison of their respective nucleotide sequences. Relying on comparisons of nucleotide sequences, however, would ignore much of the information contained in the original flow-based signal trace. In addition, existing assemblers are not optimized for 100–200 bp reads, particularly with respect to memory management due to the increased number of fragments needed to achieve equivalent genome coverage (A completely random genome covered with 100 bp reads requires approximately 50% more oversampling for its assembly to yield the same number of contiguous regions (contigs) as would result from covering it with 700 bp reads, assuming 30 bp overlap between reads.) (Lander and Waterman, 1988). To address these issues and combine the individual reads into a complete and accurate consensus genome sequence, the system uses a newly developed assembler, which applies an "overlap-align-detangle" approach to creating multiple alignments and consensus base calls for the unique regions and repeat regions of the genome.

Assembly starts with the selection of high-quality reads (as described above) to ensure that the flowgrams to be processed consist most likely of sequence data from the original sample. Before the flowgrams are used by the assembler, their signals are normalized and scaled across wells, so that they can be directly compared to each other during assembly. The overlap phase of the assembler performs an incremental all-against-all fragment comparison to identify pairwise overlaps between fragments. An initial indexing of selected seed regions of all fragments is used to quickly identify the set of candidate overlaps for a query fragment. Those candidate overlaps are assessed by directly comparing the flowgrams of the fragments. The differences between signals for each flow are calculated and evaluated, using a difference threshold that grows in proportion to the signal intensity. The overlaps are then trimmed by removing high-difference ends, and a final test is performed to determine whether the overlap passes the pair-wise stage (overlaps that are longer than 30 flows and with fewer than 3% of their flows different pass the test). A final filtering step that considers all of the passed pair-wise overlaps generates the set of overlaps sent to the alignment phase: if the longest overlap is less than 120 flows, overlaps that are 30 or more flow shorter than the longest overlap that are removed; otherwise, overlaps that are 60 or more flow shorter than the longest overlap that are eliminated.

The alignment phase of the algorithm constructs a tiled multiple alignment structure of the transitive union of all of the pair-wise overlaps reported by the overlapper. Where the pair-wise overlaps of the fragments are consistent across the fragments, the resulting alignment is equivalent to the contig alignments in conventional assemblers. Where sets of the fragments have either diverging or converging alignments, such as for the fragments on the boundaries of repeat regions, the overall alignment is broken into sections and the data structure is organized as a graph, where nodes correspond to a consistent alignment region, and edges correspond to where the alignments of subsets of fragments either diverge or converge. Each fragment in the data structure spans a path of the graph, with sections of its flowgram aligned in a node along the path.

The alignment and overlap modules work together to improve the efficiency of the software. After a query fragment's set of overlaps have been added to the multi-alignment data structure, the next query fragment is chosen by evaluating the fragments that are partially aligned in the data structure and choosing the one that would provide a large extension of an existing alignment. If no partially aligned fragments are found in the multi-alignment data structure, an unaligned fragment is chosen as the next query. Any fragment which was completely aligned by the use of other query fragments is not given to the overlap module, resulting in time savings generally proportional to the depth of coverage (since, generally, the first query fragment for a region will collect all of the other fragments to be aligned in that region and add them to the multi-alignment).

After the multi-alignment data structure has been constructed, a detangling phase evaluates the graph of alignment sections and applies an iterative set of rules to correct for any overly collapsed regions (from spurious small overlaps) or "outlier" edges in the graph (where a single read or a few reads diverge from a consistent, much larger set of reads). At the end of the detangling phase, each chain of nodes in the graph (where no branches occur along the chain) corresponds to a consistent alignment of fragments formed by the joining of the multi-alignment sections at each node. Chains whose joined alignments are longer than 500 flows are identified as contigs, and are marked for output as the result of the assembler.

The final step in the assembler is the generation of the consensus sequences for each contig. This step uses the constructed multiple alignments (which are alignments of flowgram signals, not nucleotide bases) to identify a consensus number of bases to be called for each flow. The consensus is determined by first performing a bias correction of each fragment, followed by a mean signal calculation of each alignment column. Each fragment's signals are compared to the global distribution of signals; if the fragment's signals consistently fall below or above the mean of the distributions of the signals (each signal initially identified as a 1-mer signal is compared to the mean of all 1-mer signals, and so on), the signals are scaled. After that, new means and thresholds of the signals are computed to identify the range of signals values that correspond to each *n*-mer incorporation. These thresholds are then used to base call the mean signal for each column of the contig alignment, and the contig consensus sequence is the concatenation of the base called *n*-mers of the flows.

3.2. Sequencing results

A collection of genomes of length less than 10 Mb were re-sequenced using the system described above to check the performance of this system in terms on coverage and accuracy. Clones were used whose sequence is available in GenBank. Table 1 summarizes the results obtained.

After sequencing, the raw reads were mapped against the reference genome; when assessing sequencing results, only reads that map to unique locations in the reference genome should be included. Because this process excludes repeat regions, the selected reads typically do not cover the genome completely. As the results show, the system described in this chapter achieves a high degree of coverage across the genomes with a high degree of concordance with the published genomes. Breaks in the scaffold are the result of incomplete coverage of the genome due to random chance and repeat regions that are longer than the raw sequence reads and cannot be uniquely anchored.

The assembler's results for these same bacteria are presented in Table 2. The genome assemblies are nearly as comprehensive as the mapping results, with the vast majority of bases in the assemblies correct (as measured by their

Table 1. Uniformity of coverage, achieved with the 454 sequencing system, on a number of bacterial genomes. The results shown are mapping of reads to a known genome. As "non-repeat coverage" demonstrates, the 454 process achieves a high degree of coverage across the genomes. The repeats are excluded from the mapping results, as they are not uniquely mapped with 100 bp reads

	M. genitalium	*E. coli*	*S. pneumoniae*	*B. licheniformis*
Genome size	580,069	4,639,675	2,160,837	4,222,645
Runs	0.5	3	2	3
Reades mapped	165,428	1,011,908	682,658	1,111,742
Deapth	30	22	31	26
Genomic coverage (%)	99.98	98.12	96.18	98.92
Non-repeat coverage (%)	99.9998	100.00	100.00	99.999
Consensus accuracy (%)	99.988	99.995	99.946	99.993
% bases Q40+ (%)	99.36	99.70	99.00	99.80
Q40+ accuracy (%)	99.998	99.998	99.994	99.999
Contigs (%)	17	144	220	42

Table 2. Summary of results for the 454-*de novo* assembler on several bacterial genomes

	M. genitalium	*E. coli*	*S. pneumoniae*	*B. licheniformis*
Genome size	580,069	4,639,675	2,160,837	4,222,645
Runs	0.5	3	2	3
Assembly contigs	19	140	228	98
Assembly coverage (%)	99.66	97.56	92.46	98.64
Overall accuracy (%)	99.993	99.998	99.991	99.998
Average contig size (kb)	29.5	32.4	8.8	42.5
N50 contig size (kb)	41.0	67.2	14.0	74.3
Largest contig (kb)	130	164	66	263

concordance with the published reference genome). Additionally, the assemblies generated place greater than half of the bases (the N50 contig size) into large contigs. As with the mapping results, the breaks in the contigs are the result of random chance and at the boundaries of repeats. The contigs from the assembler are output in standard file formats and readily incorporated into standard sequence viewers.

3.3. Comparative genomics

In comparative genomics, efforts up to now have typically focused on known polymorphisms or gene coding regions. Total genome comparison has long been considered an appealing, but largely unattainable, tool for discovery. The ability to perform sequence-based, whole-genome comparisons enables not only the identification of known variants, but also the identification of *de novo* mutations. The wait and cost inherent in obtaining deep sequencing of multiple genomes for comparative analysis have made the approach unrealistic in achieving results in a timely and competitive manner. Many labs have therefore come to rely on the cumbersome task of sifting through low coverage draft sequences in an effort to define a genome's critical region associated with their research question (Bhattacharyya *et al.*, 2002; Read *et al.*, 2002; Goo *et al.*, 2004). Alternative techniques, while quick and inexpensive for the identification of a known point mutation or for querying a subset of the genome, either lack the ability to detect *de novo* mutations or are costly and time intensive on a genome-wide scale (Honisch *et al.*, 2004; Hinds *et al.*, 2005; Hardenbol *et al.*, 2005; Zwick *et al.*, 2005).

In a first demonstration of whole-genome sequence comparison of multiple, cross-species genomes, four strains of *Mycobacterium* were sequenced on the 454 system, enabling a team of scientists to identify the mechanism of action of the first new tuberculosis-specific drug discovered in 40 years (Andries *et al.*, 2005). Specifically, in order to pinpoint the mechanisms of action for the drug, one drug-resistant strain of *Mycobacterium tuberculosis* (4.4 Mb) and three strains of *Mycobacterium smegmatis* (6.9 Mb) (wild-type and two drug-resistant strains) were sequenced. The mutants, R09 and R10, were grown in a controlled environment to develop drug resistance to diarylquinoline R207910. In order to generate the requisite depth of coverage across the genomes, five sequencing runs of *M. smegmatis* R09, four sequencing runs of *M. smegmatis* R10 and five sequencing runs of the *M. smegmatis* WT were performed. After the sequencing runs, the reads from the individual runs for each bacterial genome were pooled together and mapped against the consensus *M. smegmatis* genome (TIGR, 2003) using the flowgram-based mapping algorithm as described above. This process, from purified genomic DNA to assembled genomes, required seven working days for the three *M. smegmatis* genomes processed in parallel.

By mapping the reads against the reference genome and building consensus sequences, high-confidence mutations, those with a *Z*-score (as explained above) greater than four, were identified for each strain. Only the mutation positions present in *M. smegmatis* R09 and R10, but absent in the mutation list generated for the WT were further considered. Mutations present in both the WT and the mutant strains, as compared to the reference sequence, are considered to be of no significance. Since the two mutants in this study were derived directly from the WT clone, any mutations in all three microbes could not confer drug resistance as the WT clone was sensitive to the drug. These putative mutations may reflect differences between the wild-type strain and the strain sequenced by TIGR, or are the result of sequencing artifacts, either in the present sequence or in the TIGR reference sequence. This exercise in comparative genomics led to a total of 30 point mutation positions in the genomes. These 30 potential mutations were checked by comparing the raw reads against the reference sequence with BLAST (Altschul *et al.*, 1990). Further analysis rapidly reduced the 30 possible mutations to 4 valid ones that were present in both mutants, but absent in the *M. smegmatis* WT. The four mutation positions are 1,199,061, 3,969,644, 5,011,865 and 6,945,401. This information was used to identify the ATP synthase gene of *M. tuberculosis* and *M. smegmatis* as the predicted target gene of the new diarylquinoline drug.

Mutation of the ATP synthase gene at position 3,969,644 confers resistance to the drug. Complementation study demonstrated that resistance to the compound could be transferred to wild-type *M. smegmatis* by cloning the mutated form of the atpE gene, as part of the ATP F0 operon, from the resistant *M. smegmatis*. Comparisons between the reference and three *M. smegmatis* genomes highlighting the mutation at position 3,969,644 are shown in Figures 12 and 13.

3.4. Ultra-deep sequencing of PCR amplicons

Several established DNA analysis technologies are based on the principles of miniaturization and parallel processing, notably DNA microarrays and various bead-based technologies; however, they all rely on detailed *a priori* sequence information and are limited to the interrogation of very short stretches of DNA, such as single base-pair changes. The sequencing technology described here, on the other hand, is unique in combining high-throughput processing with an unparalleled capability to generate *de novo* sequence reads of 100 or more bases

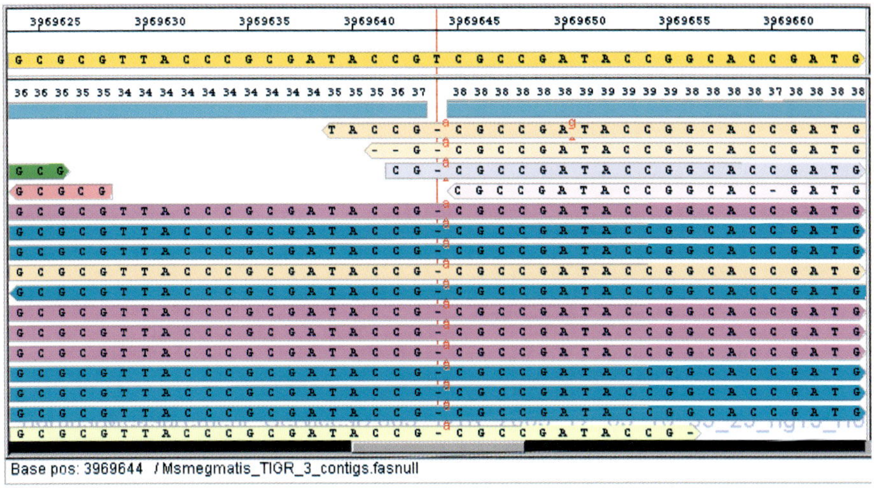

Fig. 12. An approximately 40 bp section of the genome from *Mycobacterium smegmatis* strain R09 as displayed with a viewer. The consensus sequence is displayed in yellow on top; numbers above the yellow sequence are the nucleotide position in the reference sequence. Individual reads are shown in other colors below. Numbers below the yellow consensus sequence indicate the depth of coverage at that position. Arrows on the mapped reads indicate whether the reads are forward or reverse. The T to A mutation in *Mycobacterium smegmatis* R09 is clearly identified.

in length. The only alternative, the massively parallel signature sequencing (MPSS) technology, has a reported read length of 20 bases or less, limiting its usefulness for certain applications (Brenner *et al.*, 2000). While the number of called bases per run of the 454 system represents a significant increase, the gain in the number of individual reads is even more substantial, exceeding that of traditional sequencing by 2000-fold or more. This large capacity leap opens up possibilities for new applications that are neither practical nor economical with traditional instrumentation.

A particularly attractive one, considering both the achievable throughput and read length, is the sequencing of PCR generated amplification products, here referred to as amplicons, with the specific purpose of identifying and quantifying known or novel sequence variants of low abundance. The 454 system can detect and quantify variants that may contain substitutions, insertions and deletions, simply by counting their presence. The sensitivity of detection is directly related to the level of coverage; in its current form the technology typically can detect sequence variants down to 0.1% frequency, provided at least 10,000-fold coverage (see Figure 14).

The list of potential applications for amplicon sequencing is extensive and includes sequencing of exons from disease-related genes and gene families, ribosomal RNA for microbial population studies, synthetic DNA tags used in various labeling strategies, short amplifiable DNA fragments resulting from, for example, SAGE (Velculescu *et al.*, 1995), or as described below, amplicons of viral origin relevant to drug resistance.

```
                      QUERY        3969594  cctcgggctgacgggcgatacccgagatcagcgcgttacccgcgataccgtcgccgatac 3969653
                      58735_3858_0945  4    ..................................................a......... 63
                      169835_0305_3581 94   ..................................................a......... 35
                      45183_1365_2682  19   ..................................................a......... 78
Mutant 1              196346_2551_2079 1                   ...................................a......... 45
                      299228_1092_2061 27   ..................................................a......... 86
                      132916_1686_0668 77   ..................................................a......... 18
                      193276_0501_3973 100                              .......................a......... 68
                      76215_1333_2375  78                               .......................a......... 46

                      QUERY        3969594  cctcgggctgacgggcgatacccgagatcagcgcgttacccgcgataccgtcgccgatac 3969653
                      298845_2689_0801 90   ..................................................a......... 31
                      40323_0307_2345  1                ......................................a......... 48
                      15674_2999_2644  1                ......................................a......... 48
                      149461_3070_2269 15   ............................n.....................a......... 74
Mutant 2              195177_3827_2381 39   ..................................................a......... 98
                      111775_3815_0680 8                               .......................a......... 40
                      168061_0901_0487 1                                 .....................a......... 31
                      31208_0583_1683  1                 .....................................a......... 47
                      179976_1915_1416 1                                          ............a......... 22
                      274863_3462_2981 39   ..................................................a......... 98
                      189106_2014_0516 65   ..................................................a......... 6

                      QUERY        3969594  cctcgggctgacgggcgatacccgagatcagcgcgttacccgcgataccgtcgccgatac 3969653
                      124850_2690_0974 104 ............................................................ 45
                      226456_0862_1023 105        ..................................................... 53
                      34384_0354_1505  106         ................................................... 56
Wild type             107461_2519_2710 89   ............................................................ 30
                      203421_3075_1136 1                   ............................................. 45
                      254338_0535_3866 1                   ............................................. 45
                      138303_0181_2930 19        ....................................................... 73
                      184205_3518_3516 107           ................................................... 57
                      171766_1683_0761 16   ............................................................ 75
                      113194_0257_3762 36   ............................................................ 95
                      77903_2541_0554  76   ............................................................ 17
```

Fig. 13. BLAST alignments of an approximately 60 bp section of the genome from *Mycobacterium smegmatis* mutant 1, 2 and WT. The consensus sequence (ref TIGR) and position in the reference genome are displayed on top. As in Figure 1, the T to A mutation in mutants 1 and 2, and the absence of the mutation in the wild type is clearly identified.

The 454 sequencer is capable of reading on average100 bases per template, which can be extended by bi-directional sequencing, making the effective read length close to 200 bases for each template. This is of importance when sequencing human genomic material, as the average length of a human exon is approximately 170 bases, and with a total of 200 sequenced bases 80% of all exons could be entirely covered by a single PCR-generated DNA template (Sakharkar *et al.*, 2004). Furthermore, unlike methods for interrogating a single sequence position at a time, *de novo* sequencing of 100 or more continuous bases allows the establishment of haplotype linkages of significant value while screening, for example, for drug-resistance mutations in cancer and viral specimens.

From a technical point of view, amplicon sequencing is essentially identical to the sequencing of genomic libraries with the exception that the sample preparation procedure can be greatly simplified. The preferred method relies on direct amplification of the sequencing template using hybrid primers consisting of a target specific 3′ end and a standardized adaptor sequence at the 5′ end. The double-stranded amplicons are typically purified to remove residual primers, quantified, and after appropriate dilution are ready for emulsion PCR. Alternatively, double-stranded adaptors can be ligated onto the ends of an amplicon as in the standard library preparation method. The first approach, however, requires less amplification product, is much simpler to execute, automatically

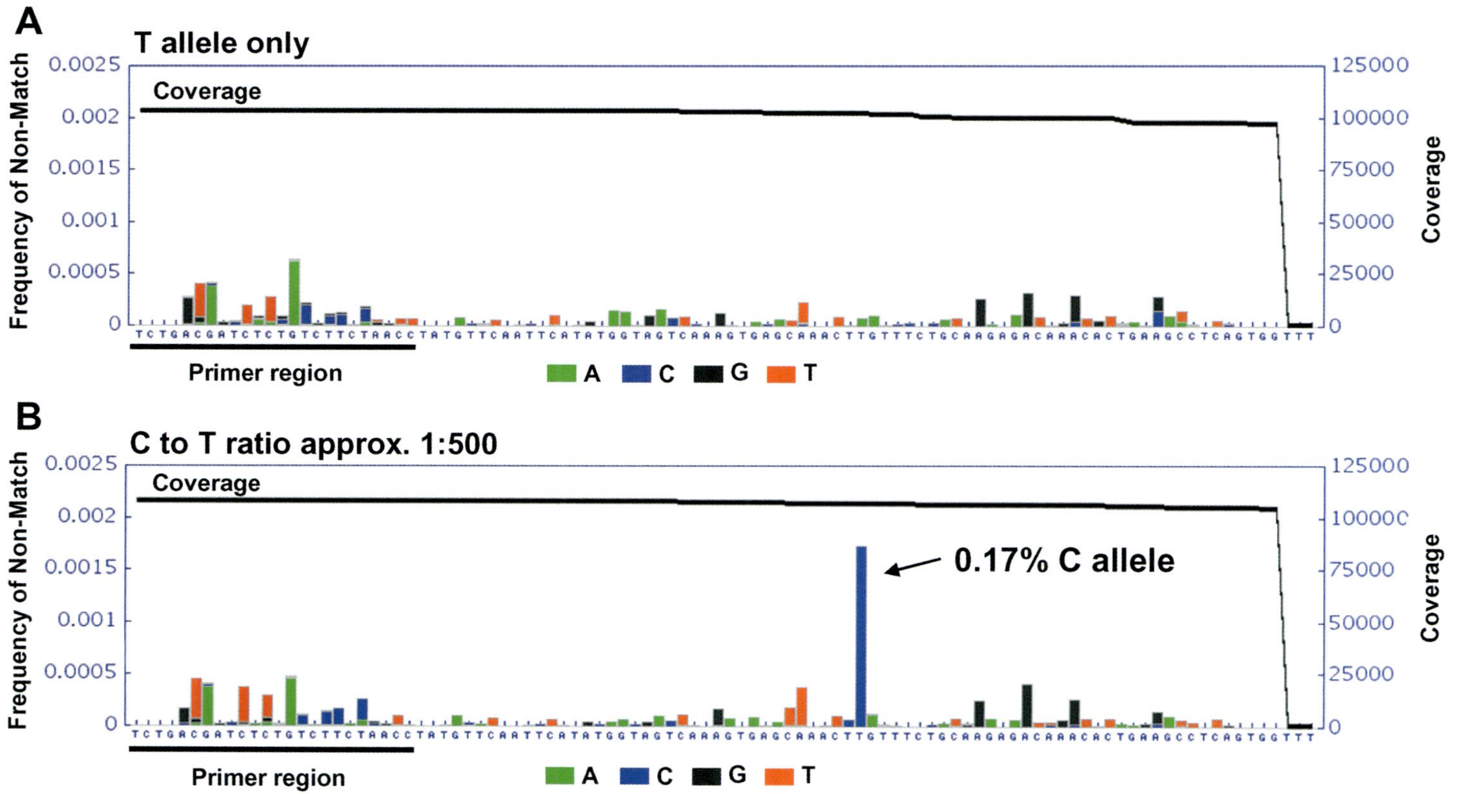

Fig. 14. A known SNP site in the HLA locus was amplified from human genomic DNA as part of a 126 bp PCR product. Amplicons representing C and T homozygous individuals were mixed in known ratios and analyzed by 454 sequencing. Each panel represents approximately 100,000 reads, which were BLAST aligned to a reference sequence shown along the *X*-axis. The left *Y*-axis indicates mutation frequency as represented by the color-coded bars, whereas the dark line indicates sequence coverage. Panel A shows reads derived from the T allele alone, whereas panel B displays a 1:500 mixture of C and T alleles and demonstrates detection of the minor allele at 0.17% frequency.

provides directionality to the primers and is well suited for multiplexing and scale-up. Conversely, the second method is labor intensive but readily allows sequencing of already existing amplicons and can be made more efficient by pooling of multiple amplicons before adaptor ligation.

The amplicon-sequencing concept was applied to the human immunodeficiency virus type 1 (HIV-1) with particular focus on the reverse transcriptase (RT) and protease genes that encode the proteins most frequently targeted by antiretroviral drugs (Simons *et al.*, 2005). Mutational profiling of clinical HIV specimens constitutes a particular challenge as individuals are infected with a mixture of viral variants termed viral quasi-species (Nowak, 1992; Visco-Comandini, 2001). Additional complexity arises when patients are treated with antiretroviral drugs, which inevitably leads to the accumulation of mutant species resistant to the drug (Chen *et al.*, 2004). With more than 20 drugs targeting RT and the protease, and an increasing number of drugs aimed at other viral components, it is becoming increasingly important to detect and continuously monitor such mutations in both patients undergoing treatment and antiretroviral-naïve patients awaiting initiation of therapy. For example, the knowledge of drug-resistant mutations will unequivocally affect the physician's choice of treatment strategy and lead to improved treatment results (Blum *et al.*, 2005; US Department of Health and Human Services: Guidelines for the Use of Antiretroviral Agents in HIV-Infected Adults and Adolescents, October 6, 2005).

A commonly used and commercially available method to screen for HIV mutations is direct Sanger sequencing of amplified PCR products. While useful to detect major sequence variants down to 10–20% abundance (Kapoor *et al.*, 2004), this approach has little utility for high-resolution analysis of complex samples, including nucleotide substitutions, as well as insertions and deletions in various ratios. Numerous non-sequence based methods have been developed for the detection of resistance mutations in HIV, but are generally limited to known mutations within known sequence contexts (Koch *et al.*, 1999). For an in-depth analysis aimed at detecting and quantifying all types of sequence variations, the only viable strategy is sequencing of individual templates. While large scale sequencing using Sanger technology is possible, it is not a realistic approach in a clinical setting due to the lack of necessary scalability and associated high costs. In addition, the necessary bacterial cloning carries the risk of bias due to sequence-based discrimination. The approach described here, on the other hand, generates hundreds of thousands of sequencing reads in parallel, and owing to the emulsion-based clonal amplification procedure guarantees the integrity of each read and preserves the molecular ratios of the original sample.

A proof-of-principle study was designed with 3 amplicons covering the entire protease gene of HIV-1 and an additional four amplicons spanning parts of RT known to harbor resistance mutations (amino acids 12-82, 84-123, 157-199 and 202-243). For simplicity, viral cDNAs previously shown by Sanger sequencing to be predominantly wild type in respect to all known major resistance mutations was used as template. To maximize usefulness of the fiber optic slide, amplicons for the seven genomic regions were pooled in equimolar ratios prior to emulsion PCR and sequenced together on a partitioned slide. In this

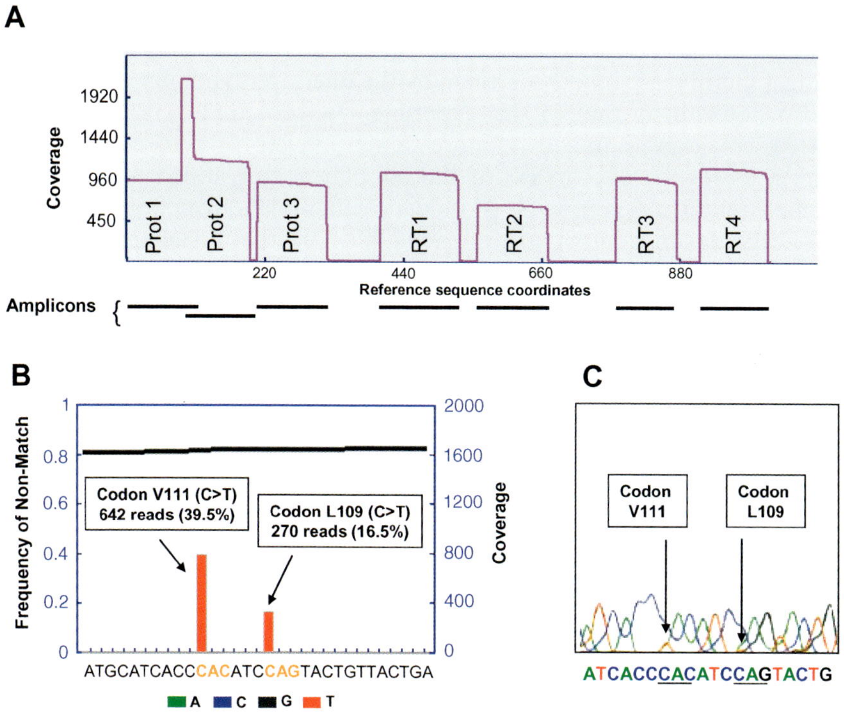

Fig. 15. (A) Seven amplicons covering the protease gene and part of RT were pooled in equimolar ratios for emulsion PCR and sequenced together on a small format fiberoptic slide ($20 \times 75\,\text{mm}^2$). The figure represents the coverage plot across the seven amplicons with approximately 1000 reads obtained for each amplicon. (B, C) Detection and quantification of mutations in a clinical pre-AZT HIV isolate was demonstrated by PCR amplification of a 156 base-pair fragment comprising RT codons 84–124. Shown are two G-to-A substitutions observed in codons L109 and V111 at 16.5% and 39.5% frequency, respectively. Sanger sequencing and automated base-calling of a 1.2 kb amplicon covering the same RT region failed to detect either mutation, although the V111 mutation is visually observed above background levels.

arrangement, each region can generate from 10,000 to 12,000 reads per loaded sample, with up to eight samples in a single run. Figure 15A demonstrates the coverage obtained across the seven pooled amplicons, with an average depth of about 1000 reads per amplicon. Preliminary experiments have demonstrated that it is possible to pool tens of amplicons, and multiplexing of thousands of amplicons is certainly conceivable and primarily limited by the precision that can be achieved in quantifying and pooling the amplicons.

For an assessment of the difference in sensitivity between Sanger and 454 sequencing, an RT-derived 1.2 kb amplicon was sequenced both directly by Sanger sequencing and, after re-amplification to generate a 156 bp template, on the 454 system. The latter method detected one G-to-A mutation at 39.5% abundance (Figure 15B), which was visually evident on the Sanger trace but was missed by the automated base-caller, and another one at 16.5%, which was not distinguishable from background signal by Sanger sequencing (Figure 15C). Clearly, even relatively abundant minor alleles at risk of being overlooked by the Sanger method are readily detected by the fiber optic slide-based sequencing. It is also noteworthy that in this experiment approximately 1500 reads were generated, coverage found to be more than sufficient to detect mutations occurring in as few as 1% or less of the templates. For added confidence in a particular mutation, it is possible to sequence a template from both directions in order to exclude any contextual sequencing bias or other systematic errors. This is possible by simply performing emulsion PCR in two separate reactions with beads containing capture primers of opposite polarity that, if desired, can be pooled for the sequencing run.

Figure 16 shows data from an antiretroviral drug-experienced patient revealing a known RT drug resistance mutation, K103N, both in the forward direction at 3.4%, and reverse direction at 3.5%. The high reproducibility that was observed lends confidence to the authenticity of the mutation and could, in a clinical setting, be of significant importance in the choice of drug regimen.

The new sequencing technology enjoys another advantage when compared to traditional approaches of HIV sequencing in its ability to detect and discriminate between different haplotypes. This is feasible since the underlying data used in the analysis consists of individual sequencing reads, rather than a composite not amenable to de-convolution.

An illuminating example is shown in Figure 17, in which a viral sample consisting of several quasi-species was analyzed for mutations within the protease gene. A 207 bp amplicon spanning amino acids 26–84 was used as a sequencing template. The mutational plot revealed several discrepancies from the reference sequence in the region of amino acids 35–37. Using a bioinformatics haplotype alignment tool, the underlying sequence variants were identified and shown to represent four distinct sequence variants within the analyzed region: three mutant sequences present at 21%, 34% and 39% each, and a small portion of reads corresponding to the wild-type sequence. Sequencing of the same sample using conventional capillary electrophoresis instruments

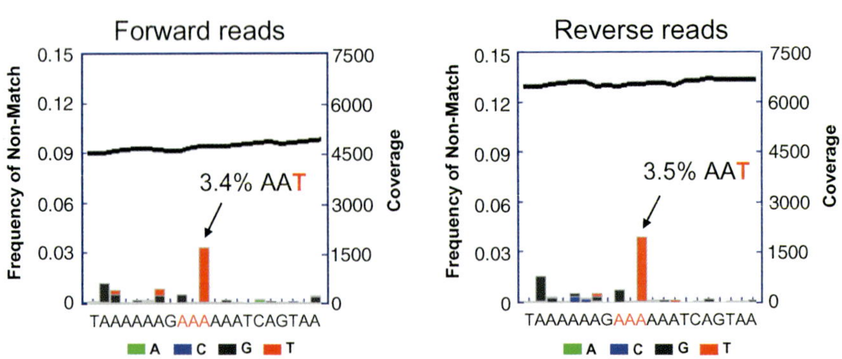

Fig. 16. Sequencing of RT from an antiretroviral drug treated patient detected the amino acid changing mutation K103N at 3.4% in the forward direction and 3.5% in the reverse direction. The mutation had previously escaped detection using standard Sanger sequencing.

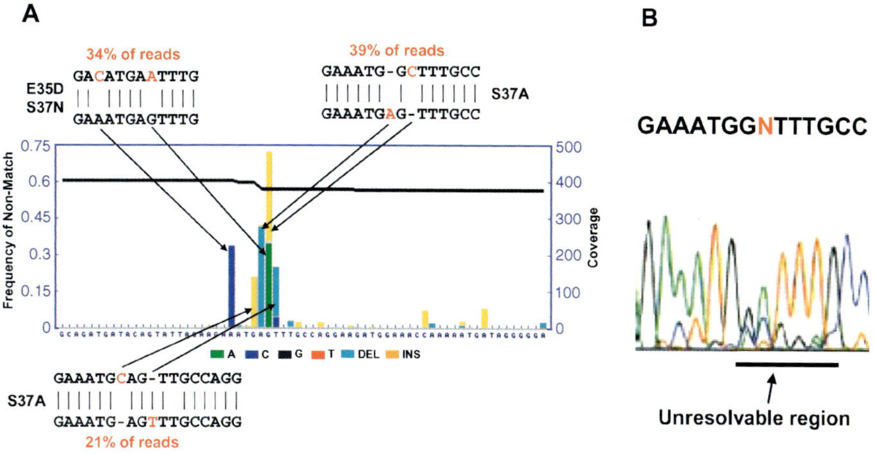

Fig. 17. Generation of individual sequencing reads allows identification of distinct subspecies, including haplotype linkages. (A) Sequencing of a 207 bp amplicon covering codons 26 through 84 of the HIV protease gene from an antiretroviral naïve patient, revealed three distinct rather abundant amino acid changing sequence variants, and a small percentage of wild-type virus (not shown). (B) Direct Sanger sequencing of an amplicon covering the same region uncovered sequence ambiguity, but could not be de-convoluted into distinct species.

highlighted an area of sequence ambiguity, but does not allow identification of individual sequence variants.

It is inherent to the data obtained by 454 sequencing that haplotype information can be gained not only across short stretches of sequence, but across the extent of entire reads. This capability is important for many applications, particularly for diagnostic sequencing of samples retrieved from HIV infected patients. In today's clinic, the use of multi-drug combinations is standard and is required for sufficient viral suppression and the prevention of HIV disease progression. To achieve successful treatment outcomes it is essential to have tools that can detect the presence, emergence and disappearance, of viral variants, especially those carrying resistance mutations. It is particularly important to know which mutations are linked within the same viral genome. While information on individual mutations can be obtained with several established methods for HIV screening, true haplotype information as produced by this new approach to sequencing may greatly improve strategic treatment decisions involving multi-drug regimens.

3.5. Scalability

At the start of its development cycle, 454 sequencing has demonstrated a very significant improvement in throughput over existing technology. The path to further performance increase will come primarily from the continued miniaturization of the fabricated sequencing substrate, allowing more wells and more sequence to be produced per unit area. However, this system is also scalable in other dimensions. Combining all of the improvements, an additional increase in throughput and a reduction in cost by two orders of magnitude may be achievable over the current performance. Scaling is based on the following considerations:

(i) Physical well density: further reduction in distance between wells will require the use of an imaging system capable of resolving smaller features. Modeling confirms that at an 8 μm pitch, background induced by the diffusion of reaction by-products out of the wells does not exceed 20%, suggesting that inter-well diffusion artifacts would not preclude a reduction in pitch by a factor of 4, equivalent to an increase in well density by a factor of 16.
(ii) Speed: a reduction in cycle time of 50% is compatible with the current well dimensions. As pitch is reduced, bead size and well depth can also be decreased, and preliminary modeling indicates that the duration of each cycle can be decreased by an additional 30%.
(iii) Yield: the percentage of wells generating high-quality sequences can be increased by improving read accuracy and cross talk correction algorithms. Improved bead deposition techniques, further correction of inter-well effects and the use of enzymatic methods to reduce chemical cross talk may allow an increase in yield to 60% of all wells.
(iv) Read lengths: by reducing carry forward and incomplete extension, improving correction algorithms and increasing the length of fragments that can be amplified and captured, sequencing 300–400 bp can be achieved, as has already been demonstrated in a laboratory setting.

4. DISCUSSION

The sequencing system described in this chapter is capable of achieving a throughput that is two orders of magnitude higher than has been attained up to now using conventional, Sanger-based, sequencing technology. This system is predicated on the idea of parallelizing all conceivable steps in the sequencing process, from sample preparation, through template amplification and sequencing, to data analysis. The methods and hardware were all developed to handle hundreds of thousands of fragments simultaneously and were made possible by significant improvements to solid-support pyrophosphate sequencing, which, in turn, have allowed the extension of read lengths beyond what had previously been reported. While the successful sequencing of reads that are 80–120 bases long has been reduced to routine practice, using 84 cycles of nucleotide additions it has been possible to achieve read lengths of 200 bases. On occasion, at 168 cycles, individual reads that are 100% accurate over greater than 400 bases have been generated. Short fragments *a priori* do not prohibit the *de novo* assembly of bacterial genomes. In fact, the larger oversampling afforded by the throughput of this system typically results in a draft sequence having fewer contigs than when sequenced with conventional Sanger sequencers.

The main goal behind the drive for low-cost high-throughput sequencing has been to decrease the cost of sequencing sufficiently to enable individualized human genome sequencing. Along the way to this goal, many potential applications of the technology are becoming available or will be enabled as soon as the bioinformatics development races to keep up with the large quantities of data and the new possibilities that inexpensive and quick sequencing allows. For example, affordable microbial sequencing, either re-sequencing for SNP identification or *de novo* sequencing of more variable strains, enables comparative genomics on strains of varying virulence, drug resistance, and host-species preference. The emulsion-based cloning employed in the 454 system, with its inherent ability to enable sequencing from single molecules in a complex mixture, opens up the possibility of massive oversampling of tag sequences or specific regions of interest and does so in a quick and cost effective manner. The detection of low abundance sequence variants in complex samples, as described above, is of considerable value in many scientific areas and holds the promise of becoming a powerful diagnostic tool in virology and oncology clinics, for instance, advancing the prospect of effective personalized medicine. The first demonstration of sequencing from complex mixtures has sensitivity below 1% in a complex mixture of HIV quasi-species present within a patient as a function of time and drug response. This level of sensitivity is barely achievable by the Sanger-based methods, and then only by cloning of fragments into bacteria. Microarray-based sequencing methods are not as sensitive or as quantitative as the direct sequencing of clonally amplified single molecules.

The applications enabled by this new sequencing technology and its usefulness to the drug discovery and development process are only beginning to be discovered. Once the technology is widely available and its power is known, additional applications will be developed. The ability to sequence *de novo* opens up a wide array of possibilities for new discovery and creative approaches to important and unaddressed problems in many areas of research and development. In the public health arena, 454 sequencing can be applied to the world-wide tracking and monitoring of the spread of specific strains of pathogenic microorganisms. In bio-defense, one can envision the rapid identification of the strain of an isolated suspected bioterrorism agent, or the identification of pathogens by sequencing complex mixtures. Sequencing the complete or partial genomes of large populations of individuals will impact our understanding of the genetic basis of human diseases. In cancer biology, rapid and inexpensive sequencing will shed light on the mutations that may give rise to cancer, help identify novel oncogenes and tumor suppressor genes and understand the basis of drug response. With the prospect of longer read lengths, whole-genome exon sequencing projects, with one amplicon per exon, can be envisioned within a foreseeable future, allowing comprehensive population-sized genetic studies and the large-scale mapping of disease susceptibility genes. Ultimately, it is not inconceivable that the scalability of this technology will enable the sequencing of individual human genomes to become part of the routine practice of medicine.

On July 20, 2006, the Max Planck Institute of Evolutionary Anthropology in Leipzig and the 454 Life Sciences Corporation announced an ambitious project to the sequence the Neandertal genome at 1 × using our advanced pyrosequencing technology. The Neandertals constitute the hominid group most closely related to currently living humans and it is anticipated that much can be learned by comparing its genome to that of the human and the chimpanzee. Of particular interest will be the regions where Neandertal is closer to chimpanzee than man, as these may indicate regions that have evolved in humans after the split from Neandertal. The project faces major challenges because an estimated 90–95% of the available DNA is microbial DNA, and furthermore the fossil DNA fragments have been degraded into small fragments with chemical modifications of the nucleobases. However, proof-of-principle experiments generated 1 million bases of Neandertal sequence owing to the single-molecule-based emulsion PCR and the high throughput of 454 picoliter sequencing. Generation of the first three billion bases of Neandertal DNA is expected to be completed within two years.

ACKNOWLEDGMENTS

The technology described in this chapter was created through the work of a large group of dedicated researchers and engineers at the 454 Life Sciences Corporation; we wish to acknowledge their extraordinary efforts. We also would like to acknowledge the organizational support and infrastructure provided by the Operations groups of the 454 Life Sciences Corporation. We want to thank Dr. Michael Kozal for providing expert advice and samples for the study of HIV, and for critical review of the manuscript. We are also indebted to Dr. Peter Verhasselt and Dr. Jean-Marc Neefs for sharing with us the results of their *M. smegmatis* work. This research was supported in part by the US Department of Health and Human Services under NIH grants 1P01HG003022-01 and 1R01HG003562-01.

REFERENCES

Altschul, S. F. *et al.* (1990). Basic local alignment search tool. J. Mol. Biol. 215, 403–410.

Andries, K. *et al.* (2005). A diarylquinoline drug active on the ATP synthase of *Mycobacterium tuberculosis*. Science 307, 223–227.

Bains, W. and Smith, G. C. (1988). A novel method for nucleic acid sequence determination. J. Theor. Biol. 135, 303–307.

Bhattacharyya, A. *et al.* (2002). Draft sequencing and comparative genomics of *Xylella fastidiosa* strains reveal novel biological insights. Genome Res. 12, 1556–1563.

Blum, R. A., Wylie, N., England, T. and French, C. (2005). HIV resistance testing in the USA – a model for the application of pharmacogenomics in the clinical setting. Pharmacogenomics 6, 169–179.

Brenner, S. *et al.* (2000). Gene expression analysis by massively parallel signature sequencing (MPSS) on microbead arrays. Nat. Biotechnol. 18, 630–634.

Chen, R., Quinones-Mateu, M. E. and Mansky, L. M. (2004). Drug resistance, virus fitness and HIV-1 mutagenesis. Curr. Pharm. Des. 10, 4065–4070.

Collins, F. S., Morgan, M. and Patrinos, A. (2003). The Human Genome Project: lessons from large-scale biology. Science 300, 286–290.

Dressman, D., Yan, H., Traverso, G., Kinzler, K. W. and Vogelstein, B. (2003). Transforming single DNA molecules into fluorescent magnetic particles for detection and enumeration of genetic variations. Proc. Nat. Acad. Sci. USA 100, 8817.

Ewing, B., Hillier, L., Wendl, M. C. and Green, P. (1998). Base-calling of automated sequencer traces using phred. I. Accuracy assessment. Genome Res. 8, 175–185.
Fagin, B., Watt, J. G. and Gross, R. (1993). A special-purpose processor for gene sequence analysis. Comput. Appl. Biosci. 9, 221–226.
Fleischmann, R. D. *et al.* (1995). Whole-genome random sequencing and assembly of *Haemophilus influenzae* Rd. Science 269, 496–512.
Ghadessy, F. J., Ong, J. L. and Holliger, P. (2001). Directed evolution of polymerase function by compartmentalized self-replication. Proc. Natl. Acad. Sci. USA 98, 4552–4557.
Goo, Y. A. *et al.* (2004). Low-pass sequencing for microbial comparative genomics. BMC Genomics 5, 3.
Hardenbol, P. *et al.* (2005). Highly multiplexed molecular inversion probe genotyping: over 10,000 targeted SNPs genotyped in a single tube array. Genome Res. 15, 269–275.
Hinds, D. A. *et al.* (2005). Whole-genome patterns of common DNA variation in three human populations. Science 307, 1072–1079.
Honisch, C. *et al.* (2004). High-throughput mutation detection underlying adaptive evolution of *Escherichia coli*-K12. Genome Res. 14, 2495–2502.
Hyman, E. (1988). A new method of sequencing DNA. Anal. Biochem. 174, 423–436.
Jacobson, K. B. *et al.* (1991). Applications of mass spectrometry to DNA sequencing. GATA 8, 223–229.
Jett, J. H. *et al.* (1989). High-speed DNA sequencing: an approach based upon fluorescence detection of single molecules. Biomol. Struct. Dynam. 7, 301–309.
Kapoor, A. *et al.* (2004). Sequencing-based detection of low-frequency human immunodeficiency virus type 1 drug-resistant mutants by an RNA/DNA heteroduplex generator-tracking assay. J. Virol. 78, 7112–7123.
Koch, N., Yahi, N., Colson, P., Fantini, J. and Tamalet, C. (1999). Genetic polymorphism near HIV-1 re-verse transcriptase resistance-associated codons is a major obstacle for the line probe as-say as an alternative method to sequence analysis. J. Virol. Methods 80, 25–31.
Lander, E. S. and Waterman, M. S. (1988). Genomic mapping by fingerprinting random clones: a mathematical analysis. Genomics 2, 231–239.
Margulies, M. *et al.* (2005). Genome sequencing in microfabricated high-density picolitre reactors. Nature 437, 376–380.
Maxam, A. M. and Gilbert, W. (1977). A new method for sequencing DNA. Proc. Natl. Acad. Sci. USA 74, 560–564.
Mehta, K., Rajesh, V. A. and Veeraswamy, S. (1993). FPGA implementation of VXIbus interface hardware. Biomed. Sci. Instrum. 29, 507–513.
Mills, D. R. and Kramer, F. R. (1979). Structure-independent nucleotide sequence analysis. Proc. Natl. Acad. Sci. USA 76, 2232–2235.
NIH News Release (2004). NHGRI seeks next generation of sequencing technologies. 14 October (http://www.genome.gov/12513210).
Nowak, M. A. (1992). Variability of HIV infections. J. Theor. Biol. 155, 1–20.
Nyren, P., Pettersson, B. and Uhlen, M. (1993). Solid phase DNA minisequencing by an enzymatic luminometric inorganic pyrophosphate detection assay. Anal. Biochem. 208, 171–175.
Ogawa, T. *et al.* (2005). Increased productivity for core labs using one polymer and one array length for multiple applications. ABRF '05: Biomolecular Technologies: Discovery to Hypotheses, Savannah, GA, 5–8 February. Poster P108-T.
Prober, J. M. *et al.* (1987). A system for rapid DNA sequencing with fluorescent chain-terminating dideoxynucleotides. Science 238, 336–341.
Read, T. D. *et al.* (2002). Comparative genome sequencing for discovery of novel polymorphisms in *Bacillus anthracis*. Science 296, 2028–20233.
Ronaghi, M. (2001). Pyrosequencing sheds light on DNA sequencing. Genome Res. 11, 3–11.
Ronaghi, M. *et al.* (1996). Real-time DNA sequencing using detection of pyrophosphate release. Anal. Biochem. 242, 84–89.
Sakharkar, M. K., Chow, V. T. and Kangueane, P. (2004). Distributions of exons and introns in the human genome. In Silico Biol. 4, 387–393.

Sanger, F., Nicklen, S. and Coulson, A. R. (1977). DNA sequencing with chain-terminating inhibitors. Proc. Natl. Acad. Sci. USA 74, 5463–5467.

Simons, J. *et al.* (2005). Ultra-deep sequencing of HIV from drug resistant patients. XIV International HIV Drug Resistance Workshop, June 7–11, Quebec City, Canada.

Tawfik, D. S. and Griffiths, A. D. (1998). Man-made cell-like compartments for molecular evolution. Nat. Biotechnol. 16, 652–656.

TIGR. (2003). Sequence obtained from TIGR http://www.TIGR.org, November .

Velculescu, V. E., Zhang, L., Vogelstein, B. and Kinzler, K. W. (1995). Serial analysis of gene expression. Science 270, 484–487.

Venter, J. C. *et al.* (2001). The sequence of the human genome. Science 291, 1304–1351.

Visco-Comandini, U., Aleman, S., Yun, Z. and Sonnerborg, A. (2001). Human immunodeficiency virus type 1 variability and long-term non-progression. J. Biol. Regul. Homeost. Agents 15, 299–303.

Zwick, M. E. *et al.* (2005). Microarray-based resequencing of multiple *Bacillus anthracis* isolates. Genome Biol. 6, R10.

Chapter 6
An Integrated System for DNA Sequencing by Synthesis

John R. Edwards, Dae Hyun Kim and Jingyue Ju

Columbia Genome Center, Columbia University College of Physicians and Surgeons, Russ Berrie Medical Science Pavilion, 1150 St. Nicholas Avenue, New York, NY 10032, USA

Contents

Abstract

The completion of the Human Genome Project has increased the need for high-throughput DNA sequencing technologies aimed at uncovering the genomic contributions to diseases. The DNA sequencing by synthesis (SBS) approach has shown great promise as a new platform for deciphering the genome. Recently, much progress has been made on the fundamental sciences required to make SBS a viable sequencing technology. One of the unique features of this approach is that many of the steps required are compatible in a modular fashion allowing for the best solution at each stage to be effectively integrated. Recent advances include emulsion-PCR based DNA template preparation, the design and synthesis of novel reporter nucleotides and new surface attachment chemistries for DNA template. The integration of these advances will lead to the development of a high-throughput DNA sequencing system in the near future.

1. INTRODUCTION

DNA sequencing is a fundamental tool for biological science. The completion of the Human Genome Project has set the stage for screening genetic mutations to

PERSPECTIVES IN BIOANALYSIS, VOLUME 2
ISSN 1871-0069 DOI: 10.1016/S1871-0069(06)02006-4

identify disease genes on a genome-wide scale (Collins *et al.*, 2003). Recent estimates seem to indicate that the number of genes is relatively constant among vertebrates (IHGS Consortium, 2004). These results point to other factors such as gene variation, regulation and alternative splicing that are involved in the fundamental differences that separate humans from other animals as well as account for many of the primary differences that make-up each individual. DNA sequencing is a primary driving force behind the search for the fundamental regulatory regions that account for these differences. Decreased cost of sequencing is critical to the comparative genomic efforts including the ultimate goals of personalized medicine based on genetic and genomic information. Accuracy, speed and size of the instrument are critical considerations for the development of new DNA analysis methods that can be used directly in the hospitals and clinical settings, for forensics or for pathogen detection in the field. Accuracy is essential for genetic mutation detection and haplotype analysis.

The Sanger dideoxy chain-termination method (Sanger *et al.*, 1977) is currently the technique of choice for large-scale DNA sequencing projects. Widely used automated versions of this method employ either four differently end-labeled fluorescent primers or terminators to generate all the possible DNA fragments complementary to the template to be analyzed. The fragments terminating with the four different bases (A, C, G, T) are then separated at single-base pair resolution on sequencing gels and identified by the four distinct fluorescent emissions (Smith *et al.*, 1986; Prober *et al.*, 1987). Application of laser induced fluorescence for DNA sequencing is a major advancement for the automated DNA sequencing technology that makes large-scale genome sequencing initiatives possible. An "ideal" set of fluorophores for four-color Sanger DNA sequencing must consist of four different fluorophores. These fluorophores should have similar high molar absorbance at a common excitation wavelength, high fluorescence quantum yields, exhibit strong and well-separated fluorescence emissions and introduce the same relative mobility shift of the DNA sequencing fragments. These criteria cannot be met optimally by the spectroscopic properties of single fluorescent dye molecules, and indeed are poorly satisfied by the initially used sets of fluorescent tags. Ju *et al.*, (1995) overcame these constraints imposed by the use of single dyes and developed fluorescence energy transfer dyes for DNA sequencing that meet the performance criteria set out as above (Ju *et al.*, 1995). The higher sensitivity offered by these new sets of fluorescent dyes also allows the direct sequencing of large-template DNA (>30 kb) with read lengths of over 700 bases per sequencing reaction, leading to significant progress in the large scale genome sequencing and mapping projects (Marra *et al.*, 1996; Lee *et al.*, 1997; Heiner *et al.*, 1998).

DNA sequencing by synthesis (SBS) is based on polymerase reaction, a key process for DNA replication inside cells. The basic concept of SBS is to use DNA polymerase to extend a primer that is hybridized to a template by a single nucleotide, determine its identity, and then proceed to the extension and detection of next nucleotide. The goal is to read out the DNA sequence serially during the polymerase reaction. This stands in contrast to Sanger sequencing in which fluorescently-labeled DNA fragments of different sizes are all generated in a single reaction and then separated and detected. SBS approaches have an advantage of easy scale-up in parallel without the need for separations. Currently available fluorescent array scanners can easily detect over 100,000 sample spots arrayed on a glass surface (Schena *et al.*, 1995). Such array scanners allow fast screening of large areas with high resolution, allowing automated detection of hundreds of thousands and even millions of samples simultaneously.

Several groups have recently reported significant advances in implementing new practical strategies for DNA sequencing. In these reports, emulsion-PCR, one commonly used technique for various biological assays including directed enzyme evolution (Tawfik and Griffiths, 1998; Ghadessy *et al.*, 2001) and genotyping (Dressman *et al.*, 2003) was used for generating template from single DNA molecules. Margulies and colleagues used emulsion-based microreactors to amplify DNA templates in a one-tube reaction for pyrosequencing (Margulies *et al.*, 2005, 2007). Beads containing the amplified templates generated from a single DNA molecule were then isolated in individual wells and reagents were flowed across the wells for the pyrosequencing reactions. Shendure and colleagues used emulsion-PCR on 1-μm beads to prepare DNA template to produce seven bases at a time, using a "sequencing by ligation" approach (Shendure *et al.*, 2005).

2. DNA Sequencing by Synthesis Methodology

The concept of DNA SBS was first revealed in 1988 with an attempt to sequence DNA by detecting the pyrophosphate group that is generated when a nucleotide is incorporated in a DNA polymerase reaction (Hyman, 1988). Pyrosequencing, which was developed based on this concept, has been explored for DNA sequencing (Ronaghi *et al.*, 1996). In this approach, each of the four dNTPs is added sequentially with a cocktail of enzymes and substrates in addition to the usual polymerase reaction components. If the added nucleotide is complementary with the first available base on the template, the nucleotide will be incorporated and a pyrophosphate will be released. The released pyrophosphate is converted to ATP by sulfurylase, and visible light is subsequently produced by firefly luciferase. If the added nucleotide is not incorporated, no light will be produced and the nucleotide will simply be degraded by the enzyme apyrase.

Pyrosequencing has been applied to single nucleotide polymorphism (SNP) detection (Ronaghi *et al.*, 1998) and genome sequencing (Margulies *et al.*, 2005). However, there are inherent difficulties in this method for determining the number of incorporated nucleotides in homopolymeric regions (e.g., a string of several Ts in a row) of the template. Additionally, each of the four nucleotides needs to be added and detected separately, which increases the overall detection time. The accumulation of undegraded nucleotides and other components could also lower the accuracy of the method when sequencing a long DNA template. Ideally, as one examines the fundamental limitations towards miniaturization, it is desirable to have a simple method to directly detect a reporter group attached to the nucleotide that is incorporated into a growing DNA strand in polymerase reaction, rather than relying on a complex enzymatic cascade.

We recently described an integrated SBS approach (Ju *et al.*, 2003) that is illustrated in Figure 1. This method relies on the use of a DNA polymerase

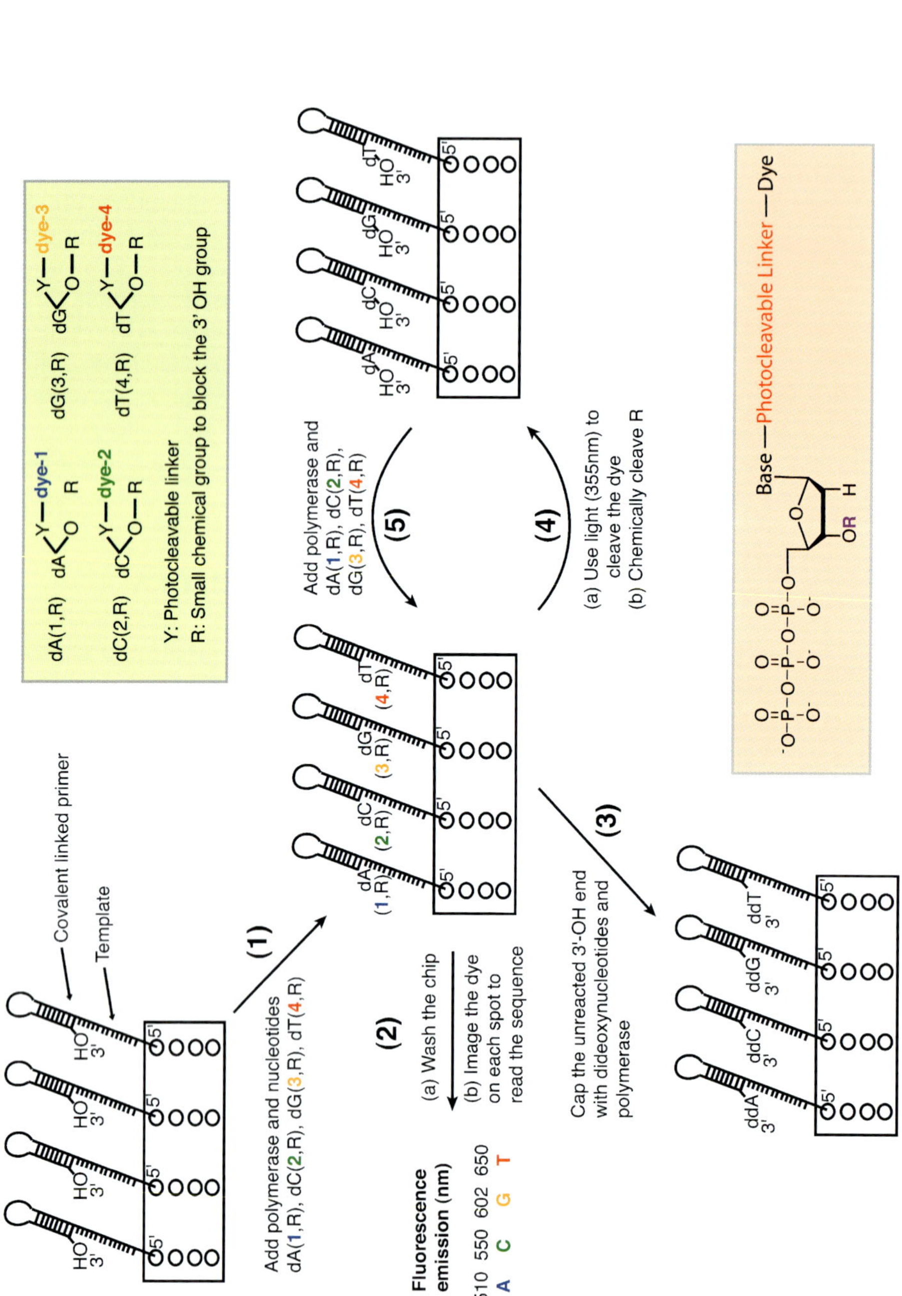

dA(1,R) dA Y—dye-1 O R
dC(2,R) dC Y—dye-2 O—R
dG(3,R) dG Y—dye-3 O—R
dT(4,R) dT Y—dye-4 O—R
Y: Photocleavable linker
R: Small chemical group to block the 3' OH group
Covalent linked primer
Template
HO 3' 5'
(1)
Add polymerase and nucleotides dA(1,R), dC(2,R), dG(3,R), dT(4,R)
(2)
(a) Wash the chip
(b) Image the dye on each spot to read the sequence
Fluorescence emission (nm)
510 550 602 650
A C G T
(3)
Cap the unreacted 3'-OH end with dideoxynucleotides and polymerase
ddA ddC ddG ddT 3'
(4)
(a) Use light (355nm) to cleave the dye
(b) Chemically cleave R
(5)
Add polymerase and dA(1,R), dC(2,R), dG(3,R), dT(4,R)
dA dC dG dT HO 3'
Base —Photocleavable Linker —Dye
OR H

reaction to read-out the DNA sequence, using novel reporter nucleotides for signal detection. After completion of the addition of each nucleotide the attached fluorescent reporter group is detected, determining the identity of the added nucleotide. The 3′ -OH moiety of each reporter nucleotide is also blocked by a functional group, which prevents the DNA polymerase from adding additional nucleotides. This blocking group needs to be easily removed, to generate a free 3′ -OH group for subsequent round of extension. This system increases the ability to accurately sequence through homopolymeric regions in the DNA template, as the addition of individual nucleotides are detected independently. In order to design an ideal system for SBS, new nucleotide analogues with the above properties must be developed.

Taking this and other factors into account, the following requirements must be met to make an entire SBS system into an efficient sequencing technology:

1. Standard cloning techniques to amplify DNA must be replaced by a high-throughput method for DNA template preparation.
2. After initial amplification, DNA templates must be physically arrayed in a format that allows each template to be probed multiple times.
3. Nucleotides must be reversible terminators (3′-OH is blocked) so that only a single nucleotide is added each step during SBS.
4. The 3′-OH blocking group used in SBS must be easily removed after detection for subsequent nucleotide addition.
5. The entire system must allow for simple washing and reagent additions between detection cycles.

Emulsion-PCR, which has been shown to have potential to address DNA template preparation for various sequencing platforms (Margulies *et al.*, 2005; Shendure *et al.*, 2005), can be readily adapted to the SBS approach shown in Figure 1. The remainder of this review is divided into two sections that describe advances in DNA attachment chemistries and in the synthesis of novel reporter

Fig. 1. In the SBS approach, a chip is constructed with immobilized DNA templates that are able to self-prime for initiating the polymerase reaction. Four nucleotide analogues are designed such that each is labeled with a unique fluorescent dye on the specific location of the base, and a small chemical group (R) to cap the 3′-OH group. Upon addition of the four nucleotide analogues and DNA polymerase, only the nucleotide analogue complementary to the next nucleotide on the template is incorporated by polymerase on each spot of the chip (step 1). After removing the excess reagents and washing away any unincorporated nucleotide analogues, a four-color fluorescence imager is used to image the surface of the chip, and the unique fluorescence emission from the specific dye on the nucleotide analogues on each spot of the chip will yield the identity of the nucleotide (step 2). After imaging, the small amount of unreacted 3′-OH group on the self-primed template moiety will be capped by excess ddNTPs (ddATP, ddGTP, ddTTP and ddCTP) and DNA polymerase to avoid interference with the next round of synthesis (step 3). The dye moiety will then be cleaved by light (~355 nm) and the R protecting group will be removed chemically to generate a free 3′-OH group with high yield (step 4). The self-primed DNA moiety on the chip at this stage is ready for the next cycle of the reaction to identify the next nucleotide sequence of the template DNA (step 5) (Ju *et al.*, 2003).

nucleotides. Each new advance further enables SBS to be developed into a viable DNA sequencing technology.

2.1. DNA attachment chemistries on surfaces

A variety of attachment chemistries have been used for the immobilization of DNA on surfaces. These chemistries have been driven by a wide range of applications including gene expression analysis using microarrays, chip based genotyping and SBS methods. For SBS, the primary requirement in the development of DNA immobilization chemistry is that the coupling reactions must produce high yields under conditions that are compatible with routine handling of DNA. These coupling conditions must not interrupt the phosphodiester bonds that comprise the DNA phosphate backbone, and should also not modify the primary amines found in guanines, cytosines and adenines.

One commonly used method is to attach streptavidin to a solid surface and then bind biotinylated DNA molecules to the immobilized streptavidin. While the streptavidin–biotin interaction is quite strong (Weber *et al.*, 1989), the hydrogen bonding interaction may prove problematic in multiple rounds of SBS. Streptavidin is also a large tetrameric protein, and its size will limit the number of available binding sites on the solid surface, limiting the number of DNA binding sites on the surface, resulting in decreased density of DNA molecules on the surface and decreased achievable read-length. Ideally, DNA would be attached covalently to the surface to eliminate any loss during multiple washing steps between nucleotide additions. The development of a chemoselective coupling chemistry for the immobilization of DNA on a solid surface is essential for accurate gene-expression measurement (Schena *et al.*, 1995) and polymorphism or mutation detection (Wang *et al.*, 1998; Debouck and Goodfellow, 1999). Because covalent coupling chemistries have been shown to typically lead to more stable DNA arrays than non-covalent chemistries, a variety of covalent coupling methods have been used for DNA immobilization on a solid surface (Beier and Hoheisel, 1999; Adessi *et al.*, 2000; Lindroos *et al.*, 2001). However, an additional improvement of the coupling chemistry for immobilizing DNA on a surface is required to achieve high selectivity and coupling efficiency. One ideal property required for the functional groups to be coupled (one from the DNA and the other from the surface) is the stability of the groups in aqueous conditions, which are typically needed to perform the coupling reaction.

We have explored Click Chemistry (Seo *et al.*, 2003, 2004), specifically azide–alkyne cycloaddition, to immobilize DNA on a glass chip for SBS (Figure 2). An amino modified glass surface is reacted with a bifunctional linker containing an NHS-ester and an alkyne on either end to functionalize the surface. Azido-labeled PCR product (created from a PCR reaction using an azido-labeled primer) is then attached to the surface using a copper catalyzed 1,3 dipolar azide–alkyne cycloaddition reaction. Alkaline conditions are then used to remove the unattached DNA strand, leaving only a single stranded DNA on the surface.

We have shown that DNA templates can be spotted in high density using the Click Chemistry with standard microarray spotters (Figure 3). In order to

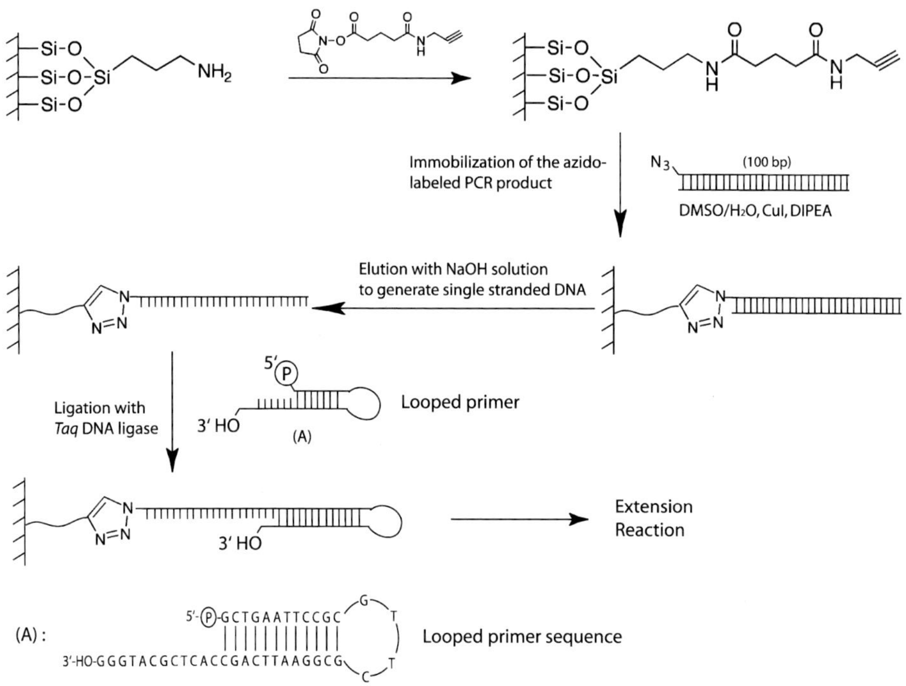

Fig. 2. Immobilization of an azido-labeled PCR DNA product on an alkynyl-functionalized surface and a ligation reaction between the immobilized single-stranded DNA template and a loop primer to form a self-priming DNA moiety on the chip. The sequence of the loop primer is shown in (A).

prevent dissociation of the primers and template during washing cycles of SBS, we ligated a looped primer directly to the template (Figure 2). This looped primer was designed such that the primer sequence self-hybridizes in a very efficient manner, making a universal primer adapter ideal for SBS. The loop sequence was carefully chosen to increase the stability of the hairpin structure (Nakano *et al.*, 2002), and has been shown to efficiently prime templates for SBS reactions (Seo *et al.*, 2005).

2.2. Novel reporter nucleotides

In order to design the reporter nucleotides used in the SBS extension reaction, it is important to examine the structure of the polymerase enzyme complexed with a DNA template, a primer and an incoming nucleotide in polymerase reaction. The 3D structure of the ternary complexes (Pelletier *et al.*, 1994) comprising a rat DNA polymerase, a DNA template-primer and a dideoxycytidine triphosphate (ddCTP) is shown in Figure 4.

What is apparent from this structure is that the 5-position of the cytosine points away from the catalytic pocket of the enzyme, while the 3′-position of the ribose ring in ddCTP is near the active amino acid residues of the polymerase and is therefore very crowded. Any group that is attached at the 3′-position of

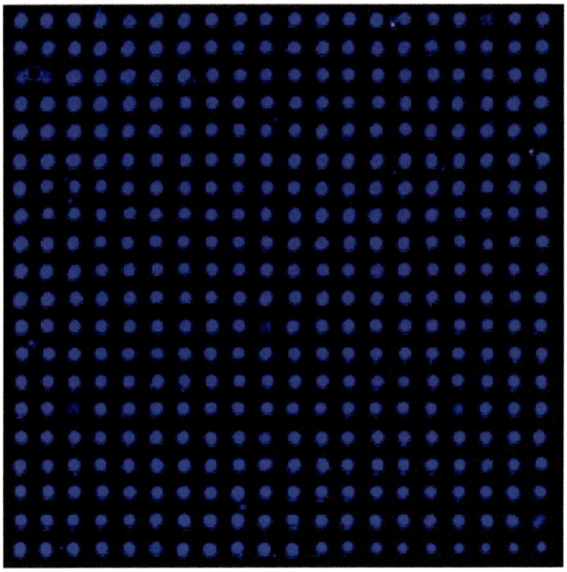

Fig. 3. Azido-labeled fluorescently modified DNA spotted on an alkyne surface using a standard microarray spotter demonstrating that the Click Chemistry used to covalently bind DNA templates to a solid surface can be easily scaled up with conventional high-throughput array spotting techniques.

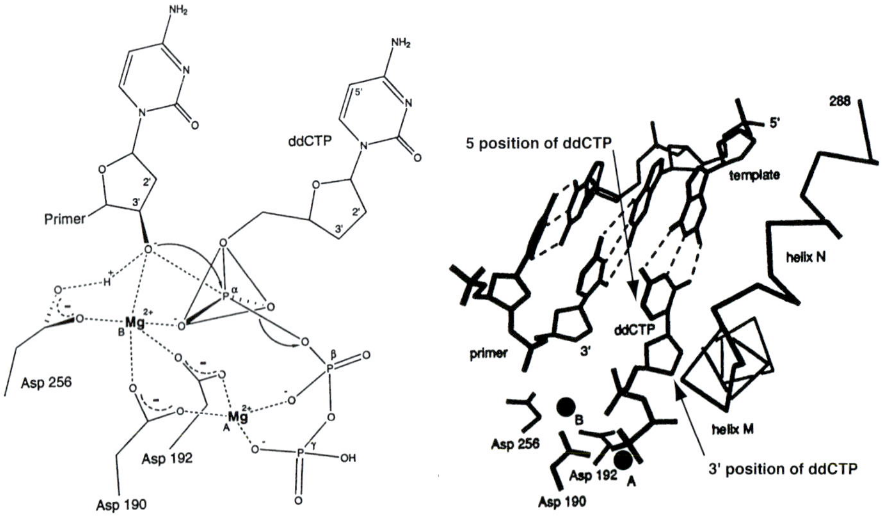

Fig. 4. The 3D structure of the ternary complexes comprising a rat DNA polymerase, a DNA template-primer and a dideoxycytidine triphosphate (ddCTP). The left side of the illustration shows the mechanism for the addition of ddCTP and the right side of the illustration shows the active site of the polymerase in the context of the polymerase–DNA complex. Note that the 3′-position of the dideoxyribose ring is very crowded, while ample space is available at the 5-position of the cytidine base.

the sugar must be small as to not interfere with the polymerase reaction. Large bulky dye molecules have been attached at the 5-position of pyrimidines and the 7-position of purines and used in enzymatic incorporation reactions such as in Sanger dideoxy-sequencing (Zhu *et al.*, 1994; Rosenblum *et al.*, 1997; Duthie *et al.*, 2002). We thus reasoned that if a unique fluorescent dye is attached to 5-position of the pyrimidines (T and C) and 7-position of purines (G and A) through a cleavable linker, and a small chemical moiety is used to cap the 3′-OH group, the resulting nucleotide analogues should be able to incorporate into the growing DNA strand as terminators. Upon removing the fluorophore and the 3′-OH capping group, the polymerase reaction will proceed to incorporate the next nucleotide analogue and detect the next base, as shown in Figure 1.

2.2.1. Nucleotide reporter groups for SBS

The reporter groups on the nucleotides for SBS must be easily detectable and then removed efficiently after detection to ensure maximum efficiency during each SBS cycle. Fluorescent dyes have been widely used as labels of nucleotides including the Sanger sequencing reactions. Braslavsky *et al.*, (2003) explored the use of photobleaching to eliminate the fluorescent signal in between nucleotide additions, without actually removing the fluorophores directly for SBS. However, studies have shown that the photobleached fluorophores remain with the DNA template and interfere with the DNA polymerase activity during the incorporation of each subsequent nucleotide. Thus, the fluorophores on the nucleotides need to be removed efficiently between each cycle of SBS.

Several different labile attachment chemistries for reporters have been used, including disulfide linkages (Mitra *et al.*, 2003) and photocleavable linkers (Ju *et al.*, 2003; Li *et al.*, 2003). The disulfide group can be chemically cleaved using 2-mercaptoethanol after the nucleotide incorporation and detection. However, the disulfide bond can be reversed and becomes destabilized under certain conditions (Pleasants *et al.*, 1989; Huyghues-Despointes and Nelson, 1992). Photocleavable linkers provide an effective and rapid method for removing the fluorophores from the nucleotide by using high intensity photons as reagents. We have developed a set of such photocleavable fluorescent nucleotide analogues (Seo *et al.*, 2005), using 2-nitrobenzyl group as the photocleavable linker to attach the fluorophore to each base at the 5-position of the pyrimidines and the 7-position of the purines (Figure 5).

These nucleotide analogues have been shown to be good substrates for incorporation into DNA by the commonly used DNA polymerase *Thermo Sequenase*, and have also been shown to incorporate efficiently in a primer extension assay. The products from the polymerase extension reaction using the four photocleavable fluorescent nucleotide analogues described in Figure 5 and the photocleavage products are shown in Figure 6. The fluorescent dyes attached to these nucleotides are removed quickly and efficiently under near-UV irradiation in nearly quantitative yield, and these photocleavage conditions are compatible for use with DNA samples (Seo *et al.*, 2005), showing no DNA damage during repeated exposure to photolysis (Figure 7).

As a demonstration of the feasibility of using these nucleotides for SBS, the above-described Click Chemistry for DNA immobilization has been combined

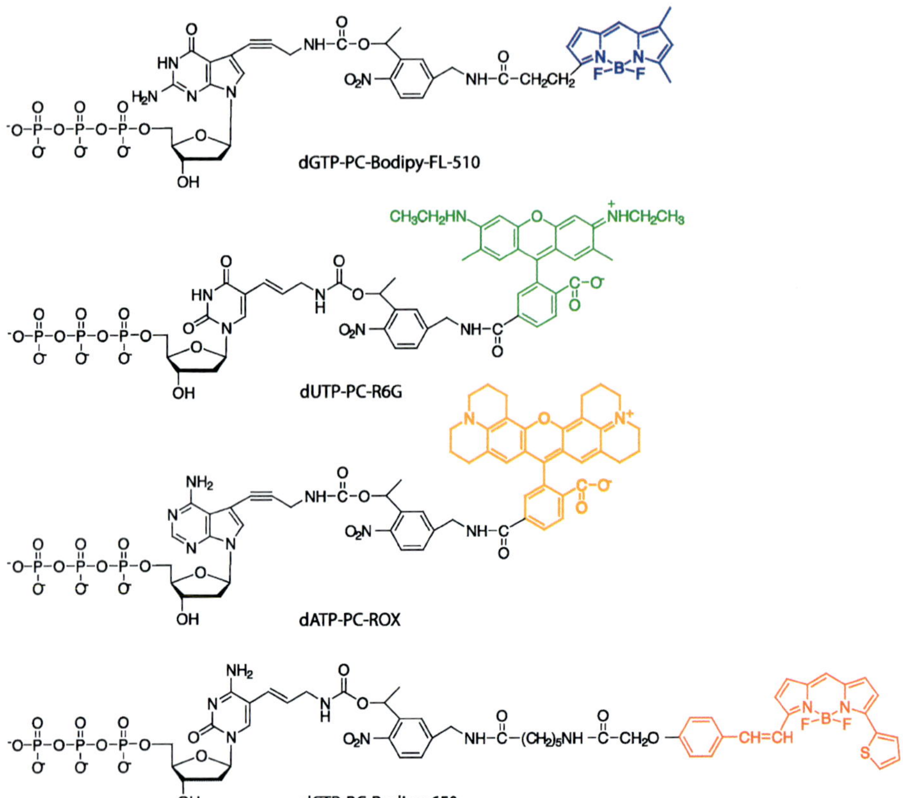

Fig. 5. Structures of four nucleotide analogues labeled through a photocleavable linker (PC) using four fluorophores with distinct fluorescent emissions, dGTP-PC-Bodipy-FL-510 ($\lambda_{abs(max)} = 502$ nm; $\lambda_{em(max)} = 510$ nm), dUTP-PC-R6G ($\lambda_{abs(max)} = 525$ nm; $\lambda_{em(max)} = 550$ nm), dATP-PC-ROX ($\lambda_{abs(max)} = 585$ nm; $\lambda_{em(max)} = 602$ nm) and dCTP-PC-Bodipy-650 ($\lambda_{abs(max)} =$ 630 nm; $\lambda_{em(max)} = 650$ nm).

with the universal looped primer to perform a series of extension reactions on a chip surface using all four photocleavable fluorescent nucleotide analogues. The principal advantage offered by the use of a self-priming moiety, as compared to using separate primers and templates, is that the covalent linkage of the primer to the template in the self-priming moiety completely prevents any possible dissociation of the primer from the template, even under vigorous washing conditions. Furthermore, the possibility of mispriming is considerably reduced,

Fig. 6. Products of DNA extension reaction and photolysis generated in solution phase to characterize the four different photocleavable fluorescent nucleotide analogues (dUTP-PC-R6G, dGTP-PC-Bodipy-FL-510, dATP-PC-ROX and dCTP-PC-Bodipy-650). After each extension reaction, the DNA extension product is analyzed by MALDI-TOF MS measurement to verify that it is the correct extension product (see Figure 7). Photolysis is then performed to produce a DNA product that is used as a primer for the next DNA extension reaction.

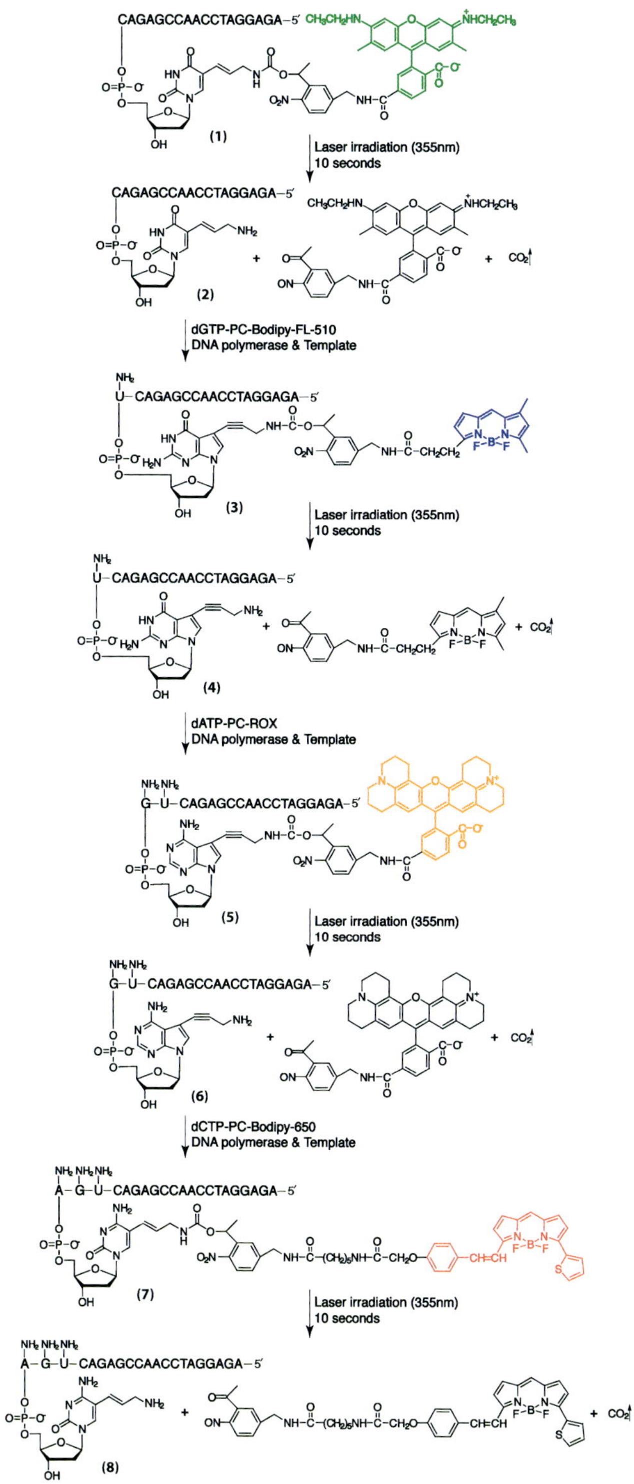
CAGAGCCAACCTAGGAGA–5′
(1)
Laser irradiation (355nm)
10 seconds
(2)
dGTP-PC-Bodipy-FL-510
DNA polymerase & Template
U–CAGAGCCAACCTAGGAGA–5′
(3)
Laser irradiation (355nm)
10 seconds
(4)
dATP-PC-ROX
DNA polymerase & Template
G–U–CAGAGCCAACCTAGGAGA–5′
(5)
Laser irradiation (355nm)
10 seconds
(6)
dCTP-PC-Bodipy-650
DNA polymerase & Template
A–G–U–CAGAGCCAACCTAGGAGA–5′
(7)
Laser irradiation (355nm)
10 seconds
(8)

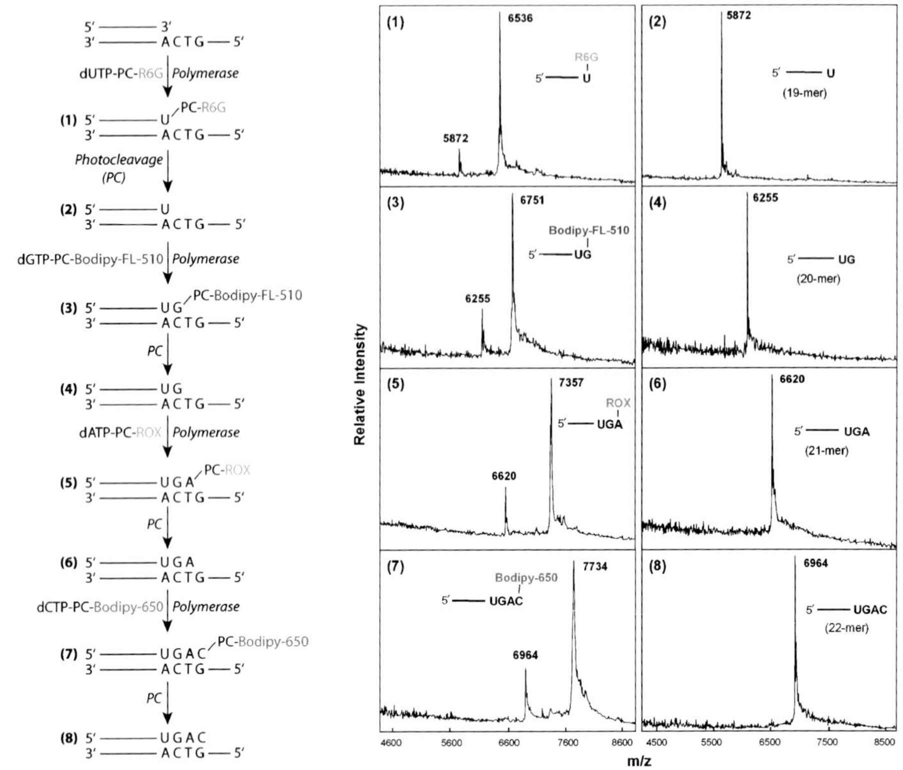

Fig. 7. The polymerase extension scheme (left) and MALDI-TOF MS spectra of the four consecutive extension products and their photocleavage products (right). Primer extended with dUTP-PC-R6G (1), and its photocleavage product 2; product 2 extended with dGTP-PC-Bodipy-FL-510 (3), and its photocleavage product 4; product 4 extended with dATP-PC-ROX (5), and its photocleavage product 6; product 6 extended with dCTP-PC-Bodipy-650 (7), and its photocleavage product 8. After 10 s of irradiation with a laser at 355 nm, photocleavage is complete with all the fluorophores cleaved from the extended DNA products. After each nucleotide is incorporated, the MS spectrum of each extension product is obtained to verify whether the incorporation was successful. In the spectra (1, 3, 5, 7) a small photocleavage product is visible due to the laser irradiation used in MALDI-TOF MS analysis. After photolysis, the spectra (2, 4, 6, 8) are obtained demonstrating complete photocleavage of each fluorophore from the extended DNA products.

and a universal loop primer can be used for all the templates allowing enhanced accuracy and ease of operation.

The four-color SBS sequencing results (Seo *et al.*, 2005) are shown in Figure 8. The structure of the self-priming DNA moiety is shown schematically in the upper panel, with the first 12-nucleotide sequence immediately after the priming site. The sequencing reaction on the chip was initiated by extending the self-priming DNA using dATP-PC-ROX (complementary to the T on the template) and Thermo Sequenase DNA polymerase. After washing, the extension of the primer by a single fluorescent nucleotide was confirmed by observing an orange signal (the emission signal from ROX) in a four-color fluorescent microarray scanner

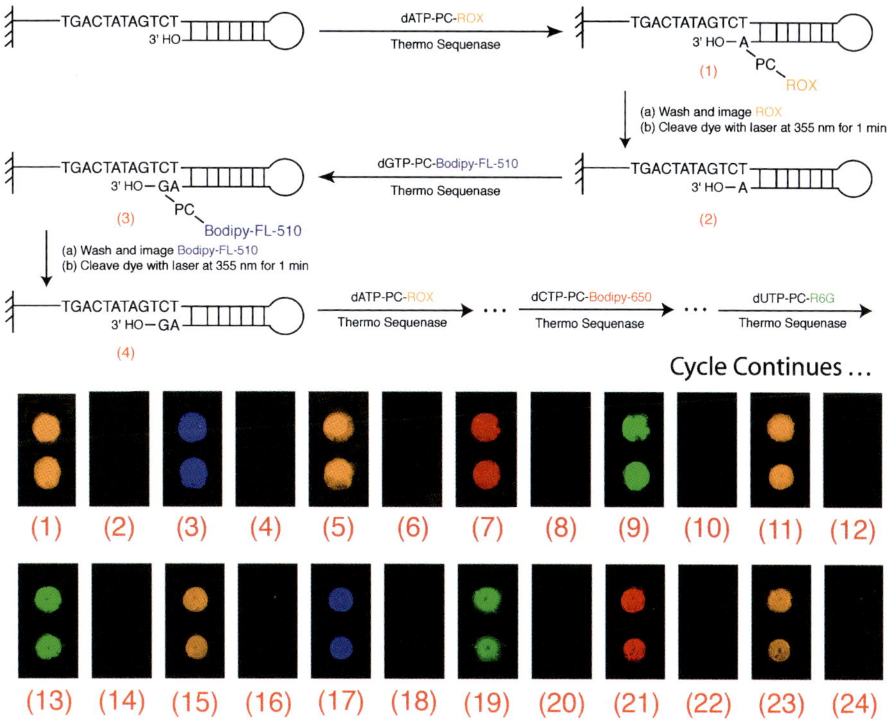

Fig. 8. Schematic representation of SBS on a chip using four PC fluorescent nucleotides (Upper panel) and the scanned fluorescence images for each step of SBS on a chip (Lower panel). (1) Incorporation of dATP-PC-ROX; (2) photocleavage of PC-ROX; (3) incorporation of dGTP-PC-Bodipy-FL-510; (4) photocleavage of PC-Bodipy-FL-510; (5) incorporation of dATP-PC-ROX; (6) photocleavage of PC-ROX; (7) incorporation of dCTP-PC-Bodipy-650; (8) photocleavage of PC-Bodipy-650; (9) incorporation of dUTP-PC-R6G; (10) photocleavage of PC-R6G; (11) incorporation of dATP-PC-ROX; (12) photocleavage of PC-ROX; (13) incorporation of dUTP-PC-R6G; (14) photocleavage of PC-R6G; (15) incorporation of dATP-PC-ROX; (16) photocleavage of PC-ROX; (17) incorporation of dGTP-PC-Bodipy-FL-510; (18) photocleavage of PC-Bodipy-FL-510; (19) incorporation of dUTP-PC-R6G; (20) photocleavage of PC-R6G; (21) incorporation of dCTP-PC-Bodipy-650; (22) photocleavage of PC-Bodipy-650; (23) incorporation of dATP-PC-ROX and (24) photocleavage of PC-ROX.

(Figure 8 [1]). After detection of the fluorescent signal, the surface was irradiated at 355 nm using an Nd-YAG laser to cleave the fluorophore. The surface was then washed, and a negligible residual fluorescent signal was detected, confirming the complete photocleavage of the fluorophore (Figure 8 [2]). This was followed by incorporation of the next fluorescent nucleotide complementary to the subsequent base on the template. The entire process of incorporation, detection and photocleavage was performed multiple times using the four photocleavable fluorescent nucleotide analogues to identify 12 successive bases in the DNA template. Thus, two conditions for a future SBS system have been satisfied – these newly developed nucleotide analogues have been shown to be excellent substrates for the

DNA polymerase, and the fluorophore group could be cleaved efficiently from the nucleotide using near UV irradiation. These results are important with respect to enhancing the speed of each cycle in SBS for high-throughput DNA analysis.

2.3. Blocking of 3′-hydroxyl groups

Another critical requirement to sequence DNA unambiguously using SBS methods is a suitable chemical moiety to cap the 3′-OH of the nucleotide so that it terminates the polymerase reaction after addition of a single nucleotide. This stepwise addition then allows the identification of the incorporated nucleotide at each step. A 3′-OH capping group on each of the nucleotides also permits all four nucleotides to be present together during the SBS extension reaction, resulting in a significant decrease in the number of cycles needed for sequencing. The 3′-OH capping group then also needs to be labile, and be able to be efficiently removed to regenerate the 3′-OH to permit the polymerase reaction to proceed to the next round. Conversely, the stepwise addition of nucleotides with a free 3′-OH group would present inherent difficulties in the detection of the sequence of homopolymeric regions.

The principal challenge posed by this requirement is the incorporation ability of the 3′-modified nucleotide by DNA polymerase into the growing DNA strand. Several groups have focused on the design and synthesis of nucleotides that have a photocleavable fluorophore on the 3′-position, as a simple way to cap the 3′-OH directly with the reporter group (Metzker *et al.*, 1994; Welch and Burgess, 1999). The rationale of this scheme is that after the fluorophore is removed, the 3′-OH group would be regenerated, and allow subsequent nucleotide addition. However, the incorporation by DNA polymerase of such a nucleotide with a photocleavable fluorescent dye on the 3′-position into a growing DNA strand has not been successfully reported so far. This is primarily due to the difficulty of DNA polymerases recognizing nucleotides with the 3′-position modified with large fluorophores. As noted earlier (Figure 4), the 3′-position on the sugar ring of a nucleotide is very close to the amino acid residues in the active site of the DNA polymerase. Thus, any bulky modification at this position will sterically hinder the DNA polymerase and prevent the nucleotide from being incorporated.

A second challenge critical to the overall efficiency and final achievable read length of SBS is the efficient removal of the 3′-OH capping group once the fluorescence signal is detected. Any DNA strand that has a remaining 3′-OH blocking group will inhibit the polymerase reaction and therefore lose its contribution to detect the next base in the template. Furthermore, the subsequent removal of the 3′-OH blocking group in subsequent SBS rounds would liberate that molecule for further addition steps, contributing to asynchronous "noise". Since each cycle of SBS sequencing essentially requires the complete removal of the 3′-OH capping group, a rapid and highly efficient process is required. It is important to use a small functional group that provides no hindrance to the DNA polymerase, while also stable enough to withstand DNA extension reaction conditions, and able to be removed easily and rapidly to regenerate a free 3′-OH under specific conditions.

Recently, we have developed a photocleavable fluorescent nucleotide with an allyl group capping the 3′-OH. These nucleotide analogues have been shown to act as substrates for a mutant DNA polymerase (Ruparel *et al.*, 2005). Figure 9 shows the synthetic scheme for the preparation of this novel nucleotide analogue, 3′-*O*-allyl-dUTP-PC-Bodipy-FL-510. Our selection of an allyl group is based on the fact that an allyl moiety, being relatively small and inert, would not provide significant hindrance for the polymerase reaction, and therefore would allow the incoming 3′-*O*-allyl modified nucleotide analogue to be accepted as a substrate by DNA polymerase. The entire cycle of a polymerase reaction using 3′-*O*-allyl-dUTP-PC-Bodipy-FL-510 as a reversible terminator is depicted in Figure 10. The extension product 11 obtained using 3′-*O*-allyl-dUTP-PC-Bodipy-FL-510 and a DNA polymerase was purified using HPLC and analyzed using MALDI-TOF MS. The base in the template immediately adjacent to the priming site was 'A'. Thus, if 3′-*O*-allyl-dUTP-PC-Bodipy-FL-510 was accepted by the polymerase as a terminator, the primer would extend by one base and then the reaction would terminate. Our results indicate that this was indeed the case.

After confirming that the extension reaction was successful, we irradiated it with near UV light at 355 nm for 10 s to cleave the fluorophore from the DNA, generating product 12. In the SBS system, this step would ensure that there would be no carryover of the fluorescence signal into the next incorporation cycle, so as to prevent the generation of ambiguous data at each step. The photocleavage product 12 was then incubated with a palladium catalyst system in aqueous solution to perform deallylation. The deallylated DNA product 13 carrying a free 3′-OH group was purified by reverse phase HPLC and then used as a primer in a second DNA extension reaction to prove that the regenerated 3′-OH was capable of allowing the polymerase reaction to continue. For the extension reaction, we used a photocleavable fluorescent nucleotide dGTP-PC-Bodipy-FL-510 and Thermo Sequenase DNA polymerase. The extension product 14 was irradiated as above, for 10 s to generate photocleavage product 15

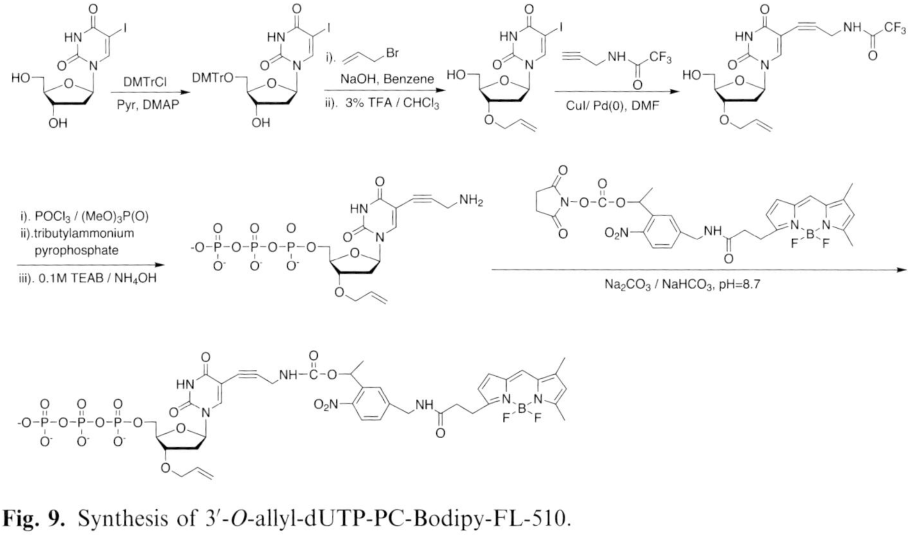

Fig. 9. Synthesis of 3′-*O*-allyl-dUTP-PC-Bodipy-FL-510.

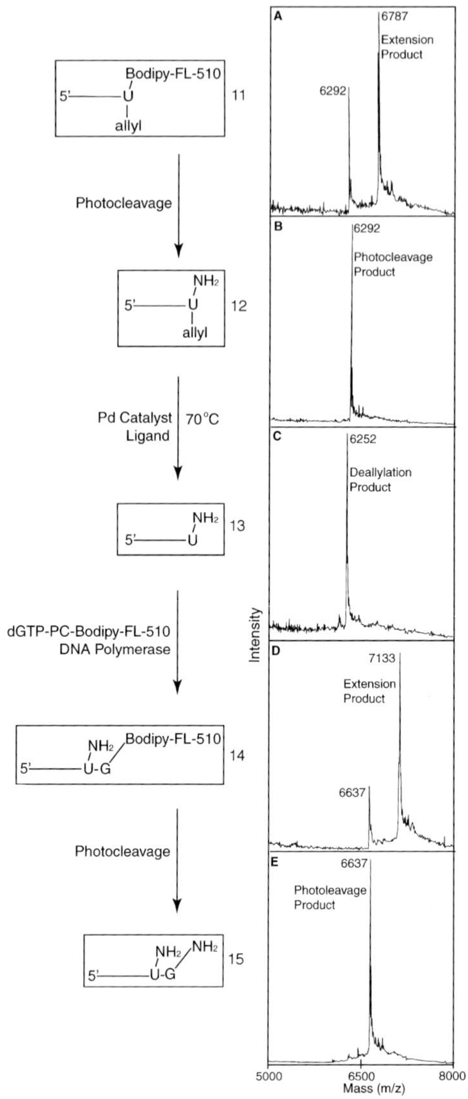

Fig. 10. MALDI-TOF MS results for each step of a polymerase reaction cycle using 3′-*O*-allyl-dUTP-PC-Bodipy-FL-510 as a reversible terminator. (A) Peak at *m/z* 6787 corresponding to the primer extension product 11 obtained using 3′-*O*-allyl-dUTP-PC-Bodipy-FL-510 and the 9°N Polymerase. (B) Peak at *m/z* 6292 corresponding to the photocleavage product 12. (C) Peak at *m/z* 6252 corresponding to the photocleavage product without the allyl group 13 obtained after incubation with the catalyst and ligand at 70°C. (D) Peak at *m/z* 7133 corresponding to the extension product 14 from the purified deallylated product using dGTP-PC-Bodipy-FL-510 and Thermo Sequenase DNA polymerase. (E) Peak at *m/z* 6637 corresponding to the photocleavage product 15.

and hence complete an entire reversible termination and extension cycle. The deallylation reaction was shown to achieve near quantitative yield under mild reaction conditions in an aqueous environment, providing an effective method to modulate the 3′-OH group of the nucleotides for SBS.

3. CONCLUSION

A substantial number of advances have been made toward the goal of making DNA SBS a viable technology for genomic research. This includes the rapid large-scale amplification of genomic libraries through emulsion-PCR, new developments in DNA attachment chemistries that allow increased array densities and novel reporter nucleotides as reversible terminators for polymerase reaction. These nucleotide analogues allow the enzymatic addition of a single nucleotide, direct detection to determine its identity, efficient removal of the reporter fluorophore and the 3′-OH blocking group to allow subsequent nucleotide additions. The integration of these developments will lead SBS to be developed into a high-throughput DNA sequencing platform for the era of personalized medicine.

ACKNOWLEDGMENTS

This work was supported by National Institutes of Health Grants P50 HG002806 and R01 HG003582, and the Packard Fellowship for Science and Engineering.

REFERENCES

Adessi, C., Matton, G., Ayala, G., Turcatti, G., Mermod, J. J., Mayer, P. and Kawashima, E. (2000). Solid phase DNA amplification: characterization of primer attachment and amplification mechanisms. Nucleic Acids Res. 28, e87.

Beier, M. and Hoheisel, J. D. (1999). Versatile derivatisation of solid support media for covalent bonding on DNA-microchips. Nucleic Acids Res. 27, 1970–1977.

Braslavsky, I., Hebert, B., Kartalov, E. and Quake, S. R. (2003). Sequence information can be obtained from single DNA molecules. Proc. Natl. Acad. Sci. USA 100, 3960–3964.

Collins, F. S., Green, E. D., Guttmacher, A. E. and Guyer, M. S. (2003). A vision for the future of genomics research. Nature 422, 835–847.

Debouck, C. and Goodfellow, P. N. (1999). DNA microarrays in drug discovery and development. Nat. Genet. 1, 48–50.

Dressman, D., Yan, H., Traverso, G., Kinzler, K. W. and Vogelstein, B. (2003). Transforming single DNA molecules into fluorescent magnetic particles for detection and enumeration of genetic variations. Proc. Natl. Acad. Sci. USA 100, 8817–8822.

Duthie, R. S., Kalve, I. M., Samols, S. B., Hamilton, S., Livshin, I., Khot, M., Nampalli, S., Kumar, S. and Fuller, C. W. (2002). Novel cyanine dye-labeled dideoxynucleoside triphosphates for DNA sequencing. Bioconjug. Chem. 13, 699–706.

Ghadessy, F. J., Ong, J. L. and Holliger, P. (2001). Directed evolution of polymerase function by compartmentalized self-replication. Proc. Natl. Acad. Sci. USA 98, 4552–4557.

Heiner, C. R., Hunkapiller, K. L., Chen, S. M., Glass, J. I. and Chen, E. Y. (1998). Sequencing multimegabase-template DNA with BigDye terminator chemistry. Genome Res. 8, 557–561.
Huyghues-Despointes, B. M. and Nelson, J. W. (1992). Stabilities of disulfide bond intermediates in the folding of apamin. Biochemistry 31, 1476–1483.
Hyman, E. D. (1988). A new method of sequencing DNA. Anal. Biochem. 174, 423–436.
International Human Genome Sequencing Consortium (2004). Finishing the euchromatic sequence of the human genome. Nature 431, 931–945.
Ju, J., Li, Z., Edwards, J. R. and Itagota, Y. (2003). Massive parallel method for decoding DNA and RNA. US Patent No. 6 664 079.
Ju, J., Ruan, C., Fuller, C. W., Glazer, A. N. and Mathies, R. A. (1995). Fluorescence energy transfer dye-labeled primers for DNA sequencing and analysis. Proc. Natl. Acad. Sci. USA 92, 4347–4351.
Lee, L. G., Spurgeon, S. L., Heiner, C. R., Benson, S. C., Rosenblum, B. B., Menchen, S. M., Graham, R. J., Constantinescu, A., Upadhya, K. G. and Cassel, J. M. (1997). New energy transfer dyes for DNA sequencing. Nucleic Acids Res. 25, 2816–2822.
Li, Z., Bai, X., Ruparel, H., Kim, S., Turro, N. J. and Ju, J. (2003). A photocleavable fluorescent nucleotide for DNA sequencing and analysis. Proc. Natl. Acad. Sci. USA 100, 414–419.
Lindroos, K., Liljedahl, U., Raitio, M. and Syvänen, A. (2001). Minisequencing on oligonucleotide microarrays: comparison of immobilisation chemistries. Nucleic Acids Res. 29, e69.
Margulies, M., Egholm, M., Altman, W. E., Attiya, S., Bader, J. S., Bemben, L. A., Berka, J., Braverman, M. S., Chen, Y. J., Chen, Z., Dewell, S. B., Du, L., Fierro, J. M., Gomes, X. V., Godwin, B. C., He, W., Helgesen, S., Ho, C. H., Irzyk, G. P., Jando, S. C., Alenquer, M. L., Jarvie, T. P., Jirage, K. B., Kim, J. B., Knight, J. R., Lanza, J. R., Leamon, J. H., Lefkowitz, S. M., Lei, J., Li, K. L., Lohman, H., Lu, V. B., Makhijani, K. E., McDade, M. P., McKenna, E. W., Myers, M., Nickerson, E., Nobile, J. R., Plant, R., Puc, B. P., Ronan, M. T., Roth, G. T., Sarkis, G. J., Simons, J. F., Simpson, J. W., Srinivasan, M., Tartaro, K. R., Tomasz, A., Vogt, K. A., Volkmer, G. A., Wang, S. H., Wang, Y., Weiner, M. P., Yu, P., Begley, R. F. and Rothberg, J. M. (2005). Genome sequencing in microfabricated high-density picolitre reactors. Nature 437, 376–380.
Margulies,, M., Jarvie,, T. P., Knight,, J. R. and Simons,, J. F. (2007). The 454 Life Sciences picoliter sequencing system. In: K. R. Mitchelson (Ed), New High Throughput Technologies for DNA Sequencing and Genomics (pp. 151–186). Elsevier, Amsterdam.
Marra, M., Weinstock, L. A. and Mardis, E. R. (1996). End sequence determination from large insert clones using energy transfer fluorescent primers. Genome Res. 6, 1118–1122.
Metzker, M. L., Raghavachari, R., Richards, S., Jacutin, S. E., Civitello, A., Burgess, K. and Gibbs, R. A. (1994). Termination of DNA synthesis by novel 3′-modified-deoxyribonucleoside 5′-triphosphates. Nucleic Acids Res. 22, 4259–4267.
Mitra, R. D., Shendure, J., Olejnik, J., Krzymanska-Olejnik, E. and Church, G. M. (2003). Fluorescent in situ sequencing on polymerase colonies. Anal. Biochem. 320, 55–65.
Nakano, M., Moody, E. M., Liang, J. and Bevilacqua, P. C. (2002). Selection for thermodynamically stable DNA tetraloops using temperature gradient gel electrophoresis reveals four motifs: d(cGNNAg), d(cGNABg), d(cCNNGg), and d(gCNNGc). Biochemistry 41, 14,281–14,292.
Pelletier, H., Sawaya, M. R., Kumar, A., Wilson, S. H. and Kraut, J. (1994). Structures of ternary complexes of rat DNA polymerase beta, a DNA template-primer, and ddCTP. Science 264, 1891–1903.
Pleasants, J. C., Guo, W. and Rabenstein, D. L. (1989). A comparative study of the kinetics of selenol/diselenide and thiol/disulfide exchange reactions. J. Am. Chem. Soc. 111, 6553–6558.
Prober, J. M., Trainor, G. L., Dam, R. J., Hobbs, F. W., Robertson, C. W., Zagursky, R. J., Cocuzza, A. J., Jensen, M. A. and Baumeister, K. (1987). A system for rapid DNA sequencing with fluorescent chain-terminating dideoxynucleotides. Science 238, 336–341.

Ronaghi, M., Karamohamed, S., Pettersson, B., Uhlén, M. and Nyren, P. (1996). Real-time DNA sequencing using detection of pyrophosphate release. Anal. Biochem. 242, 84–89.

Ronaghi, M., Uhlén, M. and Nyren, P. (1998). A sequencing method based on real-time pyrophosphate. Science 281, 363–365.

Rosenblum, B. B., Lee, L. G., Spurgeon, S. L., Khan, S. H., Menchen, S. M., Heiner, C. R. and Chen, S. M. (1997). New dye-labeled terminators for improved DNA sequencing patterns. Nucleic Acids Res. 25, 4500–4504.

Ruparel, H., Bi, L., Li, Z., Bai, X., Kim, D. H., Turro, N. J. and Ju, J. (2005). Design and synthesis of a 3′-*O*-allyl photocleavable fluorescent nucleotide as a reversible terminator for DNA sequencing by synthesis. Proc. Natl. Acad. Sci. USA 102, 5932–5937.

Sanger, F., Nicklen, S. and Coulsen, A. R. (1977). Sequencing with chain-terminating inhibitors. Proc. Natl. Acad. Sci. USA 74, 5463–5467.

Schena, M., Shalon, D., Davis, R. W. and Brown, P. O. (1995). Quantitative monitoring of gene expression patterns with a complementary DNA microarray. Science 270, 467–470.

Seo, T. S., Bai, X., Kim, D. H., Meng, Q., Shi, S., Ruparel, H., Li, Z., Turro, N. J. and Ju, J. (2005). Four-color DNA sequencing by synthesis on a chip using photocleavable fluorescent nucleotides. Proc. Natl. Acad. Sci. USA 102, 5926–5931.

Seo, T. S., Bai, X., Ruparel, H., Li, Z., Turro, N. J. and Ju, J. (2004). Photocleavable fluorescent nucleotides for DNA sequencing on a chip constructed by site-specific coupling chemistry. Proc. Natl. Acad. Sci. USA 101, 5488–5493.

Seo, T. S., Li, Z., Ruparel, H. and Ju, J. (2003). Click chemistry to construct fluorescent oligonucleotides for DNA sequencing. J. Org. Chem. 68, 609–612.

Shendure, J., Porreca, G. J., Reppas, N. B., Lin, X., McCutcheon, J. P., Rosenbaum, A. M., Wang, M. D., Zhang, K., Mitra, R. D. and Church, G. M. (2005). Accurate multiplex polony sequencing of an evolved bacterial genome. Science 309, 1728–1733.

Smith, L. M., Sanders, J. Z., Kaiser, R. J., Hughes, P., Dodd, C., Connell, C. R., Heiner, C., Kent, S. B. and Hood, L. E. (1986). Fluorescence detection in automated DNA sequence analysis. Nature 321, 674–679.

Tawfik, D. S. and Griffiths, A. D. (1998). Man-made cell-like compartments for molecular evolution. Nat. BioTech. 16, 652–656.

Wang, D. G., Fan, J. B., Siao, C. J., Berno, A., Young, P., Sapolsky, R., Ghandour, G., Perkins, N., Winchester, E., Spencer, J., Kruglyak, L., Stein, L., Hsie, L., Topaloglou, T., Hubbell, E., Robinson, E., Mittmann, M., Morris, M. S., Shen, N., Kilburn, D., Rioux, J., Nusbaum, C., Rozen, S., Hudson, T. J., Lipshutz, R., Chee, M. and Lander, E. S. (1998). Large-scale identification, mapping and genotyping of single-nucleotide polymorphisms in the human genome. Science 280, 1077–1082.

Weber, P. C., Ohlendorf, D. H., Wendoloski, J. J. and Salemme, F. R. (1989). Structural origins of high-affinity biotin binding to streptavidin. Science 243, 85–88.

Welch, M. B. and Burgess, K. (1999). Synthesis of fluorescent, photolabile 3′-*O*-protected nucleoside triphosphates for the base addition sequencing scheme. Nucleosides Nucleotides 18, 197–201.

Zhu, Z., Chao, J., Yu, H. and Waggoner, A. S. (1994). Directly labeled DNA probes using fluorescent nucleotides with different length linkers. Nucleic Acids Res. 22, 3418–3422.

Single-Molecule Sequencing

Chapter 7

Single-Molecule Fluorescence Microscopy and its Applications to Single-Molecule Sequencing by Cyclic Synthesis

Benedict Hebert[1] and Ido Braslavsky[2]

[1]*Department of Physics, McGill University, Rutherford Physics Building 228, 3600 University Street, Montreal, Quebec H3A 2T8, Canada*
[2]*Department of Physics and Astronomy, Clippinger 251B, Ohio University, Athens, OH 45701, USA*

Contents

PERSPECTIVES IN BIOANALYSIS, VOLUME 2
ISSN 1871-0069 DOI: 10.1016/S1871-0069(06)02007-6

Abstract

Single-molecule DNA sequencing (SMDS) had been proposed well before genomic research had advanced to the point where the DNA sequences of a few human individuals became available. Skepticism arose as to whether or not there was a need to replace methods that had been proven to be productive by a new technology. However, DNA information from thousands of individuals is needed to connect genomic information to the function it serves. Direct extensions of current methods are expected to be still much too expensive and slow to collect the amount of DNA and RNA sequence information that is required to enter the next phase in genomic research. Single-molecule techniques show great promise, as the next generation of DNA sequencing methods will allow the required amount of sequence information to be gathered in a timely and inexpensive manner. While several SMDS methods are under development, currently only single-molecule sequencing by cyclic synthesis advanced to the point where sequence information is produced in a massively parallel way directly from single DNA molecules. This sequencing technology relies on incorporation of fluorescently labeled nucleotides by DNA polymerase into complementary strands of DNA that are immobilized to a surface. The individual DNA strands are separated by a few microns and can be monitored as independent entities. The fluorescent signal of each incorporated labeled nucleotide is then sequentially detected using fluorescent microscopy. Because each DNA molecule is sequenced separately there is no need for synchronization between different molecules. Tens of millions of molecules can be sequenced in parallel in single small reaction volume, and thus this method readily produces high-throughput sequencing at a minimal cost. Currently this technique produces short-reading lengths, which make it suitable to re-sequencing applications in which a reference sequence is given. A single reference genome can serve as a template for the thousands of genomes produced by the short DNA fragments. These data can be used to find rare mutations and genetic heterogeneity in multiple target environments with great accuracy, high rates, and low cost. The ability to extract a massive amount of sequence information will equip cancer research with a powerful tool needed to defeat genetic diseases. In this chapter, different aspects of SMDS by cyclic synthesis will be discussed.

1. INTRODUCTION

Routine studies of individual genomes are central to the investigation of genetic variability and genetic susceptibility to diseases, but the inability to rapidly and cost-effectively sequence large amounts of DNA is a major hindrance to this goal. The recent completion of the human genome project in 2001 (Lander *et al.*, 2001; Venter *et al.*, 2001) has necessitated upwards of \$300 M in investments in two years time, and the estimated cost and time of sequencing a human genome today is set anywhere between \$10–and \$25 M in a year, still very far from the \$1000 genome objective (Chan, 2005). However, a paradigm shift has occurred recently whereby, in order to understand the function of DNA, it is not enough to produce the full sequence of a few individuals but rather we need the effort to sequence an immense amount of genome so as to relate variations in sequence and expression profiles, i.e. RNA re-sequencing, to the function of the genes. Therefore, *de novo* sequencing has been overshadowed by the potential for fast and inexpensive re-sequencing. Finding heterogeneities and inter-genomic variations will be the engine for new discoveries in the function of DNA (Bentley, 2004; Rogers and Venter, 2005).

While long read lengths are critical in *de novo* sequencing, they are less important in re-sequencing applications. With a length of as short as 16 bases (van Dam and Quake, 2002), sequences can be uniquely identified and mapped onto a template sequence and thus a method that provides a massive amount of short read lengths will be as affective as a method that produces the same amount of sequence with longer read lengths. It is expected that new and revolutionary methods will improve on Sanger sequencing in the main areas of cost and throughput, while some might also increase read lengths. Excellent reviews of the new techniques were recently published (Shendure *et al.*, 2004; Chan, 2005). This chapter will focus mainly on aspects of one of these methods: single-molecule sequencing by cyclic synthesis.

Single-molecule sequencing is a goal that has been pursued for almost two decades as a possible candidate to replace the ubiquitous Sanger method (Jett *et al.*, 1989) Different schemes have been proposed to achieve this goal, for example: (1) using exonuclease on flow-stretched labeled DNA and to detect the fluorescent product down stream (Augustin *et al.*, 2001; Werner *et al.*, 2003), (2) stretching DNA molecules in nanofabricated devices and to read fluorescent tags at the output (Chan *et al.*, 2004), (3) recording the ionic current through nanochannels while single DNA is thread through it (Meller *et al.*, 2000), (4) following the synthesis of DNA in real time by local confinement of illumination (Levene *et al.*, 2003), and (5) monitoring fluorescently labeled nucleotide incorporation on single DNA molecule step-by-step in cycle extensions (Braslavsky *et al.*, 2003). From all of the above, the demonstration that sequence information can be obtained from single DNA molecules by cyclic synthesis (Braslavsky *et al.*, 2003) leads to the development of the first working scheme for large scale single-molecule sequencing (Harris *et al.*, 2007 to be published).

DNA sequencing by cyclic synthesis (SBS) differs from the Sanger method, which relies on length separation of amplified DNA strands that terminate with a particular color according to the last base in the chain. Instead, in SBS the synthesis itself is monitored by various methods, such as pyrosequencing (Leamon *et al.*, 2003), or in polony sequencing (Mitra *et al.*, 2003). These methods monitor many reactions in parallel and thus accelerate sequencing rate and reduce cost. Out of all the cycle-extension approaches, single-molecule sequencing has the highest sequence information density, i.e. the number of sequence reads per unit area. Polymerase colony sequencing (Mitra *et al.*, 2003) has a density of about 1–2 polonies/mm^2, whereas picotiter plates (Leamon *et al.*, 2003) have a density of up to 480 wells/mm^2. The theoretical limit on density in single-molecule sequencing is the diffraction limit of light. For 670 nm emission, this limit is $\lambda/2$, or 335 nm, which entails a three orders of magnitude increase in density over picotiter plates, assuming a one-micron separation is allowed between molecules. Further more, monitoring several fields of view with a single camera introduces a major increase in throughput and opens the way for parallel sequencing of tens of millions of single DNA strands. Each DNA strand is read for about 25 bases, thus generating sequences that can readily be aligned to a reference sequence. Single-molecule sequencing is also the only cyclic-sequencing method that does not require the incorporations of nucleotides to be synchronous on all strands, a most important factor that limits read lengths in other schemes (Mitra *et al.*, 2003) and can be used to reduce error rates since reactions can terminate before the occurrence of side effects, such as misincorporation.

In this chapter, we will begin by introducing the advantages of single molecule imaging, and the theory behind the imaging systems and methods that are used in single-molecule sequencing by synthesis (SBS). We follow with an examination of the sequencing method itself and several variants that have been proposed in the last few years. We will then discuss the data analysis methodology and the sources of errors in base calling. We conclude with an overview of the applications and the performance of the technique.

2. BACKGROUND

2.1. Single-molecule detection

Single-molecule studies have had a major impact on several disciplines because of their ability to look among the smallest elements of nature, and distinguish between the ensemble average and individual behavior of the molecules (Michalet *et al.*, 2003; Bustamante *et al.*, 2004; Cecconi *et al.*, 2005). From analytical chemistry to biology, new information can be gathered by studying discrete behaviors of single molecules and generating distributions of observables quantities that are masked in ensemble averaging.

The ergodic hypothesis of statistical mechanics tells us that the average over time of a physical quantity from a single member of an ensemble is equivalent to the average over the ensemble at a given time. However there are several limitations to the applicability of this hypothesis. First the system must be homogeneous, which it often is not, especially in biological application where the cell-to-cell, protein-to-protein, or more generally molecule-to-molecule variation is simply too significant. Second, the sampling in space and time must be sufficient for the equivalency to be viable. Ensemble measurements can be used to determine the average value of a physical quantity but cannot generally be used to determine the distribution of that quantity. Studying the fluctuations in single-molecule temporal trajectories can yield detailed information about the dynamic processes, kinetics and kinematics of the molecules (Flomenbom *et al.*, 2005). An apparent paradox in single-molecule experiments is that experimentalists try their best to image a single molecule, and then they must observe tens to hundreds of them to extract useful information. This is due to the uncontrollable fluctuations in the experimental observables, such as emission intensity and emission spectrum of the fluorophores (Macklin *et al.*, 1996). Also, the observation of hundreds of single-molecule trajectories leads to the creation of distributions and the understanding of statistical properties. These experiments entail the analysis of the trajectory by itself and of the ensemble of trajectories. Nevertheless, while parameters such as relative distance between protein parts assessed by single-molecule FRET are influenced by fluctuations and need averaging to be precisely estimated even when careful control of the environment is implemented (Ha *et al.*, 1999; Rhoades *et al.*, 2003), some other observables are more robust. An example of such an observable is the presence of a fluorescent molecule that can be clearly determined with fluorescent microscopy (Nie and Zare, 1997). In SMDS by fluorescent microscopy, it is the presence of the fluorescent nucleotide that is monitored and thus the signal is relatively robust.

Fundamental limitations in the temporal resolution of single-molecule experiments stem from the intrinsic qualities of the fluorophore and the sensitivity of the detector. The absorption and emission lifetimes of a fluorophore are on the order of 10 ns, meaning that each molecule can emit up to 100 million photons in a second. This sets a lower limit for the efficiency of the detector. Occasionally, the molecule will transit to a dark state for some time – typically a few milliseconds – a phenomenon that limits the maximum rate of observation of a single fluorophore (Ambrose *et al.*, 1994). Fluorescence competes with several other deactivation channels and photochemical reactions that can lead to photodestruction of the signal molecule. This photobleaching phenomenon limits the maximum number of photons that can be integrated by the detector. Photobleaching is not a completely understood phenomena but the common thought is that fluorophores, in the dark (triplet) state, tend to interact with free oxygen and produce toxic singlet oxygen (Chen *et al.*, 2003), which in turn attacks the dye itself, but also damages other molecules like the DNA.

There are several excellent reviews on the various single-molecule observation methods (Nie and Zare, 1997; Xie and Trautman, 1998; Kulzer and Orrit, 2004). The fluorescence signal from single molecules is readily detected by photomultipliers, avalanche photo-diodes (APD), or high sensitivity cooled charge-coupled-device (CCD) cameras (Ambrose *et al.*, 1994), but the difficulty in detecting single molecules with high signal to noise ratios lies in the presence of optical background. The key challenge is to reduce the background interference, which may arise from Raman scattering, Rayleigh scattering, and impurity fluorescence. A confocal size volume ($\sim$1 fL) contains approximately $1–3 \times 10^{10}$ solvent molecules, $0.5–1 \times 10^8$ electrolyte molecules, and a large number of impurity molecules (Nie and Zare, 1997). To observe the minute amount of light given off by the single fluorophores over the optical background, different methods are successfully used to minimize the illuminated volume and thus reduce the background without reducing the signal from the molecule (Laurence and Weiss, 2003).

Some examples include, (1) near-field illumination utilizes a metal coated sharp optical fiber to confine the illumination volume (Xie and Dunn, 1994), (2) laser scanning microscopy in the confocal geometry considerably reduces out-of-focus light by spatial filtering with a pinhole in the image plane (Sheppard and Shotton, 1997), (3) two photon microscopy reduces the effective illumination volume because the intensity to excite the molecule by two simultaneous photons is high enough only at the focus (Mertz *et al.*, 1995), (4) zero-mode waveguides confine the illumination to small holes in a metal layer (Levene *et al.*, 2003), and (5) total internal reflection microscopy (TIRM) uses the evanescent field as a source to illuminate fluorophores in a thin layer near dielectric surfaces (Funatsu *et al.*, 1995; Tokunaga *et al.*, 1997; Dickson *et al.*, 1998). As a method of choice for surface-bound molecules, which is suitable to SMDS, we will elaborate on the TIRM approach.

2.2. Total internal reflection

TIRM is a technique used to look at fluorescence from a sample located within the first few hundred nanometers of the surface (Figure 1). There are several good reviews that describe this method (e.g. Axelrod, 1989, 2001; Tokunaga *et al.*, 1997; Ambrose *et al.*, 1999). Here, we briefly describe the TIR method and its application to DNA sequencing. When light strikes an interface going from a high refractive index medium to a low refractive index medium at an angle greater than the critical angle θ_c, it undergoes a total internal reflection. The critical angle is given by Snell's law:

$$\theta_c = \sin^{-1}\left(\frac{n_2}{n_1}\right)$$

where $n_{1(2)}$ is the refractive index of the first (second) medium, and $n_1 > n_2$. In the lower refractive index medium, there is an exponentially decaying electromagnetic field called the "evanescent wave". The evanescent wave excites fluorescent molecules within about 150 nm of the surface, and its intensity at the surface can be higher than the intensity of the incident beam (Ambrose *et al.*, 1999).

The fluorescence from the surface-bound molecules that are illuminated by the evanescent field, is detected by a microscope objective, through fluorescence filters by high sensitivity cooled CCD cameras. As only the vicinity of the surface is illuminated, there is a dramatic reduction of the noise from the bulk fluids, and surface-bound single molecules can be monitored with high signal to noise (Yildiz *et al.*, 2003).

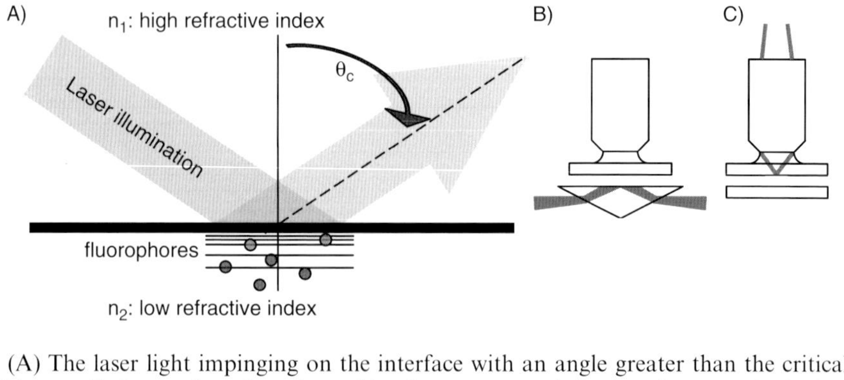

Fig. 1. (A) The laser light impinging on the interface with an angle greater than the critical angle (θ_c) is totally internally reflected, resulting in an exponentially decaying wave in the low refractive index medium. (B) Prism-based TIRM. (C) Objective based TIRM.

TIRM has the potential to generate single-molecule images even in the presence of free dye in the solution because molecules diffuse in and out of the evanescent wave region, creating a background blur, while those that are bound close to the surface become stable bright features (Funatsu *et al.*, 1995; Hebert *et al.*, 2007 to be published). TIRM is also very useful for *in vivo* imaging, for example, the studies of the basolateral membrane of the cell. Since the membrane is only about 5 nm thick, it is completely immersed in the TIRM field, as are all the transmembrane proteins and their molecular partners (Mathur *et al.*, 2000). It is important to note that there is no scanning involved in TIRM. The whole field of view is illuminated with the evanescent wave and is imaged using a cooled CCD camera.

Hence there is no illumination volume *per se* as occurs in confocal or two-photon microscopy. However, it is not possible to exceed the diffraction limit and thus there is still a convolution of the image that occurs in the optics (the objective), which means that a point particle will still appear as a Gaussian blur in the image. This phenomenon is actually helpful in designing algorithms to automatically find the features in a TIRM image, which because of their Gaussian nature their precise position can be determined down to few nanometers (Yildiz and Selvin, 2005) and even efficiently tracked in time. Another advantage of TIRM is that the location of the molecule is known, since it is attached to the surface and it is only the interface which is illuminated, therefore there are no complex focusing issues as might occur in confocal microscopy. There are several experimental geometries used to achieve TIRM near a dielectric interface in wide-field microscopy. Prism-based and through-objective TIRM (see Figure 1B and C) have been studied extensively, and each has its own advantages (Ambrose *et al.*, 1999).

Through-objective TIRM (Figure 2) requires the use of high numerical objectives. In addition to this requirement, the objective should be built from low fluorescence materials as the illumination is delivered through it. A geometric advantage is that it leaves the sample free from one side, so that fluid manipulation is simple.

The collection efficiency and the maximum angle of illumination of the objective in through-objective TIRM are characterized by the numerical aperture (NA). This number, usually 1.4–1.65, is a measure of how wide a cone of light the objective can gather or illuminate, and the greater the NA the wider the cone of light (Figure 2). The numerical aperture is equal to the refractive index of the objective lens material (n) times the sine of the maximum angle of illumination (θ_a), as given by $NA = n \cdot \sin(\theta_a)$. Hence a larger NA objective is desirable to obtain a greater angle of incidence in through-objective TIRM. For example, the refractive index of medium is 1.33–1.37, while the refractive index of glass (BK7) is 1.52. Thus, for objective built from glass the numerical aperture $n_1 \sin(\theta) > n_2$ thus, one needs $NA > 1.37$ in order to achieve objective type evanescent illumination. Even though it is possible to illuminate with an evanescent wave using a 1.4 NA objective, the margins are narrow and pure evanescent illumination is difficult to achieve given the delicate alignment. Fortunately, 1.45 NA are available from a few microscope companies, usually called TIRF objectives, as they are particularly well suited to TIR through the

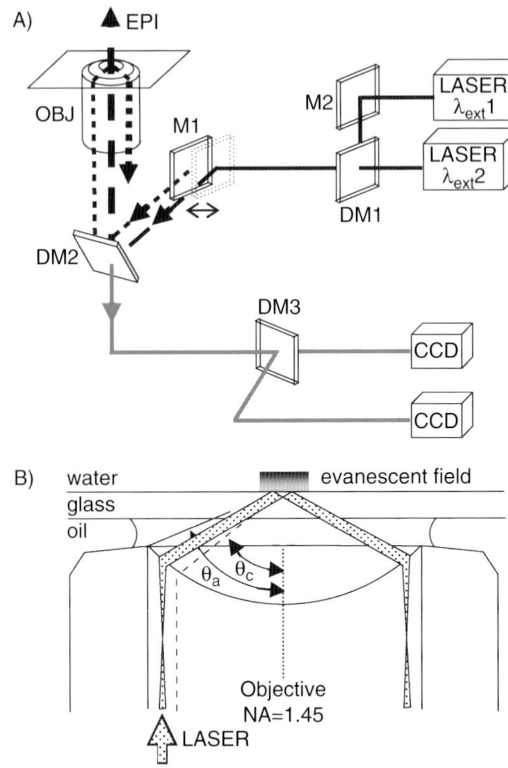

Fig. 2. Schematic drawing of the microscope used for single-molecule imaging of fluorescent molecules employing objective-type TIRFM. (A) Aligning the illumination to the appropriate angle is accomplished by translating a single mirror (M1). Multiple laser lines are combined using a dichroic mirror (DM1), for example, a diode pumped frequency-doubled Nd:YAG laser (532 nm) and a helium neon red laser (633 nm). A second dichroic mirror (DM2) introduces the laser into the objective lens (OBJ). The fluorescence is split in two (or more) channels using a dichroic mirror (DM3) and is detected by CCD cameras through appropriate fluorescence filters (see Tokunaga *et al.*, 1997 for further details). (B) Schematic drawing of objective-type TIRM (prismless TIRFM). The incident laser beam is focused on the back focal plane of the objective lens with a numerical aperture (NA) of 1.45. The term θ_a (72°) is the angle corresponding to this NA (1.52 sin(θ_a); 1.52 is the reflective index of glass), and θ_c (62°) is the critical angle of the glass-water interface (1.33 sin90° = 1.52 sin(θ_c), while 1.33 is the refractive index of water. When the incident beam is positioned to propagate along the objective edge between θ_a and θ_c, the beam is totally internally reflected producing an evanescent field at the glass–water interface (1/e penetration depth of about 150 nm). Modified with permission from figure in: Tokunaga *et al.* (1997). Copyright (1997), reprinted with permission from Elsevier.

objective applications. These extra few degrees of illumination increase the margin by a factor of 3 and thus make the alignment a relatively easy task. There are 1.65 NA objectives on the market, but they require the use of toxic oils and high refractive index glass. Thus, for most applications the 1.45 NA objectives seem to be the most efficient choice.

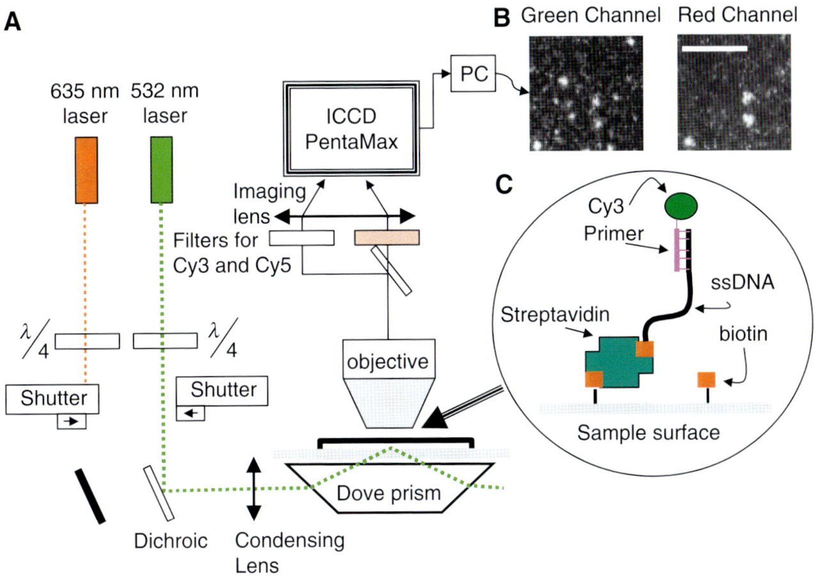

Fig. 3. Schematic drawing of prism-type TIRM. (A) Schematic drawing of the optical setup. The green laser illuminates the surface in a TIR mode while the red laser is blocked. Both Cy3 and Cy5 fluorescence spectra are recorded independently by an intensified charge-coupled device. (B) Single-molecule images are obtained by the system. The two images show colocalization of Cy3- and Cy5-labeled nucleotides in the same template (scale bar 10 μm). (C) Schematic of primed DNA templates attached to the surface of a microscope slide via streptavidin–biotin. Adapted from a figure originally published in Braslavsky *et al.* (2003). Copyright (2003), reprinted with permission from National Academy of Sciences (USA).

Prism-based TIRM can be implemented with any objective (see Figure 3). Since the imaging is made through the aqueous sample, some aberrations are introduced unless a water immersion objective is used (Peterman *et al.*, 2004). TIR is found to be an easy method to implement as no scanning is involved and the reduction in illumination depth enables one to observe surface-bound molecules with a high signal to noise.

It is possible to purchase off-the-shelf systems, but except for the objective the construction of this system is relatively simple (different configurations are illustrated in Figures 2 and 3). While surface illumination reduces noise from objects in the solution away from the surface, it does not reduce the noise from surface-bound impurities. The TIR evanescent wave will illuminate any entities on the surface, thus fluorescent dyes that adhere non-specifically to the surface will introduce noise. It is possible to reduce this noise by coating the surface with a thin metal layer that quenches the fluorescence in its very vicinity on the scale of 10 nm (Axelrod, 2001), but this implementation will also quench the signal if the molecules of interest are close to the surface as well. In the next section, we will discuss how it is possible to further reduce the noise in the system by using the FRET.

2.3. FRET theory

Förster resonant energy transfer (or fluorescent resonant energy transfer), FRET, is the energy transfer mechanism between two fluorescent dyes through long-range dipole–dipole interactions (Förster, 1948). The donor is excited at its specific excitation wavelength and this excited-state energy is transferred non-radiatively to the acceptor dye that becomes excited, while the donor returns to the ground state. The acceptor dye rapidly looses some energy through vibrational and rotational modes, and thus the energy match with the donor is lost, meaning that this energy cannot be returned to the donor. The acceptor dye eventually returns to the ground state, this time through a radiative process whereby a photon will be emitted. FRET can only happen when the two fluorescent dyes are in close proximity, usually less than 10 nm and the probability of energy transfer is strongly dependent on the inter-dye distance (Figure 4). Thus FRET is often used as a "molecular ruler", for example, to measure the

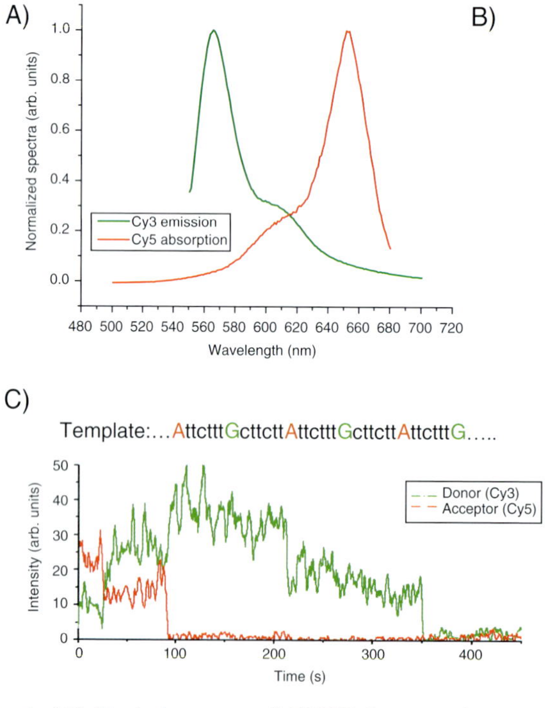

Fig. 4. (A) Typical spectra of FRET donor and acceptor molecules. In this example, the emission spectrum of Cy3 is shown to overlap the absorption spectrum of Cy5, so that FRET can occur between the two dyes. (B) Two labeled nucleotides inserted in double stranded DNA can make a single FRET pair. (C) Example of FRET between two donor dyes and two acceptor dyes. U-Cy5 and C-Cy3 are incorporated against A and G in the DNA template, the donor emission is partially quenched while the acceptors are emitting. As the acceptors bleach in single steps, the donor emission rises. Eventually the donors also undergo bleaching.

distance between two active sites on a protein that have been labeled, and therefore monitoring conformational changes through the amount of FRET between the dyes (Ha, 2001; Rhoades *et al.*, 2003; Xie *et al.*, 2004).

The orientation of the molecules in the illumination field and relative to each other is a factor that plays a role in the efficiency of the FRET as well. While usually averaged out by fast tumbling of the molecules, this orientation dependence can be of importance when incorporating fluorescent nucleotides into double-stranded DNA that has a pitch of 36° between two bases, or one turn in 10 bases (Watson and Crick, 1953; Ha *et al.*, 1996). The applications for single-molecule FRET have multiplied in the past decade which are described in several good reviews (Selvin, 2000; Ha, 2001). One important recent development is alternating-laser excitation (ALEX) of single molecules (Kapanidis *et al.*, 2005), which uses only the donor excitation wavelength that provides distance information through FRET, and uses acceptor excitation and combines this information with the donor excitation to report on relative donor–acceptor stoichiometry. Alternating both excitation wavelengths on the millisecond, microsecond, and nanosecond timescales can reveal information on structure and interaction of diffusing molecules, studies of gene transcription, and fast dynamical processes.

The crucial aspect of FRET, in its application to SMDS, is the confinement of the acceptor excitation light. Beside FRET, the smallest excitation volume that has been reported to date is $50\,\mathrm{nm} \times 50\,\mathrm{nm} \times 10\,\mathrm{nm}$ using a nanofabricated zero-mode waveguide (Levene *et al.*, 2003). This corresponds to an illuminated volume of $2.5 \times 10^{-5}\,\mu\mathrm{m}^3$, which is still more than an order of magnitude larger than the excitation volume provided by FRET, which is about $5 \times 10^{-7}\,\mu\mathrm{m}^3$. Furthermore, special equipment is required to fabricate and introduce engineered metal surfaces. Metal films can also quench the dye molecules and interfere with the detection of the molecules near the surface. In order to utilize the small excitation volume provided by FRET, the challenge is to make sure that the dyes are in close enough proximity to transfer energy. This requirement can be satisfied with single DNA molecules when donor- and acceptor-labeled nucleotides are inserted into the same DNA up to 20 bases apart.

The methods of TIRM combined with FRET provide an unparalleled increase in the signal to noise ratio of single-molecule observation. In the next section, we describe the motivation and different strategies behind the application of these techniques to SMDS. The use of FRET in such a setting will be described in more details in Section 3.4.1.

3. DNA SEQUENCING BY CYCLIC SYNTHESIS

3.1. Motivation

The advantages and feasibility of single-molecule detection at the glass–water interface using TIRM make a strong case for its use in single DNA sequencing. Current Sanger sequencing methods require a large amount of DNA to be replicated and then each of the sequencing runs is performed on one sequence at the time, a lengthy and expensive route. The alternative that DNA sequencing by cyclic synthesis offers is the sequencing of millions of fragments in parallel, and in the case of SMDS by cyclic synthesis no duplication of the DNA is needed at all. This combination would not only make whole genome sequencing far cheaper, it would also make it a lot faster. This would allow for rapid sequencing of numerous genomes and generate useful statistical comparisons.

There have been recent improvements to the ubiquitous Sanger sequencing (Sanger *et al.*, 1977), either by new methods such as massively parallel signature sequencing (MPSS) (Brenner *et al.*, 2000; Lu *et al.*, 2005) or by evolutionary approaches attempting to reduce the volumes of necessary reagents within the limits of conventional Sanger sequencing (Smailus *et al.*, 2005). These approaches have been moderately successful in lowering the overall cost per base. More recently, applications of pyrosequencing in picoliter reactors (Margulies *et al.*, 2005) have increased the throughput over current Sanger sequencing technologies by 100-fold. A close relation to "single molecule SBS" is the "amplified DNA SBS", which relies on the same principles of observation by TIRM, but requires amplification of the DNA templates. This gives a robust signal but requires additional preparation steps; the need for the templates to be synchronized might introduce duplication bias, which may limit the ultimate density of DNA targets on the surface. SMDS offers a simple sample preparation that does not require DNA amplification and holds the promise to obtain higher density of templates on the surface, both features increase the throughput. Single-molecule sequencing also removes the constraint of synchronicity encountered in other recent sequencing schemes (Kartalov and Quake, 2004; Lu *et al.*, 2005; Margulies *et al.*, 2005), in which ensemble measurements of DNA synthesis require all the strands to incorporate a given nucleotide at the same time in order to avoid de-phasing of the molecules. These advantages make SMDS by cyclic synthesis a very worthwhile pursuit. The basic scheme of SMDS by cyclic synthesis is composed of a few steps:

(1) DNA is sheared and cut into short fragments.
(2) These fragments are elongated by a common DNA tail.
(3) The DNA fragments are immobilized onto a glass surface that contains primers that match the common DNA tail.
(4) All bound fragments are then sequenced in parallel by –
    (a) polymerase extension of one base with a fluorescently labeled nucleotide;
    (b) detection by TIRM of multiple fields of view to record incorporation events on tens of millions of DNA fragments;
    (c) removal of the dye molecule; and
    (d) return to 4a with a different nucleotide.
(5) The data of each sequence is compared to a known sequence and aligned with it.
(6) Data analysis from this alignment reveals the sequence information in the target DNA.

In the next paragraphs, we discuss different aspects of this procedure. In Section 3.2 we describe the surface treatment needed to attach the single DNA molecules onto the surface, and in Section 3.3 we discuss aspects of the polymerase kinetics relevant to single-molecule sequencing. Lastly in Section 3.4 we describe different sequencing strategies.

3.2. Surface treatment

The observation of single fluorescent molecules requires a very high signal to noise ratio, and since the signal from single molecules is limited, one needs to reduce background noise to a minimum. Hence the surface on which the single DNA strands are to be attached for sequencing needs to be extensively cleaned, compatible with the anchoring method and have a low affinity to labeled nucleotides. Several good cleaning protocols are available (Kim *et al.*, 1998). For example, in previous work (Braslavsky *et al.*, 2003) a version of the RCA protocol (Kern and Vossen, 1978; Lee and Raghavan, 1999; Unger *et al.*, 1999) was used, in which glass slides were boiled in a mixture of ammonia and hydrogen peroxide followed by an extensive wash with purified water. The microscope slides were subsequently stored under purified water. After they have been thoroughly cleaned, the slides are prepared for the attachments of DNA molecules.

In order to visualize the DNA target and repeated incorporations in sequencing by cyclic synthesis, each template has to be immobilized in a definite location so that it can be matched between various image acquisition cycles. DNA will spontaneously stick to glass at a pH of about 5.5 (the isoelectric point of DNA), but we require a more specific and deterministic way to anchor the templates on the glass surface. The goal is to attach DNA to the surface while keeping it available for incorporations; therefore, it should not lie flat onto the surface and should preferably be connected at one of its ends. There are a few known protocols to attach DNA specifically to the surface, either covalently or through naturally occurring "glues" such as biotin and streptavidin, which have one of the largest free energies of association yet observed for non-covalent binding of a protein and small ligand in aqueous solution. The common basis to all these methods is that the DNA, either the template or the primer, is modified by some chemical moiety at its end. For the template it could be the 3′ or 5′ end, while in case of primer immobilization the modification must be at the 5′ end such that the 3′ end is available for incorporations. As an example of DNA attachment and surface treatment, we will elaborate on polyelectrolyte surfaces with template immobilization using streptavidin, which we used in previous work (Braslavsky *et al.*, 2003).

The initial RCA cleaning procedure leaves hydroxyl groups on the glass surface, which are de-protonated at the pH used here, and so they leave negative charges on surface. However this surface charge density is low, so it cannot provide enough electrostatic shielding against non-specific adsorption of tagged nucleotides. To increase this density, the build up of polyelectrolyte layers has been used (Decher, 1997; Kartalov *et al.*, 2003) and is illustrated in Figure 5. Polyelectrolytes are polymers whose chains contain charged functional groups. By building successive layers of polyelectrolytes on the surface, Kartalov *et al.* (2003) demonstrated that they can tune the charge density and to cover any inhomogeneities on the surface that might become sites for non-specific

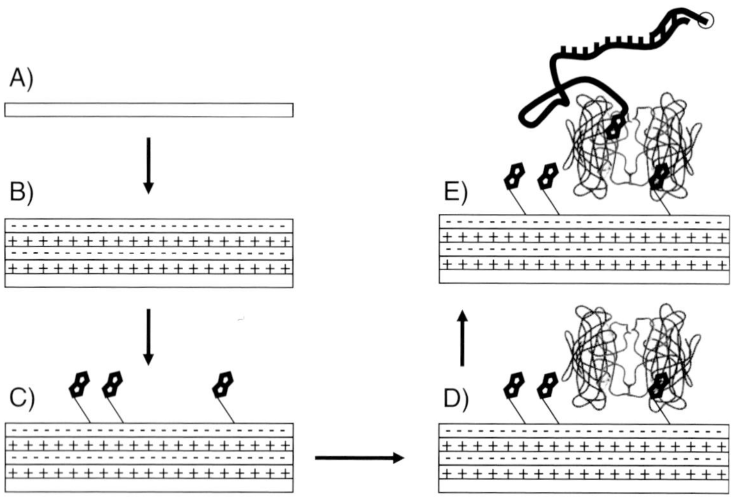

Fig. 5. 3The glass surface (A) preparation includes laying out multiple layers of electrolytes (B), and attachment of biotin to the surface (C). Streptavidin binds to the biotin layer (D), and biotinylated DNA can subsequently be attached to the surface (E). Detailed explanation is given in Kartalov *et al.* (2003). Copyright (2003), reprinted with permission from Biotechniques.

attachment. They have used positively charged polyethyleneimine (PEI) and negatively charged polyacrylic acid (PAcr). The first layer of positively charged PEI binds electrostatically to the negatively charged glass surface. The second layer, composed of negatively charged PAcr binds to PEI for the same reasons. The polymeric nature of the polyelectrolyte multi-layer results in increased charge density for each adsorbed layer. This surface was designed to efficiently reject labeled nucleotides as it has a high negative surface charge. The next step is to attach biotin ligands to the outer layer using biotin-amine (EZ-Link, Pierce), followed by the attachment of streptavidin. This treatment results in a streptavidin coated surface to which biotinylated DNA templates can be attached. While this surface treatment was successfully applied in single-molecule sequencing experiments (Braslavsky *et al.*, 2003), it was found that the quality of the surface is degraded over the cycles of incorporation, possibly due to the oxygen scavenger chemistry. Other surfaces treatments that allow extensive washes and covalent anchoring of the DNA can also be implemented (Sobek and Schlapbach, 2004), for example, Seo *et al.* (2005) anchored azido-labeled PCR products onto an alkynyl-functionalized surface. Such alternative surface treatment and a direct attachment of the DNA to the surface was successfully implemented in single-molecule sequencing for multiple cycles without apparent reduction in the surface quality over time (Harris *et al.*, 2007 to be published).

3.3. Polymerase kinetics

Current framework models for DNA polymerases (Johnson, 1993; Keller and Brozik, 2005) summarize the functions of the polymerase during the incorporation cycle. This framework is based on structural information such as the Klenow fragment structure (Beese *et al.*, 1993), on ensemble kinetics measurements such as steady and pre-steady kinetics (Kuchta *et al.*, 1987; Fiala and Suo, 2004), and on single-molecule investigations such as force dependent kinetics (Maier *et al.*, 2000; Wuite *et al.*, 2000). Despite the differences in sequence and origins, all DNA polymerases share a common structure: palm, thumb, and fingers. The polymerase resides at the end of the primer and upon docking of complementary nucleotide to the base template, it undergoes a conformation change that locks the nucleotide within the polymerase and enables bond formation with the backbone. Soon after, the polymerase opens up, releases a pyrophosphate and steps one base along the primer to the next incorporation site. Many different DNA and RNA polymerases exist (Goodman and Tippin, 2000) with different roles such as replication, repair, and error-prone polymerases that are able to overcome missing bases, and also increase genomic output by randomizing part of the genome encoding the genes of the immune system. For sequencing by cyclic synthesis, high fidelity and the ability to incorporate the particular label nucleotide required by the substrate are the desired polymerase capabilities. Exonuclease activity, by which the DNA is degraded by the enzyme, should be suppressed in order to retain labeled nucleotides that have been incorporated. While the inter-play between the polymerization and exonuclease activity of the enzyme results in an error rate that approaches one in 10^8–10^{10} bases, many polymerases with no exonuclease activity still discriminate efficiently against an incorrect base.

Most natural DNA polymerases have been found to be capable to incorporate bulky fluorescent nucleotide analogues, but with slower kinetics than their unlabeled counterparts. This is probably due to a charge difference and a steric interference when compared to the natural substrates (Zhu and Waggoner, 1997). The steric interaction is particularly problematic when several labeled nucleotides are to be inserted sequentially (Braslavsky *et al.*, 2003). For example, a mutant of the Klenow fragment of *E. coli* Pol I that does not have exonuclease activity has been found to be very efficient in incorporating fluorescently tagged nucleotides (Brakmann and Nieckchen, 2001; Brakmann, 2004); however, it does not readily incorporate several labeled nucleotides sequentially, for most attached dyes. Overcoming this problem is critical to the exonuclease single-molecule sequencing strategy (Werner *et al.*, 2003), however it is less critical to sequencing by cyclic synthesis. The limitation of the consecutive incorporation of labeled nucleotides can be removed by using cleavable dyes, in which the bulky fluorescent molecule is removed after detection. Further discussion on cleavable dyes is presented in Section 3.4.4. Directed evolution of novel polymerases (Goodman and Reha-Krantz, 1997; Brakmann, 2004; Holmberg *et al.*, 2005) can be used to develop more efficient polymerases for incorporation of labeled nucleotides. Such a polymerase should retain high fidelity while allowing incorporation of the particular fluorescent labeled nucleotides at the same time.

In the next section, we will explore a few sequencing strategies, which all have in common the use of polymerase for incorporation of labeled nucleotides into DNA templates and differ in the illumination, nucleotide substrate, and detection modes used.

3.4. Sequencing strategies

Several different approaches have been developed for use of fluorescence in SMDS. We have presented the theory behind TIRM, which confines the illumination light to within 150 nm of the surface, and FRET, which further confines the excitation region around the donor and provides excellent signal to noise ratios in single-molecule experiments. Here, we will describe in more details their application to single-molecule sequencing, and explain some of the more recent ideas on how to use fluorescence in DNA sequencing. Sequencing strategies using FRET illumination, either by cyclic synthesis or real-time mode, and the use of non-FRET illumination, cleavable dyes, and cleavable terminators are also described in detail.

3.4.1. Cyclic synthesis using FRET

The advantage of the FRET/TIRM combination over conventional wide-field TIRM is analogous to the haystack showing you exactly where the needle is, without having to look for it. The confinement of the acceptor excitation zone to a sphere of approximately 5 nm around the donor makes it unlikely to have a false positive signal (for a discussion of error, see Section 5) due to background noise or non-specific sticking to the surface. In FRET sequencing by cyclic synthesis (Braslavsky *et al.*, 2003), the common donor/acceptor pair Cy3/Cy5 has been used to demonstrate the feasibility of this technique. The general scheme is as follows: the first labeled nucleotide to be incorporated contains a donor fluorophore (Cy3), and successive nucleotides labeled with an acceptor fluorophore (Cy5) are cyclically washed in (see Figure 6). The acceptor fluorescence is detected by exciting the donor, and the acceptors thus fluoresce only if they are in the vicinity of the donor.

The noise from a non-specific attachment of labeled nucleotides to the surface has virtually disappeared because the effective illumination region is only a few nanometers. Since a non-cleavable dye was used, the elimination of the signal after detection has been achieved by bleaching the acceptor directly with the acceptor-specific laser illumination while the donor is left unharmed. Thus the use of a labeled nucleotide, as a donor combined with further incorporations of nucleotides carrying acceptor dyes, enabled the demonstration that sequence information can be obtained from single DNA molecule (Section 4.2.1 will describe single-molecule traces typical of this method). Nevertheless, this method has a few drawbacks that need to be addressed in order to accomplish this as a high-throughput method. First, the acceptor molecules are bleached, but they are not physically removed and thus further consecutive incorporations are severely compromised. Second, the donor eventually bleaches because of repeated illumination in this scheme. Third, even if both of the previous problems were solved, the limitation of the FRET excitation to a range of 5 nm would impose a limit of the read length of about 15 bases, which is too short to be aligned uniquely to a reference sequence.

In order to retain the advantage of FRET in SMDS by cyclic synthesis without the disadvantages, the donor should not be incorporated into the DNA,

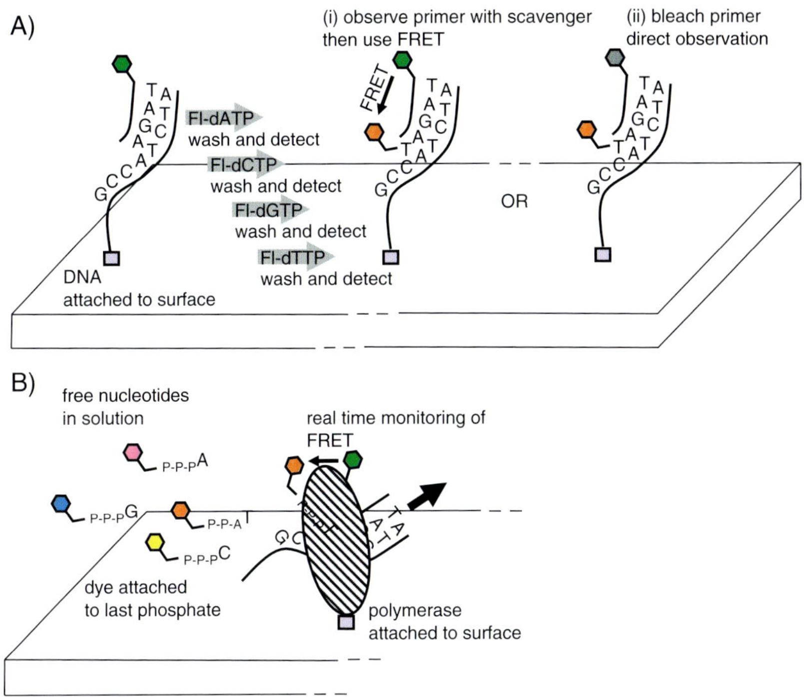

Fig. 6. Illustration of the SMDS by synthesis using FRET. (A) After observing the labeled primer, one can either use an oxygen scavenger to observe subsequent incorporations through FRET (i), or observe the incorporated fluorescent nucleotide directly (ii). Millions of DNA fragments are anchored to the surface of a glass slide and all the fragments are sequenced in parallel. (B) Real-time monitoring of the incorporation can be achieved if all types of nucleotides are present, with a label on the last of the three phosphates. The polymerase will lock on the nucleotide long enough for observation and the dye will automatically be cleaved off upon complete incorporation.

should be very stable or replaceable and would still need to be present in the vicinity of the incorporated acceptor-labeled nucleotide. A possible solution to this problem could be to label the polymerase with a donor fluorophore (Schneider and Rubens, 2001). The polymerase naturally finds its way to the 3′ position of the primer, exactly where the incorporation occurs. Thus, after washing all the reagents from the reaction chamber, reintroducing a polymerase with a donor attached to it will target the donor excitation to the right place. This would overcome all the problems posed before. It would act as a replaceable source that would not interfere with the incorporations and would not limit reading length.

Additionally, the use of robust photostable dyes would be an improvement on the sequencing by a cyclic synthesis scheme. Recently, quantum dots have been shown to act as good donors in FRET situations between a quantum dot and a fluorescent dye (Hohng and Ha, 2005). The authors have reproduced the known behavior of a DNA Holliday junction by comparing their quantum dot FRET data to conventional FRET data and obtaining the same dwell time distribution for low and high FRET states. In single-molecule sequencing, this would present the advantage of having a very long-lived donor because quantum dots are very photostable, and thus present the possibility of longer read lengths. A drawback to the use of quantum dot usage is their extensive blinking behavior (Nirmal *et al.*, 1996). This fluorescence intermittency has the potential to introduce frequent errors as false negative because the donor would be in an "off" state. The quantum dots are much bigger than conventional Cyanine dyes, so they probably will not be used directly as a label for a nucleotide. They could be used either as a label for the polymerase or possibly by fixing the quantum dot to the surface and attaching a single DNA molecule to it, with subsequent acceptor-carrying nucleotide incorporations; though this application is usable only if distance of the acceptor is kept within few nanometers from the quantum dot.

In the sequencing by cyclic synthesis method, the reaction is paused after each incorporation event. This method bears a huge advantage in throughput as the pause in activity enables the collection of information from tens of millions of fragments. The pause can be as long as needed to gather this information, which could take anywhere from several minutes to an hour with a rate that is dictated by the number of DNA fragments that are imaged per field of view, and the rate of imaging each field of view. Another SBS scheme in which no pause is required is the real-time mode which will be described next.

3.4.2. Real-time imaging

In real-time SBS, all nucleotides are present together in the reaction solution and the synthesis process is monitored constantly. Each nucleotide is labeled with a different dye. In order to enable sequential incorporation, the label is located on the last of the three phosphates and is cleaved off during the incorporation. With this method, one needs to follow the activity of the enzyme on the sub-millisecond time frame which makes it relatively hard to scale up to a massively parallel technique as only one field of view can be monitored. On the other hand, since the reaction runs freely and leaves behind unmodified DNA, it might produce long read lengths – far longer than what is achievable today by conventional Sanger methods. It might thus serve as a *de novo* sequencing method. While sequencing by cyclic synthesis could be performed at the single-molecule level or using amplified template molecules, this method has to be operated at the single-molecule level as there is no way to synchronize the incorporations at all.

One realization of the real-time SBS method could be achieved through immobilization of the polymerase labeled with the donor dye, as described previously (see Figure 6). While FRET delivers an advantage in the signal to noise and light confinement that it provides, especially because the real-time incorporation scheme is used in the presence of free-labeled nucleotide in the solution, it poses the problem of sustaining the donor dye unbleached for long periods of observation. Although this might be solved by labeling the polymerase with a quantum dot, which are photostable but have the drawback of extensive blinking (Nirmal *et al.*, 1996).

Another scheme for the realization of the real-time imaging employed zero-mode waveguides (Levene *et al.*, 2003). This innovative technique uses the evanescent illumination inside small 50 nm holes in metal films to locally illuminate a polymerase site as described above, and thus follows the synthesis process of single molecules in real time without FRET. Even though the illumination volumes are bigger than FRET, they remain sufficiently small to observe single molecules in high concentrations of free dye in solution. Since this method also avoids the problem of sustaining the donor dye unbleached, it holds the promise of achieving long read frames. However, the error rate might be high in this scheme because the integration time is small. Also, quenching of the fluorescence by the metal film could be a factor that increases the error rates, and it still has to be proven that this method can produce a significant amount of sequence information. In the next section, we return to the cyclic scheme and describe a non-FRET implementation of fluorescence microscopy to DNA sequencing.

3.4.3. *Non-FRET imaging*

In the case where a low density of free dye is present in the solution, direct imaging of the incorporated molecules using TIRM is a feasible option. The challenge in this case is to reduce the density of non-specific surface absorption to a minimum. In this scheme, the fluorescent dye is excited by the illuminating laser field, and not by a close donor dye, so that any fluorescent molecule in the field of view will emit, including non-specifically bound labeled nucleotides and other auto-fluorescent impurities. This might introduce false positives because both the pixel size of the imaging device and the convolving point spread function of the objective are much bigger than the local area taken by a single DNA molecule. Thus, any impurity or non-specific attachment of a labeled molecule within this region around a template would count as an incorporation event. Careful treatment of the surface can reduce the non-specific absorption of dye molecule to the surface. Recent experiments using this scheme have been successful in limiting the amount of non-specific binding and thus avoiding the drawbacks of the FRET illumination scheme (Harris *et al.*, 2007 to be published). Also, the optical resolution poses a limit on the minimal spot size but not on the accuracy in determining the location of the fluorophore. A new method called FIONA (Yildiz and Selvin, 2005) permits the determination of a fluorophore position down to about 2 nm.Following signals even enables one to identify two molecule positions by following a shift in the location of the spot using single-molecule high-resolution imaging with photobleaching (SHRImP) (Gordon *et al.*, 2004). These methods could be used to distinguish between a real event and a false positive event and reduce the random overlap problem to an acceptable level.

3.4.4. *Cleavable linkers*

Besides the experimental imaging considerations, there are also the molecular biology factors that need to be taken into account. The DNA polymerase is a very sophisticated enzyme capable of incorporating the correct nucleotide with less than one error in 10^5–10^6 bases (without exonuclease activity) and is an exemplary case of the integration of naturally occurring biological protein to the molecular biotechnology toolbox. However, in DNA sequencing by fluorescence, the bulky labeled nucleotide might not present such a challenge in itself to incorporate, but more importantly presents severe steric interferences for the incorporation of subsequent nucleotides. In sequential incorporations, the yield of incorporation reduces by a factor of 5 compared to incorporations of a labeled nucleotide adjacent to a non-labeled nucleotide (Braslavsky *et al.*, 2003). Although some dyes can be used as a label for consecutive incorporation (Brakmann and Nieckchen, 2001), other dyes cause the polymerase to throttle on multiple consecutive incorporations (Zhu and Waggoner, 1997).

For this reason, many research groups have focused their attention on designing nucleotides with cleavable dyes. By leaving a minimal residue on the nucleic acid, the steric interference is removed and the polymerase is able to incorporate the following nucleotide very efficiently. Two main approaches have materialized, the first of which is the inclusion of a disulfide (S–S) bond in the linker between the nucleic acid and the dye (Shimkus *et al.*, 1985; Mitra *et al.*, 2003, 2004). After incorporation, the disulfide bond can be broken by incubation with a reducer such as DTT. The second approach is the insertion of a photocleavable bond (PC) in the linker, which can be broken by UV radiation (Li *et al.*, 2003; Seo *et al.*, 2005). The advantageous use of cleavable dyes in single-molecule sequencing has been recently demonstrated (Harris *et al*,, 2007 to be published) with a yield of approximately 98% at each incorporation step. At this level of incorporation yield, more than 65% of the initial templates are sequenced to a length of more than 20 bases and thus establish this method as a practical DNA sequencing technique. This last set of experiments represents the first working scheme of SMDS – a goal that was pursued for the past 15 years by many groups.

Another aspect of DNA sequencing by cyclic synthesis is the homopolymer problem. When labeled nucleotides are washed into the reaction cell for incorporation, consecutive sites are available in each homopolymer template, such as an "AAAAAA" sequence. This might result in a few incorporations at a single site. While it is possible in principle to resolve the number of incorporation by intensity transitions (Park *et al.*, 2005) or by bleaching behavior (Gordon *et al.*, 2004), it becomes a more delicate process as the digital nature of the detection is compromised, i.e. the molecule is present or not. It is also hard to distinguish the number of incorporations by the total fluorescence due to quenching, or by the number of bleaching steps since they are sometimes hard to resolve and also require long illumination periods that might slow down the imaging process and also might be harmful for the sample. The fact that labeled nucleotides do not readily incorporate sequentially due to steric effect is an advantage for the homopolymer problem as the polymerase rapidly chokes and thus long homopolymeric runs do not entirely incorporate. Nevertheless an elegant method to cope with this problem is presented in the next section.

3.4.5. *Cleavable terminators*

Sanger sequencing utilizes 2′,3′-dideoxynucleotide triphosphates (ddNTPs), molecules that differ from deoxynucleotides by having a hydrogen atom attached to the 3′ carbon rather than an OH group. These molecules terminate DNA chain elongation because they cannot form a phosphodiester bond with the next deoxynucleotide, therefore these ddNTPs are called terminators. The homopolymer problem, which has been described in the last section, can be solved by using cleavable terminators. If the termination group can be cut after incorporation and imaging, this would allow for the incorporation of a single labeled nucleotide at a time, no matter if it is a repeat, or not. There have been recent reports of capping the 3′-OH group of an incoming nucleotides by a chemical moiety, which causes the polymerase reaction to terminate after the nucleotide is incorporated into the DNA strand (Ruparel *et al.*, 2005). The capping group can be subsequently removed to generate a free 3′-OH, and the polymerase reaction can re-initialize. It has been successfully demonstrated that fluorescently labeled nucleotides equipped with a cleavable chain terminator are active (Ruparel *et al.*, 2005). While cleavable terminators are a promising tool for SMDS, they still need to be experimentally checked at the single-molecule level to be validated as a suitable alternative. In particular, if the fluorescent dye itself is also cleaved, two cleaving stages are thus required and any type of chemistry step needs to be verified for compatibility with the other ingredients and its influence on performance. Nonetheless, another potential advantage of the cleavable terminators method is that it opens the possibility for incorporation of multiple labeled nucleotides in one step by multi-color labeling, a scheme which will be discussed in the next section.

3.4.6. *Multi-color versus one-color imaging*

In sequencing by cyclic synthesis one can implement either a single-color strategy in which all nucleotides are labeled with the same dye and each type is introduced independently into the reaction chamber, or to implement a multicolor scheme where each nucleotide species is labeled with a different dye and thus all nucleotide varieties can be introduced and imaged simultaneously. The foremost advantage of multi-color imaging in single-molecule sequencing is the reduction of the number of "wash and detect" cycles (see Figure 6): there is only one incorporation wash for four nucleotides. This might speed up the data acquisition process because current image splitting technology allows for wavelength specific, four-way splitting of the emitted light into four separate channels, each representative of a single nucleotide variety. As only one imaging cycle is needed, the increase in throughput is four-fold. Moreover, a possible advantage is that all nucleotides are present in the reaction and this might reduce the mis-incorporation rate. However, there are potential drawbacks associated with this method, the first of which being that, although it is possible to implement, splitting the signal in four separate channels increases the detection complexity as all colors need to be simultaneously focused accurately. Moreover, this scheme entails either real-time incorporation monitoring or the use of cleavable terminators because all the possible nucleotides are present, and therefore successive incorporations can occur. As real-time monitoring has its own drawbacks and cleavable terminators introduce additional cleaving steps, the potential advantage might be compromised compared to a simpler version with single dye for single-molecule sequencing purposes.

4. DATA ANALYSIS

The sequencing of DNA using single-molecule fluorescence calls for careful experimental design and subtle parameter tweaking, simply to be able to observe the incorporation of single nucleotides into the DNA template. The goal is to collect the sequence information from each molecule by itself. As multiple fields of view are imaged in order to monitor incorporations on millions of templates simultaneously, techniques that precisely monitor the position of the molecules should be addressed. The sequence information from each molecule should then be aligned to the reference sequence. For long enough sequences, it is possible to align the found sequences to the reference even if there is disagreement or "error". This "error" could come from either a real error in the sequencing or from the data under analysis – i.e. the mutations, polymorphism, or heterogeneity that the re-sequencing reveals. In order to have enough statistics to provide a meaningful picture of the DNA sequence, an over-sampling is required, which averages out random error and reveals the sequence content of the sample. As the amount of strands that are sequenced at the same time is enormous, this is not a strong limitation on the method. In this section we will elaborate on some aspects of the data analysis, starting from an example to signal analysis that is used to align the position of the molecule in time, then an example for extracting the sequence information from each molecule by FRET, and lastly a discussion on aligning the sequences to the template.

4.1. Spatial correlations

In order to return to the position of a molecule with high precision after probing other fields of view, one must either use a nanometer-positioning stage that can travel several millimeters, or use the single molecule itself as accurate fiduciary markers for repositioning. Here, we describe an example of the analysis of CCD images to extract the positions of the molecules within an image and the alignment of the images in time.

The images are first processed using a spatial band-pass filter to smooth the images and subtract background fluorescence. Coordinates of the resolved intensity spots in the filtered image were determined by locating their centroids using both intensity and eccentricity of the spots as rejection criteria to discriminate real features from noise (Crocker and Grier, 1996). A correlogram is generated by shifting the two coordinate sets relative to one another, and counting the number of correlated features at each spatial lag. It is assumed that two positions are correlated if they fall within a certain pre-set radius from each other. Fluorescently tagged proteins, DNA molecules, and other particles can be tracked in time using such methods for locating the position of particles (Crocker and Grier, 1996; Braslavsky *et al.*, 2001; Yildiz *et al.*, 2003; Babcock *et al.*, 2004; Hebert *et al.*, 2005). To illustrate this method, we describe the following experiment. DNA polymerase and a matched species of labeled nucleotide were incubated in the flow cell for 5 min and subsequently washed out.

The surface was imaged and the positions of the fluorescent molecules that appeared on the surface were correlated with the positions of the DNA molecules that were detected beforehand (see Figure 7A). When the images are superimposed, a high correlation between the primer position and the nucleotide position was found for the correct match i.e. when dUTP-Cy3 matches the available template base, A (see Figure 7C). For mismatch incorporation no peak in the correlogram is detected, (see Figure 2 in Braslavsky *et al.*, 2003). The correlogram reveals sub-pixel shifts between the images as they averaged over many molecules. This information is used to monitor a particular pixel position over time and to determine the incorporation events and thus the sequence of the DNA template attached to that particular point. The next section will discuss the extraction of the sequence information from the fluorescence data.

4.2. Data collection – base calling

Once the fields of view have been aligned using the correlograms, each molecule is detected by a few pixels of the CCD camera. After each incorporation reaction, the presence of a labeled molecule is detected by the intensity, shape, and location of the fluorescence signal at that spot. According to this signal, it can be decided automatically whether or not a nucleotide has been incorporated. The data collection of the fluorescence signal depends on the sequencing

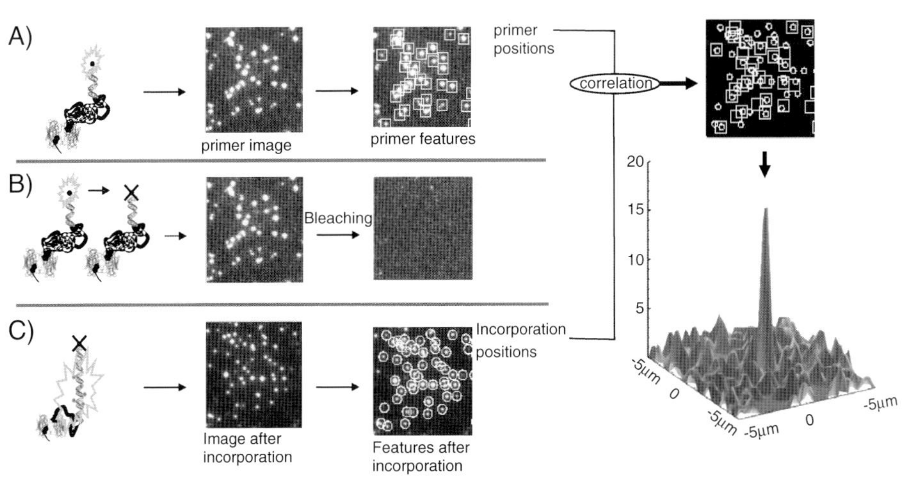

Fig. 7. Correlation between the positions of the DNA template (A), and the position of incorporation events (C). To avoid false positive signals, the primer label is bleached in between these two observations (B). Modified from Braslavsky *et al.* (2003). Copyright (2003), reprinted with permission from National Academy of Sciences (USA).

scheme. In real-time methods, a continuous stream of data on the millisecond timescale is needed. In cyclic sequencing schemes, a single or a few exposures are needed with integration times of about 100 ms to determine the presence of a fluorescent molecule. The optimal detection integration time is influenced by factors such as bleaching time of the molecules and signal to noise. The goal is to observe the molecule in as short a time as possible to reduce the thousands of field-imaging times, without bleaching the molecule and while keeping the signal to noise high, by extracting the maximum numbers of photons from a molecule. In the next section, we will elaborate on the example of single DNA molecule signals in sequencing experiments that use FRET to determine incorporation events (Braslavsky *et al.*, 2003).

4.2.1. Intensity traces

In this section, we will describe the signal collection from a FRET experiment with some additional details. As discussed previously, the background noise can be suppressed by the use of single-pair FRET as a highly localized excitation source to monitor the incorporation of nucleotides in the templates. The first labeled nucleotide to be incorporated contains a donor fluorophore (Cy3), and successive nucleotides are labeled with an acceptor fluorophore (Cy5). The acceptor fluorescence is detected by exciting the donor, and the acceptors thus fluoresce only if they are in the vicinity of a donor. The noise from a non-specific attachment of labeled nucleotides to the surface becomes very small, because the effective illumination region is only a few nanometers. In this example, the fluorescence dyes are not cleavable, hence photobleaching is used to null the acceptor fluorescence. After each incubation and FRET signal detection, the surface is illuminated with the acceptor specific excitation laser to bleach the acceptor but leave the donor unharmed. To efficiently visualize this process throughout the whole sequencing experiment, the authors used intensity traces at the primer locations for both Cy3 and Cy5 signals to calculate the FRET efficiency (Figure 8).

Alternate illumination can also be used to compare the signal from FRET to the signal from the Cy5 fluorophore directly. Some other uses of alternate illumination have been described in the literature (Kapanidis *et al.*, 2005). Since the field of view shifts slightly between each reagent exchange, one has to be careful to shift the location of the intensity trace for each image set according to the peak of the correlation function. Also, because of the uneven illumination field from TIRM, one has to subtract a local background as opposed to a general noise subtraction for the whole field of view. In essence, the average intensity over a 3×3 pixel region around the location of the primer constitutes the raw signal from the single molecule, and from that is subtracted an average over a 5×5 pixel region (excluding the central 3×3 region) that constitutes the local background. Here, it is assumed that the density of the DNA templates is low enough that the 5×5 region around the primer location does not contain another DNA molecule.

The FRET efficiency is calculated as $I_a/(I_a + I_d)$, where I_d and I_a are the average intensities of the donor (Cy3) and the acceptor (Cy5), respectively. The FRET

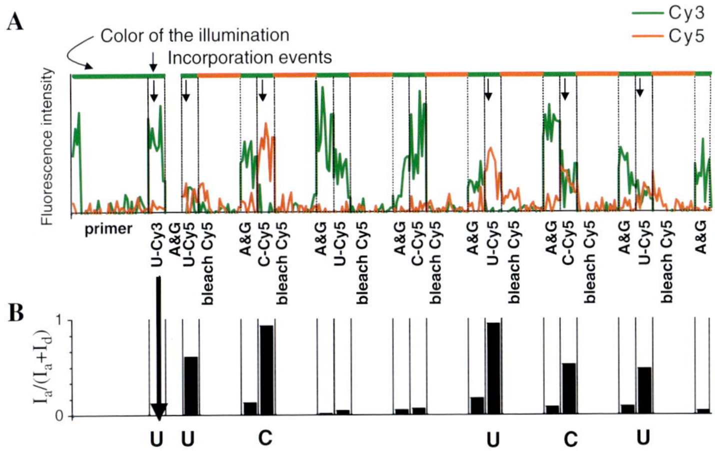

Fig. 8. Sequencing single DNA molecules with FRET. (A) Intensity trace from a single template molecule through the entire session. The green and red lines represent the intensity of the Cy3 and Cy5 channels, respectively. The label at each column indicates the last nucleotide to be incubated, and successful incorporation events are marked with an arrow. (B) FRET efficiency as a function of the experimental epoch. Reprinted from Braslavsky *et al.*(2003). Copyright (2003), reprinted with permission from National Academy of Sciences (USA).

efficiency has a higher signal to noise than quantitation of channel alone because it combines information from both fluorophores while simultaneously normalizing the relative intensities. The particular trace shown in Figure 8 reads out the correct sequence fingerprint for the template used (AAGAGA). Note the skip after the first G. This demonstrates that the sequencing scheme is asynchronous, an important feature that distinguishes sequencing at the single-molecule level from the ensemble averaging inherent in macroscopic schemes. Thus, when an incorporation reaction is incomplete on a particular template molecule, it can be successfully completed in a later cycle without producing false information, or interfering with data from other DNA templates in the field of view. While using a complete trace is very useful to determine the sequence content of the template, it has a few drawbacks. For example, long illumination times in the FRET trace mode increase the risk of bleaching, even in the presence of an oxygen scavenger, which complicates the data analysis. A simpler method, relying on the information that is deduced from the trace mode, is discussed next.

4.2.2. Single-image data collection

After careful characterization of the single-molecule signal in the experiments, one can assess what the detection probability of a molecule in one exposure will be compared to a more elaborate scheme of detection. This single-image scheme can be implemented as a simple and fast method of detection, since the digital readouts of single-color sequencing (presence or absence of a fluorescent molecule) are much simpler to analyze. Recent experiments have shown that such a collection mode is efficient and results in a reliable reading with a fast and simple data collection (Harris *et al.*, 2007 to be published).

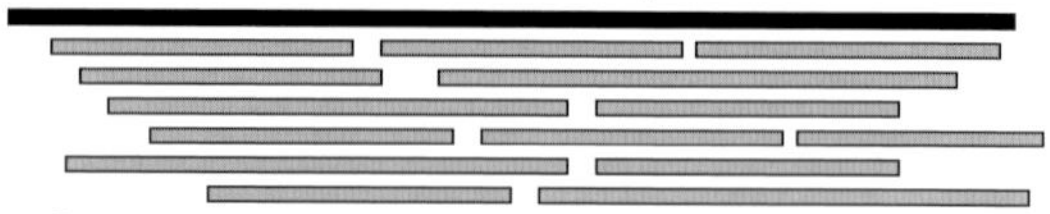

Fig. 9. Short-sequenced fragments have to be aligned with the consensus genome sequence using computer algorithms to allow detection of point mutations, insertions/deletions, and amplifications.

4.3. Aligning the sequences

Once the short fragments have been read, they have to be aligned to a reference sequence. Sequence alignment has become one of the most common tasks in bioinformatics, with applications ranging from phylogenetic analyses to identification of conserved domains and protein structure prediction. The alignment of the sequence fragments over the consensus DNA sequence is done using various computer algorithms (Notredame, 2002). Because of limited read length and error rates, any DNA sequencing scheme requires a certain amount of oversampling, if only to provide sufficient regions of overlap between the reads to assemble the genome. The short DNA fragments that are sequenced using single DNA molecules are too small to be assembled as a genome for *de novo* sequencing. Instead, alignment of these sequences with a known template (Figure 9) allows the detection of point mutations, insertions/deletions, and amplifications. Detection of rare mutations and single nucleotide polymorphisms requires a high level of coverage of the genome, and a minimized error rate. A more in-depth look at error sources and experimental caveats follows in the next section.

5. ERROR SOURCES IN BASE CALLING

Determining the base type in SMDS by fluorescence is conceptually easy: the presence of the fluorescence signal at a primer location during any given step of the sequencing cycle is indicative of an incorporation of that base in the DNA template. However, in practice, deciding whether an incorporation event has happened is not trivial. We have to consider the rate of occurrence of false-positive and false-negative signals.

False-positive signals occur when there is random correlation of a dye signal with the primer location in non-FRET single-molecule sequencing, which can be due to non-specific binding of a labeled nucleotide close to the DNA template, within the size of a pixel or so.

These can also occur because of a mis-incorporation of the labeled nucleotide by the DNA polymerase. All false-positive signals will indicate that a nucleotide has been inserted when in fact there should be none, and hence it will introduce an error in the sequence for that particular DNA template. False-negative signals originate when a nucleotide is inserted but no fluorescent signal is detected. This could be due to defective reagents such as unlabeled nucleotides,

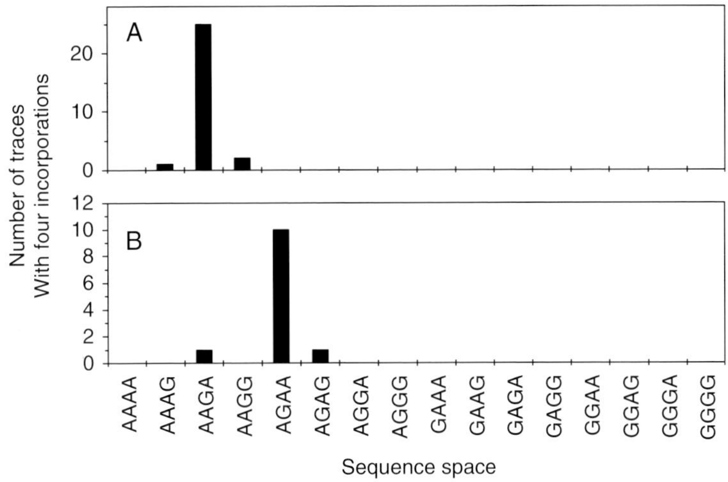

Fig. 10. Histogram of sequence space for 4-mers composed of A and G. All traces that reached at least four incorporations are included. (A) Results for template 1 (actual sequence fingerprint: AAGA). (B) Results for template 2 (actual sequence fingerprint: AGAA). Reprinted from Braslavsky *et al.* (2003). Copyright (2003), reprinted with permission from National Academy of Sciences (USA).

or a labeled nucleotide whose attached dye has bleached during the donor observation that precedes the FRET imaging. In addition, dye blinking and out of focus imaging can be sources of false-negative signals. However, the asynchronous feature of single-molecule sequencing allows one to discriminate against false-signal information for each template by virtue of statistics. For example, the sequence fingerprinting experiment described in Figure 8 was also performed with an independent template DNA sequence (Braslavsky *et al.*, 2003). Comparing the measured sequences to the set of all possible 4-mer sequences shows that the correct sequences for two templates can be discriminated with a 97% confidence level (see Figure 10).

In the re-sequencing application, the reading lengths are unique when they are longer than 16–20 bases (van Dam and Quake, 2002). Thus, when reading lengths of 20 bases or more are generated, the sequences can be aligned with a known reference sequence (Figure 9). When a high coverage of the reference sequence is obtained, it is possible to average the sequences, and thus find mutations or disagreements with the library sequence. By increasing the coverage, or sequencing depth, one can find rare mutations even in noisy raw sequence data. Some other factors can reduce error rate, for example, (1) misincorporation results in a mismatch at the end of the primer and this template will probably be terminated and thus filtered out from the template pool, (2) random overlap will look like a single addition in the alignment process, a rare event in gene sequences as it cause a shift in the reading frame, and thus can be filtered out in some cases, and (3) since the location of each molecule is known, it is possible, in principle, to sequence the same molecule twice, a procedure that would dramatically decrease the error rate.

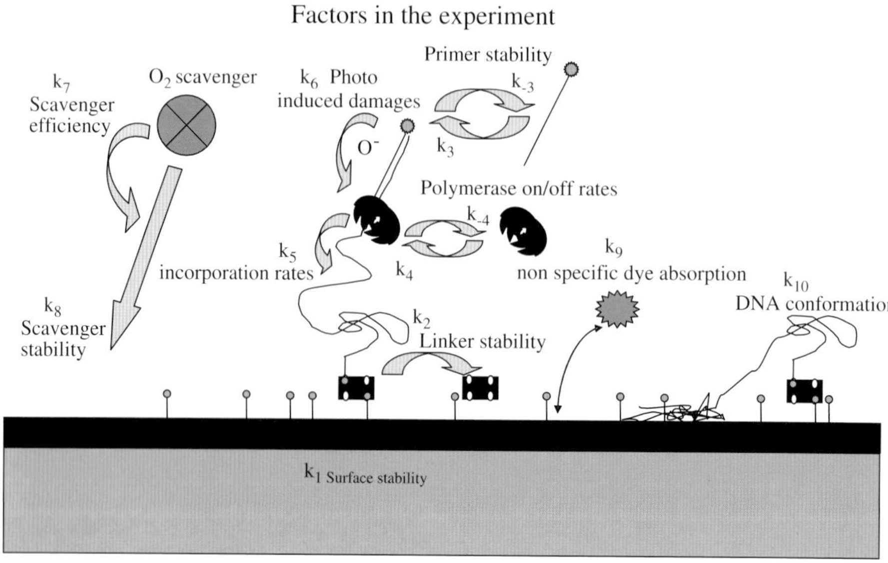

Fig. 11. Several important time constants play a role in determining the minimum reagent concentrations necessary and the error sources in the experiments.

In SMDS, each molecule contains unique information that is critical and thus one would like to examine the same molecule for the full experiment duration. The important constants for stability are not the equilibrium constants, but rather the off-rate parameters, because when the molecule leaves the anchoring position, further examination cannot be completed. Hence, parameters such as stability of the template, kinetics of incorporation, and others need to be optimized in order to increase read length, reduce error rates, and ensure robustness of the system.

Some of the potential processes that are of concern in SMDS are illustrated in Figure 11. We explain a few of these concerns below –

- The stability of the substrate: what is the lifetime of the multi-polyelectrolyte layers or other surfaces?
- The stability of the connector of the DNA to the surface, such as biotin streptavidin.
- The kinetics of incorporation of labeled nucleotides: the bulky labeled nucleotides are a possible bottleneck for the polymerase activity – a cleavable nucleotide increases the yield tremendously.
- The stability of the primer/DNA hybridization.
- The photo-induced radicals can be a source of damage to the DNA, to the dye (bleaching) and to other ingredients in the flow cell.
- The oxygen scavenger system can reduce the formation of oxygen radicals, but fluctuations in the performance of the scavenger solution can influence the sequencing operation. It might also degrade the surface.
- Non-specific sticking of the fluorescent molecules produces reading errors. It might be addressed by careful surface preparation and suitable wash solutions.

While each of these factors has to be optimized in order to achieve the required high yields, none of them pose a fundamental limit. For example, it is known that the mutation G over T occurs in high rates naturally (Kunkel, 2004) because there is very little local perturbation of the helix, and more importantly, the global conformation of the duplex is unaffected. Similar results have been reported for the A–C mispairing. Since the incorporation of the labeled nucleotide slows down incorporation rates for steric reasons, steric hinderance will also slow the incorporation of mismatched nucleotides to the point of insignificant error rates. Additionally, since synchronization is not a requirement in single-molecule sequencing, the incorporation does not have to be driven to close to 100% incorporation at every cycle and thus short cycles can reduce the probability of the incorporation of wrong bases. In the next section, we will discuss the anticipated performance of SMDS by cyclic synthesis.

6. PERFORMANCE

The performance of SMDS relies on serial scanning of multiple fields of view, each can contain approximately 20,000 single strands. The limit here will be the time it takes to scan a field of view, say on the order of 0.2 s per field of view. At this rate, scanning 5000 fields of view would take approximately 15 min. With 20,000 molecules per field of view and with incorporation into 40% of the templates per incorporation cycle it will translate to monitoring of 10^8 molecules at a rate of approximately 40,000 base/s. This scheme is useful when the reading lengths are about 20 bases, or longer. The reading length is heavily dependent on the ability of the polymerase to incorporate the fluorescent nucleotide on the DNA template. The single incorporation yield should be on the order of 97% to have a significant total yield, and current experiments have exceeded such yields (Harris *et al.*, 2007 to be published). The reading speed of the device will depend on the DNA density that is compatible with the experimental setup and on the number of fields of view that are imaged. The previous estimate of 10^8 target molecules is reasonable because such a high number of templates can be attached to a microscope slide with minimum preparation. It is interesting to note that if the average number of bases per template is larger then 30, then the equivalent of an entire human genome can be attached to one slide and re-sequenced in one experiment. At each incorporation step, about 40 Mb are incorporated on the slide with approximately 100 μL of reaction solution.

The reading speed will probably mostly be camera limited, and at a rate of 40,000 bases/s, this amounts to 3 Gb of sequence information per day. The reagent costs will be significantly reduced, but the startup equipment might still be expensive, thus the cost per base will then be determined by the reading speed and total sequence output over the long term. After the protocols for this technology have settled down, a globally cheaper instrument when compared to current robotics, can be built with microfluidics (Kartalov and Quake, 2004), which will further reduce reagent cost and will be compatible with other 'lab-on-a-chip' components such as single-cell lysis (Hong *et al.*, 2004). This would allow the creation of affordable instruments for private investigators in research laboratories, or even the relatively routine use of this technology in medical clinics.

7. APPLICATIONS

SMDS has the potential to revolutionize the genome-sequencing world by making it simpler, cheaper, and faster. By gathering the information from many different individual genomes, there is hope to discover and understand the function and variation of genes, and how they relate to diseases. For example, cancer is ultimately a disease of the genes. Identifying the entire collection of genetic aberrations in all tumor types will help discover molecular mechanisms responsible for uncontrolled cell growth and tumor metastasis (Kaiser, 2005). Many other diseases have a strong genetic component to them and usually several genes are involved in a single illness. By sequencing the genomes of individuals affected by a certain class of disease, it would be possible to find a common genetic cause to them. Also, several infectious diseases could be detected by sequencing short DNA or RNA viral strand in the blood of an individual. The detection of this viral signature would also immediately reveal the identity of the infecting agent and allow for rapid treatment of the infection.

More recently, it has been discovered that small RNA (sRNAs) can regulate transcription and protein abundance (Vaughn and Martienssen, 2005), and small interfering RNA (siRNA) have been used to suppress protein expression in place of studies using traditional knockouts. Traditional sequencing approaches have low throughput and have been limited in the number of sRNAs they could characterize. Only a few thousand had been identified, and yet ongoing improvements to Sanger sequencing has allowed over a million to be recently discovered. The applications of single-molecule methods to sRNA sequencing would allow for this to be done in multiple organisms at minimal cost. Moreover, the RNA profiling of stem cells, before and after differentiation, could help elucidate the various differentiation pathways of pluripotent cells. Given this information, one could eventually engineer stems cells to differentiate into the tissue of their choice, for the purpose of replacing damaged or diseased tissues in patients.

8. CONCLUSIONS

SMDS by cyclic synthesis is a promising new technique that minimizes cost and enhances throughput over current Sanger sequencing methods. The ability to sequence millions of bases in parallel at very high density and high data rates, without the constraint of synchronous incorporations, establishes this method as a viable option for massive DNA re-sequencing applications. Significant reductions in reagent use, combined with minimal sample preparation, contribute to lower the cost and time of the re-sequencing, as well as virtually eliminating the amplification biases. The microfluidic implementation of this method could reduce, even further, the cost of the reagents and of the device as a whole. Further, the use of FRET as a local illumination source in single-molecule sequencing by fluorescence is useful for reducing noise and false-positive signals from unspecific binding of nucleotides, and is applicable in other situations where a tightly confined excitation light is desirable. The use of cleavable fluorescent markers substantially increases the read lengths in single-molecule sequencing as steric interactions between adjacent dyes are eliminated. Further increase in read length is anticipated by optimizing reaction conditions and by choice of the DNA polymerase used. In the FRET scheme of sequencing, the lifetime of the donor is a key factor in limiting the read length; however, the use of a quantum dot as the donor might alleviate this problem.

Single-molecule sequencing technology is already at a working state, and fine-tuning of the technique will bring its performance to cost and throughput levels that would make this the method of choice for biomedical applications. This technology could allow high-throughput gene re-sequencing and with it the discovery of rare genetic aberrations, including point mutations, insertions/deletions, and amplifications. Recent experiments have shown that the high coverage afforded by parallel sequencing reveals mutations as rare as 1% (Harris *et al.*, 2006 to be published). The ability to reveal genetic inhomogeneities in small tumor samples with minimal preparation will be important for cancer research. Whole human genome re-sequencing directly from genomic DNA purified from 100 cell equivalents, without amplification, would be possible with this technology. Ten-fold genome coverage could be achieved in days, reducing re-sequencing costs by three orders of magnitude over traditional Sanger sequencing. Entire case and control groups could be studied for the discovery and detection of biomarkers for drug efficacy and adverse drug reactions. In a future where ever-present gene functional analysis and human disease gene identification are poised to assume a growing role, SMDS will hopefully provide "personal genomics" at an affordable price.

ACKNOWLEDGMENTS

We would like to acknowledge Timothy Harris from Helicos BioSciences and Stephen Quake from Stanford University for their helpful comments.

REFERENCES

Ambrose, W. P., Goodwin, P. M., Martin, J. C. and Keller, R. A. (1994). Single-molecule detection and photochemistry on a surface using near-field optical-excitation. Phys. Rev. Lett. 72(1), 160–163.

Ambrose, W. P., Goodwin, P. M. and Nolan, J. P. (1999). Single-molecule detection with total internal reflection excitation: comparing signal-to-background and total signals in different geometries. Cytometry 36(3), 224–231.

Augustin, M. A., Ankenbauer, W. and Angerer, B. (2001). Progress towards single-molecule sequencing: enzymatic synthesis of nucleotide-specifically labeled DNA. J. Biotechnol. 86(3), 289–301.

Axelrod, D. (1989). Total internal-reflection fluorescence microscopy. Method. Cell Biol. 30, 245–270.

Axelrod, D. (2001). Total internal reflection fluorescence microscopy in cell biology. Traffic 2(11), 764–774.

Babcock, H. P., Chen, C. and Zhuang, X. W. (2004). Using single-particle tracking to study nuclear trafficking of viral genes. Biophys. J. 87(4), 2749–2758.
Beese, L. S., Derbyshire, V. and Steitz, T. A. (1993). Structure of DNA-polymerase-I Klenow fragment bound to duplex DNA. Science 260(5106), 352–355.
Bentley, D. R. (2004). Genomes for medicine. Nature 429(6990), 440–445.
Brakmann, S. (2004). Optimal enzymes for single-molecule sequencing. Curr. Pharm. Biotechnol. 5(1), 119–126.
Brakmann, S. and Nieckchen, P. (2001). The large fragment of *Escherichia coli* DNA polymerase I can synthesize DNA exclusively from fluorescently labeled nucleotides. Chem. Biochem. 2(10), 773–777.
Braslavsky, I., Amit, R., Ali, B. M. J., Gileadi, O., Oppenheim, A. and Stavans, J. (2001). Objective-type dark-field illumination for scattering from microbeads. App. Optics 40(31), 5650–5657.
Braslavsky, I., Hebert, B., Kartalov, E. and Quake, S. R. (2003). Sequence information can be obtained from single DNA molecules. Proc. Natl. Acad. Sci. USA 100(7), 3960–3964.
Brenner, S., Johnson, M., Bridgham, J., Golda, G., Lloyd, D. H., Johnson, D., Luo, S. J., McCurdy, S., Foy, M., Ewan, M., Roth, R., George, D., Eletr, S., Albrecht, G., Vermaas, E., Williams, S. R., Moon, K., Burcham, T., Pallas, M., DuBridge, R. B., Krichner, J., Fearon, K., Mao, J. and Corcoran, K. (2000). Gene expression analysis by massively parallel signature sequencing (MPSS) on microbead arrays. Nat. Biotechnol. 18(6), 630–634.
Bustamante, C., Chemla, Y. R., Forde, N. R. and Izhaky, D. (2004). Mechanical processes in biochemistry. Annu. Rev. Biochem. 73, 705–748.
Cecconi, C., Shank, E. A., Bustamante, C. and Marqusee, S. (2005). Direct observation of the three-state folding of a single protein molecule. Science 309(5743), 2057–2060.
Chan, E. Y. (2005). Advances in sequencing technology. Mutat. Res.-Fundam. Mol. Mech. Mutag. 573(1–2), 13–40.
Chan, E. Y., Goncalves, N. M., Haeusler, R. A., Hatch, A. J., Larson, J. W., Maletta, A. M., Yantz, G. R., Carstea, E. D., Fuchs, M., Wong, G. G., Gullans, S. R. and Gilmanshin, R. (2004). DNA mapping using microfluidic stretching and single-molecule detection of fluorescent site-specific tags. Genome Res. 14(6), 1137–1146.
Chen, T.-S., Zeng, S.-Q., Zhou, W. and Luo, Q.-M. (2003). A quantitative theory model of a photobleaching mechanism. Chinese Phys. Lett. 20, 1940–1943.
Crocker, J. C. and Grier, D. G. (1996). Methods of digital video microscopy for colloidal studies. J. Colloid Interf. Sci. 179(1), 298–310.
Decher, G. (1997). Fuzzy nanoassemblies: toward layered polymeric multicomposites. Science 277(5330), 1232–1237.
Dickson, R. M., Norris, D. J. and Moerner, W. E. (1998). Simultaneous imaging of individual molecules aligned both parallel and perpendicular to the optic axis. Phys. Rev. Lett. 81(24), 5322–5325.
Fiala, K. A. and Suo, Z. (2004). Pre-steady-state kinetic studies of the fidelity of *Sulfolobus solfataricus* P2 DNA polymerase IV. Biochemistry 43(7), 2106–2115.
Flomenbom, O., Klafter, J. and Szabo, A. (2005). What can one learn from two-state single-molecule trajectories?. Biophys. J. 88(6), 3780–3783.
Förster, T. (1948). Intermolecular energy migration and fluorescence. Ann. Phys. 2, 55–75.
Funatsu, T., Harada, Y., Tokunaga, M., Saito, K. and Yanagida, T. (1995). Imaging of single fluorescent molecules and individual ATP turnovers by single myosin molecules in aqueous-solution. Nature 374(6522), 555–559.
Goodman, M. and Reha-Krantz, L. (1997). Synthesis of fluorophore-labeled DNA. World Patent Publication Number WO97/39150.
Goodman, M. F. and Tippin, B. (2000). The expanding polymerase universe. Nat. Rev. Mol. Cell Biol. 1(2), 101–109.
Gordon, M. P., Ha, T. and Selvin, P. R. (2004). Single-molecule high-resolution imaging with photobleaching. Proc. Natl. Acad. Sci. USA 101(17), 6462–6465.
Ha, T. (2001). Single-molecule fluorescence resonance energy transfer. Methods 25(1), 78–86.
Ha, T., Enderle, T., Ogletree, D. F., Chemla, D. S., Selvin, P. R. and Weiss, S. (1996). Probing the interaction between two single molecules: fluorescence resonance energy

transfer between a single donor and a single acceptor. Proc. Natl. Acad. Sci. USA 93(13), 6264–6268.
Ha, T. J., Ting, A. Y., Liang, J., Caldwell, W. B., Deniz, A. A., Chemla, D. S., Schultz, P. G. and Weiss, S. (1999). Single-molecule fluorescence spectroscopy of enzyme conformational dynamics and cleavage mechanism. Proc. Natl. Acad. Sci. USA 96(3), 893–898.
Harris, T. D., Buzby, P. R., Babcock, H. P., Beer, E., Braslavsky, I., Causey, M., Colonell, J. I., DiMeo, J., Efcavitch, J. W., Gill, J., Healy, J., Ickes, R., Jarosz, M. V., Karsh, W., Lapen, D., Steinmann, P., Ulmer, K. M., Weber, A., Weiss, H. and Xie, Z. (2007, to be published). Single molecule DNA sequencing.
Hebert, B., Braslavsky, I. and Quake, S. R. (2007, to be published). Single molecule measurements of DNA synthesis with individual base resolution.
Hebert, B., Costantino, S. and Wiseman, P. W. (2005). Spatio-temporal image correlation Spectroscopy (STICS) theory, verification, and application to protein velocity mapping in living CHO cells. Biophys. J. 88(5), 3601–3614.
Hohng, S. and Ha, T. (2005). Single-molecule quantum-dot fluorescence resonance energy transfer. Chem. Phys. Chem. 6(5), 956–960.
Holmberg, R. C., Henry, A. A. and Romesberg, F. E. (2005). Directed evolution of novel polymerases. Biomol. Eng. 22(1–3), 39–49.
Hong, J. W., Studer, V., Hang, G., Anderson, W. F. and Quake, S. R. (2004). A nanoliter-scale nucleic acid processor with parallel architecture. Nat. Biotechnol. 22(4), 435–439.
Jett, J. H., Keller, R. A., Martin, J. C., Marrone, B. L., Moyzis, R. K., Ratliff, R. L., Seitzinger, N. K., Shera, E. B. and Stewart, C. C. (1989). High-speed DNA sequencing – an approach based upon fluorescence detection of single molecules. J. Biomol. Struct. Dyn. 7(2), 301–309.
Johnson, K. A. (1993). Conformational coupling in DNA-polymerase fidelity. Annu. Rev. Biochem. 62, 685–713.
Kaiser, J. (2005). National Institutes of Health – NCI gears up for cancer genome project. Science 307(5713), 1182.
Kapanidis, A. N., Laurence, T. A., Lee, N. K., Margeat, E., Kong, X. X. and Weiss, S. (2005). Alternating-laser excitation of single molecules. Accounts Chem. Res. 38(7), 523–533.
Kartalov, E., Unger, M. and Quake, S. R. (2003). A poly-electrolyte surface interface for single molecule fluorescence studies of DNA polymerase. Biotechniques 34(3), 505–510.
Kartalov, E. P. and Quake, S. R. (2004). Microfluidic device reads up to four consecutive base pairs in DNA sequencing-by-synthesis. Nucleic Acids Res. 32(9), 2873–2879.
Keller, D. J. and Brozik, J. A. (2005). Framework model for DNA polymerases. Biochemistry 44(18), 6877–6888.
Kern, W. and Vossen, J. (1978). Thin Film Processes, Academic Press, New York.
Kim, J. S., Granstrom, M., Friend, R. H., Johansson, N., Salaneck, W. R., Daik, R., Feast, W. J. and Cacialli, F. (1998). Indium–tin oxide treatments for single- and double-layer polymeric light-emitting diodes: the relation between the anode physical, chemical, and morphological properties and the device performance. J. Appl. Phys. 84(12), 6859–6870.
Kuchta, R. D., Mizrahi, V., Benkovic, P. A., Johnson, K. A. and Benkovic, S. J. (1987). Kinetic mechanism of DNA-polymerase-I (Klenow). Biochemistry 26(25), 8410–8417.
Kulzer, F. and Orrit, M. (2004). Single-molecule optics. Annu. Rev. Phys. Chem. 55, 585–611.
Kunkel, T. A. (2004). DNA replication fidelity. J. Biol. Chem. 279(17), 16895–16898.
Lander, E. S., Linton, L. M., Birren, B., Nusbaum, C., Zody, M. C., Baldwin, J., Devon, K., Dewar, K., Doyle, M., FitzHugh, W., Funke, R., Gage, D., Harris, K., Heaford, A., Howland, J., Kann, L., Lehoczky, J., LeVine, R., McEwan, P., McKernan, K., Meldrim, J., Mesirov, J. P., Miranda, C., Morris, W., Naylor, J., Raymond, C., Rosetti, M., Santos, R., Sheridan, A., Sougnez, C., Stange-Thomann, N., Stojanovic, N., Subramanian, A., Wyman, D., Rogers, J., Sulston, J., Ainscough, R., Beck, S., Bentley, D., Burton, J., Clee, C., Carter, N., Coulson, A., Deadman, R., Deloukas, P., Dunham, A., Dunham, I., Durbin, R., French, L., Grafham, D., Gregory, S., Hubbard, T., Humphray, S., Hunt, A., Jones, M., Lloyd, C., McMurray, A., Matthews, L., Mercer, S., Milne, S., Mullikin, J. C., Mungall, A., Plumb, R., Ross, M., Shownkeen, R., Sims, S., Waterston, R. H., Wilson, R. K., Hillier, L. W., McPherson, J. D., Marra, M. A., Mardis, E. R., Fulton, L. A.,

Chinwalla, A. T., Pepin, K. H., Gish, W. R., Chissoe, S. L., Wendl, M. C., Delehaunty, K. D., Miner, T. L., Delehaunty, A., Kramer, J. B., Cook, L. L., Fulton, R. S., Johnson, D. L., Minx, P. J., Clifton, S. W., Hawkins, T., Branscomb, E., Predki, P., Richardson, P., Wenning, S., Slezak, T., Doggett, N., Cheng, J. F., Olsen, A., Lucas, S., Elkin, C., Uberbacher, E., Frazier, M., Gibbs, R. A., Muzny, D. M., Scherer, S. E., Bouck, J. B., Sodergren, E. J., Worley, K. C., Rives, C. M., Gorrell, J. H., Metzker, M. L., Naylor, S. L., Kucherlapati, R. S., Nelson, D. L., Weinstock, G. M., Sakaki, Y., Fujiyama, A., Hattori, M., Yada, T., Toyoda, A., Itoh, T., Kawagoe, C., Watanabe, H., Totoki, Y., Taylor, T., Weissenbach, J., Heilig, R., Saurin, W., Artiguenave, F., Brottier, P., Bruls, T., Pelletier, E., Robert, C., Wincker, P., Rosenthal, A., Platzer, M., Nyakatura, G., Taudien, S., Rump, A., Yang, H. M., Yu, J., Wang, J., Huang, G. Y., Gu, J., Hood, L., Rowen, L., Madan, A., Qin, S. Z., Davis, R. W., Federspiel, N. A., Abola, A. P., Proctor, M. J., Myers, R. M., Schmutz, J., Dickson, M., Grimwood, J., Cox, D. R., Olson, M. V., Kaul, R., Raymond, C., Shimizu, N., Kawasaki, K., Minoshima, S., Evans, G. A., Athanasiou, M., Schultz, R., Roe, B. A., Chen, F., Pan, H. Q., Ramser, J., Lehrach, H., Reinhardt, R., McCombie, W. R., de la Bastide, M., Dedhia, N., Blocker, H., Hornischer, K., Nordsiek, G., Agarwala, R., Aravind, L., Bailey, J. A., Bateman, A., Batzoglou, S., Birney, E., Bork, P., Brown, D. G., Burge, C. B., Cerutti, L., Chen, H. C., Church, D., Clamp, M., Copley, R. R., Doerks, T., Eddy, S. R., Eichler, E. E., Furey, T. S., Galagan, J., Gilbert, J. G. R., Harmon, C., Hayashizaki, Y., Haussler, D., Hermjakob, H., Hokamp, K., Jang, W. H., Johnson, L. S., Jones, T. A., Kasif, S., Kaspryzk, A., Kennedy, S., Kent, W. J., Kitts, P., Koonin, E. V., Korf, I., Kulp, D., Lancet, D., Lowe, T. M., McLysaght, A., Mikkelsen, T., Moran, J. V., Mulder, N., Pollara, V. J., Ponting, C. P., Schuler, G., Schultz, J. R., Slater, G., Smit, A. F. A., Stupka, E., Szustakowki, J., Thierry-Mieg, D., Thierry-Mieg, J., Wagner, L., Wallis, J., Wheeler, R., Williams, A., Wolf, Y. I., Wolfe, K. H., Yang, S. P., Yeh, R. F., Collins, F., Guyer, M. S., Peterson, J., Felsenfeld, A., Wetterstrand, K. A., Patrinos, A. and Morgan, M. J. (2001). Initial sequencing and analysis of the human genome. Nature 409(6822), 860–921.

Laurence, T. A. and Weiss, S. (2003). How to detect weak pairs. Science 299(5607), 667–668.

Leamon, J. H., Lee, W. L., Tartaro, K. R., Lanza, J. R., Sarkis, G. J., deWinter, A. D., Berka, J. and Lohman, K. L. (2003). A massively parallel PicoTiterPlate based platform for discrete picoliter-scale polymerase chain reactions. Electrophoresis 24(21), 3769–3777.

Lee, K. T. and Raghavan, S. (1999). Etch rate of silicon and silicon dioxide in ammonia-peroxide solutions measured by quartz crystal microbalance technique. Electrochem. Solid State Lett. 2(4), 172–174.

Levene, M. J., Korlach, J., Turner, S. W., Foquet, M., Craighead, H. G. and Webb, W. W. (2003). Zero-mode waveguides for single-molecule analysis at high concentrations. Science 299(5607), 682–686.

Li, Z. M., Bai, X. P., Ruparel, H., Kim, S., Turro, N. J. and Ju, J. Y. (2003). A photo-cleavable fluorescent nucleotide for DNA sequencing and analysis. Proc. Natl. Acad. Sci. USA 100(2), 414–419.

Lu, C., Tej, S. S., Luo, S. J., Haudenschild, C. D., Meyers, M. C. and Green, P. J. (2005). Elucidation of the small RNA component of the transcriptome. Science 309(5740), 1567–1569.

Macklin, J. J., Trautman, J. K., Harris, T. D. and Brus, L. E. (1996). Imaging and time-resolved spectroscopy of single molecules at an interface. Science 272(5259), 255–258.

Maier, B., Bensimon, D. and Croquette, V. (2000). Replication by a single DNA polymerase of a stretched single-stranded DNA. Proc. Natl. Acad. Sci. USA, 97(22), 12,002–12,007.

Margulies, M., Egholm, M., Altman, W. E., Attiya, S., Bader, J. S., Bemben, L. A., Berka, J., Braverman, M. S., Chen, Y.-J., Chen, Z. T., Dewell, S. B., Du, L., Fierro, J. M., Gomes, X. V., Godwin, B. C., He, W., Helgesen, S., Ho, C. H., Irzyk, G. P., Jando, S. C., Alenquer, M. L. I., Jarvie, T. P., Jirage, K. B., Kim, J. B., Knight, J. R., Lanza, J. R., Leamon, J. H., Lefkowitz, S. M., Lei, M., Li, J., Lohman, K. L., Lu, H., Makhijani, V. B., McDade, K. E., McKenna, M. P., Myers, E. W., Nickerson, E., Nobile, J. R., Plant, R., Puc, B. P., Ronan, M. T., Roth, G. T., Sarkis, G. J., Simons, J. F., Simpson, J. W.,

Srinivasan, M., Tartaro, K. R., Tomasz, A., Vogt, K. A., Volkmer, G. A., Wang, S. H., Wang, Y., Weiner, M. P., Yu, P. G., Begley, R. F. and Rothberg, J. M. (2005). Genome sequencing in microfabricated high-density picolitre reactors. Nature 437, 376–380.

Mathur, A. B., Truskey, G. A. and Reichert, W. M. (2000). Atomic force and total internal reflection fluorescence microscopy for the study of force transmission in endothelial cells. Biophys. J. 78(4), 1725–1735.

Meller, A., Nivon, L., Brandin, E., Golovchenko, J. and Branton, D. (2000). Rapid nanopore discrimination between single polynucleotide molecules. Proc. Natl. Acad. Sci. USA, 97(3), 1079–1084.

Mertz, J., Xu, C. and Webb, W. W. (1995). Single-molecule detection by two-photon-excited fluorescence. Optics Lett. 20(24), 2532–2534.

Michalet, X., Kapanidis, A. N., Laurence, T., Pinaud, F., Doose, S., Pflughoefft, M. and Weiss, S. (2003). The power and prospects of fluorescence microscopies and spectroscopies. Annu. Rev. Biophys. Biomol. Struct. 32, 161–182.

Mitra, R. D., Shendure, J., Olejnik, J., Edyta Krzymanska, O. and Church, G. M. (2003). Fluorescent *in situ* sequencing on polymerase colonies. Anal. Biochem. 320(1), 55–65. *Erratum in*: Anal. Biochem. (2004) 328(2), 245.

Nie, S. M. and Zare, R. N. (1997). Optical detection of single molecules. Annu. Rev. Biophys. Biomol. Struct. 26, 567–596.

Nirmal, M., Dabbousi, B. O., Bawendi, M. G., Macklin, J. J., Trautman, J. K., Harris, T. D. and Brus, L. E. (1996). Fluorescence intermittency in single cadmium selenide nanocrystals. Nature 383(6603), 802–804.

Notredame, C. (2002). Recent progress in multiple sequence alignment: a survey. Pharmacogenomics 3(1), 131–144.

Park, M., Kim, H. H., Kim, D. and Song, N. W. (2005). Counting the number of fluorophores labeled in biomolecules by observing the fluorescence-intensity transient of a single molecule. Bull. Chem. Soc. Japan 78(9), 1612–1618.

Peterman, E. J. G., Sosa, H. and Moerner, W. E. (2004). Single-molecule fluorescence spectroscopy and microscopy of biomolecular motors. Annu. Rev. Phys. Chem. 55, 79–96.

Rhoades, E., Gussakovsky, E. and Haran, G. (2003). Watching proteins fold one molecule at a time. Proc. Natl. Acad. Sci. USA 100(6), 3197–3202.

Rogers, Y. H. and Venter, J. C. (2005). Genomics – massively parallel sequencing. Nature 437(7057), 326–327.

Ruparel, H., Bi, L. R., Li, Z. M., Bai, X. P., Kim, D. H., Turro, N. J. and Ju, J. Y. (2005). Design and synthesis of a 3′-*O*-allyl photocleavable fluorescent nucleotide as a reversible terminator for DNA sequencing by synthesis. Proc. Natl. Acad. Sci. USA 102(17), 5932–5937.

Sanger, F., Nicklen, S. and Coulson, A. R. (1977). DNA sequencing with chain-terminating inhibitors. Proc. Natl. Acad. Sci. USA 74(12), 5463–5467.

Schneider, T. D. and Rubens, D. (2001). High speed parallel nucleic acid sequencing. World Patent Publication Number WO 01/16375.

Selvin, P. R. (2000). The renaissance of fluorescence resonance energy transfer. Nat. Struct. Biol. 7(9), 730–734.

Seo, T. S., Bai, X. P., Kim, D. H., Meng, Q. L., Shi, S. D., Ruparelt, H., Li, Z. M., Turro, N. J. and Ju, J. Y. (2005). Four-color DNA sequencing by synthesis on a chip using photocleavable fluorescent nucleotides. Proc. Natl. Acad. Sci. USA 102(17), 5926–5931.

Shendure, J., Mitra, R. D., Varma, C. and Church, G. M. (2004). Advanced sequencing technologies: methods and goals. Nat. Rev. Genet. 5(5), 335–344.

Sheppard, C. J. R. and Shotton, D. M. (1997). Image formation in the confocal laser scanning microscope. In: Confocal Laser Scanning Microscopy (pp. 15–31). BIOS Scientific Publishers Ltd., Springer-Verlag, New York.

Shimkus, M., Levy, J. and Herman, T. (1985). A chemically cleavable biotinylated nucleotide – usefulness in the recovery of protein DNA complexes from avidin affinity columns. Proc. Natl. Acad. Sci. USA 82(9), 2593–2597.

Smailus, D. E., Marziali, A., Dextras, P., Marra, M. A. and Holt, R. A. (2005). Simple, robust methods for high-throughput nanoliter-scale DNA sequencing. Genome Res. 15(10), 1447–1450.

Sobek, J. and Schlapbach, R. (2004). Substrate architecture and function. Pharm. Disc. (Microarray Technol.) 15, 32–44.

Tokunaga, M., Kitamura, K., Saito, K., Iwane, A. H. and Yanagida, T. (1997). Single molecule imaging of fluorophores and enzymatic reactions achieved by objective-type total internal reflection fluorescence microscopy. Biochem. Biophys. Res. Commun. 235(1), 47–53.

Unger, M., Kartalov, E., Chiu, C. S., Lester, H. A. and Quake, S. R. (1999). Single-molecule fluorescence observed with mercury lamp illumination. Biotechniques 27(5), 1008–1013.

van Dam, R. M. and Quake, S. R. (2002). Gene expression analysis with universal n-mer arrays. Genome Res. 12(1), 145–152.

Vaughn, M. W. and Martienssen, R. (2005). It's a small RNA world, after all. Science 309(5740), 1525–1526.

Venter, J. C., Adams, M. D., Myers, E. W., Li, P. W., Mural, R. J., Sutton, G. G., Smith, H. O., Yandell, M., Evans, C. A. and Holt, R. A. *et al.* (2001). The sequence of the human genome. Science 291(5507), 1304–1351.

Watson, J. D. and Crick, F. H. C. (1953). Molecular structure of nucleic acids. Nature 171, 737–738.

Werner, J. H., Cai, H., Jett, J. H., Reha-Krantz, L., Keller, R. A. and Goodwin, P. M. (2003). Progress towards single-molecule DNA sequencing: a one color demonstration. J. Biotechnol. 102(1), 1–14.

Wuite, G. J. L., Smith, S. B., Young, M., Keller, D. and Bustamante, C. (2000). Single-molecule studies of the effect of template tension on T7 DNA polymerase activity. Nature 404(6773), 103–106.

Xie, X. S. and Dunn, R. C. (1994). Probing single-molecule dynamics. Science 265(5170), 361–364.

Xie, X. S. and Trautman, J. K. (1998). Optical studies of single molecules at room temperature. Annu. Rev. Phys. Chem. 49(1), 441–480.

Xie, Z., Srividya, N., Sosnick, T. R., Pan, T. and Scherer, N. F. (2004). Single-molecule studies highlight conformational heterogeneity in the early folding steps of a large ribozyme. Proc. Natl. Acad. Sci. USA 101(2), 534–539.

Yildiz, A., Forkey, J. N., McKinney, S. A., Ha, T., Goldman, Y. E. and Selvin, P. R. (2003). Myosin V walks hand-over-hand: single fluorophore imaging with 1.5-nm localization. Science 300(5628), 2061–2065.

Yildiz, A. and Selvin, P. R. (2005). Fluorescence imaging with one manometer accuracy: application to molecular motors. Acc. Chem. Res. 38(7), 574–582.

Zhu, Z. R. and Waggoner, A. S. (1997). Molecular mechanism controlling the incorporation of fluorescent nucleotides into DNA by PCR. Cytometry 28(3), 206–211.

Chapter 8

Rapid DNA Sequencing by Direct Nanoscale Reading of Nucleotide Bases on Individual DNA chains

James Weifu Lee[1] and Amit Meller[2]

[1]*Oak Ridge National Laboratory, Oak Ridge, TN 37831-6194, USA*
[2]*The Department of Physics and Biomedical Engineering, Boston University, 44 Cummington Street, Boston, MA 02215, USA*

Contents

Abstract

We present and discuss two concepts for rapid DNA sequencing by direct nanoscale reading of nucleotide bases on individual DNA molecules. Although these two concepts are based on substantially different principles, they both rely on the fabrication of nanoscale devices using state of the art technologies. The first method, which has been recently invented at Oak Ridge National Laboratory (ORNL), is based on a systematic nanoelectrode-gated tip-to-tip electron-tunneling molecular-detection concept. According to this concept, it should be possible to obtain genetic sequence information by probing through a DNA molecule base by base at a nanometer

PERSPECTIVES IN BIOANALYSIS, VOLUME 2
ISSN 1871-0069 DOI: 10.1016/S1871-0069(06)02008-8

scale, as if looking at a strip of movie film. The nanoscale reading of DNA sequences is envisioned to take place at a nanogap (gate) defined by a pair of nanoelectrode tips as a DNA molecule moves through the gate base by base. The rationale is that sample molecules, such as the four different nucleotide bases, each with a distinct chemical composition and structure, should produce a specific perturbation effect on the tunneling electron beam across the two nanoelectrode tips. A sample molecular structure (base) could thus be detected when it enters the gate. This approach could lead to a single-molecule DNA sequencing technology that does not require the PCR amplification process and could, at least, be thousands of times faster than the current technology (Sanger's "dideoxy" protocol-based capillary electrophoresis systems). Theoretically, this new approach has the potential to perform DNA sequencing at a maximal rate of about 1,000,000 bases (~1 Mb) per second per detection gate. This method can be paralleled using parallel arrays of multiple nanoelectrode-detection gates, thus magnifying the readout throughput by additional orders of magnitudes, achieving estimated maximal rates of possibly hundred millions bases (100 Mb) per second per device. The second method, developed at Harvard in collaboration with LingVitae AS, is based on the conversion of the natural DNA to another form of DNA in which each nucleotide is substituted with a group of 16 nucleotides (each base type is substituted with a unique sequence of 16 bases). Then fluorescently tagged oligonucleotides 16-mers, matching that of the converted DNA, are hybridized to the DNA and the molecule is electrophoretically fed through a nanoscale pore fabricated in thin solid-state film. The pore is used to sequentially remove the hybridized oligonucleotides, one by one, while the flashes of light in different colors arising from the attached fluorophores are detected. Moreover, according to the invention made by the Harvard group, the fluorophores attached to the 16-mer oligonucleotides are designed to be quenched until they reach the face of the nanopore, thus eliminating the undesired background fluorescence. This method is highly suitable for massive paralleled readout using high-density arrays of nanopores, which are simultaneously probed using a single imaging device. The readout throughput thus can be magnified by several orders of magnitudes, achieving estimated readout rates of ~1–10 Mb/s.

1. INTRODUCTION

The pioneering completion of the first reference human genome sequence (Venter *et al.*, 2001) has marked the commencement of an era in which genomic variations directly impact drug discovery and medical therapies. This new paradigm has created an imminent need for cheap and ultra fast methods for DNA sequence analysis. Not only research labs will highly benefit from the rapid sequencing thousands of individual genomes during the development or clinical test phases of new drugs, but also medical practitioners will be able to routinely analyze individuals' DNAs in a clinical setting before subscribing drugs, and check them against on-line data bases in which genomic information relevant to any drug is documented. To realize ultra fast and cheap DNA sequencing, new and revolutionary technologies are needed to replace the classical Sanger's "dideoxy" protocol (Shendure *et al.*, 2004). These technologies should address two main bottlenecks: First, sample size should be reduced to the minimum possible, making it possible to read the sequence from a single DNA molecule or just a few copies. Second, readout speed should be increased by several orders of magnitude compared with the current state of the art techniques. Here we focus on two emerging methods for high-throughput DNA sequencing at the single molecule level, which employ two fundamentally different probing mechanisms (nanoelectrode-gated tip-to-tip electron-tunneling-molecular-conductance and nanopore-guided single molecule fluorescence). Both methods rely however on a common principal, namely the sequential threading of DNA through nanofabricated solid-state device.

The first method is based on a systematic nanoelectrode-gated electron-tunneling molecular-detection concept, which has recently been invented at Oak Ridge National Laboratory (ORNL) (Lee and Thundat, 2001, 2004). According to this concept (Figure 1), it should be possible to obtain genetic sequence information by probing through a DNA molecule, base by base, as it transports through a nanometer-scale molecular-detection gap (gate). Theoretically, this new approach has the potential to perform DNA sequencing at a maximal rate of about 1,000,000 bases (~1 Mb) per second per detection gate.

This method can be paralleled using parallel arrays of multiple nanoelectrode-detection gates, thus magnifying the readout throughput by additional orders of magnitudes, achieving estimated maximal rates of possibly hundred millions bases (100 Mb) per second per device. Therefore, successful development of such a potentially revolutionary DNA sequencing technology could significantly contribute to analytic nanosciences, genomics studies, and nanomedicine.

An equally exciting new method has been developed at Harvard University in collaboration with Lingvitae AS (Norway). It is based on an ultra fast optical readout of Design DNA polymers (Lexow, 2004; Meller *et al.*, 2005). This method

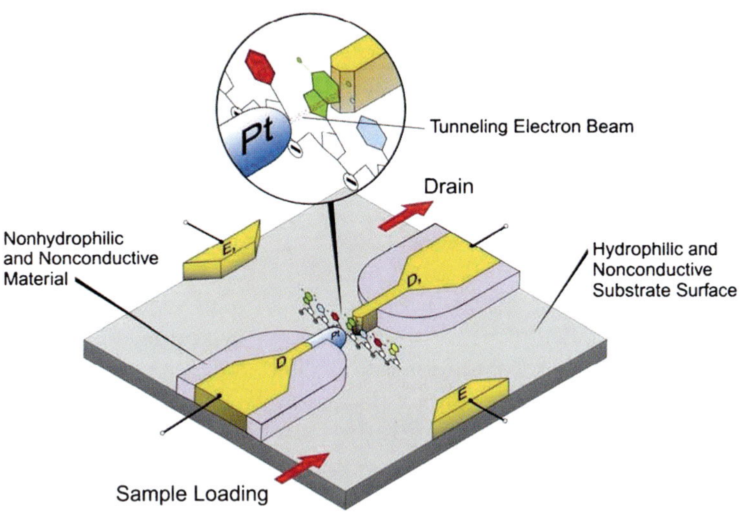

Fig. 1. Concept of DNA sequencing by nanoscale reading through programmable electrophoresis and nanoelectrode-gated tip-to-tip electron-tunneling conductance detection. The nanoscale reading of DNA sequences takes place at the nanogap (gate) defined by a pair of nanoelectrodes (D and D′) as a DNA molecule moves through the gate base by base. The electrophoresis electric field controlling movement is provided by a pair of macroelectrodes (E and E′) aligned with the nanometer detection gate on the sample plate. The sides of the detection electrodes (D and D′) are protected by nonhydrophilic and nonconductive material. The shape of this protective material is designed to produce a practical detection gate and channels for sample loading and draining.

relies on the conversion of the natural DNA to another form of DNA, called design polymers, in which each nucleotide in the original DNA is substituted with a group of nucleotides (each base type is substituted with a unique sequence of 3–16 nt.). Then fluorescently tagged oligonucleotides, matching that of the converted DNA, are hybridized to the DNA and the molecule is electrophoretically fed through a nanoscale pore fabricated in thin solid-state film. The pore is used to sequentially peel off oligonucleotides, one by one, from the design polymer while the flashes of light in different colors arising from the attached fluorophores are detected. Most strikingly, the optical readout method employed here can be paralleled through the fabrication of high-density nanopore arrays (Meller *et al.*, 2005), thus magnifying the readout throughput of this method by several orders of magnitude, achieving estimated readout rates of ~1 Mb/s or more.

2. DNA Sequencing by Nanoelectrode-Gated Electron-Tunneling Conductance Spectroscopic Molecular Detection

2.1. The concept and its origin

As illustrated in Figure 1, one of the most crucial components of this technology is the nanometer detection gate, which comprises two sharp tips of nanoelectrodes (D and D′) pointing toward each other on a nonconductive (e.g., SiO_2) plate. The spacing (nanogap size) between the two tips must be precisely tuned to about 1.5–5 nm to generate the capability for efficient passage of a single DNA chain and for base detection by tunneling-conductance spectroscopic measurement. It is now possible to fabricate this nanoelectrode-detection system using electron-beam lithography and ORNL's patented precision electrolytic nanofabrication technique.

This nanoelectrode-gated molecular-detection concept has an origin from an earlier ORNL invention (ORNL Invention Disclosure ID 0772; U.S. Patent No. 6,447,663 B1), which demonstrated that nanoelectrodes separated by nanometer distance can be fabricated using a combination of electron-beam lithography and programmable electrolytic nanofabrication (Lee and Greenbaum, 1999). This concept of precision electrolytic nanofabrication was first conceived by J.W. Lee in 1997, after experimentally demonstrating that metallic platinum atoms can be deposited precisely onto a reducing site of a nanometer biomolecule by reduction of water-soluble metal compounds such as hexachloroplatinate ($[PtCl_6]^{2-}$) using electrons generated from a photosynthetic process (Lee *et al.*, 1994, 1998). Atom-by-atom growth of metallic platinum had been demonstrated by the use of single-turnover flashes (Greenbaum, 1988) and more than 20 species of metal compounds were experimentally surveyed for the metal deposition at room temperature and neutral pH (Lee *et al.*, 1998). This idea (Lee and Greenbaum, 1999) for fabricating nanoelectrodes by precision electrolytic metal deposition subsequently received experimental support in the work of Morpurgo *et al.* (1999).

When the distance between two nanoelectrode tips was reduced to a few nanometers, significant electron-tunneling conductance across the nanogap was detected under aqueous conditions (Morpurgo *et al.*, 1999). During our experimental demonstration of manipulation of a nanometer gap size defined by a pair of nanoelectrodes (Figure 1) by using our programmable pulsed electrolytic metal deposition, we also measured the electric conductance across the nanometer gap in an aqueous solution and achieved consistent results. Inspired by the earlier findings, J.W. Lee first considered the possibility of using this type of nanoelectrode-gated electron-tunneling conductance measurement to detect single molecules such as DNA during that period. Subsequently, J.W. Lee and his ORNL colleague T.G. Thundat together developed a comprehensive DNA-sequencing nanotechnology concept (Lee and Thundat, 2001) with a series of innovations that include novel applications of a fine-tuned nanometer gap for passage of a single DNA molecule, thin-layer microfluidics and/or electrophoresis for sample loading and delivery, programmable electric fields for precise control of DNA movement, and detection of DNA nucleotide bases by nanoelectrode-gated electron-tunneling conductance measurements.

In an aqueous solution, the width of a single-stranded DNA molecule is likely to be about 1–2 nm (including some bound water molecules), while that of a double-stranded DNA is probably about 2–3 nm. Therefore, a gap size of about 1.5–5 nm should be sufficient for the passage of a DNA chain and for detection by tunneling-spectroscopic measurement across the two nanoelectrode tips. When a DNA chain enters the nanometer detection gate base by base, it will perturb the tunneling current by its screening-conduction effect. Because the chemical compositions and structures of the four distinct nucleotides (thymine, adenine, cytosine, and guanine) are different, the screening-conduction effect of each on the tunneling characteristics should be different as well. This is the rationale underlying our envisioned use of nanoelectrode-gated tip-to-tip transverse-tunneling conductance measurements to detect DNA nucleotide sequences. With this detection method, we believe it is possible to read DNA sequences directly on a single DNA molecule as it passes through the detection gate base by base, as illustrated in Figure 1. This nanotechnology also contains the capabilities of loading, separating, and feeding DNA sample molecules into the nanoelectrode-detection gate (D and D′) using micropipetting, microfluidics, and/or electrophoresis techniques. Precise control of DNA movement is another essential feature of this invention. To provide reliable detection of DNA sequences, it is important to control the DNA movement.

As illustrated in Figure 2, two programmable and perpendicular electric fields are used: an electrophoresis field that is provided by a pair of electrophoresis electrodes (E and E′) aligned with the nanometer detection gap on the sample plate, and a "holding" electric field perpendicular to the sample plate that is applied by two parallel conductive plates (H′ and H) located above and beneath the sample plate, respectively. Since a DNA molecule possesses negative charges (phosphate groups), it will move toward an anode under the influence of the electric field. To provide sufficient time and stability for the nanoelectrodes to detect a DNA nucleotide (base) at the nanogap, the electrophoresis field is pulsed and stopped for the detection period after the base enters the detection gate (Figure 3).

To hold the DNA molecule and to prevent any potential molecular drift, the "holding" electric field perpendicular to the sample plate is applied by two

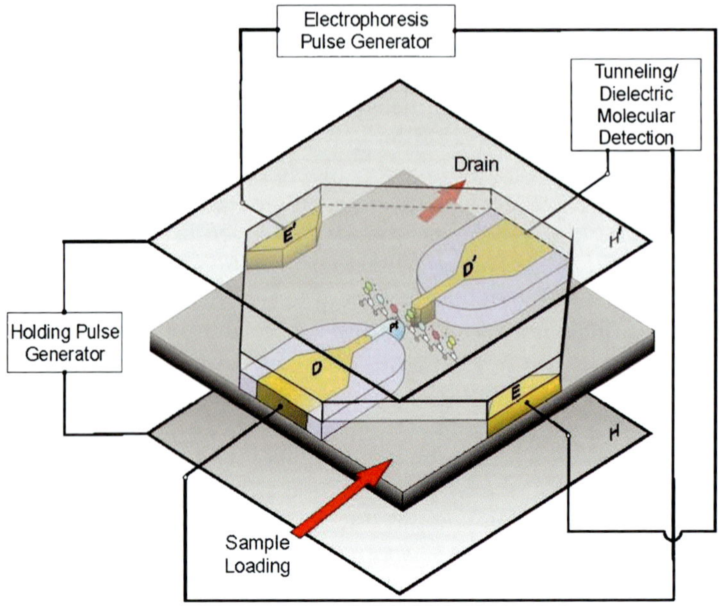

Fig. 2. A conceptual design of the nanotechnology system for nanoscale reading of DNA sequences through programmable electrophoresis and nanoelectrode-gated tip-to-tip transverse-tunneling conductance detection. The nanoscale reading by the transverse-tunneling conductance detection takes place at the nanogap (gate) defined by a pair of nanoelectrodes (D and D′) as a DNA chain transports through the gate base by base. The precise control of DNA transport could be achieved through the use of two perpendicular and coordinated programmable electric fields: an electrophoresis field provided by a pair of macroelectrodes (E and E′) on the sample plate and a perpendicular holding electric field provided by two conductive plates (H′ and H) located above and beneath the sample plate.

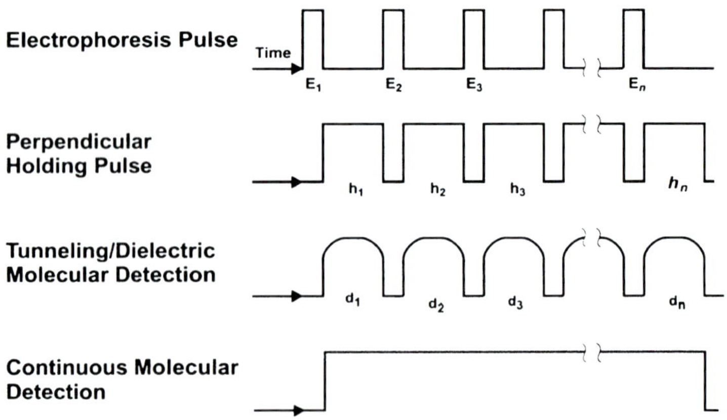

Fig. 3. Synchronization and coordination of electrophoresis, perpendicular braking and holding electric fields and tip-to-tip electron tunneling/dielectric molecular detection. The synchronization and coordination is envisioned to be achieved through application of a special computer-interfaced instrument system.

parallel conductive plates (H′ and H) located above and beneath the sample plate (Figure. 2). Because DNA has a negatively charged chain of phosphate groups, it is possible to retain it on the surface of the sample plate when the holding electric plate (H) beneath the sample plate is positively charged. If this holding electric field is strong enough, one should be able to hold a single-stranded DNA molecule with its phosphate groups down on the surface of the sample plate with its nucleotides pointing upward by pinning at the nanoelectrode gate (nanogap) as desired for detection by the transverse-tunneling electron beam. This could help to remove some randomness of the DNA molecular conformation at the gate so that a reliable and reproducible reading of the DNA sequence may be achieved. To achieve a coordinated process technology, the actions of the electrophoresis, perpendicular holding, and molecular detection are synchronized and coordinated as illustrated in Figure 3. The step size of the DNA transport can be controlled by the duration and amplitude of an electrophoresis pulse in conjunction with the perpendicular holding electric field. Precise control of these two electric fields [their sign (direction), amplitude, and duration] is achieved through use of a programmable voltage source that can be interfaced with a computer.

2.2. Potential speed of the envisioned nanoelectrode-gated DNA-sequencing system

The envisioned nanoelectrode-gated tip-to-tip electron-tunneling molecular-detection system (Figure 3) could perform DNA sequencing with ultrafast speed by direct nanoscale reading of nucleotide bases on an individual DNA chain. Both dc and/or modulated ac bias voltages could be applied across the nanometer molecular-detection gap (gate) for the nanoelectrode-gated tip-to-tip electron-tunneling molecular-conductance spectroscopic measurements. Since the control of DNA movement is achieved by use of two perpendicular electric fields that are provided by a pair of macroelectrodes (E and E′) and a pair of conductive plates (H and H′) with pulsed dc voltage, it is likely better to apply an ac voltage between the nanogap electrodes (D and D′) for tunneling conductance measurement. This will minimize the effect of dc voltage used for control of DNA movement on the sensing current measured across the nanogap. Therefore, the tip-to-tip tunneling current I is preferably an alternating current. Since the frequency of the input ac voltage (V) can be in a megahertz (MHz) range, the reading of a nucleotide base by nanoelectrode-gated tunneling conductance detection could be completed within a microsecond. By use of the programmable electric fields described above (Figures 2 and 3), it is possible to move a DNA molecule through the detection gate at a speed as fast as about 1 base per nanosecond. Therefore, this approach theoretically could have a maximal sequencing rate of about 1,000,000 bases per second per detection gate. This method can also be paralleled using parallel arrays of multiple nanoelectrode-detection gates in an integrated device, thus magnifying the readout throughput by additional orders of magnitudes, achieving estimated maximal rates of possibly hundred millions bases per second per device. With all considerations of practical operations including sample loading, it is estimated that this nanotechnology system should be able to perform DNA sequencing on an individual DNA molecule at a speed at least about 3,000 times faster than that available through the current DNA-sequencing technology.

Furthermore, because this is a single-molecule DNA-sequencing approach, it does not require PCR amplification. Consequently, this technology could have profound significance not only in increasing the speed and reducing the costs of DNA sequencing, but also may open new possibilities, including direct nanoscale reading of DNA base sequence extracted from a single cell.

2.3. Unique features of the nanoelectrode-gated molecular-detection concept

The envisioned nanoelectrode-gated transverse-tunneling conductance detection for nanoscale reading of DNA sequence is different from that of the "electrophysiology-like" detection postulated by Daniel Branton and his colleagues (Kasianowicz *et al.*, 1996; Akeson *et al.*, 1999). By measuring the ionic current passing through single-ion channels such as α-hemolysin (α-HL) in a lipid bilayer membrane using the electrophysiology-like approach, they demonstrated that an electric field can drive single-stranded DNA and RNA molecules through a 1.5-nm membrane pore (Kasianowicz *et al.*, 1996). Branton and his colleagues then postulated that by measuring the transient blockades of the ion current across the lipid bilayer membrane when a single-stranded DNA or RNA molecule passes through a protein (hemolysin)-ion channel embedded in the membrane, one might be able to obtain the genetic sequence information of the nucleic acid molecule (Akeson *et al.*, 1999).

Their nanopore approach depended on the electrophysiology-like measurement of the transient blockades of the ion current through a 1.5- to 2.5-nm pore by DNA nucleotides. The nanoelectrode-gated DNA-sequencing technology described in this chapter uses a transverse-tunneling electron beam provided by a pair of sharp nanoelectrode tips to directly probe the nucleotide sequence when a DNA chain passes through the nanoelectrode detection gate base by base (Figure 1). Because the nanoelectrode-gated electron-tunneling spectroscopic detection approach does not depend on the transient blockades of the ion current, we do not have to use the very small detection gap sizes (1.5–2.5 nm) required by the electrophysiology-like approach to limit the counter-ion flow. Therefore, a wider range of detection gap sizes (1.5–5 nm) can probably be used in the nanoelectrode-gated electronic tunneling detection approach to ensure speedy passage of an individual DNA chain, which is likely favored for rapid nanoscale reading of DNA sequences. Furthermore, the electron-tunneling detection gate comprises two precision tips of nanoelectrodes (D and D′) pointing toward each other on a nonconductive (e.g., SiO_2) plate (Figure 1). The patented precision electrolytic nanofabrication technique enables the ORNL research team to fine-tune a nanoelectrode detection gate at atomic scale to any specific gap size to achieve the desired results.

More recently, it has been demonstrated that DNA translocation through the 1.5-nm α-HL pore can be driven by an electrostatic potential drop through the pore in a range of 20–300 mV (Meller *et al.*, 2001; Bates *et al.*, 2003). The electrical driving force was estimated to be in the range of 5–100 piconewtons (pN). That is, an electrical driving force as small as 5 pN can pull a single-stranded DNA molecule through the α-HL nanopore. Therefore, it is clearly achievable to deliver a DNA molecule and drive it through our envisioned nanometer detection gate base by base using programmable electric fields. Furthermore, manipulation of DNA secondary structures such as double-stranded DNA "hairpin" molecules has also been demonstrated by use of external electric fields in conjunction with a nanopore (Vercoutere *et al.*, 2001). These research advances also indicate that it is completely possible to move a single DNA chain through the envisioned nanoelectrode detection gate, which could have a gap size in a wider range from 1.5 to 5 nm (Figure 1). Therefore, the proposed use of programmable, pulsed, and perpendicular electric fields (Figures 2 and 3) to control the transport of sample molecules through the detection gate makes sense for the nanoelectrode-gated electronic DNA-sequencing approach.

Furthermore, the design of our envisioned nanoelectrode-gated detection system can offer following three advantages over a conventional scanning tunneling microscope (STM) system (Bai, 1995; Bonnell, 2001) with regard to DNA sequence detection:

(1) The tunneling electron beam across a pair of sharp nanoelectrode tips is likely to be more focused (sharper) than an STM tunneling beam that is from a tip to a conductive plate (plane). Therefore, the nanoelectrode-gated detection system (in a tip-to-tip geometry, Figure 1) could potentially have a higher sensitivity than that of the tip-to-plane geometry.
(2) When a DNA sample is placed on an STM sample plate, the molecule often has variable orientations, conformations, and multiple nonspecific contacts with the substrate – all of which are potential sources of uncertainty in STM detection. In our envisioned nanoelectrode-gated molecular-detection system, these potential sources of uncertainty are removed by design. Each nucleotide enters the detection gate with nearly a uniform orientation: its phosphate group held down preferably with its nitrogenous base pointing up. Such an orientation is perfect for detection by the transverse-tunneling beam across the two nanoelectrode tips.
(3) In STM, by design and because of mechanical noise, the tip-to-substrate distance often varies between experiments. In our envisioned nanoelectrode-gated detection system, only a DNA chain moves through the detection gate base by base. After being fine-tuned for optimized detection sensitivity, the distance between the two nanoelectrode tips is constant (fixed), likely resulting in tunneling molecular detection that is more reproducible.

2.4. Theoretical analysis for the nanoelectrode-gated electronic detection

Some preliminary theoretical analysis in establishing the feasibility of using nanoelectrode-gated tunneling measurements to detect DNA sequences has recently been performed at ORNL. When the distance between the nanoelectrode tips is within about 10 nm, it is known that significant electron tunneling across the nanogap can occur upon application of a biased tunneling voltage (*V*). When a DNA molecule moves into the nanogap, the tip-to-tip tunneling conductance characteristics will change because of the "screening-conduction" effect of a given molecule on the tunneling electrons. In this case, the electronic transport process across the detection gap (gate) is more likely a combination of electron tunneling and electronic conduction through the body of the sample molecule. Because the chemical compositions and structures of the four nucleotide bases are different, the screening-conduction effect of each distinct nucleotide base on the tunneling current (*I*) and tunneling characteristics (such as the *I*– *V* and *dI*/*dV*– *V* curves) will likely differ. Previous studies at ORNL have experimentally demonstrated detection and characterization of a single biomolecule (PSI reaction center) by tunneling-spectroscopic measurements with an STM. For example, the *I*– *V* curve of an isolated photosystem I reaction center has quite unique diode-like characteristics along its axis pointing from the reaction-center pigment P700 to the terminal electron acceptor F_{AB} at the reducing end of the complex (Lee *et al.*, 1995, 1996). Therefore, it is possible to use nanoelectrode-gated tunneling-spectroscopic measurements to detect nucleotides that pass through the detection gate. With some standard DNA molecules of known sequence, this detection system could be calibrated. A unique electron-tunneling-molecular-conduction characteristic profile could then be established for each distinct nucleotide base and be used as a fingerprint to identify the base. Therefore, by detecting the difference in tunneling current (*I*) and/or tunneling characteristics (*I*– *V* and *dI*/*dV*– *V* curves) for each nucleotide passing through the detection gate, direct nanoscale reading of DNA sequences on a single molecule could be achieved (Figure 1). Since the tip-to-tip tunneling electrons likely emerge from a single (or a few) atom(s) of one nanoelectrode tip and tunnel through the nanogap to the tip of the other nanoelectrode for the shortest possible distance, the size of the tunneling electron beam is likely to be within a few angstroms (a fraction of a nanometer), which is sufficiently fine to make precise reading of the DNA sequence as a DNA chain passes through the detection gate base by base.

Recently using supercomputer facilities, the ORNL research team has performed a quantum mechanics computation study for the envisioned nanoelectrode-gated electron-tunneling nucleotide detection. The ORNL computational result showed that the differential conductance of nucleotides A, T, G, and C are well separated (detectable) in the measurable domain (Krstić *et al.*, 2005). An independent computational study by Zwolak and Ventra (2005) arrived at essentially the same conclusion. Both of these independent computational studies indicated that it is possible to distinguish the four nucleotide bases through our envisioned nanoelectrode-gated electron-tunneling conductance measurement.

2.5. Preliminary experimental work toward proof-of-principle demonstration

A significant experimental study on detection of a polynucleotide, such as polyadenylic acid (polyA), polyguanylic acid (polyG), or polycytidylic acid (polyC) has also been recently accomplished by measuring the electron-tunneling spectroscopic conductance across a nanoelectrode gap (gate) at ORNL. In this preliminary experimental study, a small droplet of polyA, polyC, or polyC water solution was placed into a nanoelectrode detection gap with a gap size of about 10 nm to allow the formation of electrode-DNA-electrode junctions, so that the electrical transport properties of the single-stranded DNA with various bases (polyA, polyC, or polyC) can be measured under aqueous conditions. Pure water was used as a control sample. The experimental results demonstrated that each of these polynucleotide materials has a distinct *I–V* curve. Their difference is more evident in the differential conductance curves. The results to date have led to the first preliminary demonstration of detecting different types of polynucleotides using the envisioned nanoelectrode-gated electron-tunneling conductance measurements at ORNL. These results indicate that it is indeed possible to detect nucleotide bases through the proposed nanoelectrode-gated tunneling conductance measurement, although much finer and sharper nanoelectrodes will need to be fabricated and tested to confer the capability of direct reading of individual nucleotides bases under nanometer scale. More experimental studies are underway.

2.6. Possible application of design polymers to enhance nanoelectrode-gated electron-tunneling DNA sequence detection

There is a quite interesting DNA-sequence conversion capability called "design polymer" developed by LingVitage AS of Oslo, Norway (Lexow, 2004), which could probably also help the nanoelectrode-gated electron-tunneling molecular-detection sequencing technology to enter the market earlier than we previously would expect. According to the design polymers technique (described in details in the next section), natural DNA sequences can be converted into code units (design polymers) that are tailor made to make certain analysis on a single-molecule readout platform easier. For example, a design DNA polymer can be a chain of two different DNA elements representing a binary (for example, "0" = "AAA"; "1" = "CCC") information system that carries the sequence information of a natural DNA molecule. That is, a piece of natural DNA such as "ATGC" could be represented in a binary format by "00 01 10 11," which corresponds to a design polymer of "AAA AAA AAA CCC CCC AAA CCC CCC," where "AAA" = "0" and "CCC" = "1" in the example below.

00	01	10	11
AAA AAA	AAA CCC	CCC AAA	CCC CCC

The binary code representation of the DNA employed by the design polymer could make the nanoelectrode-gated electronic tunneling DNA sequence detection easier in two ways. First, it may help the nanoelectrode-gated electronic tunneling DNA sequencing technology by reducing the number of base types from four (A, T, G, C) to only two (e.g., code units made out of A and C only) because the use of the binary design polymer would require to resolve only two different base types (A and C), instead of the four natural DNA bases. Second, single base resolution is no longer required because the code units can be adapted to fit the resolution of the readout method. To achieve a single base resolution, one would have to use extremely sharp nanoelectrodes (less than about 1 nm at the tips), while such a technical requirement could be relaxed in resolving, for example, a code unit made of three bases (e.g., AAA), because the segment of a triplet is significantly longer and thus much easier to be detected than a single base (e.g., A or C, about 1 nm in size). Therefore, the application of the design DNA polymer technique could potentially help the envisioned nanoelectrode-gated electron-tunneling DNA-sequencing technology to become available earlier than we previously anticipated.

3. DNA SEQUENCING BY MASSIVELY PARALLEL OPTICAL READOUT OF NANOPORE ARRAYS AND DESIGN POLYMER

3.1. The concept and its development

The optical readout of design polymer using nanoscale pores was developed in Dr. Meller's group at the Rowland Institute at Harvard (Meller *et al.*, 2005). It is based on the discovery that ~1.5 nm pores embedded in phospholipids bilayers (or in solid state films) can be used to unzip double-stranded DNA in a well-controlled process. In particular, DNA hairpin molecules, or double-stranded DNA molecules with short single-stranded DNA can be first threaded through the 1.4-nm α-HL protein pore and then unzipped under the influence of an electric field (Meller *et al.*, 2000; Meller *et al.*, 2001; Sauer-Budge *et al.*, 2003; Mathé *et al.*, 2004; Nakane *et al.*, 2004; Mathé *et al.*, 2006), and that the unzipping time of duplexes in the range 7–20 nucleotides depends exponentially on the applied voltage, as shown in Figure 4 (Mathé *et al.*, 2004). This strong dependence on voltage has opened the possibility of slowing down and controlling the rate in which DNA molecules are threaded through a nanopore. In particular it has become possible to adjust this time to permit the simultaneous optical detection based on single fluorophore probing (~1 ms) using state of the art wide area CCD cameras.

Using this discovery, we have developed a method for fast readout of design polymers by optical means. In our method, a single nanopore, or an array of nanopores fabricated or embedded in thin film is used to spatially restrict single DNA molecules in space. Electrical field is used to thread the DNAs through the pores in a single file manner. The rate of the DNA threading is determined by the unzipping process, which can be tuned to optimize detection. Unlike other nanopore sequencing methods there is no need for direct readout of the ion-current or the electrical current from each and every pore (Deamer and Akeson, 2000). Instead, a parallel *optical* readout scheme is employed. The main advantages of the optical readout are that it circumvents problems associated with the low noise, high bandwidth, electrical current recording, and it permits simultaneous readout of hundreds of nanopores embedded in a small membrane. This method takes full advantage of the design polymer format described below.

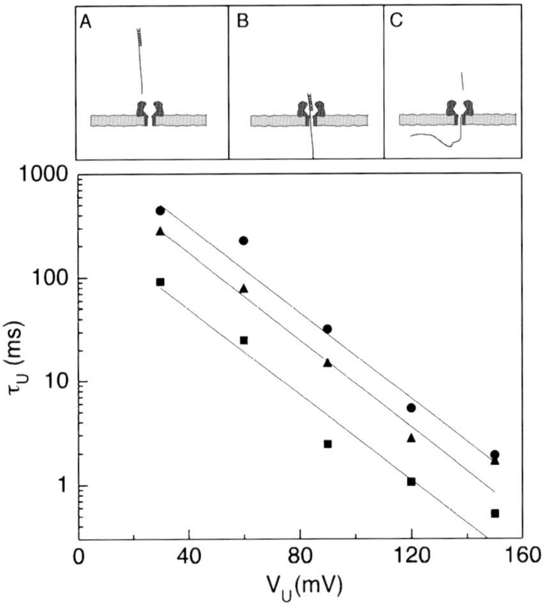

Fig. 4. Top panels: schematic illustration of the unzipping process of DNA using a membrane embedded α-hemolysin pore. (A) Double-stranded DNA with single-stranded over hang is electrophoretically threaded through the ~1.4-nm pore. (B) The ~2.2 nm double-stranded portion of the DNA cannot enter the pore, and it is unzipped at a time τ_U as shown in (C). The dependence of the characteristic unzipping time on voltage measured for DNA hairpins with 7, 9, and 10 bp (squares, triangles, and circles, respectively) is shown in the lower panel. In all cases the characteristic time can be approximated by exponentials, with a "universal" slope that does not depend on the hairpins' enthalpies, in this voltage range 30–150 mV (Mathé *et al.*, 2004). Deviations from the mono exponential behavior are present at lower voltages (data not shown).

3.2. Biochemical conversion of DNA to design polymers format

The design polymer concept, invented and patented by P. Lexow (2004) (Ling-Vitae, Norway), is a biochemical conversion method used to simplify and enhance the way in which information is stored in DNA. According to this technique, natural DNA sequences are converted into code units that make up the design polymers. The code units themselves are short (3–16 nucleotides) DNA fragments having one out of two unique sequences that we schematically label "0" or "1". Note that each unit represents a single bit of information. Thus, to represent the four bases in the DNA molecules, two bits (or two code units) are needed. The specific sequences and length of the two code units can be specified by the user prior to the conversion of the DNA, and are tailored to optimize its detection by the readout method. The biochemical conversion produces design polymer DNA molecules that are made out of a concatenation of the code units. The information is stored in the order in which the code units are concatenated. For example, adenine can be represented by the code units

"00", thymine by "01", guanine by "10", and cytosine by "11". Thus the sequence (in the native DNA) such as "ATGC" may be represented by: "00 01 10 11".

The biochemical conversion method can be used in conjunction with the nanoelectrode-gated tunneling method as described above, and with the optical nanopore detection. Not only does it reduce the four letters in the DNA to binary information, but principally it also magnifies their representation to facilitate single molecule detection. To synthesize design polymers, target DNA molecules are sheared to small pieces (less than 1 Kb) and are ligated with DNA adapters that include a recognition site for the restriction enzyme *MmeI*. This enzyme cuts 20 bp into the unknown DNA sequence, leaving a two base overhang. This overhang is used for ligation with another DNA adapter that contains a recognition site for another restriction enzyme *SfaNI*. Following this preparation stage, a cyclic conversion that includes three repeating steps starts. In each cycle the DNA is cleaved to expose three new bases in the target DNA and a DNA adapter that includes the coding units corresponding to these three bases is ligated to the other end of the DNA. The selection of only the "correct" DNA adapter is performed by PCR amplification in each cycle. Thus in each cycle three bases are removed from one end of the target DNA, and the six corresponding code units are appended on the other end. Most importantly, this process has been optimized to be highly parallel. Namely, hundreds of different DNA targets can be converted simultaneously in a single test tube (Lexow, 2004).

3.3. Features of the nanopore-guided optical readout platform

The concept of the nanopore-optical readout platform is illustrated in Figure 5. DNA will be converted to the design polymers using the biochemical conversion, and then converted from double-stranded to single-stranded DNA, as shown schematically in Figure 5A. Each base in the original DNA sequence will be represented by a unique combination of two binary code units ("0" and "1" labeled in green and red, respectively). In this case, the "0" and "1" will be defined as unique DNA sequences of 16 nucleotides each, which we call "S0" and "S1", respectively. The single-stranded design polymer will be hybridized with two types of molecular beacons that perfectly match "S0" and "S1", respectively and display minimal cross interactions (see Figure 5B). We note that one of the beacons contains a red fluorophore on its 5′ end and a quencher, "Q" at its 3′, and that the other beacon contains a yellow fluorophore at its 5′ end and the same quencher molecule at its 3′ end. The quencher molecule quenches both fluorophores. The two different colors of the fluorophores make it possible to distinguish between the two beacons. We note that in solution the molecular beacons self-quench: the four terminal bases on each end are complimentary, thus a 4-bp helix immediately forms, positioning the fluorophore in a close proximity to the quencher. This leads to a quenching of the dye. We also note that usually, upon hybridization to their targets, molecular beacons are designed to "light up" (Tyagi and Kramer, 1996; Bonnet *et al.*, 1999). However, in our case the design polymer sequence will induce the arrangement of the

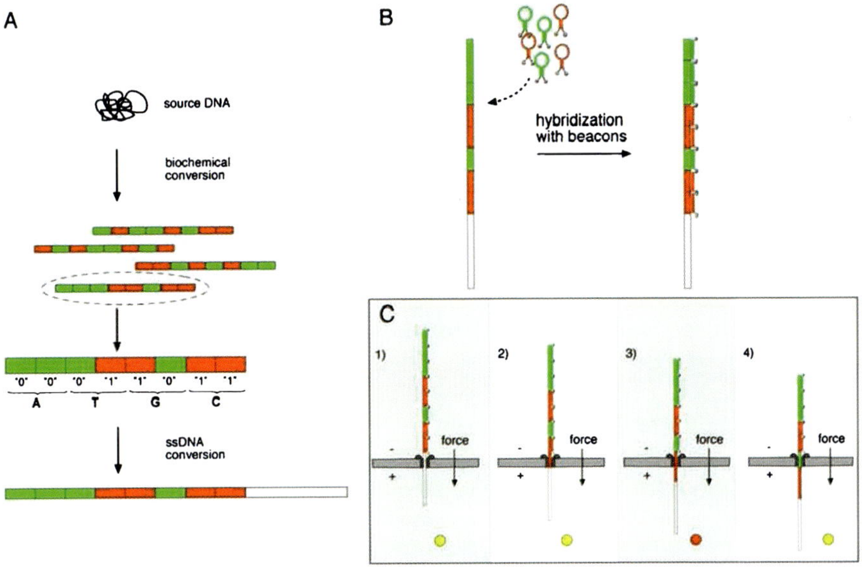

Fig. 5. DNA sequencing using optical readout of design polymers. The method involves three steps: (A) Biochemical conversion of the target DNA to design polymer format. (B) Hybridization of the design polymers with molecular beacons. (C) Optical readout of the design polymers using a nanopore-assisted method. The biochemical conversion is performed as described in the text. In this illustration the code units are labeled in green and red ("0" and "1", respectively). Many DNA fragments can be converted simultaneously. Design polymers are then converted to single-stranded format, and a poly(dA) end is added at the 3′ end of the molecules (to facilitate nanopore threading). Custom molecular beacons are hybridized with the converted design polymers in a single step (B). Finally, in step (C) we display the readout of two nucleotides (cytosine and guanine), or four code units, using the nanopore. The design polymer is threaded through the pore, while the fluorescence from the front beacon is registered; the other beacons are quenched until they reach the pore (Meller *et al.*, 2005). Each pore is capable of reading 200–1000 code units per second, and the readout can be paralleled using dense arrays of nanopores.

beacons next to each other, therefore quenchers on neighboring beacons will quench the fluorescence emission and the DNA will stay "dark" until the time that they will be removed from the DNA (excluding the first beacon). This concept is a key feature of the nanopore-optical readout method (Figure 5); it significantly reduces the fluorescence background from neighboring molecules and enables the optical detection of single molecules (Meller *et al.*, 2005).

When the molecule is introduced to the nanopore, the beacons are stripped off one by one with a time delay of ~5–10 ms. This time is tuned by the electric field intensity to optimize the signal to background levels (Mathé *et al.*, 2004, 2006). Each time that a new beacon is removed, a new fluorophore is dequenched and registered, as shown schematically in Figure 5C, by a custom made microscope. By design, the released beacon is automatically closed, thus its own fluorescence is quenched, and then it diffuses away from the pore vicinity. We also note that immediately upon the release of the first beacon, a new fluorophore, from the second beacon will light up. We estimate a readout time (for a single pore) of ~5 ms/base or 200 units per second. In a prototype setup build in Meller's lab, a custom made microscope was built for the nanopore imaging following concept described in the literature (Sabanayagam *et al.*, 2004, 2005). We used a combination of green and red diode lasers (532 and 640 nm) to illuminate the sample. Light is collected using a water immersion high numerical aperture objective, and imaged using a frame transfer cooled CCD camera equipped with electron multiplying chip. Using a custom acquisition code we are able to acquire images at > 200 fps, which correspond to an integration time of ~5 ms/image.

This method is readily extended to allow the detection of multiple pores. High-density nanopores arrays can be drilled in Si_3N_4 or in SiO_2 thin films using focused ion beam (FIB) followed by either ion or electron milling to further shrink down the pore dimension down to 1–2 nm (Li *et al.*, 2001; Storm *et al.*, 2003; Heng *et al.*, 2004; Kim *et al.*, 2006). The use of transmission electron microscope (TEM) for pore milling is particularly suitable for the fabrication of arrays, since it provide real time imaging of the processed area. Other possibilities include use of atomic layer deposition (ALD) to fine tune the diameter of fabricated nanopores (Chen *et al.*, 2004), although this technique does not provide direct feedback of the shrinking process and on pore uniformity.

3.4. Research effort toward proof-of-principle demonstration

We are currently performing feasibility studies of our optical detection method using α-HL pores embedded in phospholipid bilayer, as well as using solid state nanopores. We have designed and synthesized DNA oligos that encode a single bit (one code unit), two bits, and up to eight bits of information for the proof of concept. In parallel we have begun the fabrication of solid-state nanopore arrays in various configurations (2×2, 6×6) as has been recently shown by Kim and co-workers (Kim *et al.*, 2006). Preliminary studies indicate that these arrays perform as expected, and that larger arrays can be fabricated using similar methods. Moreover, we have discovered that our method is highly tolerable to the degree of uniformity of the nanopore dimension in any given array. Some pores may not be functional (too small or too big) but that has a little effect on the other pores in the array.

4. CONCLUSION

Two single-molecule-based DNA-sequencing nanotechnology concepts have recently been developed at ORNL and Harvard. Our studies indicate that, by employing the envisioned technologies of nanoelectrode-gated electron-tunneling spectroscopic molecular detection and/or nanopore-guided design polymer fluorescence detection, it should be possible to perform rapid DNA sequencing through direct nanoscale reading of nucleotide bases on individual DNA chains. Successful development of these DNA-sequencing nanotechnologies could have profound significance in the fields of analytic nanobiosciences, genomics studies, genome-based medical treatment, and nanomedicine.

ACKNOWLEDGMENTS

J. W. Lee wishes to thank his friends and project team members T. G. Thundat, L. Zhang, R. Kisner, P. S. Krstić, J. C. Wells, P. T. Cummings, D. Xu, X. Zhao, M. Fuentes-Cabrera, R. S. Foote, R. Zikic, and R. Jean-Pierre for their active contributions in the nanoelectrode-gated electron-tunneling DNA-sequencing research project; V. W. Pardue and D. G. Cottrell for technical illustrations; C. Thompson for secretarial assistance. This research has been supported by the U.S. National Human Genome Research Institute of the National Institutes of Health under grant numbers 1 R21 HG003578-01 and 1 R21 HG003592-01; and by the U.S. Department of Energy (DOE) Office of Science Young Scientist Award and the U.S. Presidential Early Career Award for Scientists and Engineers (to J. W. Lee). A portion of this research was conducted at the Center for Nanophase Materials Sciences, which is sponsored at Oak Ridge National Laboratory by the Division of Scientific User Facilities, U.S. Department of Energy. Oak Ridge National Laboratory is managed by UT-Battelle, LLC, for DOE under contract No. DE-AC05-00OR22725.

A. Meller wishes to thank P. Lexow of LingVitae for fruitful collaboration, and C. Sabanayagam, J. Mathé, J. Eid, V. Viasnoff, B. Hornblower, M. Kim, J. Edel, and M. Wanunu for their various contributions in the nanopore sequencing project. Insightful discussions with D. Branton, J. Golovchenko, M. Burns, and S. Ling are highly acknowledged. This work has been supported by the U.S. National Human Genome Research Institute of the National Institutes of Health under grant number HG003574-01, and by the National Science Foundation Grant NIRT-0403891.

REFERENCES

Akeson, M., Branton, D., Kasianowicz, J. J., Brandin, E. and Deamer, D. W. (1999). Microsecond time-scale resolved discrimination among polycytidylic acid, polyadenylic acid, and polyuridylic acid as homopolymers or as segments within single RNA molecules. Biophys. J. 77, 3227–3233.

Bai, C. (1995). Scanning Tunneling Microscopy and its Application. Springer Series in Surface Sceinces, Shanghai Scientific & Technical Publishers, Springer, New York.

Bates, M., Burns, M. and Meller, A. (2003). Dynamics of single DNA molecules actively controlled inside a membrane channel. Biophys. J. 84(4), 2366–2372.

Bonnell, D. A. (2001). Scanning Probe Microscopy and Spectroscopy, 2nd edition. Wiley-VCH, Inc., New York.

Bonnet, G., Tyagi, S., Libchaber, A. and Kramer, F. R. (1999). Thermodynamic basis of the enhanced specificity of structured DNA probes. Proc. Natl. Acad. Sci. 96, 6171–6176.

Chen, P., Mitsui, T., Farmer, D. B., Golovchenko, J., Gordon, R. G. and Branton, D. (2004). Atomic layer deposition to fine-tune the surface properties and diameters of fabricated nanopores. Nano. Lett. 4(7), 1333–1337.

Deamer, D. W. and Akeson, M. (2000). Nanopores and nucleic acids: prospects for ultrarapid sequencing. Tibtech 18, 147–150.

Greenbaum, E. (1988). Interfacial photoreactions at the photosynthetic membrane interface: an upper limit for the number of platinum atoms required to form a hydrogen evolving platinum metal catalyst. J. Phys. Chem. B. 92, 4571–4574.

Heng, J. B., Ho, C., Kim, T., Timp, R., Aksimentiev, A., Grinkova, Y. V., Sligar, S., Schulten, K. and Timp, G. (2004). Sizing DNA using nanometer–diameter pore. Biophys. J. 87, 2905–2911.

Kasianowicz, J. J., Brandin, E., Branton, D. and Deamer, D. W. (1996). Characterization of individual polynucleotide molecules using a membrane channel. Proc. Natl. Acad. Sci. USA 93, 13770–13773.

Kim, M.-J., Wanunu, M., Bell, D. C. and Meller, A. (2006 in press). Rapid fabrication of uniform size nanopores and nanopore arrays for parallel DNA analysis. Adv. Mater.

Krstić, P. S., Wells, J. C., Fuentes-Cabrera, M., Xu, D. and Lee, J. W. (2005). Toward electronic conductance characterization of DNA nucleotide bases. Proceedings of 2005 ChinaNano: International Conference on Nanoscience and Technology. Solid State Phenomena, Vols. 121–123 (2007) pp. 1379–1386.

Lee, I., Lee, J. W., Warmack, R. J., Allison, D. P. and Greenbaum, E. (1995). Molecular electronics of a single photosystem I reaction center: studies with scanning tunneling microscopy and spectroscopy. Proc. Natl. Acad. Sci. USA 92, 1965–1969.

Lee, J. W., Collins, R. T. and Greenbaum, E. (1998). Molecular ionic probes: a new class of Hill reagents and their potential for nanofabrication and biometallocatalysis. J. Phys. Chem. B 102(11), 2095–2100.

Lee, J. W. and Greenbaum, E. (1999). Programmable nanometer-scale metal bonding to single molecules by electrolytic deposition and depletion. ORNL Invention Disclosure ID 0772; 2002 U.S. Patent No. 6,447,663 B1.

Lee, J. W., Lee, I. and Greenbaum, E. (1996). Platinization: a novel technique to anchor photosystem I reaction centers onto a metal surface at biological temperature and pH. Biosens. Bioelectron. 11, 375–387.

Lee, J. W., Tevault, C. V., Blankinship, S. L., Collins, R. T. and Greenbaum, E. (1994). Photosynthetic water splitting: *in situ* photoprecipitation of metallocatalysts for photoevolution of hydrogen and oxygen. Energy Fuels 8(3), 770–773.

Lee, J. W. and Thundat, T. G. (2001). DNA and RNA sequencing by nanoscale reading through programmable electrophoresis and nanoelectrode-gated tunneling and dielectric detection. ORNL Invention Disclosure ID 0943; U.S. Pat. Appl. Publ. (2003), 24 pp. US 2003141189 A1; 2005 U.S. Patent No. 9,905,586 B2.

Lee, J. W. and Thundat, T. G. (2004). Separation and counting of single molecules through nanofluidics, programmable electrophoresis, and nanoelectrode-gated molecular detection. U.S. Pat. Appl. Publ. 23 pp. US 20040124084 A1.

Lexow, P. (2004). Sequencing method using magnifying tags. USA Patent No. 6,723,513 B2.

Li, J., Stein, D., McMullan, C., Branton, D., Aziz, M. J. and Golovchenko, J. A. (2001). Ion-beam sculpting at nanometre length scales. Nature 412, 166–169.

Mathé, J., Arinstein, A., Rabin, Y. and Meller, A. (2006). Equilibrium and irreversible unzipping of DNA in a nanopore. Europhys. Lett. 73, 128–134.

Mathé, J., Visram, H., Viasnoff, V., Rabin, Y. and Meller, A. (2004). Nanopore unzipping of individual DNA hairpin molecules. Biophys. J. 87, 3205–3212.

Meller, A., Mathé, J. and Eid, J. (2005). Ultra high-throughput opti-nanopore DNA readout platform. USA PCT/US05/28566.

Meller, A., Nivon, L., Brandin, E., Golovchenko, J. and Branton, D. (2000). Rapid nanopore discrimination between single polynucleotide molecules. Proc. Natl. Acad. Sci. USA 97, 1079–1084.

Meller, A., Nivon, L. and Branton, D. (2001). Voltage-driven DNA translocations through a nanopore. Phys. Rev. Lett. 86, 3435–3448.

Morpurgo, A., Marcus, C. and Robinson, D. (1999). Controlled fabrication of metallic electrodes with atomic separation. Appl. Phys. Lett. 74, 2084–2086.

Nakane, J., Wiggins, M. and Marziali, A. (2004). A nanosensor for transmembrane capture and identification of single nucleic acid molecules. Biophys. J. 87, 615–621.

Sabanayagam, C. R., Eid, J. S. and Meller, A. (2004). High-throughput scanning confocal microscope for single molecule analysis. Appl. Phys. Lett. 84, 1216–1218.

Sabanayagam, C. R., Eid, J. S. and Meller, A. (2005). Long timescale blinking kinetics of cyanine fluorophores conjugated to DNA and its effect on Forster resonance energy transfer. J. Chem. Phys. 123(224708), 1–7.

Sauer-Budge, A. F., Nyamwanda, J. A., Lubensky, D. K. and Branton, D. (2003). Unzipping kinetics of double-stranded DNA in a nanopore. Phys. Rev. Lett. 90, 238101–238104.

Shendure, J., Mitra, R. D., Varma, C. and Church, G. M. (2004). Advanced sequencing technologies: methods and goals. Nat. Rev. Genet. 5(5), 335–344.

Storm, A. J., Chen, J. H., Ling, X. S., Zandbergen, H. W. and Dekker, C. (2003). Fabrication of solid state nanopores with single-nanometre precision. Nat. Mat. 2, 537–540.

Tyagi, S. and Kramer, F. R. (1996). Molecular beacons: probes that fluoresce upon hybridization. Nat. Biotech. 14, 303–308.

Venter, J. C., Adams, M. D. and Myers, E. W. *et al.* (2001). The sequence of the human genome. Science 291, 1304–1351.

Vercoutere, W., Winters-Hilt, S., Olsen, H., Deamer, D., Haussler, D. and Akeson, M. (2001). Rapid discrimination among individual DNA hairpin molecules at single-nucleotide resolution using an ion channel. Nat. Biotechnol. 19, 248–252.

Zwolak, M. and Ventra, M. D. (2005). Electronic signature of DNA nucleotides via transverse transport. Nano Lett. 5(3), 421–424.

Chapter 9

A Single Molecule System for Whole Genome Analysis

Shiguo Zhou, Jill Herschleb and David C. Schwartz

Laboratory for Molecular and Computational Genomics, UW Biotechnology Center, Laboratory of Genetics and Department of Chemistry, University of Wisconsin, 425 Henry Mall, Madison, 53706, USA

Contents

PERSPECTIVES IN BIOANALYSIS, VOLUME 2
ISSN 1871-0069 DOI: 10.1016/S1871-0069(06)02009-X

Abstract

Optical mapping is a fully automated single molecule system for creating ordered restriction maps directly from genomic DNA molecules. The system has integrated components drawn from chemistry, physics, genetics, computer science, statistics, and engineering allowing whole genome analysis through analysis of large collections of individual DNA molecules. Such large-scale analysis is potentiated by analytes consisting of high molecular weight genomic DNA molecules that are arrayed using microfluidic devices; this step fosters high-throughput acquisition of image data. DNA "barcodes" are created using the action of restriction endonucleases on arrayed molecules from the analysis of images acquired by automated microscopy embedded within an integrated system. This system links together machine vision, barcode assembly and comparative analysis software with the massive computational power of cluster computing. After assembly of these barcodes into genome-wide physical maps, complex genome structure is revealed, characterizing genomic alterations, in addition to providing scaffolds for genome sequence assembly, or validation. Further sequence information is obtainable from barcoded DNA molecules using optical sequencing technology, delivering strings of nucleotide data from barcoded loci.

1. INTRODUCTION

Extraordinarily large molecules result from careful extraction of DNA directly from cells, and, because of their size, such molecules intrinsically potentiate new schemes for sequence acquisition either through physical mapping or nucleotide-by-nucleotide approaches (Ramanathan *et al.*, 2004, 2005). Such molecules are easily imaged by fluorescence microscopy (Morikawa and Yanagida, 1981; Schwartz and Koval, 1989) revealing polymer dynamics exploitable by manipulation schemes enabling practical approaches for molecular presentation. These features are leveraged by our single molecule platform which "barcodes" very large DNA molecules (>500 kb; 170 μm long) by sequence-specific decorations. The measurable qualities of this barcode represented as a string of visible markers along a molecule determine the extent by which a system assigns its uniqueness against other barcoded molecules in a mix of other molecules. Frequently, this "mix" includes DNA molecules originating from an entire mammalian or plant genome. Our platform utilizes single DNA molecule analytes, and our measurements discern unique molecular barcodes in the presence of noise by ensuring sufficient marker density and number, as offered by very large molecules. Individual, single DNA measurements of any kind are inherently noisy, because averages must be made through multiple measurements from either the same analyte or "identical" analytes within a population to accrue accuracy. Accordingly, the precision and reproducibility of any primary measurement in the barcoding procedure must present an acceptable level of error as gauged by the rate of reliable interpretation, which in part relies on the estimation of physical distance between successive markers (accurately placed along a molecule). As such, we create reliable barcodes using restriction endonuclease action for decoration, since bacteria have honed these enzymes into quite specific and nearly quantitative reagents. The key problem we have solved is how to assess restriction endonuclease action revealed as ordered restriction maps (barcodes) over the hundreds of thousands of single molecule barcodes required for analysis of entire plant or animal genomes. Our solution is realized as a working, fully integrated system for whole genome analysis we have dubbed – "optical mapping."

Optical mapping is now a fully integrated system for construction of genome-wide ordered restriction maps developed from the high-throughput analysis of randomly sheared genomic DNA molecules measured on a per molecule basis (see Figure 1). Because amplification or cloning is not used as part of the analysis, optical mapping is a true single molecule platform immune to errors associated with these common techniques. Briefly, the system encompasses several major functionalities described below; detailed discussion of these components follows.

1.1. Presentation of long, restriction enzyme-digested DNA molecules

We have developed a microfluidic device that works in concert with positively charged surfaces to deposit stripes consisting of elongated DNA molecules. Normally, large DNA molecules exist in solution as random coils with conformations akin to a floppy "ball of yarn," so that stable elongation of molecules becomes necessary for the high-resolution imaging of molecular barcodes. The current device boasts 48 parallel lanes for high throughput, and loads DNA solutions via capillary action onto positively charged glass surfaces. Normally relaxed DNA molecules are deposited on surfaces in an elongated state (stretched), because of the combination of capillary driven flow and their electrostatic interactions with the charged surface. Since electrostatic interactions bind molecules to the surface, enzymatic action is unhindered by common covalent bonding approaches requiring chemical linkers for the amelioration of deleterious surface interactions. Although each electrostatic bond is weak in comparison to covalent ones, the net effect of electrostatic bonding is additive, and on long molecules their sum greatly exceeds covalent bond

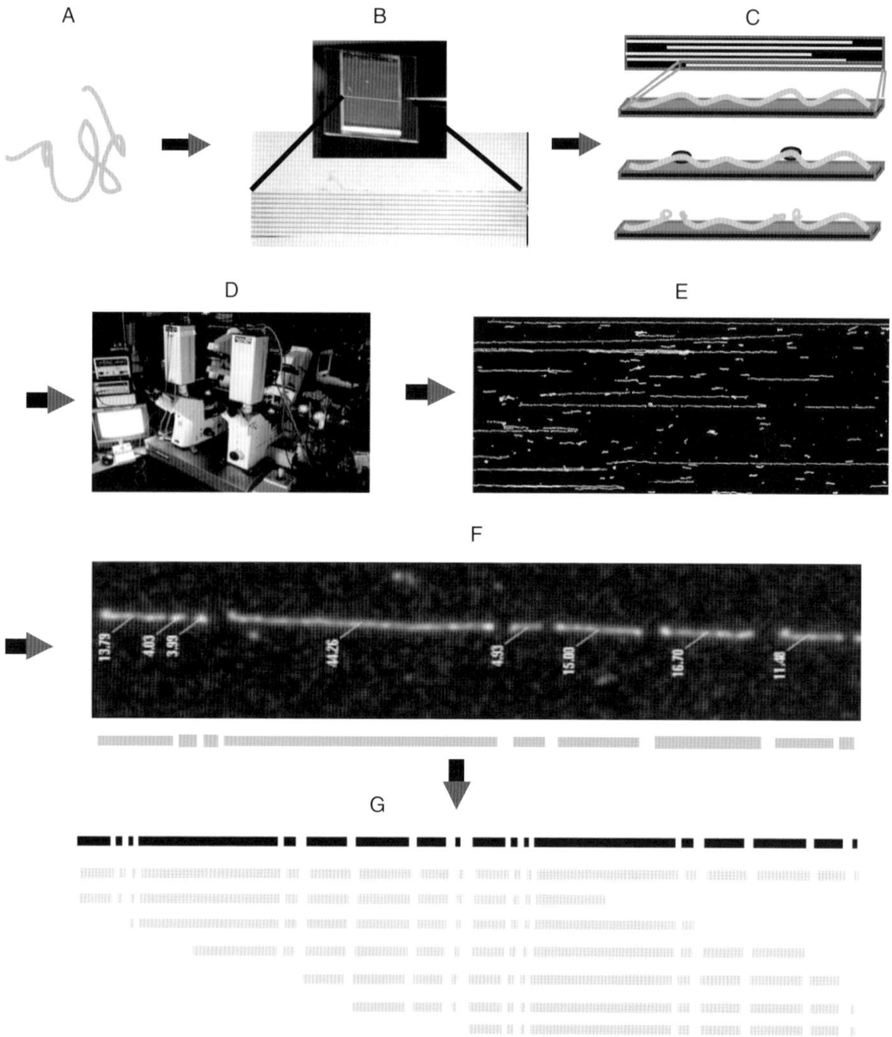

Fig. 1. The optical mapping system. (A) Genomic DNA is extracted from lysed cells. (B) DNA molecules are elongated and immobilized onto a positively charged glass surface using a microfluidic device. (C) A restriction enzyme cleaves cognate sites of the surface-bound DNA molecules. (D) The digested single DNA molecules are stained with a fluorescent dye and images are collected using a fully automated fluorescent microscope scanning system and a high resolution digital camera. (E) Single images are processed and merged together. (F) The flattened and overlapped single DNA molecule images are analyzed by machine vision software to generate single DNA molecule optical maps. (G) The single DNA molecule optical maps are formed into contigs based on overlapping restriction sites. Whole genome optical maps are constructed using ensembles of single DNA molecule maps. The dark gray barcode represents the consensus map, which includes data measurements from every single molecule map (light gray) in the underlying contig.

strengths. Accordingly, local electrostatic interactions are weak (spanning ~1 kb, compared to the thermal environment) fostering a dynamic equilibrium of "absorbed" DNA segments within a single molecule to shuttle between bound and free solution environments – only a portion of a molecule is bound at any given instant. This equilibrium supports vigorous enzymatic operations on "surface-bound" molecules enabling use of most DNA modification enzymes including restriction endonucleases. Restriction enzymes produce double-stranded "cuts" at specific cognate sequences, and these cleavage events are visualized by fluorescence microscopy as gaps (~1 μm) due to coil relaxation occurring at nascent ends; consecutive restriction fragments remain in register producing an ordered restriction map (barcode) after sizing operations.

1.2. Image acquisition, processing, and machine vision: moving from images to data files

We considered the type of "objects" we could create during the presentation stage and how to optimally image them as a single problem to be solved. This tact enabled us to critically engineer features for data acquisition in robust and operationally simple ways. As such, we developed a high-speed imaging instrument (named, "genome Zephyr"), or a single molecule "scanner," built around a standard fluorescence microscope featuring full computer control over focusing, sample positioning and digital camera functionalities, thus enabling user-free operation launched from a friendly interface. Throughput is greatly enhanced by a distributed laser illumination system offering stable, bright, monochromatic illumination to all of the Zephyr scanners in our laboratory. Essentially, the Zephyr automatically acquires strings of contiguous, overlapping micrographs, by tracking stripes of deposited DNA molecules laid down by the microfluidic system. Automatic image processing takes these images, corrects for uneven illumination, and then overlaps them (~100–150 images) maintaining proper registration into one "superimage" potentiating downstream machine vision approaches. Since we are using high-power fluorescence objectives, we realize a spatial resolution of about 300 nm. All of these features synergize critical machine vision steps that automatically identify molecules, discern "daughter" restriction fragments, and finally estimate their sizes (in kilobases), based on integrated fluorescence intensity measurements. This last step creates a restriction map for each imaged molecule, and an analyzed surface yields thousands of single molecule maps or barcodes as compact data files.

1.3. Data management, system network, map construction and analysis tools

The genome Zephyr produces a copious stream of barcode data files requiring sophisticated approaches for their management, analysis and interpretation. As such, we have integrated our imaging system (genome Zephyr and associated control and user interface subsystems) into the laboratory-wide IT infrastructure supporting all aspects of our genome analysis efforts (Figure 2). Briefly, the

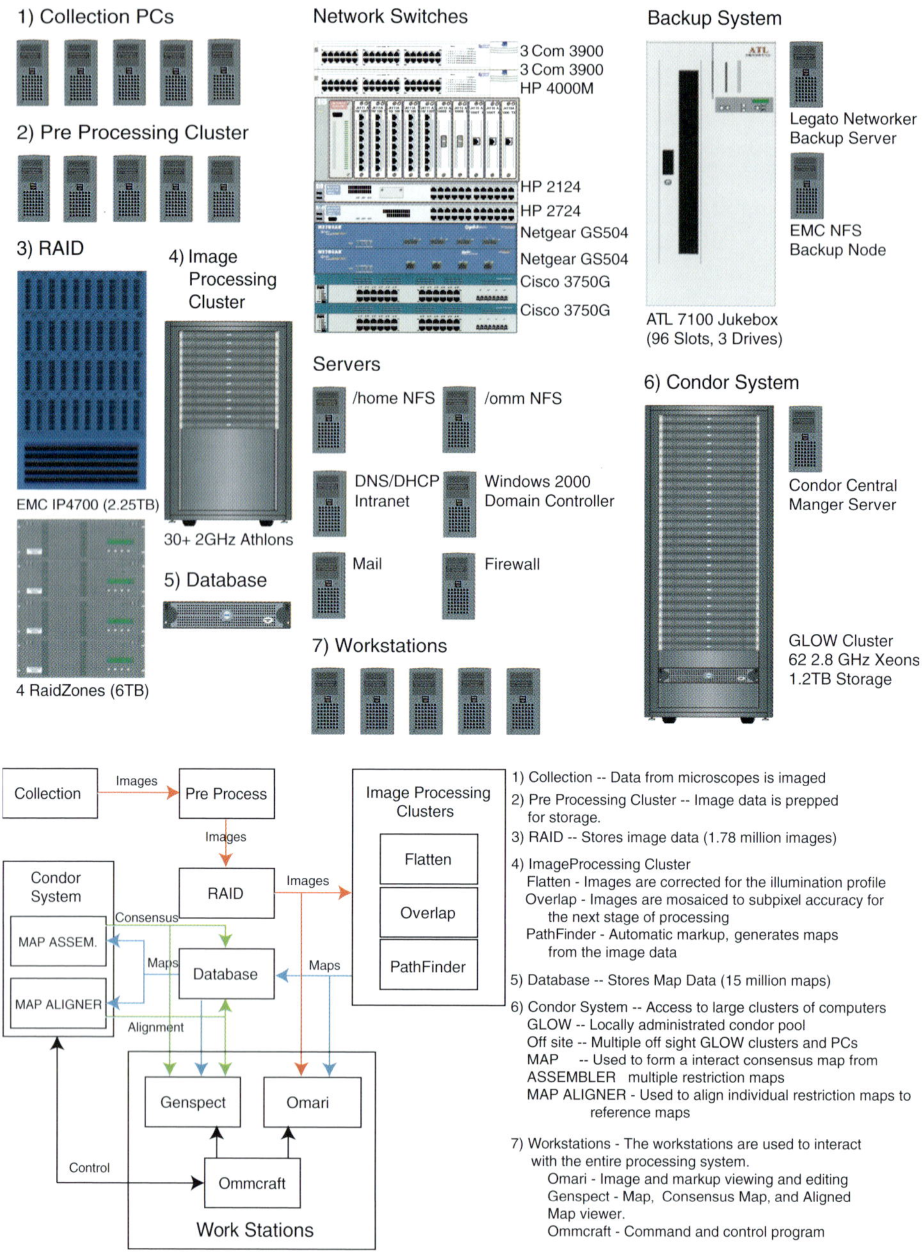

Fig. 2. Computational and network schema for optical mapping.

major components are largely controlled from a single user interface (Ommcraft), serving as a point of entry controlling and linking the majority of system functionalities. These directed operations align barcodes against each other – forming contigs (consensus maps) – or against maps found in our database, often created from the *in silico* digestion of whole genome DNA sequence data.

For this purpose, we have developed algorithms and software that work with map data in ways similar to tools created for sequence assembly and alignments (Valouev *et al.*, 2006a). Other aspects of the system include capacious storage facilities and ready access to cluster computation, making possible the rapid pairwise alignments of a large number of barcodes, necessary for their identification and localization within a given genome. These map alignment operations culminate producing deep map contigs (akin to sequence contigs, consisting of a set of overlapped and merged sequence reads) spanning entire genomes that characterize a broad size range of structural alterations (3 kb–3 Mb in size). The viewing and annotation of such findings are fully "functionalized" by a series of linked user interfaces featuring rapid presentation of image data and aligned maps or barcodes within the context of extensive annotation drawn from public databases, as well as our own.

1.3.1. Summary

Meaningful, single molecule analysis has been realized for genomic studies through the concerted exploitation of "simple" polymer phenomena. This achievement is based on our integration of seemingly disparate system components stemming from developments in microfluidics, polymer physics, novel detection schemes, software, and an IT infrastructure capable of dealing with huge single molecule data sets. As such, the unique features that optical mapping provides for whole genome analysis are as follows:

i. Since a "library" of randomly sheared genomic DNA molecules is exhaustively analyzed, prior separation of mapping substrates is not required.
ii. Maps are constructed directly from genomic DNA molecules – cloning or PCR artifacts are obviated.
iii. Genomic DNA mapping substrates are large molecules (300 kb to several megabases) enabling their maps to span repeat-rich regions.
iv. Single DNA molecule maps are "barcodes" queryable against sequence databases, revealing genomic loci and associated annotation.
v. Barcodes reveal structural homologies and differences between individuals, isolates, species, or strains.

1.4. The history of optical mapping

The optical mapping approach was first reported in 1993 (Schwartz *et al.*, 1993). In this study, yeast chromosomal DNA molecules were elongated within a flow of molten agarose developed between a cover slip and a microscope slide. Very large genomic DNA molecules were fixed in their elongated state upon gelation of the agarose matrix. A restriction enzyme was mixed with the molten agarose before gelation, and triggered by subsequent addition of magnesium, which initiated cleavage. This step enabled time lapse imaging (using digital fluorescence microscopy) of cleavage events as newly formed ends of molecules receded, creating visible gaps (movies can be viewed here: http://www.genome.org/cgi/content/full/5/1/1/DC1). Ordered restriction maps were constructed by estimating the mass of each successive "daughter" restriction fragment by integrated fluorescence intensity, and apparent contour length. The final restriction map was computed by averaging measurements taken from about 4–12 separate molecules. Overall, the final maps were very accurate, in terms of the number of fragments and their sizes, as those constructed by gel electrophoresis. The principal advance represented by this early system, aside from the use of single molecule substrates, was that it constructed ordered restriction maps, instead of less informative "fingerprints," because DNA fragment sizes and their order were determined concurrently.

Later advancements pioneered the use of charged surface modalities, in place of agarose gel, for DNA fixation, elongation, and restriction digestion. DNA solutions were applied to the edge of a "sandwich" consisting of a charged surface placed against a standard microscope slide. Surfaces were made positively charged using two different surface chemistries: polylysine (Meng *et al.*, 1995) and aminosilane (Cai *et al.*, 1995). Unlike charged surfaces for commonly used for microarray analysis, the dual requirements of molecular elongation and biochemical accessibility mandated that a critical density of charge on a surface needed to be achieved during the derivatization step (Reed *et al.*, 1998). Since arrayed molecules were imaged *after* restriction digestion on the glass surface, a large number of digested molecules could be analyzed in parallel. DNA molecules were held in place via electrostatic interactions between the negatively charged polymer and the positively charged polylysine chains, instead of being immobilized within a gel matrix. This subtle attachment scheme retained DNA during restriction digestion, yet revealed cleavage sites as visible gaps ($\sim$1 μm) due to coil relaxation occurring at newly formed ends of molecules. Consequently, imaging was performed after digestion, since restriction fragments remained stably attached to the glass surface, obviating the need for "movies" to track poorly presented molecules within a three-dimensional gel matrix. These features boosted throughput and fragments as small as 800 bp could be resolved and sized (Meng *et al.*, 1995). Overall, the surface-based mapping approach was a major breakthrough for the advancement of the optical mapping system, as the original agarose gel-based system had a resolution of $\sim$30 kb, and meager throughput.

In the mid-1990s, we further potentiated our silane-based surface modality through the development of spotted arrays consisting of deposited individual DNA molecules. Driving this development was our "co-rediscovery" of a phenomenon casually known as "the coffee-droplet-drying effect," or that unique ringed residue remaining after a coffee droplet dries on a counter leaving a spot with coffee/milk particles characteristically hugging its periphery (Deegan *et al.*, 1997; Jing *et al*, 1998). Briefly, this happens because evaporation causes convective flows that move particles from the center of the droplet toward its edge. When a DNA solution is substituted for coffee, this same convective fluid flow drags molecules across a charged surface and deposits them in a sunburst-like pattern. As such, we investigated the important variables associated with this effect – called "fluid fixation" – and these developments led to the development of the first microarray, or "chip" featuring single molecules. We then created a rather prescient spotting engine complete with an assortment of spotting tips, and this system enabled us to spot multiple samples, such as PCR amplicons (Skiadas *et al.*, 1999), and phage or cosmid clones onto a single optical mapping surface or chip for parallel enzymatic processing. These advancements pointed the way to full automation of restriction mapping through our associated development and integration of several key system components, present in today's optical mapping system: automated fluorescence microscopy – *image acquisition*, image processing vision to deal with images – *machine vision*, map construction software – *algorithms for analysis of single molecule data sets*, use of multiple computers for processing data – *cluster computing*, and a score of user interfaces that gave researcher the ability to analyze and visualize their data. It was, arguably, the first single molecule *system* for genome analysis, and it was fully automated.

We then sought to develop a new modality that would enable high-throughput map construction of entire genomes using large, genomic DNA molecules. Although, microarrays comprised of single molecules worked well for small insert clones, genomic DNA molecules proved too large and unwieldy for this approach. In part, because large DNA molecules are notoriously sensitive to shear-mediated breakage and prone to entanglement, more rigorous control of fluid flow was required to achieve optimal deposition patterns. Also, the machine vision that we developed for analysis of "small" DNA molecules did not adequately deal with molecules that could span millimeters in length. The idea was that if large genomic DNA molecules could be patterned on a surface in an oriented fashion; i.e., all molecules aligned horizontally on a surface, then this would greatly potentiate development of machine vision approaches charged with constructing restriction maps from such molecules. Accordingly, we developed a microfluidic device using soft lithography featuring a series of parallel microchannels (Dimalanta *et al.*, 2004). The device creates a spatially defined array of deposited DNA molecules, which has been fully exploited by current automated microscopy (genome Zephyr) and image analysis systems (Pathfinder).

Given this brief history, it is apparent that we have advanced the optical mapping system through fostering of synergetic developments aimed at increasing throughput, resolution, and component integration. These numerous, interlocking developments have enabled a true single molecule platform likely to see application to any analysis employing DNA molecules. The following Sections 2 and 3 present these "developments" via detailed protocols and descriptions of system components that deal with the analysis of single molecule substrates.

2. THE OPTICAL MAPPING SYSTEM

2.1. DNA preparation methods for optical mapping

The success of any optical mapping experiment depends on the availability of a high-quality DNA sample. Genomic DNA molecules that are larger than 300 kb and suitable for restriction digestion are integral parts of the experimental pipeline. Fortunately, many methods have been developed over the years that are suitable for these purposes. And while the preparation of such large substrate molecules may seem like a delicate and difficult task, the methods presented below are surprisingly robust, and allow for convenient sample transport, storage, and nearly infinite shelf life.

2.1.1. *Limitations and constraints of dealing with large DNA molecules: shearing and PFGE sample preparation*

Traditional procedures for DNA extraction involve pipetting, phenol extraction, and centrifugation. During these steps, the DNA molecules are damaged by mechanical shearing forces, and as a result are generally under ~100 kb. To get longer DNA molecules (> 500 kb), Schwartz and Cantor (1984) developed a method to isolate intact genomic DNAs for pulsed field gel electrophoresis (PFGE). Cells of any type can be embedded in low melting point agarose, a matrix that "traps" the DNA molecules, yet facilitates diffusion of large macromolecules (average pore size ~100 nm). Such immobilization allows for successive enzymatic treatment steps, while also offering protection from mechanical shearing and nuclease degradation. For microbes and plants, cell walls are first digested with protoplasting enzyme cocktails. For mammalian cells, this step is not necessary. Next, cells are completely lysed using a combination of N-laurylsarcosine and proteinase K in the presence of EDTA (0.5 M, pH 9.5). The agarose inserts can be stored for years in the EDTA solution without any detectable degradation.

2.1.2. *Extraction of DNA from PFGE inserts*

After the above treatment, DNA molecules are completely dechromatinized and suitable for restriction digestion; no further preparative steps are needed. To liberate the molecules from the protective matrix, the inserts are washed several times in TE buffer (10 mM Tris, 1 mM EDTA; pH 8.0) to remove excess EDTA and detergent. To release DNA molecules, the washed inserts are melted at 72°C for 7 min. A α-agarase solution (100 μl of TE buffer + 1 μl (1 unit) β-agarase, New England Biolabs, MA), pre-warmed to 42°C, is added to the melted inserts, and allowed to incubate at 42°C for 2 h. DNA samples are then diluted in a larger volume of TE buffer at concentrations suitable for optical mapping, typically less than 1 ng/μl. Since an internal sizing standard is required, viral or phage DNAs (such as lambda) are added to the genomic DNA solution (20 pg/μl). Samples are mounted onto optical mapping surfaces and examined by fluorescence microscopy to check molecular integrity and concentration. Long, intact DNA molecules are sheared a bit during the dilution process, yet the majority of the DNA molecules are of sufficient length (300–1,000 kb) for successful optical mapping.

2.1.3. *Direct DNA extraction via heat lysis*

Although, DNA extracted from PFGE gel inserts provides suitable substrates for optical mapping, the protocol requires some iterations to achieve a suitable DNA concentration for mapping. These iterations can be time consuming and break large DNA molecules. Consequently, a one-step procedure liquid lysate procedure was developed using heat for the lysis of cell suspensions. Here, brief heat treatment disrupts cellular integrity, while also denaturing proteins, particularly harmful nucleases. The initial heating temperature can vary from 70 to 90°C, and the initial heating time can vary from 7 min to 2 h, depending on cell type. For pathogenic bacteria, 90°C and 15–20 min heat in 0.1 M EDTA (pH 8.0) solution is used for efficient cell lysis. For mammalian cells, 70°C for 7–10 min is sufficient for complete lysis. After heat treatment, addition of proteinase K (1 mg/ml, 37°C overnight incubation) digests denatured proteins into simple amino acids. Residual proteinase K is then washed off after DNA deposition onto optical mapping surfaces.

For mammalian cells, plant nuclei and gram-negative bacteria, the above steps are sufficient for preparation of high molecular weight DNA; however, gram-positive bacteria and fungi, require a protoplasting step to remove tough cell walls, which is contoured according to cell type. For example, lysozyme is added to gram-positive bacterial samples to remove the lipopolysachharide cell wall, while Novozyme, or a mixture of cell wall lysing enzymes is used to protoplast fungal cells by digestion of chitin/glucan cell wall components.

2.2. Optical mapping surface preparation

Optical mapping surfaces are created from inexpensive glass cover slips. Rigorous cleaning and derivitazation protocols ensure uniform surface modifications that optimize molecular presentation and biochemical operations.

2.2.1. Surface cleaning

Cover slips are cleaned with strong oxidizing agents (piranha; Nano-Strip™, Cyantek Corp., Fremont, CA) to remove commercial coatings (preventing sticking) and to prepare surfaces for full protonation as a prelude to surface modifications with silane compounds (Reed *et al.*, 1998). Boiling cleaned surfaces in concentrated hydrochloric acid fully protonates surface silanol groups, which optimizes subsequent silane coupling steps.

For reasons of safety and cleanliness, a closed acid treatment system was built. All components of this system coming in contact with acids are made from Pyrex™ glass, or Teflon™; vacuum grease is not used to seal any joints. Custom Telfon™ racks were also designed to hold the cover slips securely during the cleaning process. The system is designed to process many surfaces, so that 400 cover slips (18 × 18 mm, Fisher Scientific) fit into each rack, and are held securely on three sides to anchor them in the boiling liquids. The racks are immersed in boiling Nano-Strip™ (68–75°C) for 50 min, and then rinsed extensively with high-purity, dust-free water. After six washes, the surfaces are immersed again in boiling hydrochloric acid (103°C) for 6 h, and rinsed extensively with high-purity water until a neutral pH is measured. The surfaces are removed from the Teflon™ racks one at a time, and rinsed three times in absolute ethanol. We have found that cleaned surfaces have a long shelf life at room temperature when stored under absolute ethanol in polypropylene containers.

2.2.2. Silane derivitization

Cleaned surfaces are derivatized in an aqueous solution containing trimethyl and vinyl silanes (*N*-trimethylsilylpropyl-*N*,*N*,*N*-trimethylammonium chlorides; vinyltrimethoxy-silane; Gelest Corp., PA). The trimethyl silane contains a positively charged anime group, which provides an anchor for electrostatic interactions between the DNA molecules and the optical mapping surface. The vinyl silane creates covalent cross-links between the acrylamide gel overlay and the optical mapping surface. Both silanes are bonded to the clean glass via an Si–O bond. A slotted teflon block is used to hold surfaces during the derivatization process. Thirty clean surfaces are placed into a Teflon™ block inside of a clean polypropylene (Qorpack™) container. Small amounts of trimethyl (typically 60 μl) and vinyl (typically 4 μl) silanes are carefully added to another clean polypropylene container with 250 ml of high-purity water. This mixture is shaken vigorously for several minutes. The amounts of trimethyl and vinyl silanes can be altered to obtain optimal DNA stretching, adhesion of the acrylamide overlay, and other surface chemistry properties (Cai *et al.*, 1995; Meng *et al.*, 1995; Skiadas *et al.*, 1999).

The solution is poured into the Qorpac™ polypropylene container and incubated at 65°C with gentle shaking (50 rpm) for 17.5 h. The container is then opened inside a fume hood for 1 h to thermally equilibrate. Finally, the silane solution is aspirated off and surfaces were rinsed (3 times) with high-purity water, once with ethanol, and then stored under distilled absolute ethanol. They remain usable for 4–6 weeks. Surface properties are rapidly assayed by digesting known DNA molecules (i.e., lambda DNA) to determine optimal digestion times by scoring number of observed vs. expected cleavage sites.

2.3. Microfluidic device fabrication

To control deposition of DNA molecules, a microfluidic device was developed (Dimalanta *et al.*, 2004). This device consists of microfluidic channels fabricated in polydimethylsiloxane (PDMS) by soft lithography. The microchannels are formed by molding PDMS (Sylgard™ 184, Dow Corning, Midland, MI) using a photolithographic negative master as shown in Figure 3. A master is created using standard photolithography techniques, and consists of a silicon wafer with positive-relief features made from SU-8 photoresist (MicroChem Corp., Newton, MA). Different types of masks are used – chrome and printed Mylar™ – to pattern photoresist features on the silicon master. Although, chrome masks are more expensive than the printed Mylar™ ones, they possess greater dimensional stability and thus transfer patterns with greater fidelity.

The microchannel geometry employs parallel channels, typically 100 μm wide, 8.0 μm high, and 10 mm long. The number of the microchannels on each PDMS device can be varied, and the limitations are the size of the optical mapping surface and the minimum thickness needed to hold the microchannel walls to the optical mapping surfaces without leaking (usually ≥150 μm). PDMS device replicas are easily made from a patterned silicon master. More specifically, the

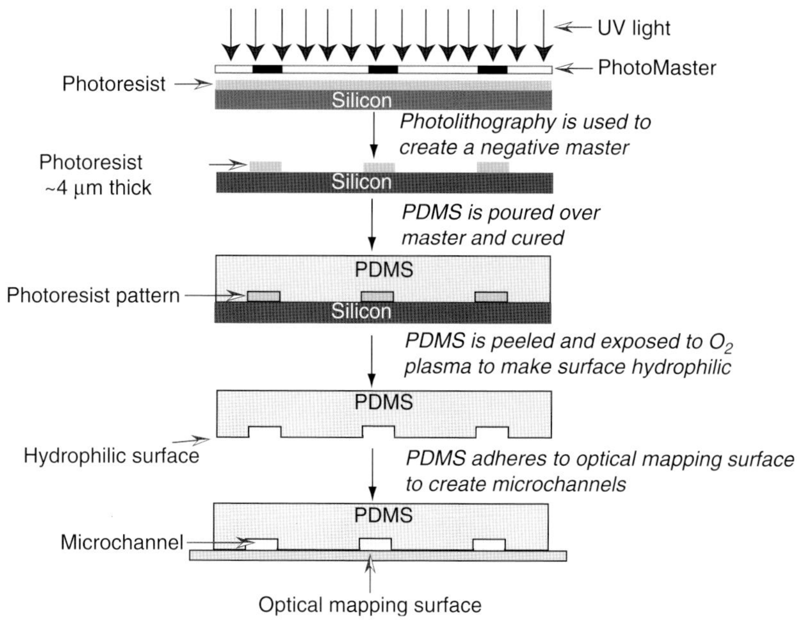

Fig. 3. Preparation of PDMS microchannels, as described in the text.

basic steps involve thorough mixing of the polymer components and catalyst, followed by degassing to remove air bubbles. The mixture is then poured onto silicon wafer master and allowed to cure at 65°C. After overnight incubation, the PDMS is fully cured and rigid enough to be peeled away from the silicon wafer. The fabricant is now hydrophobic and resistant to loading. To facilitate aqueous loadings, oxygen plasma treatment converts exposed PDMS surfaces into hydrophilic silica (Dimalanta *et al.*, 2004). Since the contact angle affects capillary action used for loading, it is controlled to range between 5 and 10°. The plasma treatment is not permanent, and the PDMS will gradually become hydrophobic in air, however storage in water extends the lifetime of hydrophilic microchannels to ~1 week.

2.4. DNA mounting, overlay, digestion, and staining

During the DNA mounting step, molecules traverse the length of the microchannels as unelongated relaxed coils. Molecule elongation and deposition takes place through a single concerted action (Jing *et al.*, 1998). Briefly, molecules ends are favored to contact the surface, and when this occurs partially absorbed molecules are anchored at one end. Because the rest of the molecule is pulled by the capillary-driven fluid flow, molecules simultaneously elongate and fully absorb to the charged surface. The degree of elongation depends on several factors that include surface charge density, fluid velocity, and molecular size. Obviously, channel geometries, rheological concerns, and device contact angles modify mentioned factors and must be tightly regulated to ensure reproducibility. This is important since the extent of molecular elongation governs restriction digestion efficiencies and the ability of machine vision to discern nascent restriction fragments. Consequently, we have found that map construction is optimized when molecules are elongated to ~70% of their B-DNA contour length. As such, the following sections fully describe mapping procedures that convey experimental details required for effective mapping.

2.4.1. Mounting/overlay

The first step of optical mapping is to load DNA solution into the microchannels. A acrylamide polymer overlay is applied after mounting that ensures retention of small restriction fragments. More specifically the steps are

i. Remove mapping surfaces from the ethanol-filled storage vessel; completely air dry on lens-paper tissue.
ii. Remove the PDMS microchannel devices from the water-filled storage vessels. Aspirate water droplets from the surface of the device; place in a plastic petri dish with channel side facing down; cut ends of each device to open channels.
iii. Lay a PDMS device (microchannel side down) onto each mapping surface. If the PDMS is not completely attached to the surface, use forceps to push the PDMS down so that the microchannel boundaries are seen intimately pressed against the glass surface.
iv. Pipette 2–3 µl of DNA solution using a wide-bore pipette tip; position the pipette tip at the entrance of the first microchannel and carefully expel the DNA solution while sliding the pipette tip along the entrance to each microchannel. The microchannels self-fill by capillary action (~5 s).
v. Aspirate excess DNA solution then carefully peel the PDMS device from the optical mapping surface.

2.4.1.1. Application of the acrylamide gel overlay.

i. Prepare a 3.3% acrylamide solution in a 1.5-ml eppendorf tube by mixing 111 µl of 29:1 (acrylamide/bisacrylamide) solution with 889 µl of high-purity water. For high salt restriction enzymes, use a 19:1 dilution: Mix 167 µl of 20% acrylamide (high purity water) with 833 µl high-purity water.
ii. Add 4–6 µl of 2% Triton-X100 solution, vortex, and degas the solution for 10–20 min.
iii. Prepare a fresh solution of 10% ammonium persulfate (APS) in water. Add 7.5 µl of 10% APS solution to the 3.3% acrylamide solution, thoroughly mix.
iv. Add 0.8 µl TEMED to 1 ml of 3.3% acrylamide solution, invert the tube a few times to mix and initiate polymerization.
v. Immediately place a 12 µl droplet of the acrylamide solution onto a mapping surface, and carefully invert the surface onto a glass microscope slide. The droplet of acrylamide will spread quickly, creating a thin and uniform coating.

vi. Place the assembly in a humidified chamber, polymerize for 20–40 min.
vii. Gently pry surface from slide using a razor blade from all four sides – peel using clean forceps. The acrylamide overlay will look like a thin transparent membrane covering the surface. Place surfaces in a humidified chamber with the overlay side facing up and add 500 μl TE buffer to each surface to wash away impurities and keep the overlay moist.

2.4.2. Digestion and staining

Choice of enzyme: Three major factors need to be considered when choosing an enzyme for optical mapping.

i. The average restriction fragment size depends on the choice of enzyme and the base composition of a given genome. Ideally, the chosen enzyme will produce products bearing an average fragment size between 10 and 30 kb to maximize the number of restriction sites per molecule, while also minimizing the occurrence of small restriction fragments less than 2 kb. Small fragments can desorb and confound discernment by machine vision.
ii. Because DNA methylation patterns are not uniformly clonal and sequence data does not represent methylation, methylation-sensitive restriction enzymes should be avoided for analysis of mammalian genomes.
iii. Enzymes that work in a NaCl-containing buffer (50–150 mM) provide optimal results, since governing electrostatic interactions are moderate.

Restriction digestion: Optimal restriction enzyme digestion conditions (digestion time and enzyme concentration) depend on the specific enzyme, porosity of the acrylamide overlay, and surface chemistry conditions. A typical protocol involves:

i. Rinse the surfaces (containing genomic DNA and an acrylamide overlay) in a moist chamber with 500 μl TE buffer. Incubate for 2–3 min and aspirate completely.
ii. Rinse with 200 μl of one time digestion buffer (containing 0.02% Triton-X) without enzyme and aspirate off the solution after 2–3 min.
iii. Add 200 μl of one time digestion buffer with enzyme (20–40U per surface). Close the humidity chamber and incubate at the proper temperature. Since digestion kinetics can vary, a preliminary time course experiment should be used to determine the optimal digestion time.
iv. After digestion, aspirate off the excess enzyme solution and wash each surface twice with 500 μl TE buffer to stop the enzymatic reaction and to remove excess salts.

Staining: Mount the surface onto a 12 μl droplet of 0.2 μM YOYO-1 solution: YOYO-1; 1,1′-[1,3-propanediylbis[(dimethyliminio)-3,1-propanediyl]]bis [4-[(3-methyl-2(3 H)-benzoxazolylidene)methyl]], tetraiodide, molecular probes, Eugene, OR, in a solution of TE buffer containing 20% β-mercaptoethanol. Seal the surface with nail polish, and incubate in a light-proof slide folder for 20 min to allow the YOYO-1 to diffuse through the overlay. The slides can be stored at 4°C overnight.

3. THE OPTICAL MAPPING SYSTEM: IMAGE ACQUISITION, PROCESSING, AND ANALYSIS

After deposited molecules are digested with a restriction enzyme and fluorochrome stained, fluorescence microscopy is used to image these results. Accordingly, whole genome analysis requires the acquisition of enormous image data sets that then must be processed and analyzed. If for example 10 times coverage (haploid human genome) is required, over 30 linear feet of DNA molecules must be imaged, at a resolution of ~250 nm, and then analyzed for the construction of maps. Assuming 100 μm wide micrograph frames, and a yield of ~10 Mb/frame, such genomic coverage is imaged within 3,000 separate frames. Given the large data sets associated with massive imaging approaches, we have automated all steps between restriction digestion and analysis of final maps. The major system components that make this possible are presented in this section.

3.1. A single molecule scanning system – "Genome Zephyr"

Image acquisition is completely automated, enabling researchers to simultaneously collect data from several collection stations without manual intervention at the rate of some 0.5 images/s (per station). Our current workstation configuration is known as the genome Zephyr (Dimalanta *et al.*, 2004) and is schematically depicted in Figure 4.

Essentially, this system is a high-throughput, high-resolution scanning system with full single molecule analysis capabilities. Briefly, an extensive suite of software (ChannelCollect; Dimalanta *et al.*, 2004) enables a commercial fluorescence microscope to rapidly focus and acquire digital images, followed by real-time processing. Since molecules span multiple frames, our software overlaps successive images with correct pixel registration to produce "superimages" comprised of upwards of 150 images, thereby efficiently tracking stripes of deposited DNA molecules laid down by the microfluidic device. The addition of laser illumination (488 nm; argon-ion laser), in place of conventional lamps, greatly reduces imaging time and provides routes for high-contrast imaging approaches that leverage "single-line" excitation. A large, water-cooled argon-ion laser is the common illumination source for 4–6 stations via a series of optical splitters and fiber optic couplings. This arrangement has proven to be economical, stable, and surprisingly flexible. What follows is a concise description of the genome Zephyr that is best followed referring to Figure 4.

Laser light enters the Zephyr station through a fiber optic coupling (A, B). Because we use a fiber optic coupling scheme, multiple microscopes are illuminated by a single source and the coherent nature of laser radiation is largely eliminated. This is necessary because destructive interference, due to any radiation coherency within the light train, will set up patterns of interference and cause shading problems. DNA molecules are stained with bis-intercalating fluorochrome, YOYO-1, and this complex is optimally excited at 491 nm. Although, 488 nm is used for illumination, typical exposure times are 150 ms.

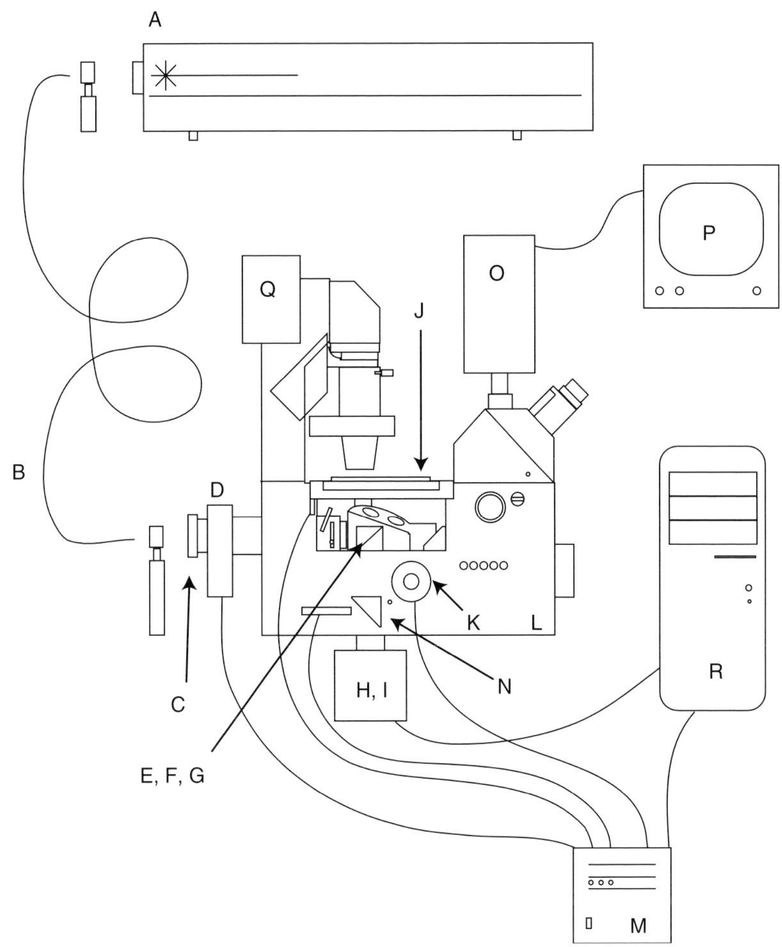

Fig. 4. Schematic illustration of a Genome Zephyr Workstation. Detailed descriptions are provided in the text. (A) Argon ion laser beam; (B) fiber optic cable; (C) rectangular aperture; (D) shutter; (E) dichroic mirror; (F) 63X objective lens; (G) emission filter; (H) holographic notch filer; (I) CCD camera; (J) X–Y stage; (K) focus motor; (L) microscope stand; (M) Ludl interface; (N) light path actuator; (L) low-light camera; (P) monitor; (Q) halogen lamp; and (R) computer.

Aside from intense illumination that allows for rapid imaging, laser illumination is spatially even and quite stable so that images exhibit consistent shading profiles. This feature facilitates downstream image processing steps that remove virtually all remnants of shading or uneven illumination. A rectangular aperture (C) spatially patterns the laser illumination, confining it to a single image frame. Without this measure, contiguous frames are prematurely photobleached, and this action renders quantitative measurements of integrated fluorescence intensity questionable.

Next, the light travels through a computer-controlled shutter (D) mounted on the rear of the microscope base. The beam hits a dichroic mirror (Omega

Optical XF2010, E), where the 488 nm light is diverted upwards toward the objective lens (Zeiss 63X Plan-Neofluar oil immersion objective, F). The YOYO-1–DNA complex emits at 509 nm (peak). This emission travels down through the objective lens, the dichroic mirror, through a long-pass emission filter (Omega Optical XF3086, G), and a holographic notch filter (Kaiser Optical Systems, HNPF-488.0AR-1.0, H) before reaching the CCD camera (I). A Roper Scientific CoolSNAPHQ^{TM} camera (1392 × 1040 pixels, Sony ICX285 chip, 12 bit digitization) is used for high-resolution image capture.

A Ludl Corp. (Hawthorne, NY) *x–y* translation stage (J), and focus motor with 50 nm resolution (K) move the next frame of data into the correct position for capture. These accessories are attached to an inverted microscope stand (Zeiss 135M; L) connected to a Ludl MAC 2000 interface (M), and controlled by a central computer (R). Together, these components control the flow of images into the analysis pipeline. Other functional components of the Zephyr include the light path actuator (N), which diverts the illumination path away from the camera so that images can be viewed through the microscope eyepiece, or through a Dage Corp (Michigan City, Michigan) SIT68GLlow-light-level video camera (O), coupled to a Sony monitor (P).

Finally, a halogen lamp (Q) allows for brightfield microscopy, and is useful for viewing set ups. All of the genome Zephyr components are controlled by a central computer interface (R) and the custom software ChannelCollect (Dimalanta *et al.*, 2004). ChannelCollect choreographs the movement of the x–y translation stage, focus motor, CCD camera shutter operation, and acquisition of digital images. The program also features an extensive user interface for viewing and modifying settings. The precise coordinates and controls for image collection are automatically generated. Complete automation enables a user to simply place a slide onto the microscope, denote a start and end coordinate pair into the program interface, and ~4,800 images are collected over the course of 150 min. Consecutive images bear a slight overlap (~20%) to ensure that molecules spanning multiple image frames are correctly overlapped and retain correct pixel registration.

3.2. Constructing single molecule restriction maps from fluorescence micrographs

3.2.1. FlatOverMerge

After acquired raw images are stored they are processed and merged together to form composite images of the original 48 microchannels of DNA deposition. This is achieved via another software program – FlatOverMerge. Although laser illumination is spatially even, further correction is necessary for accurate fluorescence intensity measurements. For this purpose, the shape and contour of the illumination field is calibrated using a photoresist-coated slide prior to sample collection (the even, spun coat of photoresist uniformly fluoresces upon illumination). ChannelCollect exports these measurements to another program, OverlapAndMerge, for additional processing. Using this calibration curve,

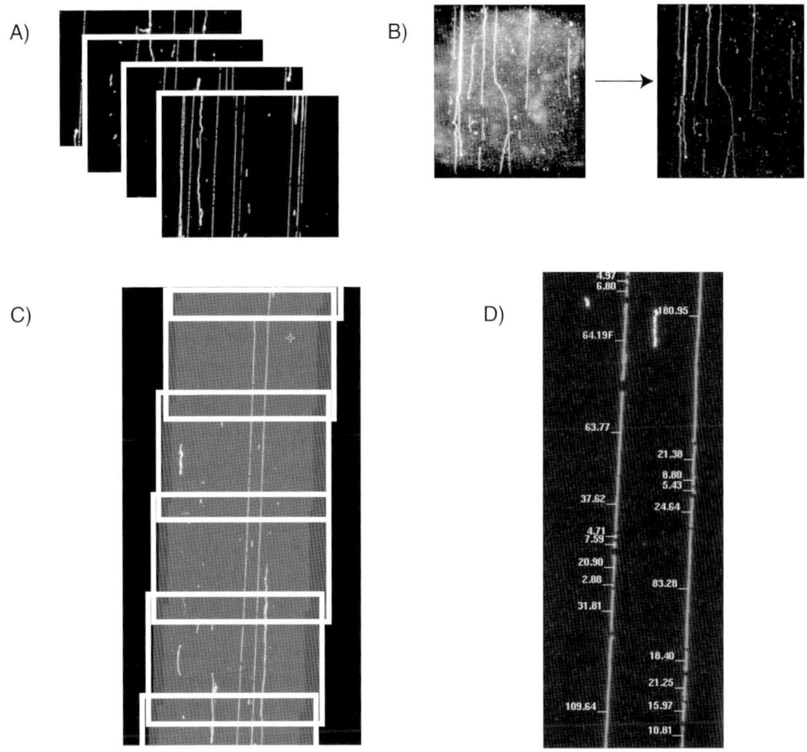

Fig. 5. Image processing. (A) Raw micrographs from CCD camera. (B) A raw image (left) is flattened by FlatOverMerge (right). (C) Images are overlapped into a composite microchannel view by FlatOverMerge. (D) Genomic DNA molecules are identified by PathFinder, and single molecule ordered restriction maps are generated.

FlatOverMerge sequentially "flattens" every image, by calculating corrections for uneven illumination on a per pixel basis. Using the slight overlap between images as a guide, it correctly orients and merges contiguous images together, forming a train of overlapped images. These images record the entire length of an imaged stripe of deposited DNA molecules. This process is illustrated in Figure 5.

3.2.2. PathFinder

PathFinder is another software program written to identify and generate single DNA molecule restriction maps from these composite images. PathFinder performs four primary functions to obtain finalized data:

i. *Segmentation*: DNA molecules are roughly isolated from background by examining image-wide pixel intensities and setting an appropriate threshold for cutoff. The precise boundaries of each DNA restriction fragment are generated by analysis of local pixel intensity gradients and an additional thresholding cutoff. Importantly, this cutoff is automatically determined.

ii. *Identification of DNA molecule backbones*: A greatly modified version of Dijkstra's algorithm is used to generate a series of lines connecting the DNA molecules identified in the segmentation step. The resulting series of connected lines traverse down the length of each DNA molecule, and generally span several contiguous images.
iii. *Identification of restriction fragments*: Restriction enzyme cuts manifest a distinctive morphology that facilitates recognition by PathFinder. Since DNA molecules are under tension on the optical mapping surfaces, the 5′- and 3′-end of each restriction fragment recoils slightly, leaving a 1–2 μm visible gap. Accordingly, the ends of DNA restriction fragments show additional fluorescence due to "DNA balls" that result from coil relaxation at these newly formed ends. An optimized series of measurements is used by PathFinder to detect these morphological features and distinguish true restriction sites from normal fluctuations in DNA stretching and reflected by variation in fluorescence intensity.
iv. *Determine the mass of each restriction fragment:* The mass of each restriction fragment is determined by its integrated fluorescence intensity value. As mentioned above, DNA molecules show some fluctuations in terms of elongation on the optical mapping surfaces. Using fluorescence intensity as a measure of mass obviates any errors due to purely length-based measurements. Internal standards are added to every experiment to facilitate accurate sizing. Typically, viral DNA of known sequence and cut pattern is added to the genomic DNA preparation at a concentration of ~20 pg/μl. Despite a series of errors stemming from a diverse set of issues, i.e., uneven fluorochrome incorporation, anomalous fluorescence intensity measurements, etc., PathFinder is able to determine the mass of genomic DNA fragments with a precision of $\sim\pm 10\%$. By averaging multiple measurements (from multiple restriction maps) at each locus, a finalized genome-wide optical map typically determines fragment sizes with an accuracy of 2–5%.

3.3. Data storage, file management and visualization

For a large optical mapping project such as a human, thousands of digital images are acquired, and storage of these images requires a sophisticated data management system (Figure 2). Typically, a single stripe of deposited DNA molecules is completely imaged by the acquisition of about 100 individual micrographs and occupies about 300 MB of disk space. As such, we process images and create maps as quickly as they are obtained to deal with the burdensome overhead associated with sizable image data sets. Furthermore, after processing and map construction, images are stored in a greatly compressed form for purposes that parallel the need to retain DNA sequence chromatograms. Storage of large image data sets has been made practical by our development of a modified JPEG format, offering near lossless compression, yet boasting a nearly a 100 fold compression rate. In part, because much of the mapping images consist of "empty pixels," a high rate of compression is achieved.

Very large data sets require databasing approaches for tracking and launching a broad range of data analysis and visualization tools. Additional functionality is accrued when the intermediate or final results of any analysis are also stored within a database. To meet these needs "OmmCraft" offers a broad range of tools under a single interface enabling users to not only access different types of files in our database, but also to invoke a suite of analysis routines and visualizations. To users, OmmCraft functions as the front end to the entire optical mapping system since it can launch local programs such as the map assembler, map aligner (described later in this chapter), image viewer "Omari," and the contig viewer "Genspect." OmmCraft also enables importation of different files formats from any accessible database. For example, OmmCraft interfaces allow Santa Cruz genome browser tracts to be readily imported and then fully utilized within this optical mapping environment. In short, OmmCraft enables users to easily manage data, perform analysis and visualization involving all aspects of optical mapping.

3.4. Optical map assembly and alignment

3.4.1. De novo map assembly

Given a clone library, physical maps are constructed by the large-scale fingerprinting or end sequencing of libraries consisting of overlapping clones. Algorithms work with these data sets to confidently identify overlapping clones for assembly into contigs spanning entire genomes. In a similar fashion, optical mapping assembles genome-wide contigs by overlapping ordered restriction maps (vs. fingerprints or sequence) constructed from individual DNA molecules (vs. clones or amplicons). In place of a clone library, optical mapping uses a single "tube" of sheared, genomic DNA; functionally, the contents of this tube is a library consisting of unique single DNA molecules. As such, the software program – Map Assembler (Anantharaman *et al.*, 1999, 1997; Lin *et al.*, 1999) – was developed to construct contigs from large, single molecule data sets. For computational economy, it assembles optical maps into a genome-wide contigs using a greedy algorithm with limited backtracking for finding an almost optimal scoring set of map contigs. All measurements have errors; however, single molecule data sets are unique in that they do not benefit from averaging at the measurement stage intrinsic to bulk approaches. Fortunately, capacious single molecule data sets enable averaging through purely computation means when measurement errors are sufficiently consistent for modeling. This means that errors do not have to be scrupulously eliminated; instead, they must simply be captured by a series of models that accurately represent them. Bayesian inference techniques work with these models to estimate the probability of overlaps, through rounds of map assembly and probability density maximization. After these error parameters are correctly estimated from the data, the best offset and alignment between a pair of maps is computed by an efficient dynamic programming algorithm. Thus, the exponential complexity of backtracking is avoided by the map assembler. The consensus map computed by map assembler is almost free of errors because the false positive error rate can be constrained below a negligible value. Although the map assembler is powerful and accurate, it cannot be readily adapted for cluster computing. Consequently, the assembly of mammalian genomes requires the systematic partitioning of large map data sets to avoid overwhelming the assembler. However, newer approaches are being developed that naturally work with cluster computing approaches and offer new routes for the assessment of genomic alterations (Valouev *et al.*, 2006a, 2006b). In addition, dramatic cost reductions in cluster computing are starting to obviate the need for these greedy algorithms, by enablement of previously "expensive" computational approaches employing exhaustive methods. Overall, these advancements promise improved throughput and new approaches dealing with experimental error.

3.4.2. Map Aligner: pairwise alignment of single DNA molecule optical maps against a reference map

The Map Aligner (software) expects local or global similarity between the two maps and works well if the two maps to be compared are very similar – alignments can be made between optical maps against other optical maps, *in silico* maps derived from sequence, and consensus maps derived from contigs constructed by the Map Assembler from optical maps. Global alignment approaches should be used when global similarity is expected, and local alignment should be used when local variation, such as insertions or deletions, are expected when comparing optical maps derived from DNA molecules with a consensus or *in silico* map derived from sequence data. For both global and local alignments, they all have gapped or ungapped options. Especially when an imperfect reference *in silico* map is used (missing some sequences from the genome sequence assembly is not unusual), gapped alignment option may give better result. Our approach for extracting multiple high-scoring alignments is based on an efficient linear scaling approach of Huang and Miller (1991). We generate confidence scores (*p*-values) using an approach similar to that used by Waterman and Vingron (1994) for sequence alignments. Given the large number of 'optical maps requiring efficient alignment with a variety of sources, we commonly use cluster computing as described below.

3.4.3. Cluster computing

A cluster computing system "Condor" (Pruyne and Livny, 1996) is used for map assembly (Figure 2). Condor is a distributed system for running computationally intensive jobs. Condor provides job queuing, job scheduling, and resource management. While similar to a traditional batch queuing system, Condor provides the additional capability of running jobs on otherwise idle desktop workstations. Owing to its checkpointing function, Condor is able to transparently move a job to a different machine when the current machine becomes actively used. Condor can also transparently redirect all I/O and system call requests back to the submitting machine, avoiding the need for a shared file system or remote user accounts. One additional benefit of Condor is that no special programming is required to use Condor's checkpoint and remote system call features. The user simply re-links their program with the Condor libraries and these facilities are made available. Finally, Condor implements "flocking," a procedure which allows jobs to run both locally and on multiple remote Condor sites, providing a massive amount of computational power.

4. APPLICATIONS OF OPTICAL MAPPING

4.1. Use of optical maps to dissect complex genome structures and facilitate sequence assembly

Physical maps reveal structural features of a genome at a resolution that roughly scales with the distance between markers. Obviously, the ultimate physical map is the complete genome sequence; however, for reasons of high cost and data sets consisting of relatively short "read" lengths, achievement of finished sequence is both relatively expensive and toilsome. Although microarray approaches readily assay single nucleotide alterations, more large-scale structural events, 500 bp to several megabases, remain difficult to assess in both "normal" populations, and especially in cancer genomes. As such, the type of physical maps (ordered restriction maps) offered by optical mapping ideally span structural features, fostering new approaches to comparative genomics (Zhou *et al.*, 2004a), sequence assembly, and the assessment of structural variation in populations. Early applications of optical mapping have prepared the way to finished sequence for a number of organisms: *Deinococcus radiodurans* (Lin *et al.*, 1999), *Escherichia coli* O157 (Lim *et al.*, 2001), *Yersinia pestis* Strain KIM (Zhou *et al.*, 2002), *Rhodospirillum rubrum* (Reslewic *et al.*, 2005), *Leishmania major* (36 chromosomes; Zhou *et al.*, 2004b), and *Thalasiosira pseudonana* (24 chromosomes; Armbrust *et al.*, 2004), and more recently mouse (Zody *et al.*, 2006). Briefly, a genome-wide optical map pinpoints errors in nascent sequence contigs, and concordances found between an optical map and sequence contigs (Zhou *et al.*, 2003). Such comparisons are a particularly effective means for validating finished sequence because libraries, PCR, and sequencing reactions are never used in the map construction. Single molecule optical maps usually span 300 kb–1 Mb, and the optical map contigs can span tens of megabases. Consequently, optical mapping is not hindered by repetitive and "unclonable" tracts of DNA, so that a genome-wide optical map spans and provides accurate distances for gaps between sequence contigs, especially difficult gaps that remain in finished and nearly finished genome assemblies.

4.2. Use of optical maps for microbial comparative genomics

The field of comparative genomics has been established in part by genome sequencing and annotation projects for a broad range of microbial species. To gain further insights, new sequencing efforts are now focusing on analyzing multiple strains or isolates within a given species. Often, the complete resequencing of several microbial strains is not completely financially feasible, and thus new whole genome approaches must be developed for high-resolution comparative genomics. Optical maps have been used for bacterial comparative genomic studies to identify and annotate chromosomal alterations between closely related bacterial strains. This approach is especially useful when one of the strains has been fully sequenced and can serve as a reference for strain comparisons. For example, two strains of *Shigella flexneri* (2a 2457 T and 301) have been fully sequenced. An optical map of an unsequenced strain of *Shigella flexneri* (serotype Y strain AMC [385Y]) was compared to the *in silico* maps of the sequenced strains. Several variants were discovered, including one novel locus implicated in serotype conversion, as well as several other loci containing IS elements and phage-related gene insertions (Wei *et al.*, 2003; Zhou *et al.*, 2004a). Using this system, genomic rearrangements and chromosomal breakpoints are readily identified and annotated against a prototypic sequenced strain.

However, the choice of restriction enzymes for generating the optical maps of unknown sequence strains is critical for successful comparative studies. For example, the *in silico* vs. *in silico* map comparisons between distantly related strains, *E. coli* K12 and *S. flexneri* serotype 2a strain 301, differ according to the choice of restriction enzyme. Some differences in the extent and location of chromosome homology were revealed using three restriction enzymes. However, most of the homologous regions were identified by all three enzymes (Zhou *et al.*, 2004a). Usually, the smaller the average fragment sizes of the strain restriction maps, the more homologies can be identified by map comparison, because more sequences will be sampled when the enzyme cuts the genome frequently. Comparing with genome sequence comparison between *E. coli* K12 and *S. flexneri* 2a 301 or 2457 T, which suggested that these two strains (301 and 2457 T) share about 80% backbone with K12, the total homology identified by map comparison is relatively low ($<50\%$). A battery of enzyme maps might help to fully and confidently discern homologies between mapped strains when the strains are not very closely related.

The ability of comparing ordered restriction maps to confidently discern genome map homologies naturally diminishes as the evolutionary distance between organisms increases. The genome structural elements, codon usage, and functional motifs represented by vastly different nucleotide composition could vary dramatically among distantly related bacterial strains. In this case, low pass shotgun sequencing (1–3 times coverage), which commonly produces a string of sequence contigs, could be combined with optical mapping to reveal large-scale genomic features that would only be apparent after additional sequence coverage and expensive finishing efforts. Essentially, these issues come down to considerations of cost and throughput. Given the recent developments to the optical mapping system, the analysis of collections of 100 strains now becomes practical at a cost of about one-tenth that of low pass sequencing, and without the need to construct libraries.

4.3. Use of optical maps for microbial identification and infectious disease diagnosis

The world of microbial pathogens displays tremendous diversity due to their intrinsic ability to adapt for survival in a constantly changing host environment.

The presence and replication of a pathogen in a certain host may or may not cause disease, depending on the host defense system as well as the antigenic capacity of the microbe. This antigenic capacity is largely determined by genes that code for cell-surface epitopes and secreted toxins. Because these genes are under such strong selective pressure, a wide range of genetic variants has emerged over the years. The ability to rapidly detect and discern these variants, combined with a detailed knowledge of toxicity and host-survival implications will have a lasting affect on both the medical and anti-bioterrorism communities. Clinicians can currently isolate and culture a wide range of infectious microbes from patients. However, if this practice is coupled with rapid tests to identify highly and/or uniquely pathogenic sub-strains, then appropriate treatment regimes can be prescribed earlier, reducing the risk of sepsis and widespread infection.

Current techniques for microbial genome identification and analysis are based on culturing, serological techniques, "chips," fingerprinting, and PCR analysis. However, these techniques suffer some significant shortcomings. For high-resolution assays, only a predetermined set of microbes and genetic loci can be identified, and scant knowledge is obtained regarding *de novo* genetic engineering efforts aimed to imbue microbes with unique pathogenicities. Whole genome techniques, such PFGE fingerprinting, DNA microarray analysis, and PCR-based subtractive hybridization allow for the detection of genomic variation among different strains or isolates. However, none of these approaches offers the ability to achieve a detailed and comprehensive whole genome view that engenders discovery and characterization of novel chromosomal alterations. As such, new approaches are required to meet these challenges to provide sufficient information on the strain identification and pathogenicity.

The optical mapping system can serve as an optical barcoding system (OBS). This system generates "barcodes" for each microbial genome from individual genomic DNA molecules. These individual genomic DNA barcodes can be used to identify microbes when they are aligned against a sequence or map database. Importantly, these alignments can provide accurate determinations within a complex mixture of microbes. Initial evaluation of this system for microbial detection and identification is shown in Figure 6. This figure shows optical barcode alignments from an artificial mixture of five different bacterial genomic DNA molecules including *E. coli* CFT073, *S. flexneri* 2a 2457 T, *Methylomonas methane*, *Enterococcus faecium*, and *Y. pestis* KIM. These barcodes were aligned against an *in silico* map database generated from 148 bacterial genome sequences found in GenBank. As shown, greater than 95% of the optical barcodes were correctly aligned against the *in silico* data from GenBank and the corresponding optical map generated from a pure culture of each microbe. Most of the genomic DNA barcodes from each bacterial strain or species hit the whole genome *in silico* or optical barcode from itself or closely related strain or species, and very few of the simulated single molecule barcodes from some bacterial strains or species hit the whole genome *in silico* or optical barcode from non-related strain or species (<5%). This result indicates that the OBS can correctly identify multiple microbial species from a complex mixture, including potential bioterrorism agents. Detailed analysis of these barcodes has also yielded

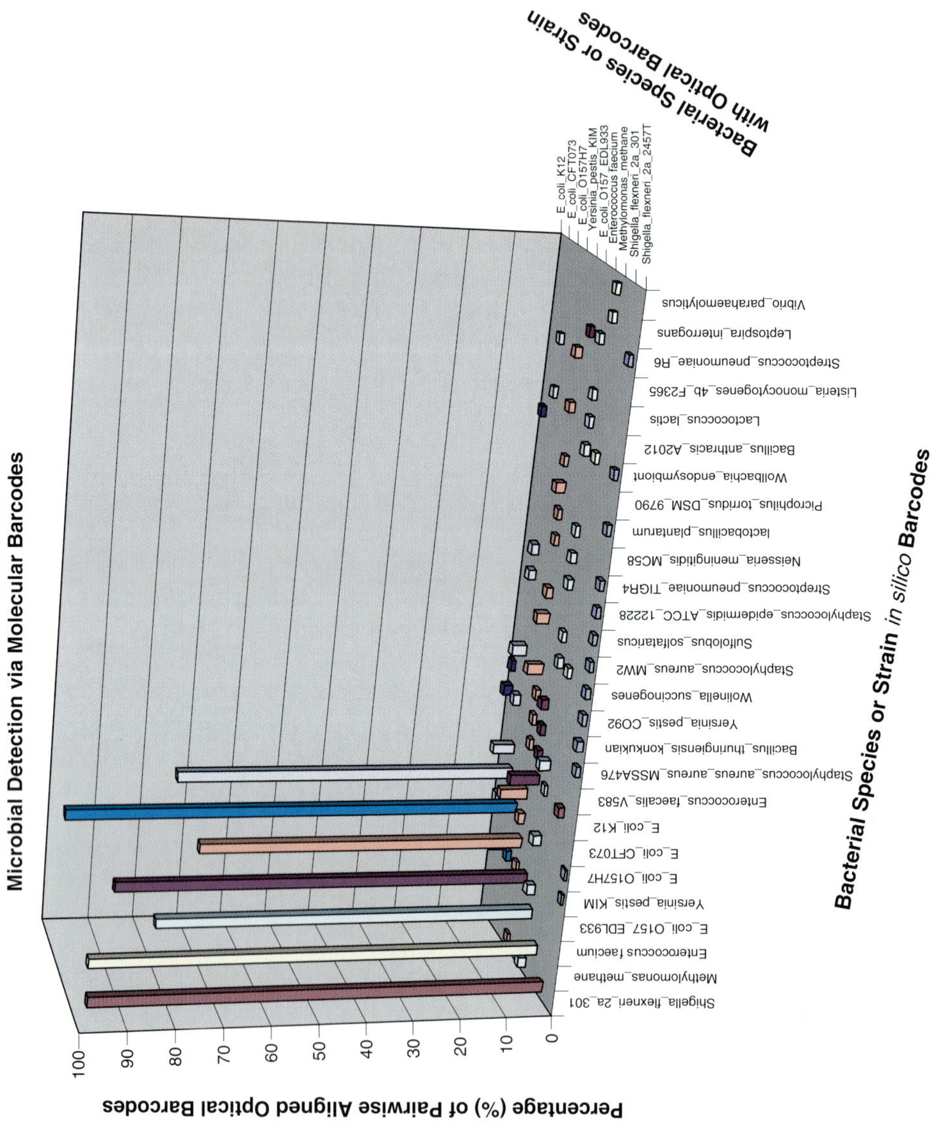
Microbial Detection via Molecular Barcodes
Percentage (%) of Pairwise Aligned Optical Barcodes
0
10
20
30
40
50
60
70
80
90
100
Bacterial Species or Strain *in silico* Barcodes
Bacterial Species or Strain with Optical Barcodes
E_coli_K12
E_coli_CFT073
E_coli_O157H7
Yersinia_pestis_KIM
E_coli_O157_EDL933
Enterococcus faecium
Methylomonas_methane
Shigella_flexneri_2a_301
Shigella_flexneri_2a_2457T
Shigella_flexneri_2a_301
Methylomonas_methane
Enterococcus faecium
E_coli_O157_EDL933
Yersinia_pestis_KIM
E_coli_O157H7
E_coli_CFT073
E_coli_K12
Enterococcus_faecalis_V583
Staphylococcus_aureus_aureus_MSSA476
Bacillus_thuringiensis_konkukian
Yersinia_pestis_CO92
Wolinella_succinogenes
Staphylococcus_aureus_MW2
Sulfolobus_solfataricus
Staphylococcus_epidermidis_ATCC_12228
Streptococcus_pneumoniae_TIGR4
Neisseria_meningitidis_MC58
lactobacillus_plantarum
Picrophilus_torridus_DSM_9790
Wollbachia_endosymbiont
Bacillus_anthracis_A2012
Lactococcus_lactis
Listeria_monocytogenes_4b_F2365
Streptococcus_pneumoniae_R6
Leptospira_interrogans
Vibrio_parahaemolyticus

detailed genomic structural alterations including rearrangements, insertions, and deletions, which in other circumstances could be interpreted as signatures of malevolent genome engineering efforts. With the recent advances in optical mapping throughput and resolution, thousands of microbes can be analyzed in a reasonable time frame, making this a valuable platform for clinical and biodefense applications.

4.4. Discovering structural alterations in mammalian genomes

The optical mapping system is a powerful platform for the discovery of genomic structural alterations, or "differences" involving kb–Mb-sized changes. Since optical maps are genome-wide ordered restriction maps, discerned structural alterations consist of missing, or additional restriction sites, insertions (novel, or enlarged restriction fragments), deletions (missing or diminished restriction fragments), inversions, duplications, translocations, aneuploidy, and complex rearrangements. When such alterations are pervasive within a population, they are characterized as structural variants, while somatic alterations, exaggerated in cancer genomes, represent lesions, or aberrations. Compared to other genome analysis techniques with the exception of costly fosmid or BAC-end sequencing, optical mapping is unique in that it simultaneously detects and structurally characterizes a broad range of genomic alterations that include: insertions, translocations, inversions, and complex genome rearrangements. As such, our current emphasis is on the analysis of mammalian genomes and cancer. For this purpose new algorithms and software have been created that automatically identify the aforementioned classes of structural genomic variation and assess their statistical confidence (Valouev *et al.*, 2006a).

Until recently, single nucleotide polymorphisms (SNPs) were though to encompass the full range of genomic diversity represented by human populations. However, recent studies have revealed that large-scale alterations are indeed numerous and common to human populations (Iafrate *et al.*, 2004; Sebat *et al.*, 2004; Tuzun *et al.*, 2005). These variants can range from small (1 kb) to extremely large (900 kb), and include insertions, deletions, and inversions. Because such common variants encompass such large chunks of the genetic landscape, they likely affect important phenotypic traits. In addition to this role in normal human diversity, structural variants have already been established as the cause of many inherited diseases (Stankiewicz and Lupski, 2002) as well as an important feature of all human cancers (Albertson *et al.*, 2003). In fact, the successful anticancer therapeutics, Gleevec, Herceptin, and Iressa have been developed against malignant chromosomal changes found in leukemia, breast cancer, and lung cancer, respectively.

Fig. 6. The percentage (%) of pairwise aligned optical barcodes or maps. Genomic barcodes were generated from a mixture of five bacterial strains. Using the software program map aligner, each barcode was aligned against 148 *in silico* reference maps from GenBank genome sequence as well as optical mapping data from each separate strain. The percentage of hits of single molecule barcodes from each test bacterial strain against all the 148 bacteria whole genome *in silico* or optical maps result from each barcode is plotted in the graph.

Despite the magnitude of these results, the annotation and discovery of structural variants has been a relatively slow process, mainly owing to the lack of high-throughput platforms that can discover and characterize these events at an acceptable level of resolution across a single individual or a population. In this regard, the optical mapping system holds great potential as a means to study structural variation in the human genome. Optical mapping can scan an entire human genome, including repetitive regions, while characterizing structural changes at ~3 kb resolution. This approach obviates clone libraries, sequencing schemes, and hybridization by directly detecting changes on single genomic DNA molecules, and provides a good bridge between sequencing analysis and cytogenetic analysis such as *in situ* hybridization. Preliminary analyses of human genomes such as cancer, stem cell, and other cell lines are exciting, and show great promises of optical mapping in the analysis of structural variations across a population of phenotypically normal individuals, as well as the discovery of complex chromosomal rearrangements in cancer genomes and clinical cases of inherited genetic disease.

5. COMPARISON OF OPTICAL MAPPING AND ALTERNATE METHODS FOR GENOME ANALYSIS

The long-standing importance of large-scale genomic variation among microbial strains, as well as recent interest in structural variation among human genomes (Iafrate *et al.*, 2004; Sebat *et al.*, 2004; Sharp *et al.*, 2005; Tuzun *et al.*, 2005) has resulted in different approaches toward evaluating these events. Optical mapping is a flexible platform for the discovery of such genomic variants.

Figure 7 illustrates a comparison of commonly used methods for discovering structural variants including genomic insertions, deletions, inversions, duplications, and translocations. As shown, only optical mapping can discover new structural variants and characterize each event at ~5 kb resolution. Alternative methods can detect larger-scale events, but offer limited utility for molecular characterization.

5.1. Microarray-based methods

Several microarray-based methods have emerged for probing genomic copy-number polymorphisms (CNPs). Array comparative genomic hybridization (CGH) (Kallioniemi *et al.*, 1992) utilizes DNA microarrays created by spotting BAC clones, or PCR/rolling circle-amplified BAC segments (Fiegler *et al.*, 2003; Smirnov *et al.*, 2004) onto a microarray slide. More recent array approaches, particularly for human genome analysis, are generated using high-density oligonucleotide chips (Lucito *et al.*, 2003; Sebat *et al.*, 2004). Genomic DNA is isolated from test and reference samples, differentially labeled, and applied to the array in a similar manner to traditional RNA expression analysis. Genomic copy-number changes are represented by spots on the array with different hybridization intensities between the test and reference samples. Events of various

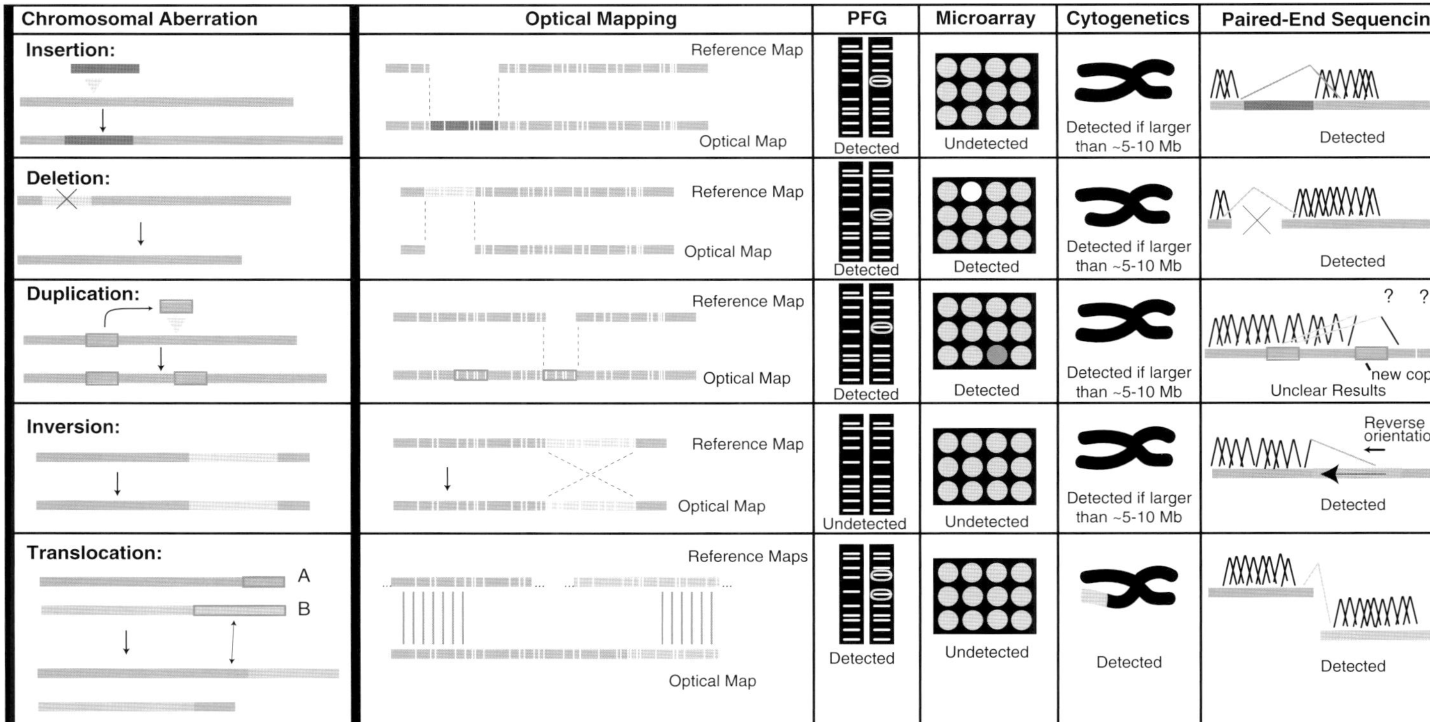

Fig. 7. Scheme for the comparison of existing methods for the discovery and characterization of structural genomic variants.

scales can be detected, depending on the resolution of the microarray used. However, the order, orientation, spacing, and genetic continuity between probes are unknown, thus limiting structural discernment of the test genome. Other structural variants that do not change DNA copy number (inversions, balanced translocations, transpositions) will remain completely undetected.

5.2. Pulsed-field gel electrophoresis

Pulsed-field gel electrophoresis (Schwartz and Cantor, 1984) is commonly used to distinguish structural variation among microbial strains on a genome-wide basis (Tenover *et al.*, 1995). Structural variants are represented by chromosomes or restriction fragments with altered electrophoretic mobilities; thus PFGE identifies the genomic regions that harbor variation, but does not provide any further evidence as to the molecular nature or sequence information involved with each event, except through toilsome Southern blot analysis.

5.3. Cytogenetics

Cytogenetic methods utilize direct visualization of intact chromosomal DNA to identify structural changes. Techniques such as metaphase chromosome banding and spectral karyotyping (SKY; Schrock *et al.*, 1996) can detect large-scale chromosomal changes that are at least 5–10 Mb in size (Shaffer, 2005). Fluorescence *in situ* hybridization (FISH; Trask *et al.*, 1991) and Fiber FISH (Heiskanen *et al.*, 1995) are higher-resolution cytogenetic approaches that can be used to confirm or identify specific loci, for which specific probes have been designed.

5.4. Paired-end sequencing

Paired-end sequencing of BAC, cosmid, or fosmid libraries is another recent method for the discovery of genomic structural variation (Volik *et al.*, 2003; Tuzun *et al.*, 2005). This approach relies upon a high-quality reference sequence. Clone libraries are generated with a uniform insert size and the end of each clone are sequenced and aligned to the reference genome. Structural variants are represented by discordant pairs of sequence alignments, displaying abnormal distance, or orientation when aligned to the test genome. This method identifies many classes of structural genomic variants. However, the high cost of library generation and sequencing limits this approach toward the analysis of large populations.

6. Optical Sequencing

De novo sequencing (genomic) requires a large number of redundant reads to mediate errors and enable complete coverage of a clone, or most of a genome.

Common factors governing effective assembly include read length and quality. Although, the availability of "finished" sequence for a rapidly increasing number of species is fueling resequencing efforts, genomic alterations associated with speciation and variation are now appreciated to involve rearrangements beyond SNPs that may defy discovery, or characterization by emerging sequencing systems. In part, potential issues stem from nascent systems offering short reads, or error rates precluding confident placement of reads within repetitive regions of genomes. As reference, repetitive sequence occupies about 50% of the human genome, and these regions are now thought to play a significant role in gene regulation. Given these concerns, we have developed a single molecule sequencing approach that may enable "renovo" (resequencing + *de novo*) sequencing operations by use of mapped or barcoded single DNA templates. We have dubbed this new system, optical sequencing, and it acquires sequence information through polymerase-mediated incorporation of fluorochrome-labeled nucleotides within optically mapped templates (Ramanathan *et al.*, 2004, 2005). The system utilizes double-stranded templates from a range of sources that include: genomic, PCR, and clone.

A key feature of the optical sequencing system is the use of mapped substrates (Figure 8 shows a schematic of the biochemical steps). Having mapped, single molecule templates offer several advantages that work together with optical sequencing to potentiate short sequence reads. First, short reads are more readily assembled into contigs because such reads are "anchored," or bear mapped positions within a genome. In this way, the number of possible overlaps is greatly limited and this relaxes requirements for long, "error-free" reads. Obviously, if strict resequencing is the endpoint, then this advantage may only hold for covering repetitive portions of a genome. Second, sequence analysis aimed at the characterization of structural alterations showing amplification, insertions, or compound events may be confounded by short, "anonymous" reads. Here, mapped templates provide unambiguous structural characterization, which can easily segue into detailed sequence characterization when reads originate from the identical mapped templates.

Although, use of double-stranded DNA molecules enables mapping, competent single-stranded templates are prepared, on mounted molecules, by a combination of *DNaseI* nicking and gapping with *T7 exonuclease* (Figure 8). Through control of nicking activity and the kinetics of gap formation, sequencing templates are formed that are long and partially double stranded in a random and patchy manner. This is an important characteristic of the substrates we use for sequencing, because long, single-stranded DNA molecules have short persistence lengths, adversely affecting elongation, which compromise presentation techniques.

Fluorochrome-labeled nucleotides are added one base at a time to an entire surface bearing mapped single molecule templates. A progressive polymerase, like *Sequenase*™ (modified T7 DNA polymerase), successively incorporates fluorochrome-labeled nucleotides when care is given to choice of linkers and fluorochrome moieties (Ramanathan *et al.*, 2005). Detection and quantitation of the number of nucleotides per cycle is accomplished through software (Peakfinder) identifying punctuates as being *bona fide* incorporations. This is

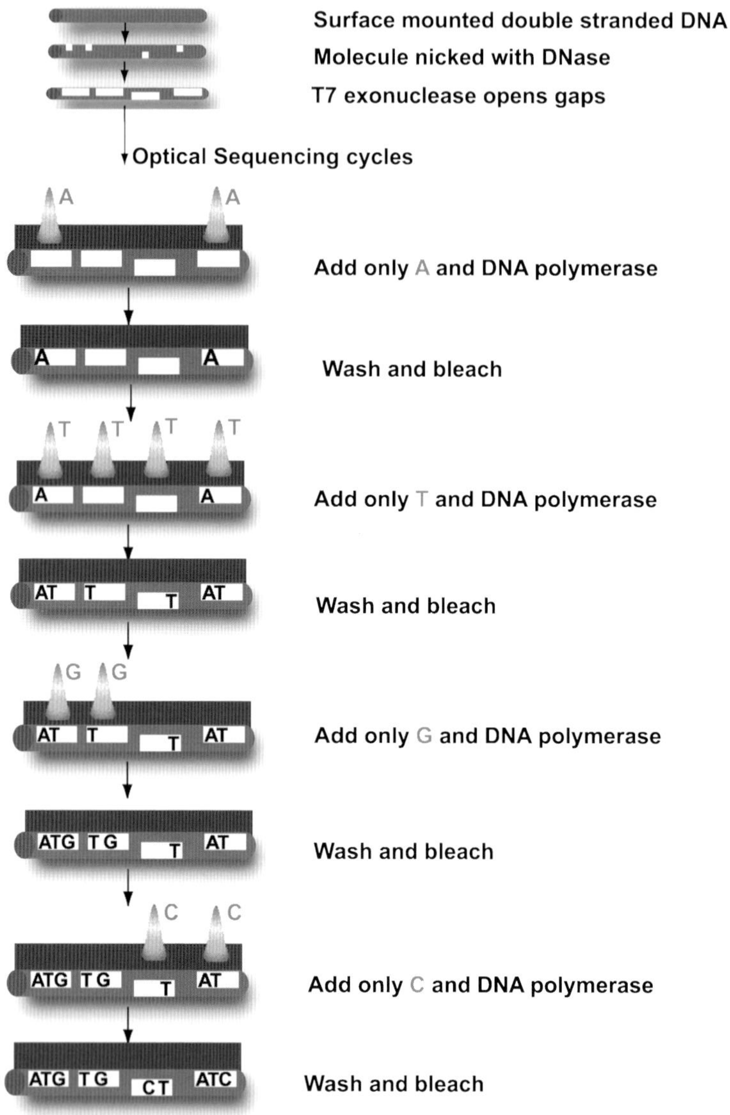

Fig. 8. Biochemical scheme of optical sequencing. Sequence acquisition from single DNA molecules elongated on optical mapping surfaces. The DNA molecules are nicked and gapped to serve as templates for DNA sequencing. Fluorochrome-labeled nucleotides (FdNTP) in the cartoon are indicated as A, T, G, and C. In each cycle, one kind of FdNTP is incorporated into the gaps, quantitated, and photobleached. This readies the template for subsequent optical sequencing cycles. Step 1: Nicking the target. Double-stranded molecules are elongated and adsorbed to an optical mapping surface. DNase is added to nick surface-elongated target DNA molecules followed by T7 exonuclease, which produces gaps and prepares the template for Sequenase v. 2.0-mediated nucleotide addition. The gaps need to be

accomplished by a routine within Peakfinder that calculates "the goodness of fit" of fluorescence intensity profiles, calculated from fluorescence imaging, against a Gaussian profile. This profile is empirically determined from analysis of fluorescent beads, small enough to approximate the point spread function of a single fluorochrome within our optical system. Finally a putative incorporation is precisely located, within a fraction of a single pixel, to a position on a specific DNA template, or background. Background punctuates are classified as noise and discounted from further analysis.

After each cycle of labeled nucleotide addition, a photobleaching step obliterates any trace of fluorescence intensity associated with *bona fide* nucleotide incorporations. Spurious fluorescent signals do not always readily photobleach, and often do not diminish in ways indicative of a limited number of fluorochromes. Photobleaching resets the fluorochrome counter within Peakfinder to a zero value, enabling the precise counting of subsequent additions. An "addition history" or tabulation of the number, location, consistency, and optical characteristics of putative incorporations (labeled nucleotides) is determined for each measured locus on a surface for all sequencing cycles. The addition history of each locus permits filtering of outliers and ranking of *bona fide* reads according to quantitative measures of "quality." From these files, short strings of novel sequence information can be ascertained and linked to known molecular loci (positions on mapped restriction fragments). In this way, sequence and map data can be combined as analysis constituting renovo sequencing.

In summary, optical sequencing is a nascent system that has achieved proof-of-principle, but requires new chemistries to be truly competitive with other emerging approaches. In this regard, advancements in our laboratory are centered on the establishment of a new single molecule analysis platform that will facilitate a barcoded molecule front end for a range of new sequencing approaches.

---

only about 20–50 bp long for sufficient sequence acquisition. Since the optical resolution of the microscope at 488-nm illumination is approximately 0.2 μm, the nicks need to be spaced approximately 1–2 μm apart (3–6 kb of B-DNA, assuming complete elongation) to efficiently distinguish different nucleotide addition loci. Step 2: Optical sequencing cycles – nick translation, gap filling, or strand displacement with labeled nucleotide. Sequenase v. 2.0 and FdNTP are added in standard buffers. (Sequenase v. 2.0 DNA polymerase has a genetically ablated 3′–5′ exonuclease activity and no strand displacement or 5′–3′exonuclease activity.) A 3′ end of a prepared gap is labeled only if the template strand contains the complementary base for a given FdNTP. Fluorescence additions are imaged and incorporated fluorochromes counted before photobleaching. The photobleaching is integral to preparing the template for the next cycle of DNA polymerase addition and, in effect, resets the fluorochrome count to zero so that the subsequent addition of another nucleotide can be counted. This also eliminates the need to label each different nucleotide with a different colored fluorochrome. The cycle is repeated for each labeled nucleotide until the desired region has been sequenced. Reprinted from Ramanathan *et al.* (2004), Copyright (2004), with permission from Elsevier.

REFERENCES

Albertson, D. G., Collins, C., McCormick, F. and Gray, J. W. (2003). Chromosome aberrations in solid tumors. Nat. Genet. 34(4), 369–376.

Anantharaman, T., Mishra, B. and Schwartz, D. (1999). Genomics via optical mapping. III: Contiging genomic DNA. Proc. Int. Conf. Intell. Syst. Mol. Biol. 18–27.

Anantharaman, T. S., Mishra, B. and Schwartz, D. C. (1997). Genomics via optical mapping. II: Ordered restriction maps. J. Comput. Biol. 4(2), 91–118.

Armbrust, E. V., Berges, J. A., Bowler, C., Green, B. R., Martinez, D., Putnam, N. H., Zhou, S., Allen, A. E., Apt, K. E., Bechner, M., Brzezinski, M. A., Chaal, B. K., Chiovitti, A., Davis, A. K., Demarest, M. S., Detter, J. C., Glavina, T., Goodstein, D., Hadi, M. Z., Hellsten, U., Hildebrand, M., Jenkins, B. D., Jurka, J., Kapitonov, V. V., Kroger, N., Lau, W. W. Y., Lane, T. W., Larimer, F. W., Lippmeier, J. C., Lucas, S., Medina, M., Montsant, A., Obornik, M., Parker, M. S., Palenik, B., Pazour, G. J., Richardson, P. M., Rynearson, T. A., Saito, M. A., Schwartz, D. C., Thamatrakoln, K., Valentin, K., Vardi, A., Wilkerson, F. P. and Rokhsar, D. S. (2004). The genome of the diatom *Thalassiosira pseudonana*: ecology, evolution, and metabolism. Science 306(5693), 79–86.

Cai, W., Aburatani, H., Stanton, V. P., Jr., Housman, D. E., Wang, Y. K. and Schwartz, D. C. (1995). Ordered restriction endonuclease maps of yeast artificial chromosomes created by optical mapping on surfaces. Proc. Natl. Acad. Sci. USA 92(11), 5164–5168.

Deegan, R. D., Bakajin, O., Dupont, T. F., Huber, G., Nagel, S. R. and Witten, T. A. (1997). Capillary flow as the cause of ring stains from dried liquid drops. Nature 389(6653), 827–829.

Dimalanta, E. T., Lim, A., Runnheim, R., Lamers, C., Churas, C., Forrest, D. K., de Pablo, J. J., Graham, M. D., Coppersmith, S. N., Goldstein, S. and Schwartz, D. C. (2004). A microfluidic system for large DNA molecule arrays. Anal. Chem. 76(18), 5293–5301.

Fiegler, H., Carr, P., Douglas, E. J., Burford, D. C., Hunt, S., Scott, C. E., Smith, J., Vetrie, D., Gorman, P., Tomlinson, I. P. and Carter, N. P. (2003). DNA microarrays for comparative genomic hybridization based on DOP-PCR amplification of BAC and PAC clones. Genes Chromosomes Cancer 36(4), 361–374.

Heiskanen, M., Hellsten, E., Kallioniemi, O. P., Makela, T. P., Alitalo, K., Peltonen, L. and Palotie, A. (1995). Visual mapping by fiber-fish. Genomics 30(1), 31–36.

Huang, X. and Miller, W. (1991). A time-efficient, linear space local similarity algorithm. Adv. Appl. Math. 12, 337–357.

Iafrate, A. J., Feuk, L., Rivera, M. N., Listewnik, M. L., Donahoe, P. K., Qi, Y., Scherer, S. W. and Lee, C. (2004). Detection of large-scale variation in the human genome. Nat. Genet. 36(9), 949–951.

Jing, J., Reed, J., Huang, J., Hu, X., Clarke, V., Edington, J., Housman, D., Anantharaman, T. S., Huff, E. J., Mishra, B., Porter, B., Shenker, A., Wolfson, E., Hiort, C., Kantor, R., Aston, C. and Schwartz, D. C. (1998). Automated high resolution optical mapping using arrayed, fluid-fixed DNA molecules. Proc. Natl. Acad. Sci. USA 95(14), 8046–8051.

Kallioniemi, A., Kallioniemi, O. P., Sudar, D., Rutovitz, D., Gray, J. W., Waldman, F. and Pinkel, D. (1992). Comparative genomic hybridization for molecular cytogenetic analysis of solid tumors. Science 258(5083), 818–821.

Lim, A., Dimalanta, E. T., Potamousis, K. D., Yen, G., Apodoca, J., Tao, C., Lin, J., Qi, R., Skiadas, J., Ramanathan, A., Perna, N. T., Plunkett, G., Burland, V., Mau, B., Hackett, J., Blattner, F. R., Anantharaman, T. S., Mishra, B. and Schwartz, D. C. (2001). Shotgun optical maps of the whole *Escherichia coli* O157:H7 genome. Genome Res. 11(9), 1584–1593.

Lin, J., Qi, R., Aston, C., Jing, J., Anantharaman, T. S., Mishra, B., White, O., Daly, M. J., Minton, K. W., Venter, J. C. and Schwartz, D. C. (1999). Whole-genome shotgun optical mapping of *Deinococcus radiodurans*. Science 285(5433), 1558–1562.

Lucito, R., Healy, J., Alexander, J., Reiner, A., Esposito, D., Chi, M., Rodgers, L., Brady, A., Sebat, J., Troge, J., West, J. A., Rostan, S., Nguyen, K. C. Q., Powers, S., Ye, K. Q., Olshen, A., Venkatraman, E., Norton, L. and Wigler, M. (2003). Representational oligonucleotide microarray analysis: a high-resolution method to detect genome copy number variation. Genome Res. 13(10), 2291–2305.

Meng, X., Benson, K., Chada, K., Huff, E. J. and Schwartz, D. C. (1995). Optical mapping of lambda bacteriophage clones using restriction endonucleases. Nat. Genet. 9(4), 432–438.

Morikawa, K. and Yanagida, M. (1981). Visualization of individual DNA molecules in solution by light microscopy: Dapi staining method. J. Biochem. Tokyo 89(2), 693–696.

Pruyne, J., Livny, M. (1996). Managing checkpoints for parallel programs. In: Proceedings of the Workshop on Job Scheduling Strategies for Parallel Processing (vol. 1162, pp. 140–154): Springer, Berlin.

Ramanathan, A., Huff, E. J., Lamers, C. C., Potamousis, K. D., Forrest, D. K. and Schwartz, D. C. (2004). An integrative approach for the optical sequencing of single DNA molecules. Anal. Biochem. 330(2), 227–241.

Ramanathan, A., Pape, L. and Schwartz, D. C. (2005). High-density polymerase-mediated incorporation of fluorochrome-labeled nucleotides. Anal. Biochem. 337(1), 1–11.

Reed, J., Singer, E., Kresbach, G. and Schwartz, D. C. (1998). A quantitative study of optical mapping surfaces by atomic force microscopy and restriction endonuclease digestion assays. Anal. Biochem. 259(1), 80–88.

Reslewic, S., Zhou, S., Place, M., Zhang, Y., Briska, A., Goldstein, S., Churas, C., Runnheim, R., Forrest, D., Lim, A., Lapidus, A., Han, C. S., Roberts, G. P. and Schwartz, D. C. (2005). Whole-genome shotgun optical mapping of *Rhodospirillum rubrum*. Appl. Environ. Microbiol. 71(9), 5511–5522.

Schrock, E., du Manoir, S., Veldman, T., Schoell, B., Wienberg, J., Ferguson-Smith, M. A., Ning, Y., Ledbetter, D. H., Bar-Am, I., Soenksen, D., Garini, Y. and Ried, T. (1996). Multicolor spectral karyotyping of human chromosomes. Science 273(5274), 494–497.

Schwartz, D. C. and Cantor, C. R. (1984). Separation of yeast chromosome-sized DNAs by pulsed-field gradient gel electrophoresis. Cell 37(1), 67–75.

Schwartz, D. C. and Koval, M. (1989). Conformational dynamics of individual DNA-molecules during gel-electrophoresis. Nature 338(6215), 520–522.

Schwartz, D. C., Li, X., Hernandez, L. I., Ramnarain, S. P., Huff, E. J. and Wang, Y. K. (1993). Ordered restriction maps of *Saccharomyces cerevisiae* chromosomes constructed by optical mapping. Science 262(5130), 110–114.

Sebat, J., Lakshmi, B., Troge, J., Alexander, J., Young, J., Lundin, P., Maner, S., Massa, H., Walker, M., Chi, M., Navin, N., Lucito, R., Healy, J., Hicks, J., Ye, K., Reiner, A., Gilliam, T. C., Trask, B., Patterson, N., Zetterberg, A. and Wigler, M. (2004). Large-scale copy number polymorphism in the human genome. Science 305(5683), 525–528.

Shaffer, L. G. (2005). American college of medical genetics guideline on the cytogenetic evaluation of the individual with developmental delay or mental retardation. Genet. Med. 7(9), 650–654.

Sharp, A. J., Locke, D. P., McGrath, S. D., Cheng, Z., Bailey, J. A., Vallente, R. U., Pertz, L. M., Clark, R. A., Schwartz, S., Segraves, R., Oseroff, V. V., Albertson, D. G., Pinkel, D. and Eichler, E. E. (2005). Segmental duplications and copy-number variation in the human genome. Am. J. Hum. Genet. 77(1), 78–88.

Skiadas, J., Aston, C., Samad, A., Anantharaman, T. S., Mishra, B. and Schwartz, D. C. (1999). Optical PCR: genomic analysis by long-range PCR and optical mapping. Mamm. Genome 10(10), 1005–1009.

Smirnov, D. A., Burdick, J. T., Morley, M. and Cheung, V. G. (2004). Method for manufacturing whole-genome microarrays by rolling circle amplification. Genes Chromosomes Cancer 40(1), 72–77.

Stankiewicz, P. and Lupski, J. R. (2002). Genome architecture, rearrangements and genomic disorders. Trends Genet. 18(2), 74–82.

Tenover, F. C., Arbeit, R. D., Goering, R. V., Mickelsen, P. A., Murray, B. E., Persing, D. H. and Swaminathan, B. (1995). Interpreting chromosomal DNA restriction patterns produced by pulsed-field gel electrophoresis: criteria for bacterial strain typing. J. Clin. Microbiol. 33(9), 2233–2239.

Trask, B. J., Massa, H., Kenwrick, S. and Gitschier, J. (1991). Mapping of human chromosome Xq28 by two-color fluorescence *in situ* hybridization of DNA sequences to interphase cell nuclei. Am. J. Hum. Genet. 48(1), 1–15.

Tuzun, E., Sharp, A. J., Bailey, J. A., Kaul, R., Morrison, V. A., Pertz, L. M., Haugen, E., Hayden, H., Albertson, D., Pinkel, D., Olson, M. V. and Eichler, E. E. (2005). Fine-scale structural variation of the human genome. Nat. Genet. 37(7), 727–732.

Valouev, A., Li, L., Liu, Y. C., Schwartz, D. C., Yang, Y., Zhang, Y. and Waterman, M. S. (2006a). Alignment of optical maps. J. Comput. Biol. 13(2), 442–462.

Valouev, A., Zhang, Y., Schwartz, D. C. and Waterman, M. S. (2006b). Refinement of optical map assemblies. Bioinformatics 22(10), 1217–1224.

Volik, S., Zhao, S., Chin, K., Brebner, J. H., Herndon, D. R., Tao, Q., Kowbel, D., Huang, G., Lapuk, A., Kuo, W. L., Magrane, G., De Jong, P., Gray, J. W. and Collins, C. (2003). End-sequence profiling: sequence-based analysis of aberrant genomes. Proc. Natl. Acad. Sci. USA 100(13), 7696–7701.

Waterman, M. S. and Vingron, M. (1994). Sequence comparison significance and poisson approximation. Statistical Sciences 9, 367–381.

Wei, J., Goldberg, M. B., Burland, V., Venkatesan, M. M., Deng, W., Fournier, G., Mayhew, G. F., Plunkett, G., III, Rose, D. J., Darling, A., Mau, B., Perna, N. T., Payne, S. M., Runyen-Janecky, L. J., Zhou, S., Schwartz, D. C. and Blattner, F. R. (2003). Complete genome sequence and comparative genomics of *Shigella flexneri* serotype 2a strain 2457t. Infect. Immun. 71(5), 2775–2786.

Zhou, S., Deng, W., Anantharaman, T. S., Lim, A., Dimalanta, E. T., Wang, J., Wu, T., Chunhong, T., Creighton, R., Kile, A., Kvikstad, E., Bechner, M., Yen, G., Garic-Stankovic, A., Severin, J., Forrest, D., Runnheim, R., Churas, C., Lamers, C., Perna, N. T., Burland, V., Blattner, F. R., Mishra, B. and Schwartz, D. C. (2002). A whole-genome shotgun optical map of *Yersinia pestis* strain kim. Appl. Environ. Microbiol. 68(12), 6321–6331.

Zhou, S., Kile, A., Bechner, M., Place, M., Kvikstad, E., Deng, W., Wei, J., Severin, J., Runnheim, R., Churas, C., Forrest, D., Dimalanta, E. T., Lamers, C., Burland, V., Blattner, F. R. and Schwartz, D. C. (2004a). Single-molecule approach to bacterial genomic comparisons via optical mapping. J. Bacteriol. 186(22), 7773–7782.

Zhou, S., Kile, A., Kvikstad, E., Bechner, M., Severin, J., Forrest, D., Runnheim, R., Churas, C., Anantharaman, T. S., Myler, P., Vogt, C., Ivens, A., Stuart, K. and Schwartz, D. C. (2004b). Shotgun optical mapping of the entire *Leishmania* major Friedlin genome. Mol. Biochem. Parasitol. 138(1), 97–106.

Zhou, S., Kvikstad, E., Kile, A., Severin, J., Forrest, D., Runnheim, R., Churas, C., Hickman, J. W., Mackenzie, C., Choudhary, M., Donohue, T., Kaplan, S. and Schwartz, D. C. (2003). Whole-genome shotgun optical mapping of *Rhodobacter sphaeroides* strain 2.4.1 and its use for whole-genome shotgun sequence assembly. Genome Res. 13, 2142–2151.

Zody, M. C., Garber, M., Adams, D. J., Sharpe, T., Harrow, J., Lupski, J. R., Nicholson, C., Searle, S. M., Wilming, L., Young, S. K., Abouelleil, A., Allen, N. R., Bi, W., Bloom, T., Borowsky, M. L., Bugalter, B. E., Butler, J., Chang, J. L., Chen, C. K., Cook, A., Corum, B., Cuomo, C. A., de Jong, P. J., DeCaprio, D., Dewar, K., FitzGerald, M., Gilbert, J., Gibson, R., Gnerre, S., Goldstein, S., Grafham, D. V., Grocock, R., Hafez, N., Hagopian, D. S., Hart, E., Norman, C. H., Humphray, S., Jaffe, D. B., Jones, M., Kamal, M., Khodiyar, V. K., LaButti, K., Laird, G., Lehoczky, J., Liu, X., Lokyitsang, T., Loveland, J., Lui, A., Macdonald, P., Major, J. E., Matthews, L., Mauceli, E., McCarroll, S. A., Mihalev, A. H., Mudge, J., Nguyen, C., Nicol, R., O'Leary, S. B., Osoegawa, K., Schwartz, D. C., Shaw-Smith, C., Stankiewicz, P., Steward, C., Swarbreck, D., Venkataraman, V., Whittaker, C. A., Yang, X., Zimmer, A. R., Bradley, A., Hubbard, T., Birren, B. W., Rogers, J., Lander, E. S. and Nusbaum, C. (2006). DNA sequence of human chromosome 17 and comparison with mouse chromosome 11. Nature 440(7087), 1045–1049.

Sequencing Validations and Analysis

Chapter 10

Sequencing Aided by Mutagenesis Facilitates the *De Novo* Sequencing of Megabase DNA Fragments by Short Read Lengths

Jonathan M. Keith,[1] David B. Hawkes,[2] Jacinta C. Carter,[3] Duncan A. E. Cochran,[4] Peter Adams,[5] Darryn E. Bryant[5] and Keith R. Mitchelson[6,7]

[1]*Institute of Molecular Bioscience and Department of Mathematics, University of Queensland, St. Lucia, Queensland 4072, Australia*

[2]*AGRF, Institute of Molecular Bioscience, University of Queensland, St. Lucia, Queensland 4072, Australia*

[3]*Leukaemia Foundation Queensland Laboratories, Queensland Institute of Medical Research, Herston, Queensland 4006, Australia*

[4]*Agen Biomedical Limited, Durbell Street, Acacia Ridge, Queensland 4110, Australia*

[5]*Department of Mathematics, University of Queensland, St. Lucia, Queensland 4072, Australia*

[6]*Capitalbio Corporation: National Engineering Research Centre for Beijing Biochip Technology, 18 Life Science Parkway, Changping District, Beijing 102206, China*

[7]*Medical Systems Biology Research Center, Tsinghua University School of Medicine, Beijing 100084, China*

Contents

PERSPECTIVES IN BIOANALYSIS, VOLUME 2
ISSN 1871-0069 DOI: 10.1016/S1871-0069(06)02010-6

Abstract

"Sequencing by synthesis" (SBS) is a rapidly emerging high-throughput low-cost sequencing technology, and is one of the front-runners in the race to the $10,000 genome. SBS reads are currently short, approximately ranging from 30 to 100–150 bp long on different technology platforms. Shotgun sequence assembly of such short reads is complicated by the presence of repeats, and this presents a major obstacle to use SBS for *de novo* sequencing of large genomes and of genomes having significant amounts of repetitive sequence. Here we propose using a radical technique called "sequencing aided by mutagenesis" (SAM) to solve many aspects of this problem. The technique involves deliberately introducing mutations into the target DNA to reduce its repetitiveness, assembling sequences of several such mutants and then inferring the target sequence from the assembled mutant sequences. We present the results of simulation SAM reassemblies of different human genomic fragments and motifs up to 600 kb long, as well as of the entire *Mycoplasma genitalium* genome. Each of these fragments could be successfully reconstructed from short reads of lengths 25, 100 or 150 bp using Phrap, demonstrating the potential of SAM as an enabling technology for short-read sequencing output.

1. INTRODUCTION

The recent delineation of the human genome and its implication for genome-wide analysis for personalized medicine is a driver of the development of sequencing technologies capable of massively increased read throughput, compared to current conventional capillary electrophoresis-based sequencers (Kling, 2003; Shendure *et al.*, 2004). Despite the efficiency of capillary-based DNA sequencers, which have a capacity to read some 600,000 bases in 24 h, several different cyclic sequencing technologies such as "sequencing by synthesis" (SBS, also called "sequencing by extension", SBE) (Leamon *et al.*, 2003; Mitra *et al.*, 2003) and "sequencing by ligation" (Shendure *et al.*, 2005) have been developed around nanoscale reactions on two-dimensional array surfaces to provide substantial depth of sequence coverage within a small reaction space, for some tens of megabases of DNA (Shendure *et al.*, 2004; Bennett *et al.*, 2005). SBS platforms employ sequencing chemistries such as single nucleotide primer extension (SnuPE) and pyrosequencing. In SnuPE, unitary base reactions add one of the four differentially labeled, reversibly terminating single nucleotides to the growing chains (Shendure *et al.*, 2004; Mitra *et al.*, 2003). These are monitored simultaneously at each localized and anchored oligonucleotide feature, and currently can read only very short runs of some 5 bp (Kartalov and Quake, 2004) of 17–34 bp with polony-FISSEQ (Mitra *et al.*, 2003; Shendure *et al.*, 2005), and up to 50 bp claimed by Solexa (www.solexa.com). These micro-fluidic array platforms are the equivalent of a million-lane sequencer that reads the sequence of each molecule at the speed of the addition reaction cycle. These short reads, in combination with the enormous parallelism of micro-arrays with millions of features and ultra-high resolution optical systems simultaneously capture signals from some 10^{6-8} different but simultaneous base-extension pyrosequencing events. These technologies may eventually provide for parallel detection of variation on $>10^8$ short unique sequences from the human genome by alignment and comparison to the known human reference sequence (Hebert and Braslavsky, 2007), i.e. massively parallel re-sequencing and genotyping applications.

1.1. Single molecule sequencing

Solexa Pty Ltd and Helicos Biosciences are both developing massively parallel array sequencing technologies in which individual alleles of DNA molecules are sequenced sequentially, without cloning. Both companies make use of novel sequencing chemistries and novel detection apparatus and are potentially able to sequence many millions of individual DNA molecules a single base at a time, simultaneously on the arrayed template molecules (Bennett *et al.*, 2005). This is expected to allow analysis of an entire human genome in a single experiment and enable the typing of known, as well as currently unknown DNA polymorphisms and provide information about patterns of linkage disequilibrium in the populations.

Helicos BioSciences has developed a parallel array technology called "true single molecule sequencing" (tSMS) technology that relies on cyclic SBS, by directly interrogating single molecules, acquiring signal information from individual labelled extension nucleotides by ultrasensitive detection at the surface plane of attachment of the molecule to the glass slide. The Helicos company has recently announced the successful sequencing of the M13 genome, including its small homopolymeric sequences. Short reads are initially anticipated from their device, although it may potentially be able to read long lengths of DNA with further development of reaction chemistries. Initial claims suggest the major role for these different single molecule sequencing technologies will be for genome re-sequencing (Hebert and Braslavsky, 2007), for fragment DNA sequencing (Poinar *et al.*, 2006) and for sequencing of short nucleic acids, e.g. micro-RNA transcripts (Henderson *et al.*, 2006).

1.2. PicoTiterPlate (Pyro)Sequencer 20

Several different "SBS" platform technologies are being developed around nanoscale reactions on solid surfaces to provide the multiplicity (depth of coverage) within a small reaction space for large DNA region cover. The technology company 454 Life Sciences (Leamon *et al.*, 2003; Margulies *et al.*, 2005, 2007) have developed a solid phase array of amplified unique genomic fragments (Ohuchi *et al.*, 1998) in picolitre volume reaction wells in a micro-fluidic analysis plate called a PicoTitre plate.

Pyrosequencing (Hyman, 1988; Ronaghi, 2001; Ehn *et al.*, 2004) is used in which native dNTPs are added to the templated-beads during successive cycles of DNA synthesis (also called SBE) typically result in single base addition events that can be monitored. Each plate view of extension signals developed on the (Pyro)Sequencer GS20 is captured by associated CCD imaging equipment. Short reads of total length 100–150 bp for each different genomic DNA fragment are seen after some 80–100 rounds of extension. However stretches of homopolymer within templates result in multiple (single) base-type extensions, which are difficult to estimate accurately. Despite these limitations, the genomes of bacteria and protozoa up to 10 Mb in length have been sequenced using the 454 Rev 1.0 whole genome sequencing system with working reads of a minimum 100 nucleotides length. The system can typically generate some 25–35 Mb of high-quality bases per run with 20 Mb per run as a minimum output.

Small genomes are sequenced to a consensus accuracy of >99.99% employing some 10–15 times over-coverage to generate this level of accuracy. Notably, the genomes that can be sequenced and assembled *de novo* without reference to a previously known genomic sequence are principally composed of unique sequences, and possess relatively few repetitious regions. This was demonstrated recently with the determination of the *de novo* sequence the 0.58 Mb genome of the bacteria *Mycoplasma genitalium* (Margulies *et al.*, 2005). The utility of the platform was also recently demonstrated with the paleogenomic *re-sequencing* of some 13 Mb of short random genomic fragments from the preserved tissues of a Siberian mammoth (Poinar *et al.*, 2006), which were then identified by comparison to elephant genomic sequence.

1.3. Non-repeat DNA sequencing

Although the sequences obtained by massively parallel SBS arrays of the genomes of bacteria *E. coli* (Shendure *et al.*, 2005) and *M. genitalium* (Margulies *et al.*, 2005) are extremely impressive, these technologies have several important limitations. Despite some 10 times genome coverage (in raw basepairs) of the 10 Mb essentially unique sequence genome of *E. coli* by Shendure *et al.* (2005), the random nature of library capture and *in vitro* amplification methods such as RCA and emulsion PCR result in only 91% of the genome having at least one time coverage. Other methods to ensure representation of "recalcitrant" genomic regions that are under-amplified by these enzyme catalysed processes must be introduced, as even a depth of sequencing cover greater than 20 times typically fails to identify the absent sequence (gaps). In addition, Margulies *et al.* (2005) noted the use of short read pyrosequencing data for *de novo* sequence reassembly of *M. genitalium* into continuous long sequence would be confounded by repeated motifs, and would necessitate hand-finishing the assembly of the numerous contigs that are generated. For both platforms, the fragmentation of the template and the short reads themselves contribute to a loss of sequence context and demand very high levels of genome coverage to provide statistical support for accurate assembly.

1.4. Limitations to the assembly of short-read data

When sequencing the complex genomes of eukaryotes, even the longer 100–150 bp reads cannot be easily reconstructed into larger contigs, unless known contiguous regions are deliberately sequenced. None of the SBS technologies described above can be used for sequencing long regions of simple sequence DNA, or homopolymer regions such as poly A, and other longer repeat regions, as they cannot intrinsically determine the length of any repeat region that is longer than the average read length. Pyrosequencing is also particularly sensitive to homopolymer elements as the quantitation of nucleotide number is not linear for elements longer than 6–7 nucleotides (Ehn *et al.*, 2004; Margulies *et al.*, 2005). For SBS technologies such as polony FISSEQ sequencing (Mitra *et al.*, 2003), the DNA polymerases employed for single molecule amplification and for sequence extension can also be potentially halted by "unamplifiable" and "unsequenceable" regions, leaving unsequenced gaps or regions of lower coverage (Shendure *et al.*, 2005). However, the nature of single nucleotide extension chemistry militates against inhibition of extension by the polymerase. Notably, pyrosequencing technology (Leamon *et al.*, 2003), FISSEQ (Mitra *et al.*, 2003) and Solexa's "Cluster DNA Amplification" (Mitchelson *et al.*, 2007) each employ a single molecule PCR-amplification process to multiply template molecules. This step could also be a contributor to the reduced representation of unamplifiable regions. The large gigabase-sized genomes of many eukaryotes, which may contain 30–60% repeated DNA, are examples of genomes that cannot be readily sequenced *de novo* using these present SBS methods.

2. PRINCIPLES OF SAM SEQUENCING

We have developed an independent technology that could overcome many of these limitations. The technique is called "sequencing aided by mutagenesis" (SAM) and involves extracting target DNA sequence information from randomly mutated variants of the target using advanced reconstruction algorithms (Keith *et al.*, 2003, 2004a, 2004b). The introduction of mutations does not destroy original sequence information, but distributes it among multiple variants. These variants lack many of the problematic features of the initial target, and are more amenable to sequencing. Because the approach involves changing the target sequence, it can address difficulties arising from any problematic sequence characteristic and relieve confusion due to repeated motifs. SAM techniques can also simplify the assembly of repeated regions by introducing "landmarks" that distinguish between different repeats. The

A

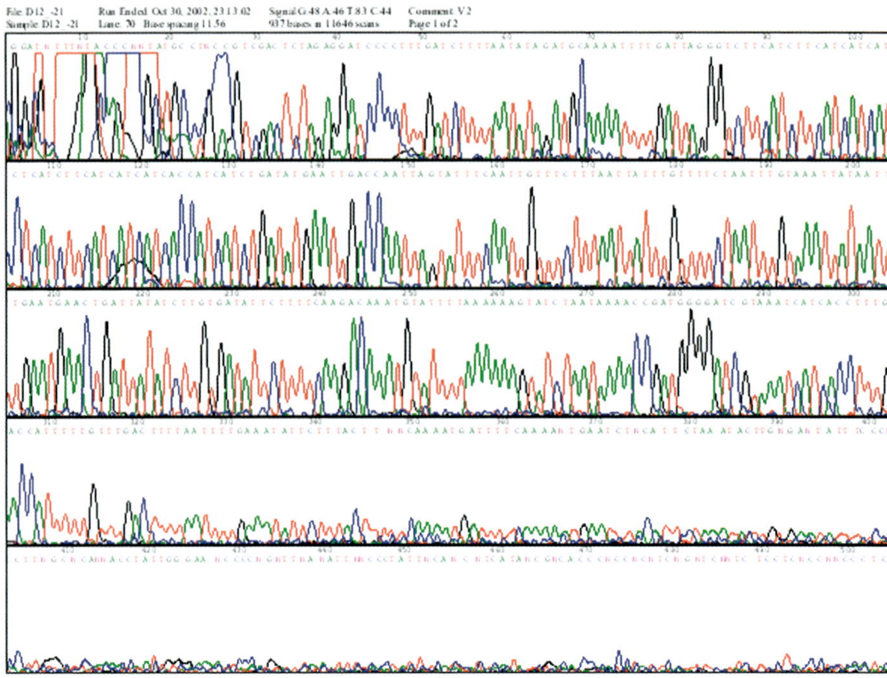

B

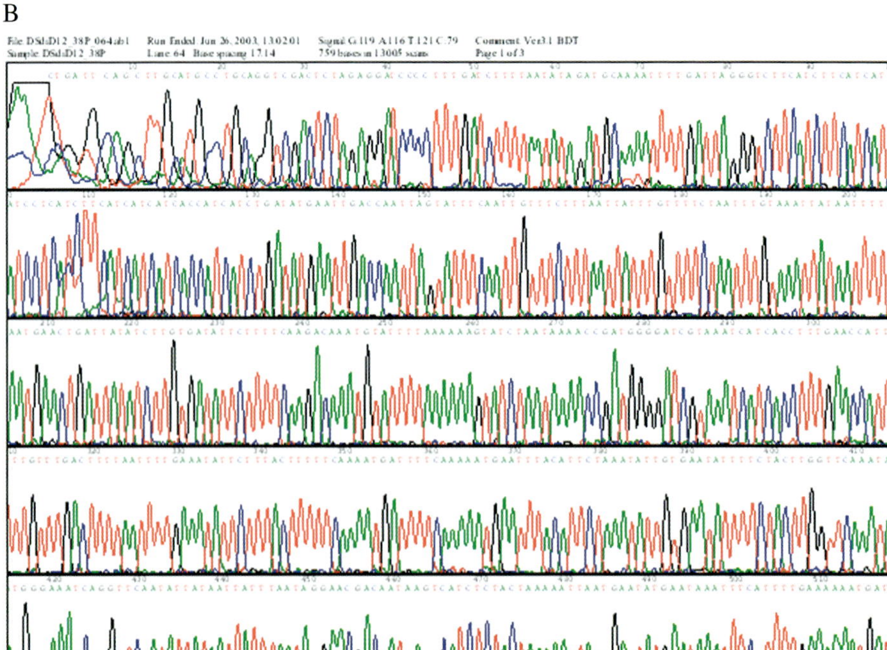

Fig. 1. Effect of dPTP mutagenesis on polymerase progression through a previously unsequenceable, repetitive AT-rich fragment of *D. discoideum* genomic DNA. (A) Wild type sequence read using ABI Big Dye terminator v 3.0. (B) Sequence of same fragment following direct mutagenesis with dPTP, and sequencing using the same kit.

introduction of landmark mutations also benefits sequence reconstruction from multiple, short overlapping reads into longer regions, necessary for *de novo* sequencing.

2.1. Mutation by nucleotide analogues

We have previously described the use of mutagenic nucleotide analogues for sequencing small intractable DNA regions (Keith *et al.*, 2004b) (see Figure 1). The nucleotide analogues are not completely random mutagens, rather preferred nucleotide transition reactions are induced (with transversions, insertions and deletions occurring very rarely) in an almost random distribution (Yu *et al.*, 1993; Zaccolo *et al.*, 1996). The different random mutations cause the mutant copies to have different sequences, which are then cloned and sequenced. Mutations introduced at very early rounds of amplification can establish "founder mutations" that occur in a significant proportion of the progeny amplimers, although these founder mutations are themselves at random loci (Figure 2). These characteristics allow the influence of founder mutations to be minimized through simple experimental design devices. The DNA sequences determined from a low number of the altered copies can then analysed using Bayesian methods to reconstruct the original wild-type sequence (Keith *et al.*, 2003). Surprisingly, this entire process has efficiencies and accuracies roughly equivalent to conventional sequencing.

2.2. Integration of SAM sequencing with SBS sequencing

We propose that these SAM techniques could be advantageously integrated with highly parallel SBS array sequencing, as it can readily generate the sequence data for SAM algorithms to reassemble large contiguous regions, even from genomic clones containing repeated motifs, homopolymer regions or recalcitrant elements. The purpose of this paper is to demonstrate the advantages of SAM sequencing methods to alleviate the limitations of SBS technology and provide a basis for improved assembly of short-read data into long contiguous sequence. We have simulated SAM sequencing reactions as would be obtained from an SBS array platform, in which reads of length 100 or 150 bp were obtained from several independently mutated copies of fragments of genomic DNA up to 600 kb long, drawn from human and other genomes. We calculated the proportion of error from our reconstructions of original (wild-type) sequence. Our analyses indicate that SAM approaches would enable the *de novo* sequencing of megabase DNA fragments from short-read sequencing procedures, including regions of repetitive sequence and base-biased sequence.

3. SIMULATED SAM SEQUENCING

SAM shotgun sequencing and reassembly was simulated using various sequences, sequence lengths, read lengths, numbers of mutants and coverage per mutant.

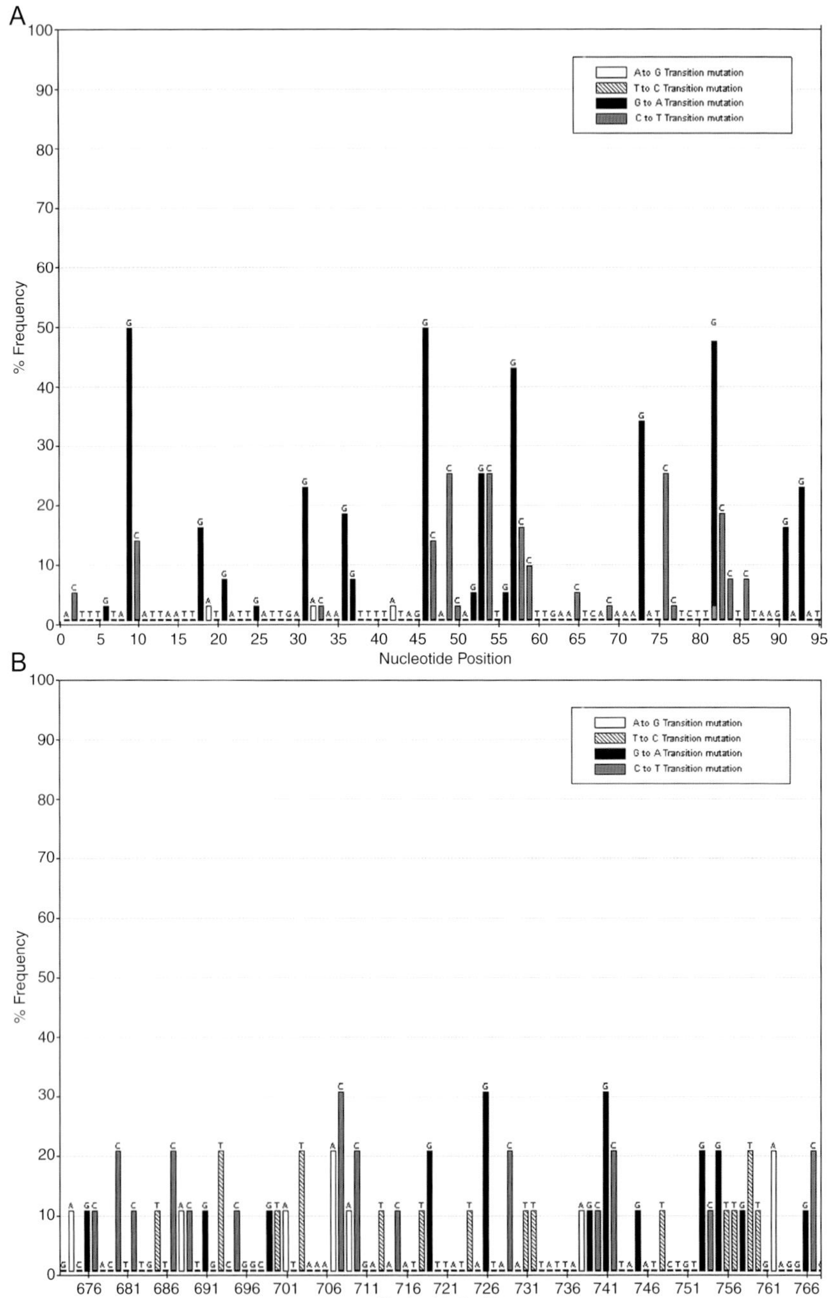

Fig. 2. Observed mutation frequencies from different combinations of mutagenic analogues. (A) Mutation frequencies observed for 45 sequences exposed to 400 μM BrdUTP. (B) Mutation frequencies observed for four sequences exposed to 200 μM dPTP and 200 μM BrdUTP during amplification. BrdUTP introduces G to A and C to T mutations that reduce the overall GC content. The addition of dPTP counteracts the overall effect by inducing A to G and T to C mutations, which increase GC content. The effect is a more random and balanced distribution of mutation, keeping the overall mutation rate at around 5.5%.

Table 1. Test sequences used for modeling

Source and notation	Fragment location	Indicative sequences	Largest number of reassembled contigs[a]
Human Chr 17 HUM-BRCA1	38,000,000–38,600,000 = 600,000 bp	BRCA1 gene 38,450,000–38,550,000	2
Human Chr 6 HUM-HLA	33,000,000–33,600,000 = 600,000 bp	HLA gene HKE2 33,365,356–33,366,689	1
Human Chr 6 HUM-subcentro	58,250,000–58,850,000 = 600,000 bp	Sub-centromeric repeats	1
Mosquito *Anopheles gambiae* Chr 3R	50,000,000–50,500,000 = 500,000 bp	AT-rich region	3
Bacteria *Mycoplasma genitalium*	0.58 Mb genome	4–10 bp homopolymers	1

[a]Maximum number of reassembled contigs refers to sequence assemblies using 10 mutants at 10-fold coverage and reads of length 150 bp.

3.1. Representative sequence motifs

Different human and other genomic DNA sequences, each of length up to ~600 kb were used. The chromosomal fragments that were analysed are indicated in Table 1.

The study sequences were chosen to represent different genomic composition characters, including human genomic regions (IHGSC, 2004) with numerous unique gene sequences interspersed with discrete regions of low complexity repeats (HUM-BRCA1, -HLA), a gene poor human genomic region with numerous mixed low complexity repeats (HUM-subcentro), and a mosquito genomic region of strong base bias (AT-rich) containing numerous short and mixed homopolymer tracts (MOS1). For comparison, the entire 0.58 Mb bacterial genome of *M. genitalium* was also analysed. Some of the confounding sequence motifs identified within these test elements are shown in the third column of Table 1, while the highest number of contigs obtained for reassembly of the full 0.6 Mb fragments using 10 mutants and 10 times coverage are shown in the last column. Importantly, these analyses suggest that SAM sequencing methods and SAM assembly allows these complex genomic elements to be assembled into either single contigs or a small number of contigs, with low sequence error from just a few independent mutants.

These motifs are illustrative of problematic sequences known or expected to prevent sequence assembly from short-read data, and include homopolymers, regions of simple repeats and strongly base-biased elements, with multiple short homopolymer regions and other regions of sequence similarity. The motifs are not exhaustive and are meant to represent some of the diverse sequences that would pose a significant challenge to conventional short-read sequencing technologies (Margulies *et al.*, 2005; Shendure *et al.*, 2005).

3.2. Initial data extraction

For each of the four sequences, six prefixes (i.e. initial contiguous subsequences) were extracted, with lengths ~100, ~200, ~300, ~400 kb, ~500 and ~600 kb.

All mutants were obtained by substituting ~10% of bases with a randomly chosen character, with no insertions and deletions. Simulation reconstructions were performed for 8, 10 or 12 mutants, with coverage per mutant of 6-fold or 10-fold, and read lengths of 150 bp. Simulated sequencing and assembly was also performed using read lengths of 100 bp, for the human HUM-BRCA1 sequences and its six prefixes, with 10 mutants and 8-fold or 10-fold coverage per mutant. Note that 150 bp is near to the upper limit of sequence read lengths now expected from SBS-array devices such as the PicoTiterPlate array, whereas 100 bp is the working length (Margulies *et al.*, 2005, 2007).

3.3. SAM assembly of simulated data

The simulations involved two stages of assembly: the first stage involving assemblies of individual mutants, and the second stage involving assemblies of the pooled contigs from the first stage. Both stages of assembly were performed using Phrap (http://www.phrap.org/; Gordon *et al.*, 1998) with suitably chosen parameters. High gap penalties were used in all assemblies. A majority rule consensus sequence was then constructed from the assembly. An "N" character was used to represent lack of consensus. We have previously described Bayesian methods that use the proportion of observed analogue-induced mutations to weight the predicted reassembled sequence (Keith *et al.*, 2004a). Although we could have used these advanced methods to reflect realistic mutation patterns of mutagenic nucleotide analogues, these measures do not seem warranted for this simulation.

4. ANALYSIS OF SAM SEQUENCING TARGET ASSEMBLIES

Results of these simulations of SAM assembly are shown in the figures below for different types of sequence motif, and for different length reads.

4.1. Assembly of contigs using 150 bp long reads

Each data point represents an average of the number of errors (i.e. miscalled bases or bases where no single character predominated) over 1–7 repeat simulations (the number of simulations is not indicated). A few simulations failed to produce results or produced clearly incorrect assemblies. These were not included in the data (with the result that some points are missing from the data set). These failed reconstructions were due to errors in stage I assembly of some mutants, which then resulted in gaps at stage II. Some assemblies of particular fragments produced two or three contigs (e.g. BRCA1 and MOS1), and for these the reported error is the lowest error for any contig. The proportion of error was calculated as described previously (Keith *et al.*, 2004a).

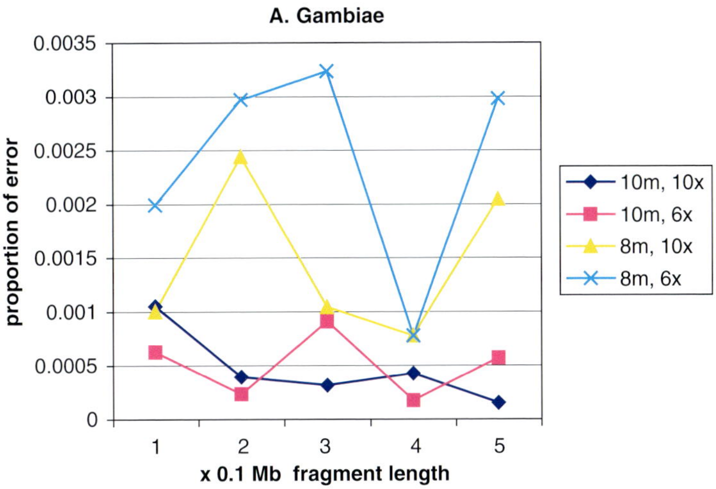

Fig. 3. The proportion of errors in the simulated reassembly of a region of *A. gambiae* Chr 3R. All mutation is at 10% and individual mutant coverage of $c = 6$-fold or 10-fold. The graph indicates the level of accuracy for all mutants of lengths ranging from 0.1 to 0.5 Mb. No knowledge of the original sequence is used during the reconstruction. The graph indicates the mean level of accuracy of the reassemblies.

4.2. AT-rich insect genomic region

Figure 3 shows the proportion of reassembly errors from SAM reconstruction of a contiguous region from chromosome 3R of the mosquito *Anopheles gambiae* (Holt *et al.*, 2002). The mosquito genome is base biased and AT rich, generally consisting of unique sequence interspersed with numerous, short 4–6 bp homopolymer tracts. Here, the original sequence could be readily reconstructed with high accuracy for fragments up to 0.5 Mb long, using 10 mutants using the SAM reassembly methodology.

4.3. Human HLA region

Figure 4A shows results from SAM reassembly of the human HLA region from a number of fragments carrying 10% random mutation and with the sequenced using 150 bp reads. We present these data obtained from the reassemblies with low levels of mutants and low sequence coverage to illustrate the minimum requirements for accurate *de novo* SAM sequencing. The overall conclusion of these experiments is that despite the introduction of 10% mutation, *de novo* sequencing of 0.5 and 0.6 Mb can be easily achieved with relatively few mutants and reasonable levels of sequence coverage, even employing short sequence reads of only 150 bp. This level of coverage is the same as reportedly employed by 454 Life Sciences for the re-assembly of non-mutated wild-type reads from the less complex 0.58 Mb genome of *Mycoplasma genetalium* (Margulies *et al.*, 2005). Note that the error in the SAM assembly was relatively independent of

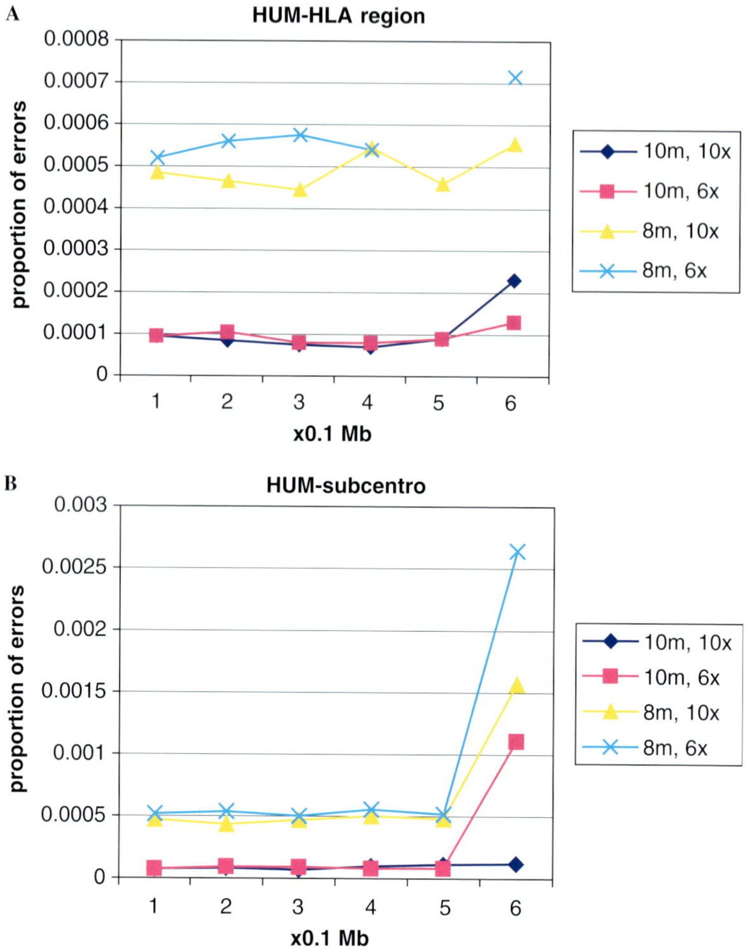

Fig. 4. (A) SAM reassembly of human chromosome 6 region from 33,000,000 to 33,600,000 inclusive. (B) SAM reassembly of 0.6 Mb fragment of Human Chr 6 from 58,250,000 to 58,850,000 inclusive, containing a subcentromeric region. These graphs show the proportion of errors in the simulated reconstructed human HLA region using various numbers of mutants for short reads of 150 bp. All mutation is at 10% and individual mutant coverage of $c = 6$-fold or 10-fold. Again no knowledge of the original sequence is used during the reconstruction. The graph indicates the level of reassembly accuracy. For example, with eight mutants of length 0.5 Mb, the average error was 0.0005, while with the addition of two more mutants (total 10) an error of ~1/10,000 was obtained.

the length of the DNA fragment between 0.1 and 0.6 Mb, and was dependent more on the number of different mutants used for the assembly.

4.4. Human sub-centromeric repeats and BRCA1 regions

Figure 4B shows a similar SAM reconstruction analysis of repeated sequence regions close to the centromere of human chromosome 6 (Guy *et al.*, 2003).

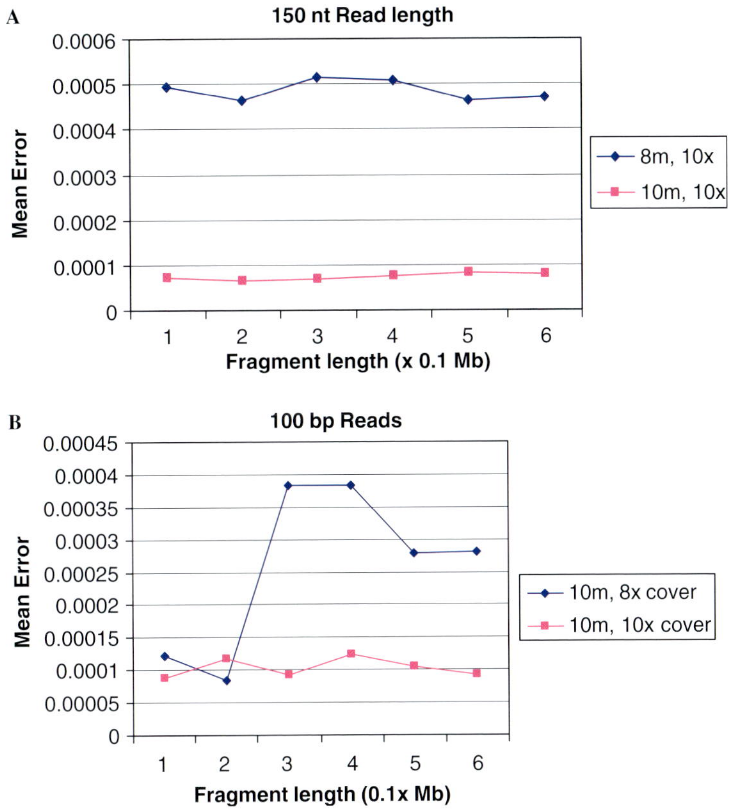

Fig. 5. SAM reassembly of human chromosome 17 region from 38,000,000 to 38,600,000, inclusive. The proportion of errors in the simulated reconstructed human BRCA1 sequence region using various numbers of mutants for short reads of (A) 150 bp and (B) 100 bp. All mutation is at 10% and individual mutant coverage of $c = 8$-fold (B) or 10-fold (A, B) as described previously. The number of mutants used for each analysis curve was either 8 mutants (A) or 10 mutants (A, B).

Here, sequence could be readily reconstructed with high accuracy for fragments up to 0.5 Mb long. For lengths of 0.6 Mb, the mean error increased slightly for reconstructions employing fewer than 10 mutants at 10-fold coverage. Results of SAM sequencing of sequence regions surrounding the BRCA1 gene of chromosome 17 are shown in Figure 5. Here again sequence could be readily reconstructed with high accuracy for fragments up to 0.5 Mb long, and again with the error increasing slightly for lengths of 0.6 Mb for reconstructions employing fewer than 10 mutants at 10-fold coverage.

4.5. Assembly with 100 bp long reads

Remarkably, the use of reads of only 100 bp had a relatively small effect on the mean error of reassembly (Figure 5B), compared to the error when reads of

150 bp were used (Figure 5A). For sets of reassemblies with reads of 150 bp there was little change in the error for fragments between 100 and 600 kb in length. Each data point is an average over 27 repeated simulations. However, for reads of 100 bp, although error remained constant for 10 mutants and 10-fold coverage over the entire 600 kb, there was a trebling of the error for lengths longer than 300 kb when 10 mutants at 8-fold coverage were analysed. The sudden increase in the proportion of errors between 200 and 300 kb fragments is probably due to the presence of reconstruction ambiguities in this part of the fragment. That this problem is not seen for reads of length 150 bp may indicate that the ambiguities are resolved for longer reads. Thus, it also appears that increasing the level of fragment coverage can improve the reconstruction of individual mutants at stage I, whereas analysis of additional mutants may reduce the proportion of error. The number of mutants necessary for error-free reconstruction of these three human genomic regions using 150 bp reads is shown in Figure 6. These data suggest that high levels of coverage ($c = 10$) and use of either 12 or 15 mutants allow reconstruction with proportions of errors less than 1/10,000. High-density PicoTiterPlate sequencing arrays (Margulies *et al.*, 2005, 2007) and polony arrays (Shendure *et al.*, 2005) can readily achieve these levels of sequencing coverage within a single experiment.

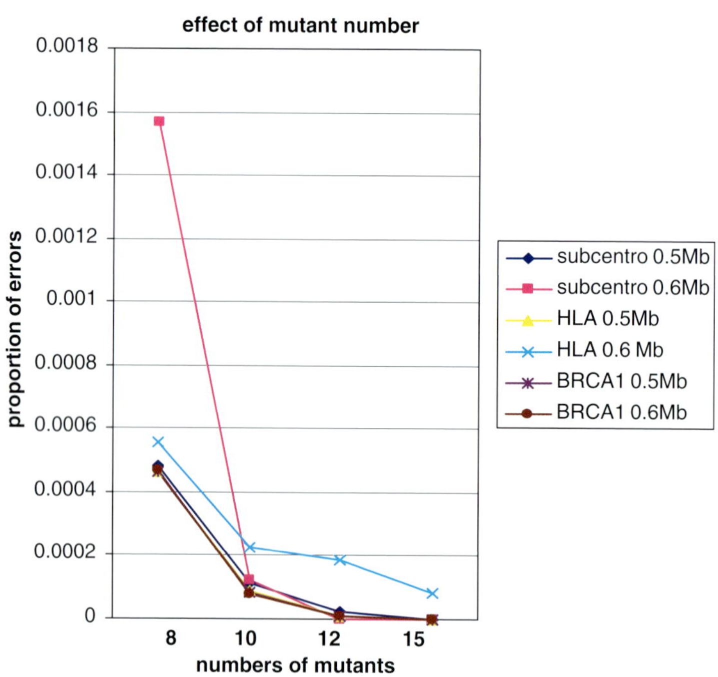

Fig. 6. The proportion of errors in the three simulated reassembled human gene regions versus the number of mutant copies analysed. All mutation is at 10% and individual mutant coverage of $c = 10$-fold. The graph indicates the level of accuracy for mutants of lengths 0.5 and 0.6 Mb. The poorer reconstruction of the 0.6 Mb HLA region fragment suggests an effect of sequence composition during stage I assembly.

Table 2. SAM reassembly of the human chromosome 17 region with 25 bp reads. The target fragments were assembled partially in several simulations. The proportion of errors for the assemblies is low.

Target length	100,000	200,000	400,000
Actual length	99,998	199,988	315,289
Mean error	0.0000533	0.000107	0.0001934

4.6. Modeling with 25 bp long reads

Table 2 shows data from simulations of the same BRCA1 genomic region, but using 25 bp reads only. Ten mutant sequences were simulated. Each base had a probability 0.1 of mutating and being replaced by one of the other three possible bases. All 25-mers were determined for all sequences – for the original sequence and for each of the 10 mutants. The 25-mers from different mutants were initially assembled separately from each other and from the original sequence.

Mini-contigs, assembled in stage I, were then assembled together in a second stage using phrap. Each mini-contig gets an equal "vote" towards this consensus, regardless of whether it came from a mutant sequence, or from the original. This process successfully reconstructed the 100,000, 200,000 and elements of 400,000 length fragments with an error of 1–3 bases per 10,000. For example, 25-mers from the 400,000 length fragment were assembled into a single contig of length 399,959, which differed from the original in only 86 bases, an error proportion of approximately 0.0002. It is likely that even longer fragments could be reconstructed using more than 10 mutants. The ability to assemble with very short reads of 25 bp, is remarkable, considering that the target is known to contain numerous mono- and di-nucleotide repeats of various lengths and is difficult to assemble using conventional assembly tools (without SAM), even from full-length reads (~500 bp).

4.7. Simulated SAM sequencing of the *M. genitalium* genome

Margulies *et al.* (2005, 2007) recently reported the comparative resequencing of the 0.58 Mb genome of *M. genitalium* (and other genomes) using pyrophosphate-based extension sequencing in picolitre-sized wells. Employing a 40-fold oversampling they achieved a re-sequencing consensus of 99.9% global genome coverage, resulting in 10 contigs and a consensus accuracy of 99.97% (mean error 0.0003). For *de novo* assembly, without reference to a known sequence they achieved an overall coverage of 96.54% and an accuracy of 99.96% (mean error 0.0004) and achieved an assembly of some 25 contigs ranging in size from 1.2 to 94.5 kb. The gaps between the contigs range in size from 10 to 2,399 bp. In contrast, modeling using SAM methods found that genome reassembly into a single 0.58 Mb contig could be achieved using 10 mutants with 10-fold genome coverage with an average error of 0.0001 (Figure 7). The assembly of a single contig could also be achieved using a lower number of

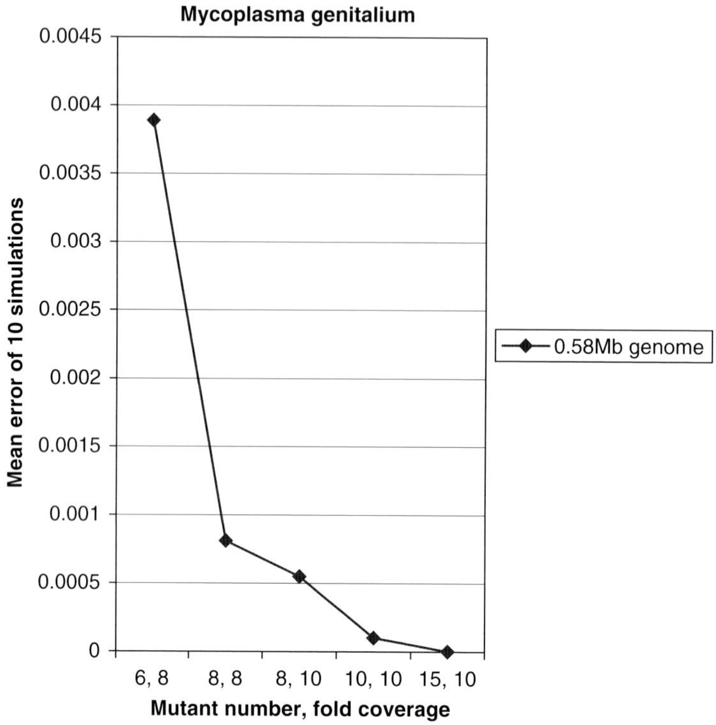

Fig. 7. SAM reassembly of the entire genome of the bacterium *M. genitalium*. The proportion of errors observed in the simulated reconstruction using various numbers of mutants with short sequence reads of 100 bp. All mutation is at 10% and modeling of sequence assembly involved using either 6, 8, 10 or 15 mutants, where the individual mutant sequence coverage was $c = 8$-fold or 10-fold.

mutants and lower level coverage, but each with greater mean error in individual base calls.

Table 3 compares *de novo* pyrosequencing of *M. genitalium* to an $8 \times 8 = 64$ coverage SAM simulation. Although the base calling error rate is higher for SAM in this simulation, a single contig genome was readily reconstructed. These data illustrate that *de novo* SAM sequencing may permit assembly of contiguous sequence, whereas pyrosequencing alone to 40 times coverage fails to generate contiguous sequence, even from this relatively simple bacterial genome. If we increased the number of mutants used in the analysis to 15, the proportion of error decreased almost two orders of magnitude to 0.000002 (Figure 7). The errors seen in our analysis were due to the chance mutation of the same nucleotide at the same position in a significant proportion of mutants, and hence the mutation frequency was seen to decrease as mutant numbers were increased. This type of error would occur during actual SAM sequencing, although it could be reduced by using methods known to minimize foundation mutation effects (e.g. using mutants from several independent SAM mutation experiments, or use of more than one type of mutagen where each introduces different mutations), and further by using Bayesian assembly methods (Keith *et al.*, 2003, 2004a).

Table 3. Comparison of the sequencing of the *M. genitalium* genome by GS20 pyrosequencing (Margulies *et al.*, 2005) to using SAM with array pyrosequencing to an 8/10000 error level

M. genitalium sequence	Cover	Contig number	Contig size	Gaps	Mean error
GS20 pyrosequencing	40 ×	25	1.2 –94.5 kb	10–2399 bp	0.0004
SAM + Array pyrosequencing	64 ×	1	0.84 Mb	None	0.0008

Pyrosequencing, and some other short-read techniques employing native nucleotides or unblocked nucleotide analogues are subject to limitations on the read quality of homopolymer regions. Figure 8 illustrates the ability of SAM sequencing to overcome limitations in homopolymer sequencing. Here the SAM mutations are seen to overcome another problem that occurs in PCR sequencing, with mis-alignment and looping out of part of the homopolymer region during strand reannealing, which results in loss of downstream sequence register. The random mutations, we believe, help to keep the tract in register, and give a more discrete determination of homopolymer length as well as allowing continued quality sequence downstream of the tract.

5. DISCUSSION

5.1. Assembly of human genomic DNA using SAM methodologies

These simulated assemblies demonstrate the potential of SAM sequencing to overcome impediments to conventional sequence assembly, and provide some estimates of the efficiency of the SAM protocols for parameters typical of SBS-array sequencing technologies. We have undertaken simulation studies on genomic sequences containing regions of sequence repetition and complexity that frequently confounds conventional assembly algorithms, examining contiguous DNA fragments of 0.5 Mb and longer. These experiments indicate the potential of SAM methods to facilitate *de novo* sequencing and reconstruction of large DNA fragments. Our calculations demonstrate that use of SAM sequencing approaches is very effective in permitting reassembly of large contiguous sequences of 0.5 Mb (or larger), with respect to both number of mutants sequenced, the depth of coverage of each mutant, and the low average proportion of error across the sequence.

5.2. Accuracy of the assemblies are relatively independent of target length

If we ignore the results for the longest sequences assembled (~600 kb) then the error rate of the assembly appears to depend only on the number of mutants,

Fig. 8. SAM sequencing of a homopolymer poly T tract on the non-coding strand of a cloned horse cDNA fragment. (A) Wild-type sequence, note disruption beyond the homopolymer region. The sequence disruption is likely due to homopolymer foldback resulting in a mixture of fragment lengths. (B) and (C) The sequences of two independent mutated clones of the cDNA. Introduction of mutations in the homopolymer region prevents random foldback, allowing downstream sequence to be read. The estimation of homopolymer length may differ markedly between wild-type and the mutated sequences.

not on the sequence, sequence length or coverage per mutant. This is not too surprising, as the sequence, the sequence length and coverage per mutant mainly affect whether the assembly is possible. For choices of these parameters outside some "safe" range, the method will fail because of mis-assemblies at stage I, or because the stage I contigs are too short to permit successful stage II assembly. The dominant influence on the proportion of miscalled or undetermined bases, given that the assembly succeeded, is the mutation level. The proportion decreases with increasing number of mutants, with an error level around 0.0001 achieved for about 10 mutants. Overall, the number of mutants required for sequence reconstruction would not need to be increased significantly for longer sequences. Certainly our data suggests the number of mutants required to achieve a given accuracy is not primarily dependent on target length, rather the required number of mutants will depend on the degree of mutation and the desired accuracy.

5.3. Can SAM sequencing aid SBS array short-read sequencing?

While the (pyro)sequencer GS20 typically generates more than 25 Mb with a Phred quality score of 20 (or more) for bases called during the sequencing of the 0.58 Mb genome of *M. genitalium* (Margulies *et al.*, 2005), the substantially lower accuracy of the short individual (feature) reads demands higher coverage for assembly of the entire genome. Homopolymer regions define one of the limits of the (semi-) quantitative pyrosequencing process (Ehn *et al.*, 2004; Ronaghi, 2001) with runs up to at least seven nucleotides able to be assessed accurately. However, alignment may require insertion of additional "padding" (bases) into different copies of individual element reads during *de novo* sequence assembly.

For simulation of SAM sequencing we have assumed perfect sequencing accuracy for each read (including all coverage) of our mutant copies. While this does not account for the reported 99.4% raw base-read accuracy observed for actual pyrosequencing output on PicoTiterPlates (Margulies *et al.*, 2005), our simulation is intended to explore the advantages of SAM sequencing in overcoming regions of low sequence complexity and homopolymer tracts that occur in eukaryote genomes. Considering our level of introduced mutation of 10%, if additional random errors such as insertions, deletions and homopolymer tract errors were introduced into our raw base reads at the same level of 0.6%, it would have little effect on the accuracy of SAM sequence reconstruction.

Church and colleagues have also reported extensive sequencing of prokaryote genomes using a PCR-colony sequencing method (Mitra *et al.*, 2003) and a related sequencing by ligation approach (Shendure *et al.*, 2005) with reads of 26 bp per amplicon. They note that during their ground-breaking resequencing of the entire ~3.3 Mb genome of an *E. coli* strain that, "despite 10 times coverage in terms of raw basepairs, only 91.4% of the genome had at least one time coverage", and further noted "substantial fluctuations in coverage were observed due to the stochasticity of the RCA step of library construction." While, their data indicates that the vast majority of the problem is due to insufficient formation of closed circles during the library construction prior to RCA, we would suggest that some residual problems could be due to sequence biases as well as some "very difficult" sequence that larger library sizes and oversampling may not fully address.

5.4. Costs and coverage for SAM sequencing

Our simulations suggest that sequence coverage required for SAM shotgun sequencing is not significantly higher than for conventional shotgun, even at moderate intensity of 10% mutation. Significantly, different sequence motifs representing problematic regions, such as sub-centromeric repeated regions and base-biased DNA, were present in the sequences used in this study. Nevertheless, all sequences could be reconstructed with errors close to 0.0001 for arrays consisting of 10 mutants sequenced to either 6-fold or 10-fold coverage. These simulations also indicate that *de novo* SAM sequencing of 0.6 Mb lengths can be readily achieved with reads of only 100 or 150 bp. Our simulation of successful sub-megabase assembly of human genomic DNA using only 25 bp reads is also indicative of the benefits of mutation sequencing for very short read lengths.

One observation is that the theoretical calculation of the number of mutants (Keith *et al.*, 2004b) required for a particular proportion of errors is consistent with errors observed in simulations of a range of different genomes and types of sequence motif. For example, a proportion of errors of ≤ 0.0001 was calculated if 10 mutants (with 10% mutation) are sequenced to 10-fold coverage. This level of accuracy was also projected with simulated SAM data from genomic fragments up to 580,074 bp from bacteria, and from different human genomic regions using reads of length 150 bp (100 bp for bacteria), and including the AT-rich fragment from *A. gambiae* Chr 3R. Further, a similar level of accuracy for SAM sequencing was projected with read lengths of 100 bp, the current read length of output from the array pyrosequencer (see Figures 5B and 7).

Our modeling also suggests that Bermuda Agreement accuracy for finished sequence can be achieved with short-read SAM sequencing, using no more than 2–3-fold higher cover than used for conventional array pyrosequencing or equivalent array-based SBS method. This observation is important because it means that the anticipated costs associated with the SAM approach are not significantly greater than for conventional short-read shotgun, while for intractable sequence regions the costs are significantly lower for SAM sequencing than for conventional shotgun approaches alone. These findings are also highly significant for array sequencing where 2–3-fold higher sequence coverage can easily be achieved at minimum cost. Importantly, the depth of coverage achievable on SBS arrays can reasonably provide data necessary for SAM assembly with errors <1 in 10,000. For example, an array with only 100,000 features and delivering 50 bp read data is sufficient to sequence 10 mutants of a 50 kb target to 10-fold coverage. Reads of 100 bp could provide similar levels of cover to a target of 100 kb in one experiment. The 454 PicoTitrePlate GS20 sequencer can reportedly cover some 25–35 Mb in a single experiment (Margulies *et al.*, 2007), while the Church laboratory (Shendure *et al.*, 2005) report 30 Mb of sequence.

The cost of array pyrosequencing using the sequencer "GS20" is about $7,500 per plate experiment, and delivers about 30 Mb of quality sequence. Table 4 extrapolates this cost onto a mammalian genome in current prices, it compares favourably against current Sanger sequencing at some 0.3–0.5 cents per base (full costing). If Sanger sequencing could achieve 0.1 c per base, the cost advantage of SAM–SBS array pyrosequencing would be removed, however the ability to recover data from recalcitrant and repeat-motif regions would remain an advantage. Some additional costs for handling and sequencing of large insert clones are included for eukaryote genome sequencing where mate-pair information is collected. Our modeling, using 10 mutants and a 10-fold coverage, suggests that Bermuda Agreement accuracy for finished *de novo* sequence can be achieved using short-read SAM sequencing, with about 2.5-fold higher

Table 4. *De novo* genomic sequencing. The cost of *de novo* GS20 pyrosequencing using the "GS20 sequencer" is about $7,500 per plate experiment, and delivers about 30 Mb (25–35 Mb) of quality sequence. The completion of the *M. genitalium* genome to 1/10000 error using SAM with array pyrosequencing would currently cost about $18,000 (2006–2007 values)

	Over sampling	Contiguous assembly	Cost per Mb of $Phred_{20}$ quality sequence data	Percentage sequence coverage (%)
Simple prokaryote genomes				
Sanger sequencing[a]	8–12 ×	Yes, 1 contig	$3,000–5,000	100
454 array pyrosequencing	40 ×	No, 25 contigs	$250	96.5
SAM + 454 pyrosequencing	100 ×	Yes, 1 contig	$250	100
Complex eukaryote genomes				
Sanger sequencing[a]	10–20 ×	Not possible	$4,500–7,500[b]	90
454 array pyrosequencing	100 ×	Not possible	$250	< 40?
SAM + 454 pyrosequencing	100 ×	Possible?	$250	> 90?

[a]At a cost of 0.3–0.5 c per base.
[b]Additional cost for mate pair and large clone handling and sequencing.

oversampling than is used for conventional GS20 pyrosequencing or equivalent short-read technology. Our coverage is some 10-fold higher than used for conventional Sanger sequencing. However, the cost differences between array pyrosequencing and Sanger sequencing are approximately 100–120-fold per megabasepair of $Phred_{20}$ sequence data, to the favour of GS20 array pyrosequencing.

We suggest that the integration of SAM and SBS technologies will advance *de novo* array-based sequencing, dramatically reducing sequencing cover needed for larger genomes to a maximum defines by SAM theory, while improving sequencing accuracy, the depth of cover and the length of sequence that can be reassembled correctly from SBS read data. Here, we chose to illustrate the effect of SAM mutation of DNA targets by *in silico* simulation of completely random mutations at any base in the target region to a level of 10%. We have previously described Bayesian methods that use the proportion of observed mutations to weight the predicted reassembled sequence (Keith *et al.*, 2003). We would also suggest that Bayesian mathematical methods could contribute to more efficient assembly of short-read SAM sequence data.

5.5. The advantages of SAM sequencing

Our second objective was to improve the efficiency of sequencing of different DNA motifs, including improving the sequencing of currently intractable DNA, repetitive regions and base-biased elements (*AT-rich* or *GC-rich*) that confound many sequencing technologies. Figure 8 illustrates that mutagenic analogue nucleotides can introduce mutations into homopolymer regions, illustrated here with a small cloned poly T (poly A) element. Although the quality of conventional Sanger sequencing is improved through introduction of mutant bases that disrupt the homopolymer tracts, this chemistry would have obvious advantage to pyrosequencing, in which the disruption of the homopolymer into shorter elements may allow more accurate determination of their total repeat length.

Our aim is to create a sequencing technology that sees little difference between these intractable DNA motifs and readily sequenceable DNA. SAM techniques also facilitate the cloning of refractory regions that are under-represented in genomic libraries (Keith *et al.*, 2004), by altering the sequence of inhibitory motifs and structures. The ability to create different library representations would aid sequencing and genome finishing, which is still a practical constraint on SBE methods (Margulies *et al.*, 2005, 2007).

5.6. Overcoming the biochemical limitations of SBS

Mitra *et al.* (2003) note "there are three biochemical sources of error that will likely determine the maximum read length attainable using the FISSEQ approach: mispriming, misincorporation and incomplete extension." Each of these errors have the compounding effect of de-phasing the extension step on individual DNA molecules within a polony, causing loss of correct signal and introduction of erroneous base addition signals at de-phased DNA templates. SBS arrays are also significantly limited by the inability of DNA polymerases to read through even short *difficult to read* DNA motifs. Indeed, the current poor uniformity of SBS arrays is probably related in part to frequently encountering short sequences that impede polymerase progress, and de-phasing would certainly cause signal failure at a (significant) proportion of features. We have also previously demonstrated that SAM sequencing methods prevent DNA polymerase slippage (Keith *et al.*, 2004b) and thus could reduce the frequency of de-phasing events. This effect is due to the introduction of a more uniform local nucleotide environment (in repeated elements).

SBS methods such as pyrosequencing that make single nucleotide additions of native dNTPs are also particularly prone to error at homompolymer regions. This is because the quantification of the strength of addition signals does not permit homopolymers longer than 6–7 bp to be distinguished accurately, and because limiting amounts of individual natural dNTPs must be used in each extension cycle to minimize the misincorporation effects that occur at higher concentrations. Hence another major drawback with pyrosequencing is the incomplete extension through long homopolymer repeats, leading to loss of register on the many copies of the template, and causing read dropout at individual templated-beads. The introduction of random mutations during the SAM process can reduce the effective homopolymer lengths into a series of shorter tracts, which are then more tractable to pyrosequencing giving either accurate quantitation of shorter (sub)tracts, or reduced bead dropout.

Although the accumulated HGP sequence is significantly greater than 8 times coverage, significant portions of the human genome still remain unable to be sequenced using current technologies (IHGSC, 2004). We suggest that development of SAM sequencing could be instrumental in helping to sequence these difficult regions, as SAM–SBS array sequencing could readily provide contiguous coverage of large fragments containing repeated sequence motifs.

REFERENCES

Bennett, S. T., Barnes, C., Cox, A., Davies, L. and Brown, C. (2005). Toward the $1000 human genome. Pharmacogenomics 6, 373–382.

Ehn, M., Nourizad, N., Bergstrom, K., Ahmadian, A., Nyren, P., Lundeberg, J. and Hober, S. (2004). Toward pyrosequencing on surface-attached genetic material by use of DNA-binding luciferase fusion proteins. Anal. Biochem. 329, 11–20.

Gordon, D., Abajian, C. and Green, P. (1998). Consed: a graphical tool for sequence finishing. Genome Res. 8, 195–202.

Guy, J., Hearn, T., Crosier, M., Mudge, J., Viggiano, L., Koczan, D., Thiesen, H. J., Bailey, J. A., Horvath, J. E., Eichler, E. E., Earthrowl, M. E., Deloukas, P., French, L., Rogers, J., Bentley, D. and Jackson, M. S. (2003). Genomic sequence and transcriptional profile of the boundary between peri-centromeric satellites and genes on human chromosome arm 10p. Genome Res. 13, 159–172.

Hebert, B. and Braslavsky, I. (2007). Single molecule fluorescence microscopy and its applications to single molecule sequencing by cyclic synthesis. In: K. R. Mitchelson (Ed), New High Throughput Technologies for DNA Sequencing and Genomics (pp. 207–244). Elsevier, Amsterdam.

Henderson, I. R., Zhang, X., Lu, C., Johnson, L., Meyers, B. C., Green, P. J. and Jacobsen, S. E. (2006). Dissecting *Arabidopsis thaliana* DICER function in small RNA processing, gene silencing and DNA methylation patterning. Nat. Genet. 38(6), 721–725.

Holt, R. A., Subramanian, G. M., Halpern, A., Sutton, G. G., Charlab, R., Nusskern, D. R., Wincker, P., Clark, A. G., Ribeiro, J. M. and Wides, R. *et al.* (2002). The genome sequence of the malaria mosquito Anopheles gambiae. Science 298, 129–149.

Hyman, E. D. (1988). A new method of sequencing DNA. Anal. Biochem. 174, 423–436.

International human genome sequencing consortium (IHGSC) (2004). Finishing the euchromatic sequence of the human genome. Nature 431, 931–945.

Kartalov, E. P. and Quake, S. R. (2004). Microfluidic device reads up to four consecutive base pairs in DNA sequencing-by-synthesis. Nucleic Acids Res. 32, 2873–2879.

Keith, J. M., Adams, P., Bryant, D., Cochran, D. A. E., Lala, G. H. and Mitchelson, K. R. (2004a). Algorithms for sequencing aided by mutagenesis. Bioinformatics 20, 2401–2410.

Keith, J. M., Adams, P., Bryant, D., Mitchelson, K. R., Cochran, D. A. E. and Lala, G. L. (2003). Inferring an original sequence from erroneous copies: a Bayesian approach. In: Chen, Y.-P. P. (Ed.), Proceedings of the 1st Asia-Pacific Bioinformatics Conference (vol. 19, pp. 23–28) APBC2003.

Keith, J. M., Cochran, D. A. E., Lala, G. H., Adams, P., Bryant, D. and Mitchelson, K. R. (2004b). Unlocking hidden genomic sequence. Nucleic Acids Res. 32, e35.

Kling, J. (2003). Ultrafast DNA sequencing. Nat. Biotech. 21, 1425–1427.

Leamon, J. H., Lee, W. L., Tartaro, K. R., Lanza, J. R., Sarkis, G. J., deWinter, A. D., Berka, J. and Lohman, K. L. (2003). A massively parallel PicoTiterPlate based platform for discrete picoliter-scale polymerase chain reactions. Electrophoresis 24, 3769–3777.

Margulies, M., Egholm, M., Altman, W. E., Attiya, S., Bader, J. S., Bemben, L. A., Berka, J., Braverman, M. S., Chen, Y. J., Chen, Z., Dewell, S. B., Du, L., Fierro, J. M., Gomes, X. V., Godwin, B. C., He, W., Helgesen, S., Ho, C. H., Irzyk, G. P., Jando, S. C., Alenquer, M. L., Jarvie, T. P., Jirage, K. B., Kim, J. B., Knight, J. R., Lanza, J. R., Leamon, J. H., Lefkowitz, S. M., Lei, M., Li, J., Lohman, K. L., Lu, H., Makhijani, V. B., McDade, K. E., McKenna, M. P., Myers, E. W., Nickerson, E., Nobile, J. R., Plant, R., Puc, B. P., Ronan, M. T., Roth, G. T., Sarkis, G. J., Simons, J. F., Simpson, J. W., Srinivasan, M., Tartaro, K. R., Tomasz, A., Vogt, K. A., Volkmer, G. A., Wang, S. H.,

Wang, Y., Weiner, M. P., Yu, P., Begley, R. F. and Rothberg, J. M. (2005). Genome sequencing in microfabricated high-density picolitre reactors. Nature 437, 376–380.

Margulies, M., Jarvie, T. P., Knight, J. R. and Simons, J. F. (2007). The 454 Life Sciences Picoliter sequencing system. In: K. R. Mitchelson (Ed), New High Throughput Technologies for DNA Sequencing and Genomics (pp. 151–186). Elsevier, Amsterdam.

Mitchelson, K. R., Hawkes, D. B., Turakulov, R. and Men, A. (2007). Overview: developments in DNA sequencing. In: K. R. Mitchelson (Ed), New High Throughput Technologies for DNA Sequencing and Genomics (pp. 1–44). Elsevier, Amsterdam.

Mitra, R. D., Shendure, J., Olejnik, J., Krzymanska-Olejnik, E. and Church, G. M. (2003). Fluorescent *in situ* sequencing on polymerase colonies. Anal. Biochem. 320, 55–65.

Ohuchi, S., Nakano, H. and Yamane, T. (1998). *In vitro* method for the generation of protein libraries using PCR amplification of a single molecule and coupled transcription/translation. Nucleic Acids Res. 26, 4339–4346.

Poinar, H. N., Schwarz, C., Qi, J., Shapiro, B., Macphee, R. D., Buigues, B., Tikhonov, A., Huson, D. H., Tomsho, L. P., Auch, A., Rampp, M., Miller, W. and Schuster, S. C. (2006). Metagenomics to paleogenomics: large-scale sequencing of mammoth DNA. Science 311, 392–394.

Ronaghi, M. (2001). Pyrosequencing sheds light on DNA sequencing. Genome Res. 11, 3–11.

Shendure, J., Mitra, R. D., Varma, C. and Church, G. M. (2004). Advanced sequencing technologies: methods and goals. Nat. Rev. Genet. 5, 335–344.

Shendure, J., Porreca, G. J., Reppas, N. B., Lin, X., McCutcheon, J. P., Rosenbaum, A. M., Wang, M. D., Zhang, K., Mitra, R. D. and Church, G. M. (2005). Accurate multiplex polony sequencing of an evolved bacterial genome. Science 309, 1728–1732.

Yu, H., Eritja, R., Bloom, L. B. and Goodman, M. F. (1993). Ionization of bromouracil and fluorouracil stimulates base mispairing frequencies with guanine. J. Biol. Chem. 268, 15,935–15,943.

Zaccolo, M., Williams, D. M., Brown, D. M. and Gherardi, E. (1996). An approach to random mutagenesis of DNA using mixtures of triphosphate derivatives of nucleoside analogues. J. Mol. Biol. 255, 589–603.

Chapter 11
Genome Sequencing and Assembly

Annette McGrath

Australian Genome Research Facility, University of Queensland, St. Lucia, Queensland 4072, Australia

Contents

Abstract

From the advent of genome sequencing and assembly in to the 1970s to the present day a great deal has been achieved in genome science. We now have the experimental means, tools and computing hardware necessary to successfully sequence and assemble large complex genomes. However challenges remain and draft genome sequences are widely known to contain misassemblies, largely due to repetitive DNA. As technology advances, new ways to generate large amounts sequence data in a short time and the increasing use of low-coverage genome scans are resulting in a greater reliance on bioinformatics tools for sequence

PERSPECTIVES IN BIOANALYSIS, VOLUME 2
ISSN 1871-0069 DOI: 10.1016/S1871-0069(06)02011-8

assembly and comparison. With these new directions in genome science, this research area will remain as an active one for some time.

1. INTRODUCTION

The foundations of the genomic era was laid in the late 1970s with the sequencing of the single-stranded bacteriophage φX174 leading to the landmark sequencing of bacteriophage lambda by Sanger *et al.* (1977, 1982). This genome was sequenced by a random shotgun sequencing approach, a technique pioneered by Sanger and colleagues that broke the genomic DNA at random and sequenced the ends of these random fragments, after being cloned into M13 vectors (Sanger *et al.*, 1980). The lambda genome is 48.5 kb in size and due to its relatively larger size, the use of computer programs were becoming more important to handle the data and to reassemble the random fragments into the genome from which they came (Staden, 1982). Progress toward sequencing of larger genomes remained stalled largely due to the lack of necessary computational resources required for assembly of the raw sequencing reads into a finished genome. The first breakthrough was with the sequencing of the first bacterial genome in 1995. The 1.83 Mb *Haemophilus influenzae* genome was determined by a group at the Institute for Genome Research (TIGR) in 1995 (Fleischmann *et al.*, 1995), closely followed by the sequencing of the genome of *Mycoplasma genitalium* (0.5 Mb) by the same group in the same year (Fraser *et al.*, 1995). The *H. influenzae* genome was determined by a shotgun sequencing approach, which provides a high level of redundancy and proved to be a cost-effective and efficient way to sequence what were regarded as large genomes at the time. The reconstruction of the *H. influenzae* genome from the shotgun sequencing reads required the creation of a new assembly program, the TIGR assembler (Sutton *et al.*, 1995).

Currently, we are now truly in the genome age with over 2000 different genome projects, from viruses to eukaryotes, either completed or in progress (http://www.genomesonline.org/). This review will focus on the laboratory approaches to the genome sequencing process and the computational tools and methodologies used to reconstruct a genome from its constituent reads and will present an overview of upcoming trends in genome sequencing and assembly.

2. APPROACHES TO GENOME SEQUENCING

2.1. Whole genome shotgun sequencing

Shotgun sequencing involves the construction of libraries of DNA fragments created from the genomic DNA of the organism of interest. This method has been popular because there is no need for prior knowledge of the relative map position of larger clones such as bacterial artificial chromosomes (BACs). The latter approach was made popular by the ordered-clone approach and is described below. In random shotgun sequencing, the genomic DNA is sheared, either enzymatically or more usually by mechanical means, to break it into fragments of a target size, for example, 3–5 kb. Figure 1 below depicts a simplified bacterial genome sequencing process.

The ends of the fragmented DNA are repaired and fragments are size selected to obtain DNA fragments of the required size that is then subcloned into, usually, a plasmid vector to create a library of clones that will be used for the shotgun sequence generation. The library is then transformed into a suitable *E. coli* host, colonies are selected and plasmid DNA is purified for sequencing. Sequencing from plasmid vectors has the advantage of being highly automatable, the template is easy to prepare and the method is very reliable producing high-quality data from the sequence reads. For most genomic shotgun

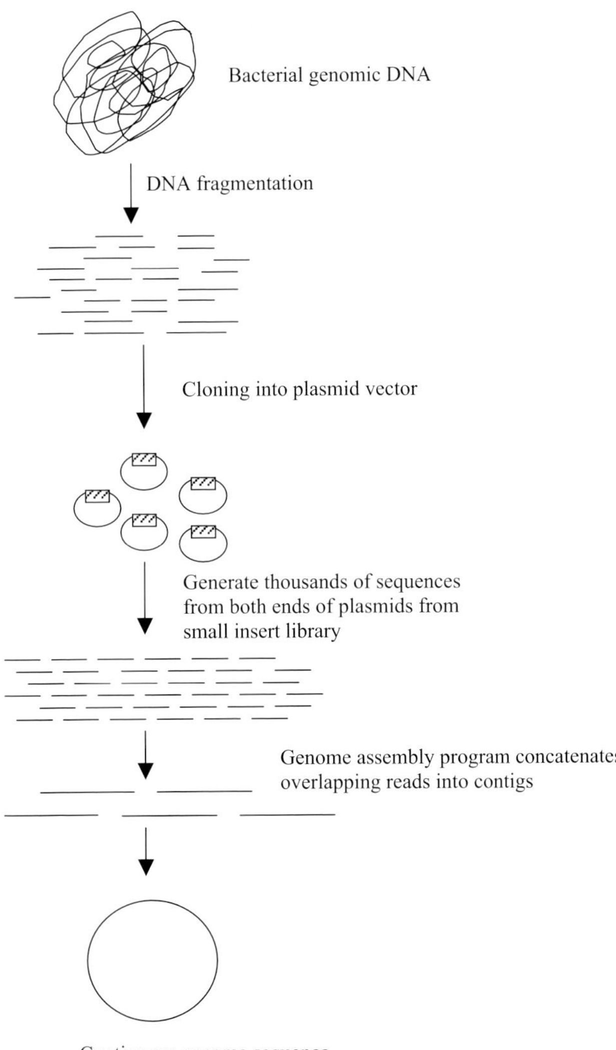

Fig. 1. Schematic representation of the shotgun sequencing and assembly process.

sequencing projects, small insert libraries are used to gain the majority of the sequencing data. These libraries typically contain inserts in the range of 2–5 kb. Libraries with shorter inserts will be of less utility to the assembly process as reads of both ends of the plasmid vector are likely to overlap and therefore, when trying to reconstruct the genome programmatically, these reads will not be as informative as reads that can help assembly a larger area in the local assembly. Medium-insert libraries are typically in the range of 5–10 kb in size and are very useful in the generation of a 'scaffold' of the contigs used in the later stages of the assembly and gap-closure phases. A typical bacterial genome project would use at least one library of small insert size and one of medium insert size. Larger eukaryotic genome projects would also typically make use of vectors that can uptake larger inserts such as fosmids (inserts sizes in the range of 30–45 kb) or BAC libraries (with insert sizes of 100–200 kb).

Initially shotgun sequence assembly projects only generated a single read from one direction of the plasmid clone. Later, a double-barreled shotgun approach was proposed (Edwards *et al.*, 1990), enhancing the original method and introducing the use of long-range linking information. Currently, sequencing technologies allow us to generate sequences between 600 and 1000 bp, depending on the running conditions on the sequencing machine. For small insert libraries therefore, the middle of the clone will be unknown. As the sequencing reads were generated from either end of the same DNA template, the reads are called mate-pair reads. The mate-pairs are known to be related and therefore information about the relative placement of the mate-pairs in the contigs can be determined, as the maximum distance between the reads in the contig can be a maximum of the upper limit of the size range which was size selected for cloning. This information is extremely useful for genome sequence assembly tools as it helps in placement of the reads, and any overlapping sequences in relation to each other.

Shotgun sequencing is a two-step process. Following the high-throughput generation of shotgun reads and sequencing assembly, the resulting contig data are rarely complete. A second more directed phase of sequencing, termed finishing, is usually required to ensure that the assembly is complete and correct. In this phase the assembly data are inspected, often manually, to identify any problematic areas. Any anomalous data, such as sequences of vector or *E. coli* host origin or chimeric clones, are identified and removed. Where necessary, regions where additional data are required to close any remaining gaps are identified and appropriate reactions are designed and performed (see Figure 2). There may also be regions of poor quality where the sequence is uncertain or insufficient reads available to link and join the contigs into the consensus sequence, or that additional data are required to confirm the sequence. Manual inspection of the consensus is also required to correct assembly or base-calling errors.

In contrast to the highly automated shotgun generation phase, the finishing process is time-consuming and expensive requiring extensive manual intervention, and was largely a manual task until recently with the advent of a number of interactive tools that help to automate the finishing process and reduce the amount of human intervention required. The most commonly used tools are Consed (Gordon *et al.*, 1998), which includes a program Autofinish, (Gordon *et al.*, 2001) to automatically predict any additional sequencing reads, and GAP4 (Bonfield

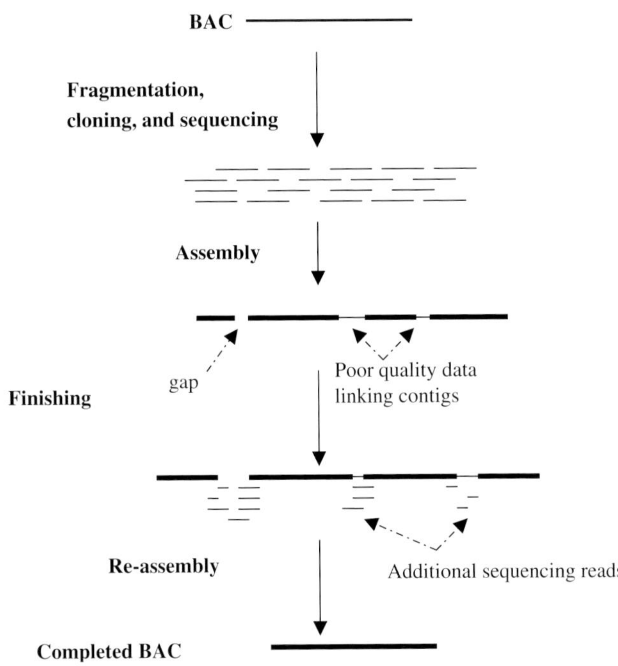

Fig. 2. The sequence finishing process of a fictional BAC clone depicting some of the problems due to incomplete coverage or poor-quality data.

et al., 1995). Both of these tools provide a graphical interface to allow human finishers to examine the assembly and its constituent reads, to identify problem areas, to remove problematic reads and make base-edits where necessary. The power of these tools is evident with the success of finishing approximately 300 published completed bacterial genomes, with approximately another 1000 in progress (http://www.genomesonline.org/). The vast majority of these small (<20 Mb) genomes have been sequenced by whole genome shotgun sequencing.

2.2. Clone-by-clone approach

The clone-by-clone approach is also known as the ordered clone approach or the hierarchical shotgun sequencing approach and was made popular by the public human genome sequencing effort undertaken by the International Human Genome Sequencing Consortium (IHGSC, 2001). A number of bacterial genomes were also sequenced using this approach including *Mycobacterium tuberculosis* (Cole *et al.*, 1998) and *Streptomyces coelicolor* (Bentley *et al.*, 2002). The clone-by-clone approach was developed at a time when whole genome shotgun sequencing of large genomes in particular was considered unfeasible due to the computing resources required and the presence of repeats in eukaryotic genomes that greatly complicate the assembly process. This approach requires that a map of overlapping large-insert clones, >50 kb, that cover the whole genome be constructed prior to undertaking a sequencing effort. This step is called physical mapping. BACs are commonly used for physical mapping as they can stably contain inserts of foreign DNA that range in size from 100 to 200 kb.

Methods to identify overlapping clones include fingerprinting clones by restriction enzyme digestion and gel electrophoresis or by hybridization mapping. Fingerprint mapping relies on the fact that large insert clones from a genomic library of the organism that share sufficient common restriction enzyme fingerprint patterns can be assumed to overlap and that the sequence of these two clones will assemble into a contiguous region. The hybridization mapping method relies on digesting chromosomal DNA with a rare cutting restriction enzyme and using these fragments as probes to order the clones from a cosmid library. The underlying theory is that pairwise comparison of the fingerprints or hybridization patterns of BACs will help infer which clones overlap to create a BAC contig map, which will cover the entire genome sequence. A schematic representation of the ordered clone approach is shown in Figure 3.

The set of overlapping BACs that cover the entire genome is called the minimal tiling path. Each of the individual BACs from the minimal tiling path of unique clones are then sequenced to a high level of redundancy by shotgun sequencing and assembled. Therefore each region of the genome is covered at least once, apart for the regions that are found in the overlaps between the BACs, which are sequenced at least twice. The overlaps must be large enough, e.g. 5 kb, to avoid difficulties caused by repeats. As the BACs are of a manageable size, commonly used tools, such as Phrap, could be used to assemble them. As the sequence assembly is local in nature, the prospect of long-range misassembly is eliminated. Sequences from overlapping BACs are then assembled together to obtain the entire genome sequence. This process may be done by hand, if the genome is sufficiently small, or if the genome is large by the use of a specialized assembler, e.g. GigAssembler (Kent and Haussler, 2001), in the case of the public human genome project. GigAssembler was built to order, orient and merge sequence data from approximately 30,000 large insert clones,

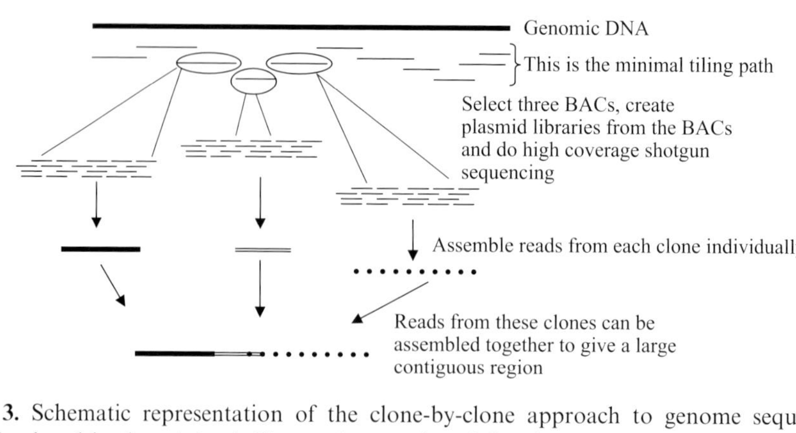

Fig. 3. Schematic representation of the clone-by-clone approach to genome sequencing. BACs placed in the minimal tiling path are selected for shotgun sequencing and finishing, assembled individually and then these assemblies are combined and overlapped to produce the consensus sequence.

while taking into account any incorrect mapping data and correcting for misassemblies in the initial contig data. The ordered clone method provides a guaranteed route to reach the whole genome sequence as each BAC assembly can be considered a local assembly and this assembly can be accurately placed on the genome, according to the initial physical map and rechecked independently by reference to the data on which this map was based.

2.3. Sequencing more complex genomes

Many of the principles that apply to bacterial genome sequencing also apply to eukaryotic genome sequencing. However, the larger scale of the task has meant that the clone-by-clone approach was initially the more favored approach. The larger genomes of many eukaryotes were all successfully determined by using the clone-by-clone approach, they include *Saccharomyces cerevisiae* (Goffeau *et al.*, 1996), *Caenorhabditis elegans* (*C. elegans* Sequencing Consortium, 1998), *Arabidopsis thaliana* (The Arabidopsis Initiative, 2000) and most significantly the approximately 3 Gb human genome (IHGSC, 2001). The IHGSC built their successful reconstruction and resolution of sequence ambiguities on the available BAC-based physical map (International Human Genome Mapping Consortium, 2001), which also allowed more detailed finishing of the euchromatic portion of the human genome and reinvestigation of regions of repetition (IHGSC, 2004).

2.4. Assembly of whole genome shotgun sequence

There was considerable doubt within the community that the whole genome shotgun method alone could be successfully applied to eukaryotic genome sequencing. Discussions in the literature were underway before this method was applied to large eukaryotic sequencing efforts with opposing views being expressed on whether such an approach would be feasible for complex genomes, and particularly the human genome (Weber and Myers, 1997; Green, 1997). Although a very successful approach for sequencing bacterial genomes, as we have seen, the larger size, polymorphic nature and much higher repetitive content of eukaryotic genomes posed significant challenges to the process of sequence assembly. Repetitive sequences in eukaryotic genomes are also longer than in bacterial genomes and can span the entire length of inserts from small insert clones thereby making correctly positioning the repeats a more difficult task than in simpler genomes. In addition, mammalian genomes have large segmental duplications, reads from which it can incorrectly assemble in one location during the assembly process. The sequence-finishing step using data exclusively from whole genome shotgun eukaryotic projects is also likely to be extensive, due to the difficulty in resolving repeats.

The first complex genome to be sequenced using a whole genome shotgun approach was that of *Drosophila melanogaster* (Adams *et al.*, 2000; Myers *et al.*, 2000). This was a draft sequence of the euchromatic portion of the genome and comprised less than 3% repetitive sequence and produced an assembly with approximately 2500 gaps and was assembled by the specifically built program, the Celera assembler. The finished *Drosophila melanogaster* genomic sequence did however make use of the clone-by-clone approach (Celniker *et al.*, 2002) and therefore the finished *Drosophila* genome can be considered to be a hybrid approach to genome sequencing and assembly. In fact, the authors indicated that if they were to redo the project from the beginning that they would use a combination of various levels of coverage from small- and medium-insert plasmid libraries and a low coverage of BAC libraries with BAC end sequencing, as well as restriction enzyme fingerprints of BAC clones in a tiling path across each arm to resolve collapsed repeats and to verify the physical map and final assembly, as well as use fluorescence *in situ* hybridization (FISH) to associate the sequence scaffolds to their chromosome of origin.

The human genome was sequenced by two groups who announced the release of the sequence simultaneously (IHGSC, 2001; Venter *et al.*, 2001). The genome sequencing efforts were performed by the two opposing techniques; that undertaken by the Celera Genomics team being a whole genome shotgun approach to generate a 5.1 × coverage while the IHGSC effort was a hierarchical shotgun sequencing approach from over 29,000 overlapping clones. Celera assembled a 5.1 × coverage from their own shotgun data and a 2 × coverage of the data from the public effort deposited in public databases to assemble the genome. Even after the publication of the genome sequence, debate raged on in the literature among representatives from both groups regarding the veracity of the whole genome shotgun approach to complex genome sequencing and whether or not the Celera assembly could be viewed as a true whole genome shotgun assembly (Waterston *et al.*, 2002, 2003; Myers *et al.*, 2002; Green, 2002; Adams *et al.*, 2003). Waterston *et al.* (2002) point to the fact that the Celera assembly of the human genome used public data, deposited in public domain databases, to aid their assembly and that therefore the Celera assembly was not an independent assembly and could not be regarded as a true test of whether or not a whole genome shotgun approach could be used to assemble a genome. They showed that the Celera assembly used public data in three ways; they used 'shredded' HGP BACs to reconstruct a perfect tiling, they used assembled HGP BAC data for gap filling between Celera contigs, and they used compartmentalized assembly strategies which reduced the problem of overlap detection as the problem was then focused onto small local regions. With the release of more complete genome builds from NCBI in the interim years allowing better comparison of the accuracy, completeness and coverage of both sequencing efforts, Celera performed a comparison to show that their assembly provided better order and orientation, and that the public effort provided an assembly with greater coverage of exact and near exact repeats (Istrail *et al.*, 2004).

Whole genome shotgun sequencing from either small insert libraries, or more commonly from a mixture of libraries of varying clones insert sizes has now been accepted as a viable option for the sequencing of complex genomes, with the recent reports of the genomes of the fungus *Aspergillus nidulans* (Galagan *et al.*, 2005), domestic dog *Canis familiaris* (Lindblad-Toh *et al.*, 2005) and chimpanzee (The Chimpanzee Sequencing and Analysis Consortium, 2005) being sequenced using only a whole genome shotgun approach. The combination approach of using a physical map and BAC sequencing as well as with whole genome shotgun has also been employed for the genomes of the Brown Norway rat (Rat Genome Sequencing Project Consortium, 2004) and the domestic chicken (International Chicken Genome Sequencing Consortium, 2004).

3. PROBLEMS INHERENT WITH GENOME ASSEMBLIES

3.1. Repetitive DNA

As noted previously, while it would be expected that a single contig representing the entire genome would be reconstructed after genome assembly, this is rarely the case, even for simple genomes due to a number of complicating factors with the underlying raw data.

The main complicating factors affecting genome assemblies are repetitive sequence and biological artifacts that result in errors in the assembly data and in incomplete or biased coverage of the genome. Ideally, the assembly program should be able to distinguish between different copies of the repetitive sequences and to place them in their correct genomic location. However this is not the case and the problem of misassembled genomes resulting from repetitive sequence is widespread and all assembly tools are affected (Salzberg and Yorke, 2005).

In general, repeats that are shorter than the average sequence read length are easier to deal with than repeats that are longer than the average read length, which is currently approximately 600–700 bp with available technology, as these repeats can be accurately placed in the correct location and orientation within the genome sequence. Also repeats that have more base differences than the sequencing error rate should also be handled effectively as they will de distinguished by the assembly program as being repeat-induced overlaps, rather than true overlaps. Longer repeats are problematic to assemble because even though they originate from different locations in the genome, the assembly program cannot distinguish between the different repeated copies, the reads are all assembled together which tends to result in collapse the assembly at that point. This is especially true in the case of tandem repeats, resulting in a (apparent) very deep coverage of this region, in comparison to the rest of the genome (Figure 4).

In this example, as both copies of the repeat are assembled together resulting in stacking at that point, the intervening reads in the region between the repeats can only be assembled together, and therefore cannot be included in the assembly. This stacking results in an incorrect genome reconstruction, and the correct reconstruction must be resolved by labor-intensive and time-consuming finishing processes. Identical repeats are the most problematic as it is impossible to distinguish between copies of the repetitive element. These types of repeats are generally rare as copies of repetitive elements accumulate different point mutations throughout time and this information can be used to distinguish between them. Repeats can be of a number of different categories, such as microsatellites, low-complexity DNA, transposons or retrotransposons such as short interspersed nuclear elements (SINEs, e.g. Alu repeats, ~300 bp in length)

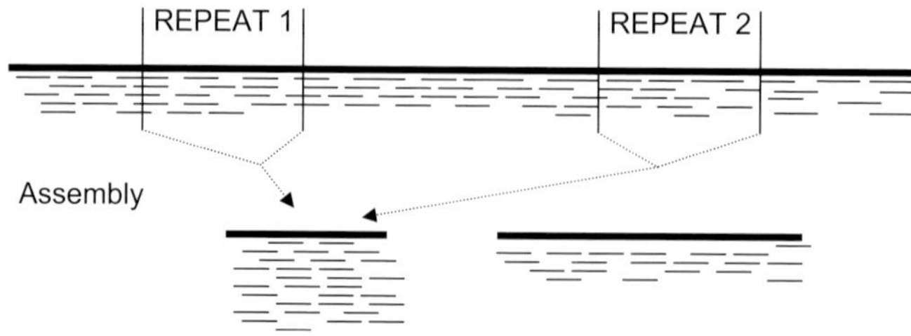

Fig. 4. Possible assembly of a region containing two repeats. The repeat sequences can assemble together producing read stacking at that point and leaving the intervening reads to assemble together. These reads cannot be extended to the left or right due to the stacking of the repeat sequence.

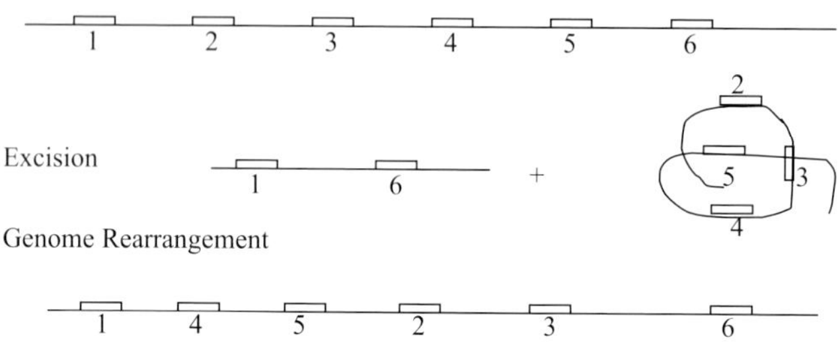

Fig. 5. Rearrangements and excision in a genome location caused by many identical copies of a repetitive sequence.

and long interspersed nuclear elements (LINEs of 500–5000 bp in length), or LTR retroposons such as long-terminal repeats of approximately 700 bases in length. Other classes of repetitive sequence that can cause problems for an assembly program include gene duplication where genes duplicate and then diverge in sequence, or long-segmental duplications that can be very long and have very similar copies of a very long portion of the genome. Misassemblies of repetitive DNA containing regions can therefore result also in the excision of a repeat in particular genome locations, as well as erroneous genome rearrangements, in addition to collapsed areas of the genome (Figure 5).

Techniques for repairing regions of the genome that contain repetitive DNA are labor intensive and depend on finding differences between the sequencing reads that are aligned together at a particular point. For example, if three constituent sequencing reads at point X contain a C base and an equal number contain a T, this could indicate that there two copies of a repeat have been assembled into one location. If two copies of a repeat are suspected, its constituent reads can be isolated and each copy of the repeat can be assembled independently and then placed into the rest of the assembly. However, while these differences may be true SNPs they may not be detectable if the shotgun sequence coverage in this region is low. This is especially true for diploid mammalian genomes as both copies of the chromosome have slightly diverged and therefore it is difficult to distinguish true polymorphisms between the haplotypes from incorrectly assembled and collapsed repeats. Salzberg and Yorke (2005) recommend that all assemblies be made available for other groups to view them and correct them if necessary. The Assembly Archive at NCBI is the largest resource available at the moment to capture draft and finished genomes (Salzberg *et al.*, 2004).

3.2. Data quality

Sequence data can also contain errors. Current sequencing technology does not produce sequence reads that are 100% correct all of the time, nor is the data of consistent quality throughput all parts of the sequence read. The data that comes from the beginning and ends of the sequence reads are most prone to error. A number of factors cause this including the fact that for longer reads, the signal is of weaker intensity due to the geometric distribution of the incorporation of ddNTPs during the sequencing reaction and as the longer fragments take longer to move through the gel, they are more diffuse than shorter fragments and more difficult to call. Variable sequence quality is not however restricted to the ends of the reads, as secondary structure effects or compressions of GC-rich region peaks can on occasion be poorly resolved in the middle of the sequence. Therefore, the base-caller may not identify the correct base or the correct number of bases and this makes sequencing data 'noisy'. Assembly programs are affected by noisy data when trying to find overlaps between fragments, as single base differences could prevent otherwise identical sequences from overlapping (true overlaps). Assembly programs therefore need to tolerate a low level of error from the raw input data. This is a delicate balance between being able to identify true overlaps in the presence of sequencing errors while not increasing the amount of repeat-induced overlaps. This strategy works where the level of sequence divergence between the repeats is greater than the sequencing error rate. Having error-free sequence data would greatly reduce the complexity of the algorithms required to assemble a genome sequence and as it would require less reads, would make the cost of sequencing much lower. The level of error in sequence data is estimated to be approximately 3% (Hill *et al.*, 2000).

To reflect the fact that the data quality in all parts of a sequence read is not the same, Green and colleagues (Ewing *et al.*, 1998; Ewing and Green, 1998) developed a system where a quantitative measure of the quality of each base call is assigned. This method estimates the error probability for each base call and is implemented in the program Phred. Since its introduction in 1998, Phred quality values have become the industry standard and it remains the most widely used method of base-call error estimation. Phred's quality values are logarithmic and are defined as $Q = -10 \log_{10} P$, where Q is the quality value and P the estimated error probability of a base call. Quality values range from 0 to 99, increasing values indicating increasing quality. A Phred quality value of 10 corresponds to a 90% chance of accuracy, or a 1 in 10 chance that the base call is incorrect. The standard that is generally required is a Phred quality of 20, Q20, which corresponds to a predicted error rate of 1% or a 99% chance of the base call being correct. For the public human genome project, the aim was that the overall sequence quality conform to the Bermuda Standard of being accurate to at least 1 bp in 10,000, or Q40 (Bentley, 1996).

Many assembler programs can use the base quality information in the assembly process. Those that do not use quality values directly can still use the quality values as a method of automatically trimming low-quality data from the sequence reads, for example, a clip being introduced in the sequence read when the average quality value falls below a predetermined value over a sliding window of a fixed width, e.g. 50 bases. Applied Biosystems have also introduced a base-caller that assigns Phred-like quality values to bases (KB base-caller), although details of the algorithm have not been released (http://docs.appliedbiosystems.com/pebiodocs/04362968.pdf).

3.3. Cloning artifacts

Other problems with the assembly of sequence data include biological features inherent to the nature of cloning and sequencing plasmid inserts. These features include the fact that the raw sequence data can contain contaminating vector or bacterial host sequence. Depending on how close the sequencing primer sequence is to the vector insert site, at least some of the insert sequence read will be of vector origin. This sequence must be removed prior to assembly or recognized as vector by the assembly program. If they are not removed, they can cause false overlaps between otherwise unrelated sequences. Furthermore, depending on the cloning strategy, there is a possibility of multiple inserts from different parts of the genome being cloned into one vector site. These chimeric clones could cause misassemblies if they are not recognized as such. Clone insert sequences which are derived completely from the bacterial host DNA must also be recognized and discarded or not considered further in the assembly. As random fragments of DNA can be inserted into the vector insert in either orientation, the orientation of the read in the genome is also not known, depending on the experimental approach.

Simplistically, the process of randomly shearing the genomic DNA of the organism and then cloning all of the random fragments into a shotgun library is a stochastic process. Owing to the random nature of the sampling, in an 8 × coverage, for example, not all genomic locations are sampled evenly. Some will be samples more than eight times while others may not be sampled at all. In addition, some regions of DNA may not be clonable at all, if for example the sequence contained in the vector is toxic to the bacterial host, these clones containing the toxic inserts, would never survive to be sampled in the sequencing process. These processes lead to incomplete coverage and will result in gaps in the coverage which can only be resolved through application of labor-intensive directed approaches during genome assembly finishing.

4. A MATHEMATICAL MODEL OF SHOTGUN SEQUENCING

Work done by Lander and Waterman (1988) helped to formulate a mathematical framework for estimating various statistical parameters associated with shotgun sequencing. Their intention was to provide a model for the physical

G = Genome length

L = read length of acquired bases

N = number of reads

T = length of overlap

Coverage $c = NL/G$

The probability that a base is not sequenced $P_0 = e^{-c}$

The number of gaps = Ne^{-c}

Total gap length = Ge^{-c}

Fig. 6. Key formulas from the Lander–Waterman mathematical framework for estimating various statistics associated with shotgun sequencing.

mapping projects that were underway at the time. While it was not intended to model shotgun sequencing they have since been used in this way, and these statistical measures were widely used when planning the public human genome sequencing project. The Lander–Waterman model has been used to determine the contig landscape of a randomly fragmented genome given the length of the genome, the average read length, the number of fragments studied and the minimum overlap used for the sequence assembly. Some of the formulas used are presented in Figure 6. It is commonly used to determine the number of sequencing reads required for a given shotgun genome coverage, and the expected number of gaps in an assembly, as functions of the number of reads sequenced.

Lander and Waterman (1988) made two important assumptions; that the sequencing reads were random and independent of each other. Their model shows that coverage of any given base in shotgun coverage is distributed according to a Poisson distribution. In Figure 6, it can be seen that coverage is defined as the total number of acquired bases divided by the expected genome size. It is therefore independent of read length, as the acquired bases are what are being reported on. However, this equation is often used to predict the number of reads that will be required to give a certain level of coverage using an average expected read length from the sequencing machines. The total number of gaps can be calculated as can the total gap length at any given coverage. Using the Lander–Waterman model at a coverage level of 10 you would expect 1 gapped region per 1,000,000 nucleotides. In practice, the values determined by Lander–Waterman are almost always an underestimate, due to non-random sampling arising in cloning biases in shotgun libraries, repeats and other low-complexity regions such as GC/AT rich regions.

5. GENOME ASSEMBLY APPROACHES AND PROGRAMS

5.1. The shortest common superstring model

Once all of the sequencing data has been generated, the task of arranging the sequences into the correct order and orientation to reconstruct the genome sequence then begins. Initial attempts to solve the sequence assembly problem formulated it as the shortest common superstring (SCS) problem, a well-studied problem in computer science. This approach models the sequencing reads as strings and from the set of reads (s_1–s_n) that comprise the dataset, the challenge is to find the SCS, S, that contains each of the sequencing reads (s_1–s_n). This SCS should be the reconstructed genome of the organism and intuitively appears to be a sensible approach to the problem. It can be argued however that the SCS does not accurately model the sequence assembly problem, as the orientation of the reads must be known and the sequence must be error and repeat-free. Furthermore, there is no efficient algorithm for solving the SCS problem. The SCS problem has been shown to be NP-hard and also MAX-SNP hard which means that this problem cannot be efficiently solved in polynomial time, nor can an approximate superstring solution be found that is almost as short as the optimal. An approximation of the SCS problem therefore needed to be found which would efficiently solve the superstring problem without increasing the size of the solution. This greedy algorithm computes all of the possible overlaps between the sequence strings and assigns a score to each of the potential overlaps. Those strings with the highest scores are then merged, the resulting combined string is returned to the set and the process is iterated until it is not possible to combine any more strings. This then provides us with the assembled consensus sequence. The greedy approximations have been constantly refined and have been shown to produce closer approximations that can be efficiently computed, resulting in a two to three fold approximation (Armen and Stein, 1996). Greedy algorithms were first used for DNA fragment assembly by Rodger Staden (1979).

The fact that sequencing and cloning technology is not 100% accurate all of the time and errors can be introduced at either stage, complicates the task of finding overlaps, as any two reads which should overlap may not be 100% identical as would be expected for perfect data. Therefore further approximations, which will allow two reads to overlap provided that the only errors between them can be explained by sequencing or cloning artifacts, need to be added to the greedy algorithm in practice, to identify the overlaps in sequences with errors. Two sequence reads with base differences can be said to overlap if the edit distance between them is less than the sequencing error rate. The edit distance is the minimum number of edits (substitutions or insertions and deletions) that need to be performed to convert sequence x into sequence y. An error tolerance or fraction, e, is provided to the algorithm. When this value is 0, the solution becomes the SCS problem. To be regarded as true overlaps, rather than repeat-induced overlaps, overlaps were expected to occur at either the start or end of a sequence read. In determining overlaps, small overlaps should also be ignored as they can be arrived at purely by chance, due to the four-letter DNA alphabet.

The earliest sequence assembly programs used the greedy algorithm with modifications, and were successfully used for the fragment assembly problems that were being sequenced at the time. Some of the modifications to the greedy algorithm include taking additional information into account such as the score of the alignment (which has been adjusted to by the assignment of positive or negative values for each of matches, mismatches, gap extension penalties and gap opening penalties), looking at the length of the overlap region or looking at the percentage of bases shared between the two sequences with a potential overlap as a fraction of the length of the overlap region. Algorithms that have used some of these measures are the TIGR Assembler (Sutton *et al.*, 1995), used to assemble the first bacterial genome sequenced, Phrap (http://www.phrap.org) and CAP3 (Huang and Madan, 1999). Both Phrap and CAP3 are still widely used.

5.2. Case-study: Phrap

Phrap is a very widely used assembly tool, particularly for BAC assembly and bacterial genome assembly. It has also been used in some modules of the assembly tools developed to handle complex eukaryotic genomes such as Atlas (Havlak *et al.*, 2004) and Phusion (Mullikin and Ning, 2003). It is used in conjunction with the base caller Phred (Ewing *et al.*, 1998) and uses Phred's quality scores in the assembly process to discriminate between repeat-induced and true overlaps, and in determining the consensus sequence. It also uses internally computed quality measures to aid the assembly process in the presence of repeats. In addition, it can use the entire length of the sequence read rather than a portion of the read that has been trimmed for quality and it reconstructs the consensus as a mosaic of the high-quality read segments rather than as a true consensus. A description of the Phrap algorithm follows. Unfortunately, despite its widespread adoption, a paper describing this algorithm has never been published and details about the algorithmic approach need to be obtained from the user documentation (http://www.phrap.org/phredphrap/phrap.html) which by the authors' admission is somewhat outdated and incomplete. Nonetheless, it is possible to illustrate the major steps involved.

All of the sequence and quality data are read and any region at the beginning or end of a read that consists almost entirely of homopolymer bases are converted to N's. The data at the beginning and ends of sequence reads are typically of poorer quality and can lead to spurious matches if they are not handled effectively. Reads are converted to upper-case characters prior to finding potential overlaps. The overlap detection process begins by finding all matching words of at least a given size, 14 at default settings, between any pair of sequences. The matching words are found by constructing a list of all of the words of the default word size in the sequences, sorting the list and scanning the sorted list for matching words. To extend the overlap between matching pairs of sequence, a band of a specified width, 14 by default, which is centered on the imaginary diagonal on a dot matrix plot of the two sequences is defined. Overlapping bands are identified and merged. Other matching segments are then identified which exceed a minimum scoring threshold, 30 by default, by a recursive Smith–Waterman-based search (SWAT), masking those segments already merged. This step allows for the detection of repeats, which have matches in different bands. The pairwise SWAT alignment scores (+1 for a match, −2 for a mismatch, gap opening penalties of −4 and gap extension penalties of −3) are adjusted to reflect the nucleotide composition of the region being scored. Biased nucleotide composition in the sequence segment causes the scores to be adjusted downwards. This is the complexity-adjusted score.

The confirmed part of the each read is then determined from the pairwise matches, corresponding to the part of the read in a SWAT alignment, which is aligned to another read, ignoring reads from the same template. Consideration is given to confirmation by reads on an opposite strand or by a different chemistry type and Phrap-adjusted quality values are then computed on the basis of this information. When more than one read is available to confirm the sequence read then the single highest quality value from all opposite strands is chosen. Strongly confirmed reads, those having matches in a reverse sense read, and chimeric and deletion reads, which are ignored, are also identified in this step.

The Phrap-adjusted quality values are used to compute LLR scores, which are used in the construction of the layout. LLR scores take base quality values into consideration and roughly equate to log likelihood ratios for the hypothesis that the reads are true overlaps to the alternative hypothesis that they are from repeats that are 95% similar. A true overlap will have a positive LLR score while an overlap due to repeats will have a negative LLR score. Nearly identical repeats will therefore not be excluded. The layout, which is the order in which the reads align relative to each other, is constructed by sorting all of the matching pairs in order of decreasing LLR score and progressively merging pairs with high scores by the greedy algorithm until no more merges can be made. The consistency of the layout is checked at the pairwise comparison level. Previously identified deletion and chimeric reads are not merged.

The contig consensus sequence is constructed as a mosaic of the highest quality parts of the individual reads in any given part of the layout. The consensus is constructed by a weighted directed graph whose nodes are selected positions in the reads. There are two types of edges: bidirectional edges between aligned positions in two overlapping reads which is assigned a weight of 0, and unidirectional edges between positions within the same read from 5′ to 3′ direction and are assigned a weight equal to the total Phred quality of the sequence between the two base positions. The highest scoring path through the graph is identified and this is used to construct the consensus. The reads are then aligned to the resulting contigs and any inconsistencies between the read sequences and the contig consensus are identified. The score for any contig position is determined from the individual reads from which that segment of the consensus originated, and the scores are adjusted downwards if there are any discrepancies with other reads.

5.3. Overlap–layout–consensus

Despite the success of the greedy algorithms, as they are concerned with building a consensus sequence from overlapping reads, they are essentially assembling in a local nature and can therefore miss some long-distance relationships between reads which can help resolve repeats in particular. These greedy assembly programs can be confused by repeats which in turn can lead to misassemblies and an incorrect consensus sequence. A further issue with the greedy algorithms is their requirements for large amounts of memory, which limits their usefulness in the assembly of complex genomes. For example, assuming an $8\times$ coverage of a genome, the greedy algorithms require up to IMB of RAM per Mb of the genome. Therefore they can be used for bacteria, BACs and some simple eukaryotic genomes but handling more complex genomes are beyond their capability. The modern genome age therefore heralded the need for a new generation of sequence assembly tools.

6. NEW GENERATION SEQUENCE ASSEMBLY TOOLS

The tools that are used to reconstruct the genomes of more complex organisms have made use of developments in the field of graph theory. A graph is a collection of edges and vertices. As we have seen, the greedy algorithm used an essentially overlap–layout–consensus method to assemble the genome, although in some programs these three steps are not always distinct or explicit. The overlap–layout–consensus method has largely been used for almost all of the assemblers written to cope with the assembly of complex mammalian genomes, such as Arachne (Batzoglou *et al.*, 2002) and the Celera Assembler (Huson *et al.*, 2001). Improvements to the overlap–layout-consensus method that allow them to be used for complex genomes are due to the use of graph theoretic methods. The formulation of the sequence assembly problem as a graph theoretic problem was initially proposed by Peltola *et al.* (1984), and restated more recently by Kececioglu and Myers (1995). The three steps were defined as *overlap*, where the reads which fit the overlap criteria are found; *layout*, where the layout between the reads that induces a layout consistent with the orientation of the reads; and *consensus*, where a multiple alignment of the consistent reads in the layout is performed.

These methods construct a graph at the layout stage from the overlaps that have been found. They regard the sequencing reads as nodes in the graph, which are connected by edges. In the context of sequence assembly, an edge represents an overlap between two reads. The layout algorithm therefore has to find the optimal sub-graph of true overlaps through the graph. During construction of the graph, redundant information is removed, by discarding reads that are fully contained in other reads and by removing transitive edges. For example if the paths $x \rightarrow y \rightarrow z$ and $x \rightarrow z$ exist, then the information in the $x \rightarrow y \rightarrow z$ path can be removed as it is redundant. The layout algorithm determines the relative location and orientation of reads with respect to each other. A Hamiltonian path through the graph visits each vertex exactly once although it does not necessarily pass through all of the edges. A contig is therefore a path in the graph. This is referred to as a Hamiltonian cycle, if the same start and end node are used, and computing this is well-known to be NP-complete.

There is a need therefore for the use of heuristics when determining the layout. Following the determination of the layout, a multiple sequence alignment of all of the overlapping reads in the contig is constructed as directed by the layout graph. The consensus sequence is constructed from the multiple alignment by majority voting, and this consensus infers the genome sequence. Some assemblers use quality values and therefore can take a more sophisticated approach to the consensus. Assembly tools constructed using this approach include the Celera Assembler and Arachne.

6.1. Case Study: Arachne

Arachne is designed specifically for large and complex eukaryotic genomes using mate-pair reads from a whole genome shotgun sequencing approach (Batzoglou *et al.*, 2002). It employs the overlap–layout–consensus approach and incorporates a number of features for large scale genome projects such as an efficient and sensitive method for finding overlaps between reads, merging reads based on forward and reverse mate-pair links, and detecting repeat contigs using the information in the forward and reverse read pair-mate inconsistencies. Interestingly, Arachne also tries to correct sequence errors prior to sequence alignment. Arachne was first described by Batzoglou *et al.* (2002) and was tested using simulated data from small and mid-sized genomes. Arachne2 was published a year later (Jaffe *et al.*, 2003) that included algorithmic adaptations to allow assembly of mammalian genomes and alterations to how gaps and misassemblies in contigs and supercontigs are handled. This was applied to the mouse genome assembly and increased the N50 contig length, N50 corresponding to the length at which 50% of the contigs are of at least that length, by 50%.

An outline of the main stages in the Arachne assembly algorithm is presented below. These steps are performed by seven main program modules. A more detailed examination of the filling gaps module is presented below for illustration of the principles of the use of graph-based algorithms in the overlap–layout–consensus method.

1. *Input data*
   The input data are paired forward and reverse reads from whole genome shotgun data, although unpaired data can also be used. Phred quality values are used by Arachne to trim the reads to remove terminal regions with poor quality and to eliminate reads containing very little high-quality sequence. Known vector sequences are also trimmed and known contaminants are removed.
2. *Overlap detection and alignment; sort and extend (performed by the alignment module)*
   A sort and extend procedure is used to compare all reads to each other. Using k, a parameter with possible values between 8 and 24, a sorted table of k-mer frequencies from the sequence reads and their complements is produced by the alignment module. Owing to the large memory requirements, phase 1 has been optimized so that it can be completed in 100 passes. High-frequency k-mers are discarded as they are likely to originate from repeated sequence. Arachne identifies all instances of read-pairs that share one or more overlapping k-mers. A three-step process is then used to compute pairwise alignments
   a. *Overlap the shared k-mers.*
   b. *Extend the shared k-mers to Alignments.*
      The k-mers are merged into longer, imperfect but gap-free alignments. Mismatches are not allowed at the ends.
3. *Refine alignment by using banded diagonal dynamic programming.*

4. *Error correction*
Sequencing errors among reads are detected by multiple alignments and corrected by the majority rule, taking Phred quality scores into account. This step largely corrects for single discrepant base differences in a read due to sequencing errors, rather than slightly different copies of a repeated sequence. Occasional insertions and deletions are also detected and the alignment corrected.
5. *Evaluation of alignments*
A penalty score is assigned to each aligned pair of overlapping reads with penalties ranging from high to low depending on the sequence quality. These are summed to produce an overall penalty score. If the penalty score exceeds 100, the alignment is rejected.
6. *Identification of paired pairs*
Complexes of paired pairs are built. A paired pair is an instance of two plasmids with similar insert size with overlaps occurring at both ends. These are extended where possible with other paired pairs. These collections of paired pairs are merged into contigs and a consensus is formed which is later treated as a large read. At this stage the contigs sizes can be up to 30k in length.
7. *Contig assembly*
Potential repeat boundaries are identified and contigs are not assembled across these boundaries. The criterion for merging reads is very conservative which is problematic if reads x and y fail to overlap due to sequencing errors. It does however rarely produce misassemblies. Arachne defines dominant and sub-reads. If read A overlaps with B, and all of the reads extending A to the left and right is the superset of those extending B, then A dominates B. A sub-read is one which is fully included within another read. A potential repeat region is where read R can be extended to the right by reads X and Y which do not overlap with each other. Arachne tries to eliminate some spurious repeat boundaries by dropping dominated reads, and performing a second round of contig merging. Sub-reads, which also introduce repeat boundaries, are ignored in the initial contig construction stage.
8. *Detection of repeat contigs*
Potentially misassembled contigs are detected by either their unusually high depth of coverage which can be assessed by using the log-odds ratio, with a repeated contig having a log-odds ratio less than 1, or they typically have conflicting links to other contigs.
9. *Creation of supercontigs*
With the repeat contigs marked, the remaining contigs are ordered and orientated using forward and reverse links from the plasmids into layouts called supercontigs, two forward–reverse links being required to create a supercontig.
10. *Filling gaps in supercontigs*
There are now a number of supercontigs with interleaved gaps, which Arachne attempts to fill with the repeat contigs.

11. *Consensus derivation and post consensus merger*
    The layout of overlapping reads is converted to a consensus sequence by converting pair-wise alignments into multiple alignments using the relative placement of reads. If an alignment is suggested from the placement that does not exist, Arache attempts to recover them. From the left-most end of the contig, it moves to the right assessing using the consistent alignments to determine which base to align to the multiple alignment. A quality-weighted vote determines which base is placed in the consensus.

6.2. Filling gaps in supercontigs

Arachne uses a Hamiltonian graph based approach at the gap filling stage. A graph, $G = (V,E)$ is defined such that V is the set of contigs. An edge E connects two contigs if they are known to overlap. Therefore any two contigs, A and B in G, are connected by a path p, the distance $d_p(A,B)$, which corresponds to the length of sequence between A and B in the path. For every pair of contigs in G, the shortest paths are stored in a new graph, $G_{paths}(V,E_{paths})$ where E_{paths} corresponds to the computed shortest paths between contigs. For every pair of contigs, A, B that are consecutive in a supercontig, Arachne attempts to fill the gap between A and B as follows:

(i) If a path has already been computed for A, B and (A,B) is in E_{paths}, then fill the gap with the contigs in the path
(ii) If this is not the case, firstly find each contig C with forward–reverse links with D so that these links are positioned between A and B in D. The set of these contigs is called the set of targets and is represented by the vector V_C.
(iii) Then G is searched for a path from A to B that only uses nodes in V_C. If a path is found in G, then this corresponds to the path between A and B. The ordered contigs contained in p can then be used to fill the gap between A and B.

6.3. Alternative approaches

An alternative to the overlap–layout–consensus approach was undertaken by Pevzner *et al.* (2001) in the construction of their assembly tool, Euler. They make use of another graph theoretic approach to determine the layout that was first proposed for use in sequencing by hybridization (SBH). SBH is a technique that involves creating an array of all possible k-tuples, as it is not known in advance which subset will be useful, and then hybridizing labeled DNA from the organism of interest to the array. The reverse complement of the sequence represented by the k-tuples will be bound. Hybridization signals are detected and the sequence can be reconstructed by finding the DNA sequence whose base sequence corresponds to the k-tuples that were detected by hybridization. The problem of reconstructing the DNA sequence is represented in graph theory terms by constructing a graph on $(k-1)$-tuples, so that each of the edges has a length of 1. The probe sequences are the nodes in the graph. Two nodes are connected if the $(k-1)$ suffix of one node matches the $(k-1)$ prefix of another node and implicitly represent all of the k-tuples.

Similar to the overlap–layout–consensus method the problem of finding a path through the graph remains. However, instead of using a Hamiltonian path, Pevzner (1989) proposed applying an Eulerian path to the SBH. An Eulerian path passes along each edge in the graph once, but may pass through each node more than once. This problem is not NP-complete and the algorithms that exist to compute this are in linear time. The Eulerian path method, as applied to SBH, was limited largely due to the fact that there was a physical limit to size of DNA that could be sequenced by this technique as the array must contain all k-tuples, and to the limits of hybridization as there may be errors in the hybridization data due to non-specific binding. In addition, there may be more than one solution to an Eulerian path and repetitive DNA can lead to confusing tangles.

Idury and Waterman (1995) extended the use of Eulerian paths to a hybrid SBH/shotgun sequencing approach. They showed that shotgun data could be broken down into overlapping k-tuples, simulating a SBH experiment. As an array of probes no longer needs to be constructed, there is no size limitation to the DNA sequence which can be reconstructed by this method. The set of all k-tuples in the shotgun data corresponds to all of the k-tuples in the original DNA, and therefore solving the path through the graph is equivalent to assembling the DNA. As noted, shotgun sequencing data does contain errors and therefore these can lead to spurious edges which greatly complicate the graph. As the initial Idury/Waterman algorithm did not scale-up well, this work was extended by Pevzner *et al.* (2001) and was implemented as the assembly program Euler.

The implementation of Euler introduced a number of measures to address some of the practical issues with DNA sequencing with the potential to confuse the program. In a novel approach, they introduced an error correction stage as the first stage to correct for errors in the raw sequence data, to prevent errors from excessively tangling the graph. This step is usually the last step in the consensus generation of the more conventional assembly programs. They also combined the information in the original reads to help guide the assembly. This leads to the formulation of the Eulerian SuperPath problem; finding the Eulerian path consistent with all read-paths, a read-path being a path in the graph that corresponds to a sequence read.

7. ASSEMBLY OF GENOMES BY COMPARATIVE MEANS

Using Sanger sequencing technologies, at present there are over 350 published completed genomes and over 1500 genome sequencing projects in progress, the majority of which are bacterial genomes. Closely related species may be sequenced to help understand the mechanism, of virulence, drug resistance, develop genetic fingerprints of strains or to study the genotypic nature of phenotypic differences between strains (Read *et al.*, 2002). With access to emerging technologies which produce millions of reads in a short run time, the number of genomes being sequenced, particularly bacterial genomes will increase greatly. In addition, due to the possibly prohibitive finishing costs and bioinformatics analyses required to complete genomes sequenced using emerging technologies, researchers may wish to use information from previously assembled sequences to guide them in the assembly process.

Another factor that will influence the use of previously sequenced data from closely related species is the low-coverage mammalian genomes currently being funded by the NHGRI (http://www.genome.gov/Pages/Research/Sequencing/SeqProposals/Lowcoveragemammalianseq.doc). The goal of the, currently 25, low-coverage mammalian genome projects is to provide a 2 × coverage of each of these genomes for comparative genomic analysis to help to identify the functional elements in the human genome. To assemble eukaryotic genomes at such a low coverage is a challenge and additional information can be gained by aligning low-coverage genome data to previously sequenced genomes. Genomic sequencing to lower levels of coverage, such as the 0.66 × pig genome study, do not provide enough information to allow an initial genome assembly to be attempted (Wernersson *et al.*, 2005).

In comparative assembly, in place of the conventional overlap–layout–consensus approach, an alignment–layout–consensus approach is used, for example in AMOS-Cmp (Pop *et al.*, 2004). In this method, the overlap phase is removed entirely and is replaced by an alignment phase, whereby the reads are aligned against a reference sequence, using a modified MUMmer (Kurtz *et al.*, 2004) algorithm and produces a layout directly from the alignment phase. The layout is further refined in the next steps to resolve repeats and handling insertions and deletions in the target genome with respect to the reference genome. A multiple alignment is then constructed to create the consensus, followed by scaffolding steps to determine the order and orientation of the contigs. The authors note that this would be a valuable tool for bacterial comparative assembly.

With the decision by the NHGRI to fund the low-coverage sequencing of 25 mammalian genomes, the problem arises of how to assemble these low-coverage genomes to gain as much biologically relevant data as possible. An approach proposed by Gnerre *et al.* (2005) uses synteny between species to improve a low-coverage assembly. At low coverage the ability of a conventional assembler to distinguish between 'true' overlaps and repeat-induced overlaps is diminished. However assemblies can be improved by aligning the reads to high-quality finished genomes from phylogenetically related organisms and using these alignments to confirm links between contigs in the lower coverage genome. This information can then be incorporated into an existing assembly, e.g. created by Arachne (Gnerre *et al.*, 2005).

8. ASSEMBLY OF SEQUENCE DATA FROM EMERGING SEQUENCING TECHNOLOGIES

Sequencing technology based on Sanger dye-termination and currently, separation by capillary electrophoresis has been the gold standard for many years now. However with the push for cheaper alternative technologies to make the $1000 genome a reality (http://grants.nih.gov/grants/guide/rfa-files/RFA-HG-05-004.html) a great deal of effort is being placed on the development of alternative sequencing technologies. See Mitchelson *et al.* (2007) in this volume for a review of some of the emerging technologies. While contemporary sequencing technologies routinely produce read lengths of 600–700 bases in length, many of the new sequencing technologies produce reads which are much shorter. For example, the pyrosequencing-based technology of the 454 Life Sciences Corporation results in reads that are currently a working length of 108 bases on average (Margulies *et al.*, 2007), although over 25 million reads at a high quality are generated per 4 h run (Margulies *et al.*, 2005). Similarly, Solexa's clonal single-molecule array technology sequencing by synthesis approach produces large numbers, up to a billion, of even shorter reads, 20 bases (http://www.solexa.com). The new sequencing technology companies are focusing on resequencing or resequencing and de novo assembly. In addition to being shorter reads, the reads that are produced from the new sequencing technologies are of a different form than those from traditional Sanger sequencing.

Data from the 454 Life Sciences Corporation is in the form of flowgrams and they have produced the Newbler assembler for *de novo* sequencing (Margulies *et al.*, 2007; http://www.454.com). This assembler follows an overlap–layout–consensus algorithm utilizing three modules: Overlapper, Unitigger and Multialigner. The Overlapper module operates in flowgram space rather than nucleotide space for the detection of overlaps, using a scalar value based on the sum of the product of the individual flowgram signal intensities over the entire putative overlap region. If this value exceeds a threshold score then a putative overlap is designated. The Unitigger, which works on the potential overlaps identified by the Overlapper; and the Multialigner, which aligns all of the read signals, both operate in nucleotide space. Chaisson *et al.* (2004) have modeled the assembly of short reads using simulated data from viruses, BACs and bacteria using the FragmentGluer algorithm as implemented in Euler+ (Pevzner *et al.*, 2004) and found that while it is feasible to assemble these reads using existing assembly tools, the genome finishing effort, may be prohibitive.

Given that the vast majority of DNA sequencing is still performed using Sanger methodology, and pyrosequencing is still a relatively new technology; can sequencing data from both technologies be combined and assembled together and how does existing assembly tools cope with data from these new technologies. Data presented by the 454 Life Sciences Corporation would suggest that it is possible to assemble flowgrams and chromatograms together using the same assembly tool (Desany *et al.*, 2005) and can improve assemblies which suffer from cloning biases such as unclonable regions, although details are not provided. Support for incorporating flowgram data into existing tools is underway, for example, with the release of the Staden package 1.6.0. However, anecdotal evidence (http://sourceforge.net/forum/forum.php?thread_id=1462155&forum_id=347718) would suggest that established assembly tools such as Phrap, do not handle the 454 Life Sciences Corporation flowgram data well.

REFERENCES

Adams, M. D., Celniker, S. E., Holt, R. A., Evans, C. A., Gocayne, J. D., Amanatides, P. G., Scherer, S. E., Li, P. W., Hoskins, R. A., Galle, R. F., George, R. A., Lewis, S. E., Richards, S., Ashburner, M., Henderson, S. N., Sutton, G. G., Wortman, J. R., Yandell, M. D., Zhang, Q., Chen, L. X., Brandon, R. C., Rogers, Y. H., Blazej, R. G., Champe, M., Pfeiffer, B. D., Wan, K. H., Doyle, C., Baxter, E. G., Helt, G., Nelson, C. R., Gabor, G. L., Abril, J. F., Agbayani, A., An, H. J., Andrews-Pfannkoch, C., Baldwin, D., Ballew, R. M., Basu, A., Baxendale, J., Bayraktaroglu, L., Beasley, E. M., Beeson, K. Y., Benos, P. V., Berman, B. P., Bhandari, D., Bolshakov, S., Borkova, D., Botchan, M. R., Bouck, J., Brokstein, P., Brottier, P., Burtis, K. C., Busam, D. A., Butler, H., Cadieu, E., Center, A., Chandra, I., Cherry, J. M., Cawley, S., Dahlke, C., Davenport, L. B., Davies, P., de Pablos, B., Delcher, A., Deng, Z., Mays, A. D., Dew, I., Dietz, S. M., Dodson, K., Doup, L. E., Downes, M., Dugan-Rocha, S., Dunkov, B. C., Dunn, P., Durbin, K. J., Evangelista, C. C., Ferraz, C., Ferriera, S., Fleischmann, W., Fosler, C., Gabrielian, A. E., Garg, N. S., Gelbart, W. M., Glasser, K., Glodek, A., Gong, F., Gorrell, J. H., Gu, Z., Guan, P., Harris, M., Harris, N. L., Harvey, D., Heiman, T. J., Hernandez, J. R., Houck, J., Hostin, D., Houston, K. A., Howland, T. J., Wei, M. H., Ibegwam, C., Jalali, M., Kalush, F., Karpen, G. H., Ke, Z., Kennison, J. A., Ketchum, K. A., Kimmel, B. E., Kodira, C. D., Kraft, C., Kravitz, S., Kulp, D., Lai, Z., Lasko, P., Lei, Y., Levitsky, A. A., Li, J., Li, Z., Liang, Y., Lin, X., Liu, X., Mattei, B., McIntosh, T. C., McLeod, M. P., McPherson, D., Merkulov, G., Milshina, N. V., Mobarry, C., Morris, J., Moshrefi, A., Mount, S. M., Moy, M., Murphy, B., Murphy, L., Muzny, D. M., Nelson, D. L., Nelson, D. R., Nelson, K. A., Nixon, K., Nusskern, D. R., Pacleb, J. M., Palazzolo, M., Pittman, G. S., Pan, S., Pollard, J., Puri, V., Reese, M. G., Reinert, K., Remington, K., Saunders, R. D., Scheeler, F., Shen, H., Shue, B. C., Siden-Kiamos, I., Simpson, M., Skupski, M. P., Smith, T., Spier, E., Spradling, A. C., Stapleton, M., Strong, R., Sun, E., Svirskas, R., Tector, C., Turner, R., Venter, E., Wang, A. H., Wang, X., Wang, Z. Y., Wassarman, D. A., Weinstock, G. M., Weissenbach, J., Williams, S. M., Woodage, T., Worley, K. C., Wu, D., Yang, S., Yao, Q. A., Ye, J., Yeh, R. F., Zaveri, J. S., Zhan, M., Zhang, G., Zhao, Q., Zheng, L., Zheng, X. H., Zhong, F. N., Zhong, W., Zhou, X., Zhu, S., Zhu, X., Smith, H. O., Gibbs, R. A., Myers, E. W., Rubin, G. M. and Venter, J. C. (2000). The genome sequence of *Drosophila melanogaster*. Science 287(5461), 2185–2195.

Adams, M. D., Sutton, G. G., Smith, H. O., Myers, E. W. and Venter, J. C. (2003). The independence of our genome assemblies. Proc. Natl. Acad. Sci. USA, 100, 3025–3026.

Armen, C. and Stein, C. (1996). A 2 2/3 approximation algorithm for the shortest common superstring problem. Proc. Combinatorial Pattern Matching 1996 Lecture Notes in Computer Science 1075, 87–101.

Batzoglou, S., Jaffe, D. B., Stanley, K., Butler, J., Gnerre, S., Mauceli, E., Berger, B., Mesirov, J. P. and Lander, E. S. (2002). ARACHNE: a whole-genome shotgun assembler. Genome Res. 12, 177–189.

Bentley, D. R. (1996). Genomic sequence information should be released immediately and freely in the public domain. Science 274, 533–534.

Bentley, S. D., Chater, K. F., Cerdeno-Tarraga, A. M., Challis, G. L., Thomson, N. R., James, K. D., Harris, D. E., Quail, M. A., Kieser, H., Harper, D., Bateman, A., Brown, S., Chandra, G., Chen, C. W., Collins, M., Cronin, A., Fraser, A., Goble, A., Hidalgo, J., Hornsby, T., Howarth, S., Huang, C. H., Kieser, T., Larke, L., Murphy, L., Oliver, K., O'Neil, S., Rabbinowitsch, E., Rajandream, M. A., Rutherford, K., Rutter, S., Seeger, K., Saunders, D., Sharp, S., Squares, R., Squares, S., Taylor, K., Warren, T., Wietzorrek, A., Woodward, J., Barrell, B. G., Parkhill, J. and Hopwood, D. A. (2002). Complete genome sequence of the model actinomycete *Streptomyces coelicolor* A3(2). Nature 417(6885), 141–147.

Bonfield, J. K., Smith, K. F. and Staden, R. (1995). A new DNA sequence assembly program. Nucleic Acids Res. 23, 4992–4999.

Celniker, S. E., Wheeler, D. A., Kronmiller, B., Carlson, J. W., Halpern, A., Patel, S., Adams, M., Mark Champe, M., Dugan, S. P., Frise, E., Hodgson, A., George, R. A., Hoskins, R. A., Laverty, T., Muzny, D. M., Nelson, C. R., Pacleb, J. M., Park, S.,

Pfeiffer, B. D., Richards, S., Sodergren, E. J., Svirskas, R., Tabor, P. E., Wan, K., Stapleton, M., Sutton, G. G., Venter, J. C., Weinstock, G., Scherer, S. E., Myers, E. W., Gibbs, R. A., Rubin, G. M. (2002). Finishing a whole-genome shotgun: Release 3 of the *Drosophila melanogaster* euchromatic genome sequence, Genome Biol. 3, research0079.1–0079.14

Chaisson, M., Pevzner, P. and Tang, H. (2004). Fragment assembly using short reads. Bioinformatics 20, 2067–2074.

Cole, S. T., Brosch, R., Parkhill, J., Garnier, T., Churcher, C., Harris, D., Gordon, S. V., Eiglmeier, K., Gas, S., Barry, C. E., III, Tekaia, F., Badcock, K., Basham, D., Brown, D., Chillingworth, T., Connor, R., Davies, R., Devlin, K., Feltwell, T., Gentles, S., Hamlin, N., Holroyd, S., Hornsby, T., Jagels, K., Krogh, A., McLean, J., Moule, S., Murphy, L., Oliver, K., Osborne, J., Quail, M. A., Rajandream, M. A., Rogers, J., Rutter, S., Seeger, K., Skelton, J., Squares, R., Squares, S., Sulston, J. E., Taylor, K., Whitehead, S. and Barrell, B. G. (1998). Deciphering the biology of *Mycobacterium tuberculosis* from the complete genome sequence. Nature 393(6685), 537–544.

Desany, B., Alvarez, P., Rounsley, S., Brockman, W., Altman, B., Benitez, M., Clark, C., Knight, J., Perbost, C., Vijayadamodar, G., Butler, J., Kleber, M., Zody, M. C., Nusbaum, C., Irzyk, G., Du, L. and Jaffe, D. B. (2005). Assembly of *Neurospora crassa* genome using 454 sequencing technology and new assembler. Genomes, Medicine and the Environment Conference. Hilton Head, SC, October 17–19 (http://www.454.com/downloads/news-events/publications/GMEC_454.pdf).

Edwards, A., Voss, H., Rice, P., Civitello, A., Stegemann, J., Schwager, C., Zimmermann, J., Erfle, H., Caskey, C. T. and Ansorge, W. (1990). Automated DNA sequencing of the human HPTR locus. Genomics 6, 593–608.

Ewing, B. and Green, P. (1998). Base-calling of automated sequencer traces using Phred. II. Error probabilities. Genome Res. 8, 186–194.

Ewing, B., Hillier, L., Wendl, M. C. and Green, P. (1998). Base-calling of automated sequencer traces using Phred. I. Accuracy assessment. Genome Res. 8(3), 175–185.

Fleischmann, R. D., Adams, M. D., White, O., Clayton, R. A., Kirkness, E. F., Kerlavage, A. R., Bult, C. J., Tomb, J. F., Dougherty, B. A., Merrick, J. M., McKenney, K., Sutton, G. G., FitzHugh, W., Fields, C., Gocayne, J. D., Scott, J., Shirley, R., Liu, L.-I., Glodek, A., Kelley, J. M., Weidman, J. F., Phillips, C. A., Spriggs, T., Hedblom, E., Cotton, M. D., Utterback, T. R., Hanna, M. C., Nguyen, D. T., Saudek, D. M., Brandon, R. C., Fine, L. D., Fritchman, J. L., Fuhrmann, J. L., Geoghagen, N. S. M., Gnehm, C. L., McDonald, L. A., Small, K. V., Fraser, C. M., Smith, H. O. and Venter, J. C. (1995). Whole-genome random sequencing and assembly of *Haemophilus influenzae* Rd. Science 269(5223), 496–512.

Fraser, C. M., Gocayne, J. D., White, O., Adams, M. D., Clayton, R. A., Fleischmann, R. D., Bult, C. J., Kerlavage, A. R., Sutton, G., Kelley, J. M., Fritchman, R. D., Weidman, J. F., Small, K. V., Sandusky, M., Fuhrmann, J., Nguyen, D., Utterback, T. R., Saudek, D. M., Phillips, C. A., Merrick, J. M., Tomb, J. F., Dougherty, B. A., Bott, K. F., Hu, P. C., Lucier, T. S., Peterson, S. N., Smith, H. O., Hutchison, C. A., III and Venter, J. C. (1995). The minimal gene complement of *Mycoplasma genitalium*. Science 270(5235), 397–403.

Galagan, J. E., Calvo, S. E., Cuomo, C., Ma, L. J., Wortman, J. R., Batzoglou, S., Lee, S. I., Batürkmen, M., Spevak, C. C., Clutterbuck, J., Kapitonov, V., Jurka, J., Scazzocchio, C., Farman, M., Butler, J., Purcell, S., Harris, S., Braus, G. H., Draht, O., Busch, S., D'Enfert, C., Bouchier, C., Goldman, G. H., Bell-Pedersen, D., Griffiths-Jones, S., Doonan, J. H., Yu, J., Vienken, K., Pain, A., Freitag, M., Selker, E. U., Archer, D. B., Peñalva, M. A., Oakley, B. R., Momany, M., Tanaka, T., Kumagai, T., Asai, K., Machida, M., Nierman, W. C., Denning, D. W., Caddick, M., Hynes, M., Paoletti, M., Fischer, R., Miller, B., Dyer, P., Sachs, M. S., Osmani, S. A. and Birren, B. W. (2005). Sequencing of *Aspergillus nidulans* and comparative analysis with *A. fumigatus* and *A. oryzae*. Nature 438, 1105–1115.

Gnerre, S., Alvarez, P., Brockman, W., Butler, J., Chin, C., Grabherr, M., Kleber, M., Mauceli, E., Chang, J., Clamp, M., Mesirov, J., Lindblad-Toh, K., Lander, E. S. and

Jaffe, D. B. (2005). Synteny-assisted assembly of genomes at low coverage. 9th International Conference on Research in Computational Molecular Biology, Boston, USA, May (http://www.broad.mit.edu/events/recomb2005/posters/posters/0775bb4c4fac5f53858a02250df4e591_abstract_v5.pdf).

Goffeau, A., Barrell, B. G., Bussey, H., Davis, R. W., Dujon, B., Feldmann, H., Galibert, F., Hoheisel, J. D., Jacq, C., Johnston, M., Louis, E. J., Mewes, H. W., Murakami, Y., Philippsen, P., Tettelin, H. and Oliver, S. G. (1996). Life with 6000 genes. Science 274(5287), 546, 563–567.

Gordon, D., Abajian, C. and Green, P. (1998). Consed: a graphical tool for sequence finishing. Genome Res. 8, 195–202.

Gordon, D., Desmarais, C. and Green, P. (2001). Automated finishing with autofinish. Genome Res. 11, 614–625.

Green, P. (1997). Against a whole-genome shotgun. Genome Res. 7, 410–417.

Green, P. (2002). Whole genome disassembly. Proc. Natl. Acad. Sci. USA 99, 4143–4144.

Havlak, P., Chen, R., Durbin, K. J., Egan, A., Ren, Y., Song, X. Z., Weinstock, G. M. and Gibbs, R. A. (2004). The Atlas genome assembly system. Genome Res. 14, 721–732.

Hill, F., Gemund, C., Benes, V., Ansorge, W. and Gibson, T. J. (2000). An estimate of large-scale sequencing accuracy. EMBO Reports 1, 29–31.

Huang, X. and Madan, A. (1999). CAP3: a DNA Sequence Assembly Program. Genome Res. 9, 868–877.

Huson, D. H., Reinert, K., Kravitz, S. A., Remington, K. A., Delcher, A. L., Dew, I. M., Flanigan, M., Halpern, A. L., Lai, Z., Mobarry, C. M., Sutton, G. G. and Myers, E. W. (2001). Design of a compartmentalized shotgun assembler for the human genome. Bioinformatics 17(Suppl. 1), S132–S139.

Idury, R. M. and Waterman, M. S. (1995). A new algorithm for DNA sequence assembly. J. Comp. Biol. 2, 291–306.

International Chicken Genome Sequencing Consortium (2004). Sequence and comparative analysis of the chicken genome provide unique perspectives on vertebrate evolution. Nature 432, 695–716.

International Human Genome Mapping Consortium (2001). A physical map of the human genome. Nature 409(6822), 934–941.

International Human Genome Sequencing Consortium (IHGSC) (2001). Initial sequencing and analysis of the human genome. Nature 409, 860–921.

International Human Genome Sequencing Consortium (IHGSC) (2004). Finishing the euchromatic sequence of the human genome. Nature 431(7011), 931–945.

Istrail, S., Sutton, G. G., Florea, L., Halpern, A. L., Mobarry, C. M., Lippert, R., Walenz, B., Shatkay, H., Dew, I., Miller, J. R., Flanigan, M. J., Edwards, N. J., Bolanos, R., Fasulo, D., Halldorsson, B. V., Hannenhalli, S., Turner, R., Yooseph, S., Lu, F., Nusskern, D. R., Shue, B. C., Zheng, X. H., Zhong, F., Delcher, A. L., Huson, D. H., Kravitz, S. A., Mouchard, L., Reinert, K., Remington, K. A., Clark, A. G., Waterman, M. S., Eichler, E. E., Adams, M. D., Hunkapiller, M. W., Myers, E. W. and Venter, J. C. (2004). Whole-genome shotgun assembly and comparison of human genome assemblies. Proc. Natl. Acad. Sci. USA 101, 1916–1921.

Jaffe, D. B., Butler, J., Gnerre, S., Mauceli, E., Lindblad-Toh, K., Mesirov, J. P., Zody, M. C. and Lander, E. S. (2003). Whole-genome sequence assembly for mammalian genomes: Arachne 2. Genome Res. 13, 91–96.

Kececioglu, J. D. and Myers, E. W. (1995). Combinatorial algorithms for DNA sequence assembly. Algorithmica 13, 7–51.

Kent, W. J. and Haussler, D. (2001). Assembly of the working draft of the human genome with GigAssembler. Genome Res. 11, 1541–1548.

Kurtz, S., Phillippy, A., Delcher, A. L., Smoot, M., Shumway, M., Antonescu, C. and Salzberg, S. L. (2004). Versatile and open software for comparing large genomes. Genome Biol. 5, R12.1–R12.9.

Lander, E. S. and Waterman, M. S. (1988). Genomic mapping by fingerprinting random clones: a mathematical analysis. Genomics 2, 231–239.

Lindblad-Toh, K., Wade, C. M., Mikkelsen, T. S., Karlsson, E. K., Jaffe, D. B., Kamal, M., Clamp, M., Chang, J. L., KulbokasIII, E. J., Zody, M. C., Mauceli, E., Xie, X., Breen, M., Wayne, R. K., Ostrander, E. A., Ponting, C. P., Galibert, F., Smith, D. R., DeJong, P. J., Kirkness, E., Alvarez, P., Biagi, T., Brockman, W., Butler, J., Chin, C. W., Cook, A., Cuff, J., Daly, M. J., DeCaprio, D., Gnerre, S., Grabherr, M., Kellis, M., Kleber, M., Bardeleben, C., Goodstadt, L., Heger, A., Hitte, C., Kim, L., Koepfli, K. P., Parker, H. G., Pollinger, J. P., Searle, S. M., Sutter, N. B., Thomas, R., Webber, C., Baldwin, J., Abebe, A., Abouelleil, A., Aftuck, L., Ait-Zahra, M., Aldredge, T., Allen, N., An, P., Anderson, S., Antoine, C., Arachchi, H., Aslam, A., Ayotte, L., Bachantsang, P., Barry, A., Bayul, T., Benamara, M., Berlin, A., Bessette, D., Blitshteyn, B., Bloom, T., Blye, J., Boguslavskiy, L., Bonnet, C., Boukhgalter, B., Brown, A., Cahill, P., Calixte, N., Camarata, J., Cheshatsang, Y., Chu, J., Citroen, M., Collymore, A., Cooke, P., Dawoe, T., Daza, R., Decktor, K., DeGray, S., Dhargay, N., Dooley, K., Dooley, K., Dorje, P., Dorjee, K., Dorris, L., Duffey, N., Dupes, A., Egbiremolen, O., Elong, R., Falk, J., Farina, A., Faro, S., Ferguson, D., Ferreira, P., Fisher, S., FitzGerald, M., Foley, K., Foley, C., Franke, A., Friedrich, D., Gage, D., Garber, M., Gearin, G., Giannoukos, G., Goode, T., Goyette, A., Graham, J., Grandbois, E., Gyaltsen, K., Hafez, N., Hagopian, D., Hagos, B., Hall, J., Healy, C., Hegarty, R., Honan, T., Horn, A., Houde, N., Hughes, L., Hunnicutt, L., Husby, M., Jester, B., Jones, C., Kamat, A., Kanga, B., Kells, C., Khazanovich, D., Kieu, A. C., Kisner, P., Kumar, M., Lance, K., Landers, T., Lara, M., Lee, W., Leger, J. P., Lennon, N., Leuper, L., LeVine, S., Liu, J., Liu, X., Lokyitsang, Y., Lokyitsang, T., Lui, A., Macdonald, J., Major, J., Marabella, R., Maru, K., Matthews, C., McDonough, S., Mehta, T., Meldrim, J., Melnikov, A., Meneus, L., Mihalev, A., Mihova, T., Miller, K., Mittelman, R., Mlenga, V., Mulrain, L., Munson, G., Navidi, A., Naylor, J., Nguyen, T., Nguyen, N., Nguyen, C., Nguyen, T., Nicol, R., Norbu, N., Norbu, C., Novod, N., Nyima, T., Olandt, P., O'Neill, B., O'Neill, K., Osman, S., Oyono, L., Patti, C., Perrin, D., Phunkhang, P., Pierre, F., Priest, M., Rachupka, A., Raghuraman, S., Rameau, R., Ray, V., Raymond, C., Rege, F., Rise, C., Rogers, J., Rogov, P., Sahalie, J., Settipalli, S., Sharpe, T., Shea, T., Sheehan, M., Sherpa, N., Shi, J., Shih, D., Sloan, J., Smith, C., Sparrow, T., Stalker, J., Stange-Thomann, N., Stavropoulos, S., Stone, C., Stone, S., Sykes, S., Tchuinga, P., Tenzing, P., Tesfaye, S., Thoulutsang, D., Thoulutsang, Y., Topham, K., Topping, I., Tsamla, T., Vassiliev, H., Venkataraman, V., Vo, A., Wangchuk, T., Wangdi, T., Weiand, M., Wilkinson, J., Wilson, A., Yadav, S., Yang, S., Yang, X., Young, G., Yu, Q., Zainoun, J., Zembek, L., Zimmer, A. and Lander, E. S. (2005). Genome sequence, comparative analysis and haplotype structure of the domestic dog. Nature 438(7069), 803–819.

Margulies, M., Egholm, M., Altman, W. E., Attiya, S., Bader, J. S., Bemben, L. A., Berka, J., Braverman, M. S., Chen, Y. J., Chen, Z., Dewell, S. B., Du, L., Fierro, J. M., Gomes, X. V., Godwin, B. C., He, W., Helgesen, S., Ho, C. H., Irzyk, G. P., Jando, S. C., Alenquer, M. L., Jarvie, T. P., Jirage, K. B., Kim, J. B., Knight, J. R., Lanza, J. R., Leamon, J. H., Lefkowitz, S. M., Lei, M., Li, J., Lohman, K. L., Lu, H., Makhijani, V. B., McDade, K. E., McKenna, M. P., Myers, E. W., Nickerson, E., Nobile, J. R., Plant, R., Puc, B. P., Ronan, M. T., Roth, G. T., Sarkis, G. J., Simons, J. F., Simpson, J. W., Srinivasan, M., Tartaro, K. R., Tomasz, A., Vogt, K. A., Volkmer, G. A., Wang, S. H., Wang, Y., Weiner, M. P., Yu, P., Begley, R. F. and Rothberg, J. M. (2005). Genome sequencing in microfabricated high-density picolitre reactors. Nature 437(7057), 376–380.

Margulies, M., Jarvie, T. P., Knight, J. R. and Simons, J. F. (2007). The 454 Life Sciences Picoliter Sequencing System. In: K. R. Mitchelson (Ed), New High Throughput Technologies for DNA Sequencing and Genomics (pp. 151–186). Elsevier, Amsterdam.

Mitchelson, K. R., Hawkes, D. B., Turakulov, R. and Men, A. E. (2007). Developments in DNA sequencing. In: K. R. Mitchelson (Ed), New Technologies for DNA Sequencing and Genomics (pp. 1–45). Elsevier, Amsterdam.

Mullikin, J. C. and Ning, Z. (2003). The phusion assembler. Genome Res. 13(1), 81–90.

Myers, E. W., Sutton, G. G., Delcher, A. L., Dew, I. M., Fasulo, D. P., Flanigan, M. J., Kravitz, S. A., Mobarry, C. M., Reinert, K. H., Remington, K. A., Anson, E. L., Bolanos,

R. A., Chou, H. H., Jordan, C. M., Halpern, A. L., Lonardi, S., Beasley, E. M., Brandon, R. C., Chen, L., Dunn, P. J., Lai, Z., Liang, Y., Nusskern, D. R., Zhan, M., Zhang, Q., Zheng, X., Rubin, G. M., Adams, M. D. and Venter, J. C. (2000). A whole-genome assembly of *Drosophila*. Science 287(5461), 2196–2204.

Myers, E. W., Sutton, G. G., Smith, H. O., Adams, M. D. and Venter, J. C. (2002). On the sequencing and assembly of the human genome. Proc. Natl. Acad. Sci. USA 99, 4145–4146.

Peltola, H., Soderlund, H. and Ukkonen, E. (1984). SEQAID: a DNA sequence assembling program based on a mathematical model. Nucleic Acids Res. 12(1), 307–321.

Pevzner, P. (1989). L-tuple DNA sequencing: computer analysis. J. Biomol. Struct. Dynam. 7, 63–73.

Pevzner, P. A., Tang, H. and Tesler, G. (2004). *De novo* repeat classification and fragment assembly. Genome Res. 14, 1786–1796.

Pevzner, P. A., Tang, H. and Waterman, M. S. (2001). An Eulerian approach to DNA fragment assembly. Proc. Natl. Acad. Sci. USA, 98, 9748–9753.

Pop, M., Philippy, A., Delcher, A. L. and Salzberg, S. L. (2004). Comparative genome assembly. Briefings Bioinform. 5, 237–248.

Rat Genome Sequencing Project Consortium (2004). Genome sequence of the Brown Norway rat yields insights into mammalian evolution. Nature 428, 493–521.

Read, T. D., Salzberg, S. L., Pop, M., Shumway, M., Umayam, L., Jiang, L., Holtzapple, E., Busch, J. D., Smith, K. L., Schupp, J. M., Solomon, D., Keim, P. and Fraser, C. M. (2002). Comparative genome sequencing for discovery of novel polymorphisms in *Bacillus anthracis*. Science 296(5575), 2028–2033.

Salzberg, S. L., Church, D., DiCuccio, M., Yaschenko, E. and Ostell, J. (2004). The genome assembly archive: a new public resource. PLoS Biol. 2(9), E285.

Salzberg, S. L. and Yorke, J. A. (2005). Beware of misassembled genomes. Bioinformatics 21(24), 4320–4321.

Sanger, F., Air, G. M., Barrell, B. G., Brown, N. L., Coulson, A. R., Fiddes, C. A., Hutchison, C. A., Slocombe, P. M. and Smith, M. (1977). Nucleotide sequence of bacteriophage phi X174 DNA. Nature 265(5596), 687–695.

Sanger, F., Coulson, A. R., Barrell, B. G., Smith, A. J. and Roe, B. A. (1980). Cloning in single-stranded bacteriophage as an aid to rapid DNA sequencing. J. Mol. Biol. 143, 161–178.

Sanger, F., Coulson, A. R., Hong, G. F., Hill, D. F. and Petersen, G. B. (1982). Nucleotide sequence of bacteriophage lambda DNA. J. Mol. Biol. 162, 729–773.

Staden, R. (1979). A strategy of DNA sequencing employing computer programs. Nucleic Acids Res. 6, 2601–2610.

Staden, R. (1982). Automation of the computer handling of gel reading data produced by the shotgun method of DNA sequencing. Nucleic Acids Res. 10, 4731–4751.

Sutton, G., White, O., Adams, M. and Kerlavage, A. (1995). TIGR assembler: a new tool for assembling large shotgun sequencing projects. Genome Sci. Technol. 1, 9–19.

The Arabidopsis Initiative (2000). Analysis of the genome sequence of the flowering plant *Arabidopsis thaliana*. Nature 408, 796–815.

The *C. elegans* Sequencing Consortium (1998). Genome sequence of the Nematode *Caenorhabditis elegans*. A platform for investigating biology. Science 282, 2012–2018.

The Chimpanzee Sequencing and Analysis Consortium (2005). Initial sequence of the chimpanzee genome and comparison with the human genome. Nature 437, 69–87.

Venter, J. C., Adams, M. D., Myers, E. W., Li, P. W., Mural, R. J., Sutton, G. G., Smith, H. O., Yandell, M., Evans, C. A., Holt, R. A., Gocayne, J. D., Amanatides, P., Ballew, R. M., Huson, D. H., Wortman, J. R., Zhang, Q., Kodira, C. D., Zheng, X. H., Chen, L., Skupski, M., Subramanian, G., Thomas, P. D., Zhang, J., Gabor Miklos, G. L., Nelson, C., Broder, S., Clark, A. G., Nadeau, J., McKusick, V. A., Zinder, N., Levine, A. J., Roberts, R. J., Simon, M., Slayman, C., Hunkapiller, M., Bolanos, R., Delcher, A., Dew, I., Fasulo, D., Flanigan, M., Florea, L., Halpern, A., Hannenhalli, S., Kravitz, S., Levy, S., Mobarry, C., Reinert, K., Remington, K., Abu-Threideh, J., Beasley, E., Biddick, K., Bonazzi, V., Brandon, R., Cargill, M., Chandramouliswaran, I., Charlab, R., Chaturvedi,

K., Deng, Z., Di Francesco, V., Dunn, P., Eilbeck, K., Evangelista, C., Gabrielian, A. E., Gan, W., Ge, W., Gong, F., Gu, Z., Guan, P., Heiman, T. J., Higgins, M. E., Ji, R. R., Ke, Z., Ketchum, K. A., Lai, Z., Lei, Y., Li, Z., Li, J., Liang, Y., Lin, X., Lu, F., Merkulov, G. V., Milshina, N., Moore, H. M., Naik, A. K., Narayan, V. A., Neelam, B., Nusskern, D., Rusch, D. B., Salzberg, S., Shao, W., Shue, B., Sun, J., Wang, Z., Wang, A., Wang, X., Wang, J., Wei, M., Wides, R., Xiao, C., Yan, C., Yao, A., Ye, J., Zhan, M., Zhang, W., Zhang, H., Zhao, Q., Zheng, L., Zhong, F., Zhong, W., Zhu, S., Zhao, S., Gilbert, D., Baumhueter, S., Spier, G., Carter, C., Cravchik, A., Woodage, T., Ali, F., An, H., Awe, A., Baldwin, D., Baden, H., Barnstead, M., Barrow, I., Beeson, K., Busam, D., Carver, A., Center, A., Cheng, M. L., Curry, L., Danaher, S., Davenport, L., Desilets, R., Dietz, S., Dodson, K., Doup, L., Ferriera, S., Garg, N., Gluecksmann, A., Hart, B., Haynes, J., Haynes, C., Heiner, C., Hladun, S., Hostin, D., Houck, J., Howland, T., Ibegwam, C., Johnson, J., Kalush, F., Kline, L., Koduru, S., Love, A., Mann, F., May, D., McCawley, S., McIntosh, T., McMullen, I., Moy, M., Moy, L., Murphy, B., Nelson, K., Pfannkoch, C., Pratts, E., Puri, V., Qureshi, H., Reardon, M., Rodriguez, R., Rogers, Y. H., Romblad, D., Ruhfel, B., Scott, R., Sitter, C., Smallwood, M., Stewart, E., Strong, R., Suh, E., Thomas, R., Tint, N. N., Tse, S., Vech, C., Wang, G., Wetter, J., Williams, S., Williams, M., Windsor, S., Winn-Deen, E., Wolfe, K., Zaveri, J., Zaveri, K., Abril, J. F., Guigo, R., Campbell, M. J., Sjolander, K. V., Karlak, B., Kejariwal, A., Mi, H., Lazareva, B., Hatton, T., Narechania, A., Diemer, K., Muruganujan, A., Guo, N., Sato, S., Bafna, V., Istrail, S., Lippert, R., Schwartz, R., Walenz, B., Yooseph, S., Allen, D., Basu, A., Baxendale, J., Blick, L., Caminha, M., Carnes-Stine, J., Caulk, P., Chiang, Y. H., Coyne, M., Dahlke, C., Mays, A., Dombroski, M., Donnelly, M., Ely, D., Esparham, S., Fosler, C., Gire, H., Glanowski, S., Glasser, K., Glodek, A., Gorokhov, M., Graham, K., Gropman, B., Harris, M., Heil, J., Henderson, S., Hoover, J., Jennings, D., Jordan, C., Jordan, J., Kasha, J., Kagan, L., Kraft, C., Levitsky, A., Lewis, M., Liu, X., Lopez, J., Ma, D., Majoros, W., McDaniel, J., Murphy, S., Newman, M., Nguyen, T., Nguyen, N., Nodell, M., Pan, S., Peck, J., Peterson, M., Rowe, W., Sanders, R., Scott, J., Simpson, M., Smith, T., Sprague, A., Stockwell, T., Turner, R., Venter, E., Wang, M., Wen, M., Wu, D., Wu, M., Xia, A., Zandieh, A. and Zhu, X. (2001). The sequence of the human genome. Science 291(5507), 1304–1351.

Waterston, R. H., Lander, E. S. and Sulston, J. E. (2002). On the sequencing of the human genome. Proc. Natl. Acad. Sci. USA 99, 3712–3716.

Waterston, R. H., Lander, E. S. and Sulston, J. E. (2003). More on the sequencing of the human genome. Proc. Natl. Acad. Sci. USA 100, 3022–3024.

Weber, J. L. and Myers, E. W. (1997). Human whole-genome shotgun sequencing. Genome Res. 7, 401–409.

Wernersson, R., Schierup, M. H., Jorgensen, F. G., Gorodkin, J., Panitz, F., Staerfeldt, H. H., Christensen, O. F., Mailund, T., Hornshoj, H., Klein, A., Wang, J., Liu, B., Hu, S., Dong, W., Li, W., Wong, G. K., Yu, J., Wang, J., Bendixen, C., Fredholm, M., Brunak, S., Yang, H. and Bolund, L. (2005). Pigs in sequence space: a 0.66X coverage pig genome survey based on shotgun sequencing. BMC Genomics 6, E70.

Chapter 12

Valid Recovery of Nucleic Acid Sequence Information from High Contamination Risk Samples – Ancient DNA and Environmental DNA

George A. Kowalchuk,[1] Jeremy J. Austin,[2] Paul S. Gooding[3] and John R. Stephen[3]

[1]*Netherlands Institute of Ecology, PO Box 40, 6666 ZG Heteren, The Netherlands*
[2]*Department of Environmental Biology, University of Adelaide, North Terrace Adelaide, SA 5005, Australia*
[3]*Agricultural Division – AGRF, Plant Genomics Centre, University of Adelaide, Hartley Grove, Waite Campus PMB 1 Glen Osmond, SA 5064, Australia*

Contents

Abstract

The introduction of molecular techniques into environmental and forensic sciences has opened up an entirely new window of observation into our world's biology and its history. However, the very sensitive nature of such methods leaves them particularly prone to error and artifact. This is especially true in the analysis of problematic samples with suboptimal nucleic acid quantity or quality, as is often the case in microbial ecological studies and ancient DNA investigations. Numerous studies have highlighted the potential sources and extent of artifact in such molecular approaches, but such cautionary messages have generally gone unheeded. Some unavoidable sources of error and artifact are inherent to molecular studies and must simply be kept in mind when interpreting the robustness of results. Other causes of error can at least be partially addressed via good laboratory practice measures, proper controls and thorough data scrutiny. Some laboratories and disciplines have been keen to adopt appropriate measures to circumvent some of the validity issues involved in

PERSPECTIVES IN BIOANALYSIS, VOLUME 2
ISSN 1871-0069 DOI: 10.1016/S1871-0069(06)02012-X

molecular studies, whereas others lag behind. Here, we identify sources of error, contamination and artifact in nucleic acid-based studies, especially as they apply to problematic material, suboptimal in sample size, source, quality or purity. In evaluating these concerns, we further offer some considerations and suggestions to help minimize sources of error and artifacts in laboratories studying molecular ecology and evolution with frequently problematic, and usually rare, sample materials.

1. INTRODUCTION

In the past two decades, the ability to extract, amplify and characterize nucleic acids from a diverse range of environmental samples and post-mortem human, animal and plant remains has revolutionized our ability to study the ecology and evolution of biotic communities past and present (Pace, 1997; Whitman *et al.*, 1998; Hofreiter *et al.*, 2001). Indeed, molecular access to ancient materials is yielding astounding breakthroughs ranging from determining mammoth ancestry (Krause *et al.*, 2006) to reconstructing and analyzing past pandemic-causing viruses (Taubenberger *et al.*, 1997; Tumpey *et al.*, 2005). Similarly, molecular analyses of novel and extreme environments, from isolated deep subsurface Antarctic lake ice (Priscu *et al.*, 1999; Christner *et al.*, 2001) to super-hot deep sea vents (Kashefi and Lovely, 2003), has revolutionized our appreciation of microbial diversity. However, the comparative ease of application of the new molecular technologies in general (e.g. kit-based DNA extraction and high-throughput DNA sequencing) relative to the technical challenges related to eliminating contamination and authentication of results means that misleading information can and often does enter the public domain (Ashelford *et al.*, 2005; Hugenholtz and Huber, 2003). The problem of incorrect and aberrant sequences in environmental and forensic studies is universal, but is particularly challenging to control for high-contamination risk samples, where more traditional types of validation (e.g. reference to museum voucher specimens or bacterial culture collections) are not possible (Rappe and Giovannoni, 2003). Proving that a nucleic acid sequence has a single biological origin (i.e. is not a chimera) and is endogenous to a single sample or specimen (i.e. is not an external contaminant) is still a major challenge, especially in the fields of ancient DNA and environmental microbial ecology.

While such issues have been recognized for some time now, recent developments that provide greater access to ancient and pristine, low-biomass samples, coupled with tremendous advances in methods of rapid data generation (Shendure *et al.*, 2004; Margulies *et al.*, 2005; Venter *et al.*, 2004), have recently magnified these concerns. For instance, researchers are pushing the limits of possibility even further in the amplification of nucleic acids from more recalcitrant ancient materials where little intact DNA remains (Yang *et al.*, 2003; Willerslev *et al.*, 2003, 2004). Similarly, recent expeditions to new and extreme environments are providing access to microbial communities that have to date been isolated from man (e.g. deep sea, deep soil and sediment, deep ice and submerged lakes), and where contamination risks are often compounded by the low biomass present in such pristine environments. Concerns of contamination also extend to such disciplines as astrobiology, where obviously it must be clearly demonstrated that detection of supposed extraterrestrial life is not the detection of contamination or artifact.

The main focus of this chapter is to address the challenges of recovering nucleic acid sequence information from especially troublesome samples, and the issues involved in subsequent data validation. In particular, we discuss the steps required from environmental sample to data submission and the potential sources of contamination, bias and artifacts during each of these steps. We should stress that our focus is helping to ensure the quality of nucleic acid-derived information in environmental research, and we do not purport to provide quality control guidelines for clinical, forensic or medical diagnostics. We further offer several recommendations and considerations that should be followed during the course of molecular environmental studies to help minimize the risks of faulty data generation and dissemination.

2. FEATURES OF HIGH CONTAMINATION AND ARTIFACT RISK SAMPLES

Two technical challenges face all studies involving high contamination risk samples, both of which can produce erroneous DNA sequences. The first involves the problem of sequence accuracy. DNA sequences may contain errors due to post-mortem damage of template DNA (Paabo *et al.*, 1990), polymerase chain reaction (PCR) jumping creating chimeric molecules (DeSalle *et al.* 1993; Wang and Wang, 1997) and formation of heteroduplex molecules during PCR that are non-systematically repaired during subsequent cloning experiments (Speksnijder *et al.*, 2001; also see Section 3). The second involves the problem of contamination by exogenous contemporary DNA.

Irrespective of the starting material, DNA from non-living material is damaged to a greater or lesser extent, a feature that must be considered in all downstream analyses. Within living cells, the structure and sequence of DNA is maintained by a variety of repair pathways (Lindahl, 1993). After cells cease to be metabolically active, DNA is degraded and modified by a range of biological and chemical processes – endogenous nucleases, microbial degradation, cross-linking, oxidation and hydrolysis. Together these result in fragmentation of the phosphodiester backbone producing shorter and shorter fragments (often no longer than 100–500 bp), a dramatic drop in copy number, and the accumulation of baseless sites (e.g. depurination), blocking lesions (e.g. oxidation of pyrimidines to 5-hydroxy-5-methyl hydantoin and 5-hydroxy hydantoin) and miscoding lesions (e.g. hydrolysis of adenine to hypoxanthine, cytosine to uracil). DNA fragmentation severely limits the size of PCR amplicons that can be obtained, especially ancient specimens, forcing larger targets to be amplified as multiple, short, overlapping fragments (Stone and Stoneking, 1998). Blocking lesions prevent the elongation of DNA strands by *Taq* polymerase, further reducing the effective copy number of DNA fragments of a specified length in ancient specimens. Miscoding lesions can cause incorrect bases to be incorporated during strand synthesis. In particular a damaged cytosine (uracil) will be copied as an A instead of G, and a damaged adenine (hypoxanthine) will be copied as C instead of T. Thus, even in the absence of contaminating DNA, the short and damaged nature of ancient DNA means that sequences obtained will contain errors.

With samples for microbial analysis, the problem is often the exact opposite; namely, the researchers wish to recover nucleic acids only from living cells, without contamination from extracellular nucleic acids present in the environment. Sample washing steps have typically been used to help alleviate this problem, but few controls have been put into place to verify the effectiveness of such measures. Alternatively, the use of RNA, which is far less stable than DNA both intra- as well as extracellularly, can help ensure that the extracted nucleic acids are predominantly from intact cells. Under the supposition that general microbial activity is proportional to cellular ribosome numbers (Wagner, 1994; Keener and Nomura, 1996), the use of ribosomal RNA, as opposed to or in addition to DNA, has gained popularity, and is generally regarded as providing additional insight into the identities of active microbial populations. However, it must be recognized that the correlation between ribosome numbers and activity only holds for cells during relatively rapid growth (Kemp *et al.*, 1993), which is rarely encountered by most microbial populations *in situ.* A number of sample features are correlated with increased risk of contamination with studies of environmental and ancient nucleic acids. The most important of these is low-target concentrations. As target numbers decrease, detection of true targets above the background of potential artifacts and contaminants becomes increasingly difficult. Other characteristics of problem samples include the presence of compounds that impede nucleic acid extraction or inhibit downstream analyses, difficulties of access, which may necessitate expensive and less than optimal sampling, and a lack of knowledge concerning the nature of the expected targets.

2.1. Sources of nucleic acids and sampling

Proper sampling of high contamination risk materials is the first of many critical steps to ensure accuracy and quality of all subsequent DNA sequence information. Whether the samples are museum specimens, *in situ* sub-fossil bone and archaeological material, or low-DNA environmental samples (e.g. ice, soil, sediment, water), the very act of sampling can introduce exogenous contaminating DNA that may be at much greater quantity and quality than in the sample itself. In order to ensure the greatest integrity of samples, field workers should wear disposable gloves at all times and change them regularly, and preferably face masks and coveralls, to limit the amount of human-associated DNA (i.e. human skin, hair, sweat and aerosols, skin and respiratory tract micro-organisms and micro-fauna) from contaminating collecting equipment and specimens. All sampling equipment should be sterile and DNA-free, as well as free from DNases and RNases. This can be achieved by purchasing supplies that are guaranteed sterile and DNA-free or pre-treating equipment with acid, bleach or UV light. Swabs of all sampling equipment should be taken prior to sampling to monitor for any pre-existing contamination. Contamination can also be monitored during sample collection. For example, when taking sediment cores, a known DNA sample can be used to spike the coring equipment (Christner *et al.*, 2005). Absence of the known contaminant in any subsequently analyzed samples strongly suggests that external contaminants were not introduced during the sampling process itself.

It has been shown that sample core surface layers can contain large numbers of introduced bacteria (Christner *et al.*, 2005). Thus, where possible, samples should be collected to allow subsequent sterilization and/or removal of the outer surface so that only the innermost sections are used for nucleic acid extraction (Rogers *et al.*, 2004). Similarly, special devices for reliable large-volume sampling can help to ensure the integrity of liquid samples, such as deep-sea water (Wommack *et al.*, 2004). These sampling and sample work-up considerations are also of paramount importance in astrobiology, where robust and clean sampling strategies must be devised far in advance and operated at a distance (Bada, 2001). Sampling design should also allow for enough sample to be collected so that independent replication of results and alternative supporting analyses can be carried out (e.g. for ancient DNA work, carbon dating, collagen analysis, amino-acid racemization; for microbial work, phospholipid fatty acid, culture or microscopic analyses). In many cases, especially for precious museum samples or unique environments, there is a necessary trade-off between obtaining the largest and most representative sample possible versus physical damage to unique and important material. Given these sampling issues of unique, rare or difficult to access samples, sample storage and curation is of utmost importance to maintain sample integrity for future or long-term analyses.

These issues have become an even greater concern in recent years as we are afforded greater access to new environments, including ancient material, pristine environments (deep sea vents, deep soils and sediments, ice-cores and buried lakes), and low-biomass environmental and newly discovered ancient materials. Microbial community analysis of ancient samples might be considered as the riskiest of all sample types, and results of studies targeting such materials have been called into question (Vreeland *et al.*, 2000; Fish *et al.*, 2002).

2.2. Nucleic acid extraction

There are three major technical issues involved during extraction of nucleic acids from low-DNA, forensic or environmental samples: (1) extraction efficiency; (2) how well the extracted nucleic acids actually represent those in the sample; and (3) chemical and nucleic acid contamination. If sample size and quality is sufficient, the efficiency of extraction may not be a serious issue, as many analyses only require small amounts of nucleic acid template. However, given the increased risk of artifacts when working with low concentrations of extracted nucleic acids (see Section 2.1), efficient nucleic acid extraction is generally desirable. Obviously, as molecular target numbers approach limits of reliable detection, such as in very small or dilute samples, extraction efficiency becomes an important issue. For highly dilute samples, ultracentrifugation devices may facilitate the concentration of large volume extractions. Also, although many samples may contain sufficient nucleic acids for proper analyses, these nucleic acids can be bound to the sample substrate or contain intermolecular cross-links that inhibit downstream analyses. Addition of PTB (*N*-phenacylthiazolium bromide) appears to break intermolecular cross-links caused by Maillard reactions, releasing more PCR amplifiable DNA during the extraction process (Poinar *et al.*, 1998). In addition, many cells, cell walls or organism structures can be highly impervious to lysis. Numerous detergents, lytic enzymes and physical disruption methods have been introduced to facilitate liberation of nucleic acids from recalcitrant samples (Griffiths *et al.*, 2000; Hurt *et al.*, 2001). However, it must be kept in mind that there is a general trade-off between the rigorousness of extraction methods and the quality of nucleic acids extracted, and decreased nucleic acid quality again increases risks of downstream artifacts. Although manufactured kits have recently become available for the routine and potentially standardized extraction of nucleic acids from diverse sample types, it must be remembered that such kits have not been rigorously validated across a full range of organisms and sample types.

If research goals include comparisons of target frequencies it is essential that extracted nucleic acid targets are in the same ratio as present in the original sample. Unfortunately, nucleic acids from different organisms, tissues or substrates are not liberated and recovered equally, and this potential for bias must be recognized when interpreting results. This issue is less serious if one is comparing like samples, as any extraction biases can be assumed to remain constant across all samples. The relative representation levels of extracted nucleic acids may be a moot point if single targets are being analyzed, but is a serious issue for complex samples and analyses.

Contamination during extraction procedures can be of two types: co-extraction of substances that may inhibit downstream analyses and the introduction of contaminant sequences during extraction procedures that can lead to the recovery of erroneous results. The former contamination type is generally the result of failure to eliminate compounds from the sample that inhibit downstream analyses, whereas the latter introduces false sequences into the sample. Co-extraction of contaminating DNA and enzymatic inhibitors is a serious issue, potentially leading to false negative results or compounding artifact risks. The range of starting material and preservation conditions, in the case of ancient DNA studies, is extremely broad including soils, sediments, rocks, vegetation, water, ice, archival museum and pathology specimens (dried ethanol and/or formalin fixed skins, hair, feather, bones and tissues), archaeological material (e.g. bone, clothing, pottery, stone tools, food residues), faeces, naturally or artificially mummified remains, coprolites (subfossil faeces) and forensic specimens (e.g. body fluids, hair, skin, stool). This diversity of material and preservation creates a range of challenges for DNA extraction and subsequent enzymatic manipulations, and often necessitates customized extraction and purification procedures. Compounds that are co-extracted with nucleic acids, such as phenolic compounds and humic acids, often require additional purification steps or extract dilution prior to downstream analyses, both of which can decrease target density and therefore increase contamination risk (e.g. by necessitating use of additional PCR cycles; see Section 3).

The concerns associated with the introduction of contaminant sequences during extraction procedures are very similar to those encountered in sample collection and storage. Again, concerns are greatest when dealing with samples containing low-target numbers and/or nucleic acid quality. Preparations of sample material prior to actual extraction procedures must be carried out in an otherwise DNA-free laboratory environment. Also, all reagents and equipment used for the extraction must be free of nucleic acid contamination. It is generally assumed that commercial kits and reagents are free of nucleic acids, but this is often not the case, and precautionary UV treatment of all materials is therefore recommended.

3. AMPLIFICATION AND/OR RECOVERY OF NUCLEIC ACIDS IN THE LABORATORY

The PCR has provided the ability to amplify minute amounts of target nucleic acids, from a complex mixture of possible targets, to concentrations that allow molecular characterization. Indeed, PCR has made ancient DNA studies possible and has revolutionized microbial ecology (Hugenholtz *et al*., 1998). Much of the power of PCR lies in its sensitivity due to the exponential amplification of representative nucleic acids. However, it is this very sensitivity that makes PCR prone to contamination artifacts. Furthermore, not all potential PCR templates amplify with the same efficiency, and PCR product ratios may not match those present in the original sample pool. PCR products may also contain errors introduced at early stages by the infidelity of DNA amplification. Lastly, PCR methodologies can produce artifact sequences by a number of mechanisms that introduce aberrant sequence information into amplicon sequences (e.g. mispriming, foldback product amplification, etc).

The problem of contamination by exogenous DNA in PCR-based studies is two-fold. First, general environmental DNA (e.g. human and animal hair, skin and sweat, airborne micro-organisms – bacteria, fungi and invertebrates, pollen, etc.) is ubiquitous and poses a threat to all studies relying on the amplification of nucleic acids. This is especially true of studies where the target amplicons are similar or indistinguishable from common DNA contaminants, such as in human and microbial studies. For instance, most museum specimens are heavily contaminated with modern human DNA, simply as a consequence of handling during collection, preservation, curation and examination. Similarly, most laboratories and equipment are contaminated with microbial DNA targets. Second, molecular biological experiments can easily generate further contamination via the phenomenon of "PCR carryover", which represents an extremely serious contamination threat. A successful PCR amplification can contain 10^{12}–10^{15} copies of a single amplicon that can rapidly become spread across clothing, laboratory surfaces, door handles, corridors and offices via aerosol droplets created when PCR products are handled. These contamination sources are highly concentrated; a single aerosol droplet can contain 10^{5}–10^{9}

amplicons, all of which are potential targets for subsequent reactions. For samples with low biomass or only small amounts of intact DNA, it becomes clear that laboratory contamination can far exceed the number of targets within a sample. Indeed, for those working with difficult templates, an irony of PCR is that the more sloppily it is performed, the greater the chance of obtaining positive results. Of course, many of these results will be falsely positive.

In addition to contamination, PCR-based methods are prone to error and artifacts associated with the amplification process. In many comparative studies, it is important to know whether the PCR products represent the template ratios present in the original sample. However, due to differences in primer accessibility, DNA template structure and the dynamics of thermocycling (Suzuki and Giovannoni, 1996), not all potential targets for PCR are amplified with same efficiency (Kanagawa, 2003). Consequently, the ratio of different PCR products may not reflect the original ratio of template sequences. This may not be an issue for studies that attempt to recover a single target sequence without regard for quantitative aspects, but greatly reduces the power of PCR for estimating relative abundances of targets in environmental samples. Quantitative PCR protocols are helping to overcome some of these issues, but even where appropriate positive controls are included, it is hard to extrapolate the results of controlled experiments to field samples of mixed templates (Bruggemann *et al.*, 2000). Although some of these issues can be minimized by the use of appropriate protocols and adjustments in amplification conditions, PCR-based studies of microbial community structure must be scrutinized carefully and are inherently compromised in their quantitative power (von Wintzingerode *et al.*, 1997).

The majority of nucleic acid procedures involve the use of specific primers that target sequences of interest. Any such amplification-based strategies will always be limited by the specificity of the primers, as primer design is based upon known sequence information. Thus, even if primer specificity is tested rigorously in the laboratory, true specificity to the desired target, and inclusiveness to the breadth of a target group, are impossible to predict in applications involving samples of unknown and complex composition.

Some level of infidelity is inherent in any system that replicates nucleic acids. On the one hand, polymerases typically used in PCR are all known to be prone to some degree of error, and the use of enzymes with very high fidelity is certainly recommended. Error rates also climb as levels of DNA damage increase in a sample. Complicating this issue is the fact that the nucleotide positions that are most prone to error introduction during amplification, or due to DNA damage, often coincide with positions most susceptible to mutation. Thus, artifact may actually mimic natural processes, complicating the search for robust markers to differentiate between phylogenetic lineages of interest (Gilbert *et al.*, 2003). A final, but critical concern in many amplification procedures is the generation of chimeric products, whose sequence originates from more than one single source molecule. Such products can be created during PCR and cloning procedures when multiple templates are used to form a single product or when heteroduplex molecules are formed, cloned, and subsequently "corrected" by the *E. coli* host (Kopczynski *et al.*, 1994; Wang and Wang, 1997; Speksnijder *et al.*, 2001).

With the utility of PCR, it is easy to forget that the first molecular microbial surveys were actually conducted without the benefit of PCR (Olsen *et al.*, 1986; Pace *et al.*, 1986). These groundbreaking studies relied on the painstaking screening of shotgun clones taken directly from environmental DNA. The ability to amplify specific rRNA genes from environmental samples by PCR provided the breakthrough for rapid discovery of vast amounts of microbial diversity (Giovannoni *et al.*, 1990; Hugenholtz *et al.*, 1998). Interestingly, new advances in the high-throughput screening of environmental libraries have brought the pre-PCR approach of microbial gene discovery back in vogue. Soil metagenomes, cloned within either large- or small-insert libraries, can be directly screened for rRNA genes (Rondon *et al.*, 2000; Liles *et al.*, 2003; Venter *et al.*, 2004). Although these strategies may provide a perspective of the representative community, free of PCR artifacts, they are still highly demanding on human and financial resources. As new screening and sequencing strategies improve, this may become more feasible for standard analyses. Of course, one must keep in mind other biases inherent in these protocols such as preferential cloning (e.g. due to insert size), and biases in DNA isolation procedures (Frostegård *et al.*, 1999; also see Section 2.2). The current introduction of massively parallel sequencing by synthesis technologies (see Section 5) may in time permit the elimination of many biases introduced by PCR and cloning steps.

4. CONSIDERATION IN LABORATORY SET-UP

All work on high contamination risk samples should be carried out in low-DNA laboratory environments. To reduce general environmental DNA contamination and eliminate PCR carryover it is essential that all work on low-DNA samples be carried out in a dedicated laboratory that is physically isolated from other molecular biology laboratories. The general guidelines for such a laboratory are: This laboratory should operate with positive air pressure, HEPA filtered air to prevent influx of micro-organisms, dust, pollen, human hair and shed skin, and overhead germicidal UV lights providing nightly UV irradiation to remove any surface DNA contamination. Laboratory surfaces and equipment should be routinely washed with bleach to remove surface DNA contamination and all equipment and consumables should be purchased new and never have entered another molecular facility. Movement of researchers should always be in a one-way direction, from the clean laboratory to the post-PCR laboratory on any one day. All researchers entering the clean laboratory should wear full body coveralls (including head, arm and leg covering), shoe covers, double gloving, dust or surgical masks and face shields to prevent human and microbial DNA from contaminating the laboratory and specimens.

These issues are fully recognized in the world of ancient DNA studies and though such facilities do not come cheap, they have resulted in the construction of purpose-built, ultra-clean laboratories, routine adoption of stringent procedures for elimination of contaminating sequences from reagents, and an acceptance of the need to validate results in multiple laboratories. Despite a growing number of reports of similar problems in interpretation of data generated by microbial ecology laboratories, similar standards have not yet been universally implemented.

4.1. Controls necessary to ensure validity

The controls and criteria required to authenticate DNA sequences from high contamination risk samples in ancient DNA labs have evolved overtime as new aspects of sequence inaccuracy and contamination have become known. These criteria are a guide only and may not be appropriate to all studies, but appropriate laboratory facilities, negative controls, replication and cloning of amplification products are a minimum set of criteria that should be met to ensure sequence accuracy and authenticity.

The criteria for valid authenticity include:

(1) A physically isolated "clean-room" pre-PCR laboratory, utilizing positive-pressure, HEPA filtered air, nightly UV irradiation of work spaces, regular cleaning with bleach, the use of protective clothing and one-way movement of personnel from clean lab to post-PCR facility. These measures will reduce or eliminate background levels of environmental DNA contamination of specimens, consumables and chemicals and PCR carryover from post-PCR laboratories to the pre-PCR clean laboratory.
(2) Multiple blank extraction controls and negative PCR controls that bracket sample extractions and PCR amplifications. These controls monitor for contamination of extraction and amplification reagents and consumables, as well as sporadic contamination during DNA extraction and PCR set-up.
(3) Replication-Repeated extraction and amplification of DNA from different sub-samples of the same specimen in the same laboratory and in a second independent laboratory is essential to detect laboratory- or experiment-specific contamination as well as confirm sequence accuracy. Errors introduced by base modifications, chimeras and heteroduplexes are unlikely to occur in exactly the same pattern in replicated extraction/amplification experiments.
(4) Cloning of PCR products and sequencing of multiple clones to detect sequence errors due to DNA damage, contamination and chimeras.
(5) Quantification of the number of starting templates. Quantification is important because PCRs that are initiated from a small number of starting templates have a high chance of containing errors due to DNA damage. If a small number of templates are present (<1000), multiple amplifications will be needed to establish a consensus sequence.
(6) In ancient DNA studies, PCR amplification efficiency should decrease with increasing target length. Failure to observe this inverse relationship is strong evidence that DNA templates contain modern contaminating DNA. Conversely, this approach could be used to differentiate DNA from living and "fossil" microbial targets.
(7) Biochemical preservation. Independent evidence for good biochemical preservation of a sample can support the authenticity of an ancient DNA sequence.

5. LOOKING TO THE FUTURE

A new generation of sequencers has only just entered the market, (see this volume, Margulies *et al.*, 2007), which will allow novel approaches to be taken to the analysis of ancient and complex DNA sources. The 454 Life Sciences Corporation "GS20" is the first to make of these machines to an impact (Poinar *et al.*, 2005), with models by Solexa, Helicos and Agencourt Personal Genomics soon to follow. These systems have a great advantage in artefact-avoidance over current technologies due to the elimination of a cloning step from most analyses. Their preference for short DNA fragments and downstream software for elimination of artefact sequences opens a doorway to ancient DNA "paleogenomics". Genomic sequencing of extinct species is a new and exciting field that promises to answer many questions in molecular evolution, adaptation, speciation and genomic evolution. Two recent studies have demonstrated the feasibility of sequencing genomes of extinct species, but they also highlight the significant problems and limitations of the approach, that will likely limit its application to high-profile species and/or exceptionally well preserved samples. These problems revolve around the issues of DNA contamination, DNA damage/degradation and problems of dealing with trace amounts of DNA and will be of extreme importance in any attempts to sequence genomes of extinct hominids.

Metagenomics approaches, either using traditional bacterial cloning techniques or the new parallel sequencing technique developed by the 454 Life Sciences Corporation, differ from the targeted PCR-based approaches in that all genomic sequence in a sample are anonymously sequenced. For ancient DNA samples this creates a significant problem because contaminating microbial, environmental and human DNA will be sequenced alongside the endogenous genomic DNA. Unambiguously separating authentic from contaminating DNA creates a major bioinformatics challenge. One solution requires an annotated genome sequence of a close relative to identify and classify the genomic sequences obtained. However, even with such a framework much sequence data may be falsely included or excluded. A second solution requires sequencing multiple individuals and using interindividual sequence similarity to identify genomic regions that are likely to be endogenous to the sample and not from contamination. The problem of contaminating DNA is not trivial. Noonan *et al.* (2005) attempted genome sequencing of two extinct cave bears. Only 1–6% of all sequences obtained could be identified as probably cave bear in origin, and a massive 60–65% had not match to any sequence in the public databases. Poinar *et al.* (2005) sequenced 28 Mb from an exceptionally well-preserved mammoth specimen, only 45% of which was identified as mammoth DNA. Clearly, contaminating DNA is a major problem and will require significantly more sequencing and bioinformatics power in order to gain sufficient coverage of any ancient genome of interest. The highly fragmented nature of DNA and high level of damage will significantly impact coverage and sequence accuracy of the genome. Metagenomics of extinct species requires multi-fold coverage to correct for sequence modifications that have accumulated during the samples history. Assembling many millions of short overlapping fragments may be difficult or impossible for large parts of the genome, particularly in repetitive regions.

Most ancient DNA specimens are not as well preserved as the mammoths, nor can such large samples be taken for bulk DNA extraction. The vast bulk of ancient DNA comes from non-frozen conditions, where residual DNA is preserved in far smaller amounts and fragment sizes and with higher microbial and other contaminating DNA content. Genome sequencing via the 454 Life Sciences Corporation approach requires relatively large amounts of DNA, something that most specimens or samples cannot provide. In some cases, larger samples may be sacrificed, but for the majority of specimens, access to metagenomics technology will require the development of new methods to amplify the trace amounts of ancient DNA from normal specimens, without producing locus or allele bias.

The immediate challenge for researchers involved in molecular microbial ecology will be to adapt their approaches to take advantage of the new technologies without becoming swamped in data, or losing sight of potential biases introduced by these new systems. It is also important to note that these machines are not technically limited to the generation of raw genomic or metagenomic data; no doubt imaginative ecologists will adapt the capacity of this equipment to reveal specific aspects of microbial community structure on a depth and scale unavailable via current profiling methods. Cautious interpretation of the output will no doubt be educated by the experiences of the past 13 years since "community scale" DNA profiling of microbiota first began (Muyzer *et al.*, 1993).

REFERENCES

Ashelford, K. E., Chuzhanova, N. A., Fry, J. C., Jones, A. J. and Weightman, A. J. (2005). At least 1 in 20 16S rRNA sequence records currently held in public repositories is estimated to contain substantial anomalies. Appl. Environ. Microbiol. 71, 7724–7736.

Bada, J. L. (2001). State-of-the-art instruments for detecting extraterrestrial life. Proc. Natl. Acad. Sci. USA 98, 797–800.

Brüggemann, J., Stephen, J. R., Chang, Y.-J., Macnaughton, S. J., Kowalchuk, G. A., Kline, E. and White, D. C. (2000). Competitive PCR-DGGE analysis of bacterial mixtures an internal standard and an appraisal of template enumeration accuracy. J. Microbiol. Methods, 40, 111–123.

Christner, B. C., Mosley-Thompson, E., Thompson, L. G. and Reeve, J. N. (2001). Isolation of bacteria and 16S rDNAs from Lake Vostok accretion ice. Environ. Microbiol. 3, 570–577.

Christner, B. C., Mikucki, J. A., Foreman, C. M., Denson, J. and Priscu, J. C. (2005). Glacial ice cores: a model system for developing extraterrestrial decontamination protocols. Icarus 174, 572–584.

DeSalle, R., Barcia, M. and Wray, C. (1993). PCR jumping in clones of 30-million-year-old DNA fragments from amber preserved termites (*Mastotermes electrodominicus*). Experientia 49, 906–909.

Fish, S. A., Shepherd, T. J., McGenity, T. J. and Grant, W. D. (2002). Recovery of 16S ribosomal RNA gene fragments from ancient halite. Nature 417, 432–436.

Frostegård, A., Courtois, S., Ramisse, V., Clerc, S., Bernillon, D., Le Gall, F., Jeannin, P., Nesme, X. and Simonet, P. (1999). Quantification of bias related to the extraction of DNA directly from soil samples. Appl. Environ. Microbiol. 65, 5409–5420.

Gilbert, M. T. P., Willerslev, E., Hansen, A. J., Barnes, I., Rudbeck, L., Lynnerup, N. and Cooper, A. (2003). Distribution patterns of postmortem damage in human mitochondrial DNA. Am. J. Hum. Genet. 72, 32–47.

Giovannoni, S. J., Britschgi, T. B., Moyer, C. L. and Field, K. G. (1990). Genetic diversity in Sargasso Sea bacterioplankton. Nature 345, 60–63.

Griffiths, R. I., Whiteley, A. S., O'Donnell, A. G. and Bailey, M. J. (2000). Rapid method for coextraction of DNA and RNA from natural environments for analysis of ribosomal DNA- and rRNA-based microbial community composition. Appl. Environ. Microbiol. 66, 5488–5491.

Hofreiter, M., Serre, D., Poinar, H. N., Kuch, M. and Paabo, S. (2001). Ancient DNA. Nat. Rev. Genet. 2, 353–359.

Hugenholtz, P., Goebel, B. M. and Pace, N. R. (1998). Impact of culture-independent studies on the emerging phylogenetic view of bacterial diversity. J. Bacteriol. 180, 4765–4774.

Hugenholtz, P. and Huber, T. (2003). Chimeric 16S rDNA sequences of diverse origin are accumulating in the public databases. Int. J. Syst. Evol. Microbiol. 53, 289–293.

Hurt, R. A., Qui, X., Wu, L., Roh, Y., Palumbo, A. V., Tiedje, J. M. and Zhou, J. (2001). Simultaneous recovery of RNA and DNA from soils and sediments. Appl. Environ. Microbiol. 67, 4495–4503.

Kanagawa, T. (2003). Bias and artifacts in multi-template polymerase chain reactions (PCR). J. Biosci. Bioeng. 96, 317–323.

Kashefi, K. and Lovely, D. R. (2003). Extending the upper temperature limit for life. Science 301, 934.

Keener, J. and Nomura, M. (1996). Regulation of ribosome synthesis. In: F. C. Neidhardt (Ed), *Escherichia coli* and *Salmonella*: Cellular and Molecular Biology(2nd Edn), pp. 1417–1431). American Society for Microbiology, Washington, DC.

Kemp, P. F., Lee, S. and LaRoche, J. (1993). Estimating the growth rate of slowly growing marine bacteria from RNA content. Appl. Environ. Microbiol. 59, 2594–2601.

Kopczynski, E. D., Bateson, M. M. and Ward, D. M. (1994). Recognition of chimeric small-subunit ribosomal DNAs composed of genes from uncultivated microorganisms. Appl. Environ. Microbiol. 60, 746–748.

Krause, J., Dear, P. H., Pollack, J. L., Slatkin, M., Barnes, I., Lister, A. M., Spriggs, H., Ebersberger, I., Pääbo, S. and Hofreiter, M. (2006). Multiplex amplification of the mammoth mitochondrial genome and the evolution of *Elephantidae*. Nature 439, 724–727.

Lindahl, T. (1993). Instability and decay of the primary structure of DNA. Nature 362, 709–715.

Liles, M. R., Manske, B. F., Bintrim, S. B., Handelsman, J. and Goodman, R. M. (2003). A census of rRNA genes and linked genomic sequences within a soil metagenomic library. Appl. Environ. Microbiol. 69, 2684–2691.

Margulies, M., Egholm, M., Altman, W., Attiya, S., Bader, J., Bemben, L., Berka, J., Braverman, M., Chen, Y., Chen, Z., Dewell, S., Du, L., Fierro, J., Gomes, X., Godwin, B., He, W., Helgesen, S., Ho, C., Irzyk, G., Jando, S., Alenquer, M., Jarvie, T., Jirage, K., Kim, J., Knight, J., Lanza, J., Leamon, J., Lefkowitz, S., Lei, J., Li, K., Lohman, H., Lu, V. B., Makhijani, K., McDade, M., McKenna, E., Myers, M., Nickerson, E., Nobile, J., Plant, R., Puc, B., Ronan, M., Roth, G., Sarkis, G., Simons, J., Simpson, J., Srinivasan, M., Tartaro, K., Tomasz, A., Vogt, K., Volkmer, G., Wang, S., Wang, Y., Weiner, M., Yu, P., Begley, R. and Rothberg, J. (2005). Genome sequencing in microfabricated high-density picolitre reactors. Nature 437, 376–380.

Margulies, M., Jarvie, T. P., Knight, J. R. and Simons, J. F. (2007). The 454 Life Sciences Picoliter Sequencing System. In: K. R. Mitchelson (Ed), New High Throughput Technologies for DNA Sequencing and Genomics (pp. 151–186). Elsevier, Amsterdam.

Muyzer, G., de Waal, E. C. and Uitterlinden, A. G. (1993). Profiling of complex microbial populations by denaturing gradient gel electrophoresis analysis of polymerase chain reaction-amplified genes coding for 16S rRNA. Appl. Environ. Microbiol. 59, 695–700.

Noonan, J. P., Hofreiter, M., Smith, D., Priest, J. R., Rohland, N., Rabeder, G., Krause, J., Detter, J. C., Paabo, S. and Rubin, E. M. (2005). Genomic sequencing of Pleistocene cave bears. Science 309, 597–600.

Olsen, G. J., Lane, D. J., Giovannoni, S. J., Pace, N. R. and Stahl, D. A. (1986). Microbial ecology and evolution: a ribosomal RNA approach. Ann. Rev. Microbiol. 40, 337–365.

Paabo, S., Irwin, D. M. and Wilson, A. C. (1990). DNA damage promotes jumping between templates during enzymatic amplification. J. Biol. Chem. 265, 4718–4721.

Pace, N. R., Stahl, D. A., Lane, D. J. and Olsen, G. J. (1986). The analysis of natural microbial populations by ribosomal RNA sequences. Adv. Microb. Ecol. 9, 1–55.

Pace, N. R. (1997). A molecular view of microbial diversity and the biosphere. Science 276, 734–740.

Poinar, H. N., Hofreiter, M., Spaulding, W. G., Martin, P. S., Stankiewicz, B. A., Bland, H., Evershed, R. P., Possnert, G. and Pääbo, S. (1998). Molecular coproscopy: dung and diet of the extinct ground sloth *Nothrotheriops shastensis*. Science 281, 402–406.

Poinar, H. N., Schwarz, C., Qi, J., Shapiro, B., MacPhee, R. D. E., Buigues, B., Tikhonov, A., Huson, D. H., Tomsho, L. P., Auch, A., Rampp, M., Miller, W. and Schuster, S. C. (2005). Metagenomics to paleogenomics: Large-scale sequencing of mammoth DNA. Science 311, 392–394.

Priscu, J. C., Adams, E. E., Lyons, W. B., Voytek, M. A., Mogk, D. W., Brown, R. L., McKay, C. P., Takacs, C. D., Welch, K. A., Wolf, C. F., Kirshtein, J. D. and Avci, R. (1999). Geomicrobiology of subglacial ice above Lake Vostok, Antarctica. Science 286, 2141–2144.

Rappé, M. S. and Giovannoni, S. J. (2003). The uncultured microbial majority. Ann. Rev. Microbiol. 57, 369–394.

Rogers, S. O., Theraisnathan, V., Ma, L. J., Zhao, Y., Zhang, G., Shin, S.-G., Castello, J. D. and Starmer, W. T. (2004). Comparisons of protocols for decontamination of environmental ice samples for biological and molecular examinations. Appl. Environ. Microbiol. 70, 2540–2544.

Rondon, M. R., August, P. R., Bettermann, A. D., Brady, S. F., Grossman, T. H., Liles, M. R., Loiacono, K. A., Lynch, B. A., MacNeil, I. A., Minor, C., Tiong, C. L., Gilman, M., Osburne, M. S., Clardy, J., Handelsman, J. and Goodman, R. M. (2000). Cloning the soil metagenome: a strategy for accessing the genetic and functional diversity of uncultured microorganisms. Appl. Environ. Microbiol. 66, 2541–2547.

Shendure, J., Mitra, R. D., Varma, C. and Church, G. M. (2004). Advanced sequencing technologies: methods and goals. Nat. Rev. Genet. 5, 335–344.

Speksnijder, A. G. C. L., Kowalchuk, G. A., De Jong, S., Kline, E., Stephen, J. R. and Laanbroek, H. J. (2001). Micro-variation artifacts introduced by PCR and cloning of closely related 16S rDNA sequences. Appl. Environ. Microbiol. 67, 469–472.

Stone, A. C. and Stoneking, M. (1998). MtDNA analysis of a prehistoric Oneota population: implications for the peopling of the new world. Am. J. Hum. Genet. 62, 1153–1170.

Suzuki, M. T. and Giovannoni, S. J. (1996). Bias caused by template annealing in the amplification of mixtures of 16S rRNA genes by PCR. Appl. Environ. Microbiol. 62(2), 625–630.

Taubenberger, J. K., Reid, A. H., Krafft, A. E., Bijwaard, K. E. and Fanning, T. G. (1997). Initial genetic characterization of the 1918 "Spanish" influenza virus. Science 275, 1793–1796.

Tumpey, T. M., Basler, C. F., Aguilar, P. V., Zeng, H., Solorzano, A., Swayne, D. E., Cox, N. J., Katz, J. M., Taubenberger, J. K., Palese, P. and Garcia-Sastre, A. (2005). Characterization of the reconstructed 1918 Spanish influenza pandemic virus. Science 310, 77–80.

Venter, J. C., Remington, K., Heidelberg, J. F., Halpern, A. L., Rusch, D., Eisen, J. A., Wu, D. Y., Paulsen, I., Nelson, K. E., Nelson, W., Fouts, D. E., Levy, S., Knap, A. H., Lomas, M. W., Nealson, K., White, O., Peterson, J., Hoffman, J., Parsons, R., Baden-Tillson, H., Pfannkoch, C., Rogers, Y. H. and Smith, H. O. (2004). Environmental genome shotgun sequencing of the Sargasso Sea. Science 304, 66–74.

von Wintzingerode, F., Göbel, U. B. and Stackebrandt, E. (1997). Determination of microbial diversity in environmental samples: pitfalls of PCR-based rRNA analysis. FEMS Microbiol. Rev. 21, 213–229.

Vreeland, R. H., Rozenwieg, W. D. and Powers, D. W. (2000). Isolation of a 250 million-year-old halotolerant bacterium from a primary salt crystal. Nature 407, 897–900.

Wang, G. C.-Y. and Wang, Y. (1997). Frequency of formation of chimeric molecules as a consequence of PCR co-amplification of 16S rRNA genes from mixed bacterial genomes. Appl. Environ. Microbiol. 63, 4645–4650.

Wagner, R. (1994). The regulation of ribosomal RNA synthesis and bacterial cell growth. Arch. Microbiol. 161, 100–106.

Whitman, W. B., Coleman, D. C. and Wiebe, W. J. (1998). Prokaryotes: the unseen majority. Proc. Natl. Acad. Sci. USA 95, 6578–6583.

Willerslev, E., Hansen, A. J., Binladen, J., Brandt, T. B., Gilbert, M. T. P., Shapiro, B., Bunce, M., Wiuf, C., Gilichinsky, D. A. and Cooper, A. (2003). Diverse plant and animal genetic records from Holocene and Pleistocene sediments. Science 300, 791–795.

Willerslev, E. A., Hansen, A. J., Rønn, R., Brand, T. B., Barnes, I., Wiuf, C., Gilichinsky, D., Mitchell, D. and Cooper, A. (2004). Long-term persistence of bacterial DNA. Curr. Biol. 14, 9–10.

Wommack, K. E., Williamson, S. J., Sundbergh, A., Helton, R. R., Glazer, B. T., Portune, K. and Cary, S. C. (2004). An instrument for collecting discrete large-volume water samples suitable for ecological studies of microorganisms. Deep-Sea Res. I(51), 1781–1792.

Yang, D. Y., Eng, B. and Saunders, S. R. (2003). Hypersensitive PCR, ancient human mtDNA and contamination. Hum. Biol. 75, 355–364.

Subject Index